Understanding Host-Microbiome Interactions - An Omics Approach

Ravindra Pal Singh • Ramesh Kothari
Prakash G. Koringa • Satya Prakash Singh
Editors

Understanding Host-Microbiome Interactions - An Omics Approach

Omics of Host-Microbiome Association

Springer

Editors
Ravindra Pal Singh
Saurashtra University
UGC-CAS Department of Biosciences
Saurashtra University
Rajkot, Gujarat
India

Ramesh Kothari
Saurashtra University
UGC-CAS Department of Biosciences
Saurashtra University
Rajkot, Gujarat
India

Prakash G. Koringa
Anand Agricultural University
College of Vet. Sc. and Animal Husbandry
Anand Agricultural University
Anand, Gujarat
India

Satya Prakash Singh
Saurashtra University
UGC-CAS Department of Biosciences
Saurashtra University
Rajkot, Gujarat
India

ISBN 978-981-10-5049-7 ISBN 978-981-10-5050-3 (eBook)
DOI 10.1007/978-981-10-5050-3

Library of Congress Control Number: 2017952944

Printed on acid-free paper

This Springer imprint is published by Springer Nature
The registered company is Springer Nature Singapore Pte Ltd.
The registered company address is: 152 Beach Road, #21-01/04 Gateway East, Singapore 189721, Singapore

Preface

Microbial communities and their hosts (human, plant, and animal) have been coevolved and coadapted under diverse environmental selective pressures over millions of years. These hosts rely on integrated interactions of specific microbiome for their successful survival. The evolution of the meta-omics (metagenomics, meta-transcriptomics, and meta-proteomics) technologies in sequencing is fostering a detonation of interest in how the gut and rhizospheric microbiomes impact physiology and propensity to disease. These advancements in technologies have recently been provided with deeper understanding of the complexity of the host-microbe association in terms of phylogeny and function connectivity. Among other host microbiomes, meta-omics technologies have significantly been carried out on human gut and plant rhizospheric microbiomes. Those studies have found that small fraction of the whole microbial communities are always associated with the host, and, assisting in host's physiology, such microbial communities sometimes are referred to as a core microbiome. The core microbial communities are now being used for modulating the host microbiome in order to reduce the incidence of diseases and improving the health of the hosts. However, much information of these interesting associations in different research articles is available from the Western world and has limited availability from Asian counties. In this book, thus, we have collected several chapters from prominent scientists from Asian counties which will be available to students at graduate and postgraduate levels.

Presently, a majority of scientists working on microbiology are trying to understand phylogeny and functional aspects of different host microbiomes. However, rapid development of these meta-omics technologies makes it hard to retrieve all the required informations from diverse research reports. The book, thus, will be an excellent resource to get updated information of different host microbiomes in terms of their community composition and interactive functions. As it demands tremendous and dedicated effort, we are extremely thankful to all the authors for their prompt responses and their contributions. I extend my earnest appreciation to Mr. Kumar of Springer for his valuable support to facilitate completion of the task.

Rajkot, India — Ravindra Pal Singh
Rajkot, India — Ramesh Kothari
Anand, India — Prakash G. Koringa
Rajkot, India — Satya Prakash Singh

Contents

Part III Plant Microbiome

Part IV Animal Microbiome

Contributors

Kruti G. Dangar UGC-CAS Department of Biosciences, Saurashtra University, Rajkot, India

Oliver Deusch The WALTHAM Centre for Pet Nutrition, Leicestershire, UK

Pravin Dudhagara Department of Biosciences (UGC-SAP), Veer Narmad South Gujarat University, Surat, India

Indra Ramjibhai Gadhvi Department of Marine Science, Maharaja Krishnakumarsinhji Bhavnagar University, Bhavnagar, Gujarat, India

Anjana Ghelani Department of Life Sciences, Hemchandracharya North Gujarat University, Patan, India

Shubhrima Ghosh Enzyme and Microbial Biochemistry Laboratory, Department of Chemistry, Indian Institute of Technology Delhi, Hauz Khas, New Delhi, India

Reeta Goel Department of Microbiology, CBSH, G.B.Pant University of Agriculture and Technology, Pantnagar, Uttaranchal, India

Sangeeta D. Gohel Department of Biosciences (UGC-CAS), Saurashtra University, Rajkot, Gujarat, India

R. Hemamalini Enzyme and Microbial Biochemistry Laboratory, Department of Chemistry, Indian Institute of Technology Delhi, Hauz Khas, New Delhi, India

Kevin R. Hughes The WALTHAM Centre for Pet Nutrition, Leicestershire, UK

Bhagya Iyer Institute of Science, Nirma University, Ahmedabad, Gujarat, India

Lata Jain ICAR-National Institute of Biotic Stress Management, Baronda, Chhattisgarh, India

Anjali Joshi Department of Biosciences (UGC-CAS), Saurashtra University, Rajkot, Gujarat, India

Chaitanya G. Joshi Department of Animal Biotechnology, College of Veterinary Science & Animal Husbandry, Anand Agricultural University, Anand, Gujarat, India

Mital J. Kaneria Department of Biosciences (UGC-CAS), Saurashtra University, Rajkot, Gujarat, India

Suganya Kannan Department of Microbial Technology, School of Biological Sciences, Madurai Kamaraj University, Madurai, Tamil Nadu, India

Amanpreet Kaur Department of Biotechnology, TIFAC-Centre of Relevance and Excellence in Agro and Industrial Biotechnology (CORE), Thapar University, Patiala, India

S.K. Khare Enzyme and Microbial Biochemistry Laboratory, Department of Chemistry, Indian Institute of Technology Delhi, Hauz Khas, New Delhi, India

Juma Kisuse Department of Bioscience and Biotechnology, Faculty of Agriculture, Kyushu University, Fukuoka, Japan

Prakash G. Koringa Department of Animal Biotechnology, College of Veterinary Science and Animal Husbandry, Anand Agricultural University, Anand, Gujarat, India

Vishal Kothari Department of Biotechnology, Junagadh Agricultural University, Junagadh, Gujarat, India

Charmy Kothari Department of Biotechnology, Christ College, Rajkot, Gujarat, India

Ramesh K. Kothari UGC-CAS Department of Biosciences, Saurashtra university, Rajkot, Gujarat, India

Govindan Krishnamoorthy Department of Microbial Technology, School of Biological Sciences, Madurai Kamaraj University, Madurai, Tamil Nadu, India

P U Krishnaraj Department of Agricultural Microbiology, College of Agriculture, Vijayapura University of Agricultural Sciences, Dharwad, Karnataka, India

Vinay Kumar ICAR-National Institute of Biotic Stress Management, Baronda, Chhattisgarh, India

Anil Kumar Department of Biotechnology, TIFAC-Centre of Relevance and Excellence in Agro and Industrial Biotechnology (CORE), Thapar University, Patiala, India

Anju Kunjadiya Center for Interdisciplinary Studies in Science and Technology (CISST), Sardar Patel University, Anand, Gujarat, India

Balu Alagar Venmathi Maran Fisheries Science Institute, Chonnam National University, Yeosu, Republic of Korea

Zoe V. Marshall-Jones The WALTHAM Centre for Pet Nutrition, Leicestershire, UK

Murugan Marudhamuthu Department of Microbial Technology, School of Biological Sciences, Madurai Kamaraj University, Madurai, Tamil Nadu, India

Chandra Shekar Mootapally Department of Marine Science, Maharaja Krishnakumarsinhji Bhavnagar University, Bhavnagar, Gujarat, India

Department of Animal Biotechnology, College of Veterinary Science & Animal Husbandry, Anand Agricultural University, Anand, Gujarat, India

Jiro Nakayama Department of Bioscience and Biotechnology, Faculty of Agriculture, Kyushu University, Fukuoka, Japan

Neelam Mustakali Nathani Department of Life Sciences, Maharaja Krishnakumarsinhji Bhavnagar University, Bhavnagar, Gujarat, India

UGC-CAS Department of Biosciences, Saurashtra university, Rajkot, Gujarat, India

Prabha Palanichamy Department of Microbial Technology, School of Biological Sciences, Madurai Kamaraj University, Madurai, Tamil Nadu, India

Rupal D. Pandya UGC-CAS Department of Biosciences, Saurashtra University, Rajkot, India

Bhupendra Singh Panwar Department of Animal Biotechnology, College of Veterinary Science and Animal Husbandry, Anand Agricultural University, Anand, Gujarat, India

Malik Ahmed Pasha Department of Agricultural Microbiology, College of Agriculture, Vijayapura University of Agricultural Sciences, Dharwad, Karnataka, India

Amrutlal K. Patel Hester Biosciences Limited, Ahmedabad, Gujarat, India

Rajesh Patel Department of Life Sciences, Hemchandracharya North Gujarat University, Patan, India

Paresh Poriya Department of Marine Science, Maharaja Krishnakumarsinhji Bhavnagar University, Bhavnagar, Gujarat, India

Nirali M. Raiyani UGC-CAS Department of Biosciences, Saurashtra University, Rajkot, India

Shalini Rajkumar Institute of Science, Nirma University, Ahmedabad, Gujarat, India

Kalpna D. Rakholiya Department of Biosciences, Institute of Biotechnology, Saurashtra University, Rajkot, Gujarat, India

Jalpa Rank Department of Biosciences (UGC-CAS), Saurashtra University, Rajkot, Gujarat, India

Chetana Roat Department of Microbiology and Biotechnology, University School of Sciences, Gujarat University, Ahmedabad, India

Pritesh H. Sabara Department of Animal Biotechnology, College of Veterinary Science and Animal Husbandry, Anand Agricultural University, Anand, Gujarat, India

Meenu Saraf Department of Microbiology and Biotechnology, University School of Sciences, Gujarat University, Ahmedabad, India

Amit K. Sharma Department of Biosciences (UGC-CAS), Saurashtra University, Rajkot, Gujarat, India

Satya P. Singh Department of Biosciences (UGC-CAS), Saurashtra University, Rajkot, Gujarat, India

Ravindra Pal Singh Department of Biosciences (UGC-CAS), Saurashtra University, Rajkot, Gujarat, India

Department of Biological Chemistry, John Innes Centre, Norwich, UK

Ravindra Soni Department of Agricultural Microbiology, College of Agriculture, Indira Gandhi Krishi Vishva Vidyalaya, Raipur, Chhattisgarh, India

M. Sudhakara Reddy Department of Biotechnology, TIFAC-Centre of Relevance and Excellence in Agro and Industrial Biotechnology (CORE), Thapar University, Patiala, India

G.S. Sutaria Main Dry Farming Research Station, Junagadh Agricultural University, Rajkot, Gujarat, India

Deep Chandra Suyal Department of Microbiology, CBSH, G.B.Pant University of Agriculture and Technology, Pantnagar, Uttaranchal, India

Jalpa R. Thakkar Department of Animal Biotechnology, College of Veterinary Science and Animal Husbandry, Anand Agricultural University, Anand, Gujarat, India

Foram J. Thakrar Department of Biosciences (UGC-CAS), Saurashtra University, Rajkot, Gujarat, India

Rameshwar Tiwari Enzyme and Microbial Biochemistry Laboratory, Department of Chemistry, Indian Institute of Technology Delhi, Hauz Khas, New Delhi, India

Ruchi Trivedi Department of Animal Biotechnology, College of Veterinary Science and Animal Husbandry, Anand Agricultural University, Anand, Gujarat, India

V.D. Vora Main Dry Farming Research Station, Junagadh Agricultural University, Rajkot, Gujarat, India

Corrin V. Wallis The WALTHAM Centre for Pet Nutrition, Leicestershire, UK

About the Editors

Ravindra Pal Singh is a research assistant at the Department of Biochemistry, John Innes Centre, Norwich, UK. He obtained his Ph.D. from CSIR-CSMCRI, India, in 2013. During his scientific career, he has received several prestigious scholarships, such as an outstanding postdoctoral fellowship provided by Tel Aviv University, Israel, in 2013, a JSPS postdoctoral fellowship in 2013, and a Dr. D. S. Kothari postdoctoral fellowship provided by the UGC, India, in 2016. He has published 28 articles in different scientific journals and received a patent while working in the field of microbial ecology and bacteriology. He can be contacted at ravindrapal.1441@gmail.com.

Ramesh Kothari is currently working as a professor at the Department of Biosciences, Saurashtra University, Rajkot, Gujarat, India. He received a Ph.D. in microbiology from the Saurashtra University, Rajkot, in 2003. Soon after, he joined Christ College Rajkot where he served as an assistant professor for 14 years. He completed his postdoctoral research at Rutgers University, Newark, USA. His research work focuses on bioremediation and metagenomics. He has more than 41 international peer-reviewed research publications to his credit.

Prakash G. Koringa has been an assistant professor at Anand Agricultural University (AAU), Gujarat, India, since 2004. He obtained his Ph.D. from the Department of Animal Biotechnology, College of Veterinary Sciences and Animal Husbandry, AAU, Anand, Gujarat. His work focuses on rumen metagenomics and cecal microbiome in broilers and their application for the host health. He has published several research papers in peer-reviewed journals as well as a number of book chapters.

Satya P. Singh is currently working as professor and head at the UGC-CAS Department of Biosciences, Saurashtra University, Rajkot, Gujarat, India. He completed his master's in microbiology from the G. B. Pantnagar University of Agriculture and Technology, Pantnagar, India, and carried out his doctoral research at Griffith University, Brisbane, Australia. Prof. Singh has worked at the National Food Research Institute, Tsukuba, Japan, as a visiting scientist and also visited Yangon University, Myanmar, as visiting professor. Prof. Singh has been working on the diversity, phylogeny, and enzymatic characteristics of the halophilic/haloalkaliphilic bacteria, actinomycetes, and archaea of various saline ecosystems. He has published 82 research papers and contributed 17 book chapters in various books.

Part I

Next Generation Sequence Technology

The Omics Era and Host Microbiomes

1

Ravindra Pal Singh and Ramesh Kothari

Abstract

Advancement in 'omics' technologies permits the quantitative monitoring of the plethora of biological molecules in natural systems in a high-throughput manner. Such technologies allow determination of the variation between different biological states (RNA, DNA and protein) on a genomic scale. From a fundamental knowledge perspective, no single omics technique can completely disentangle the complexities of host microbiomes. As a result, several 'omics' platforms have been developed in order to better understand the systems biology of host-microbiome interactions (Zhang et al. 2010). For instance, metagenomics, meta-transcriptomics, meta-proteomics and metabolomics methods provide information on the metagenome. Overall changes in the mRNA or proteins levels of the host microbiome can be determined dynamic changes of all classes of molecules within a microbiome over a given time period. Integration of different layers of information obtained from multi-omics approaches are required in order to paint a meaningful canvas of functional and dynamic interactions of host-microbiome communities. To date, the application of more than one -omics technology have been applied to different host microbiomes, including in plant and humans (Tlaskalova-Hogenova et al. 2011; Fernandez et al. 2013; Tkacz and Poole 2015; Addis et al. 2016; Lareen et al. 2016). Application of multi-omics approaches to the microbiome of these hosts unravels their essential functions, which are key throughout the host's life cycle; thus, associated microbial communities are sometimes referred to as the 'secondary genome' of the host (Siboni et al. 2008). This perspective chapter briefly describes what has

R.P. Singh (✉)
Department of Biosciences, Saurashtra University, Rajkot 360005, Gujarat, India

Department of Biological Chemistry, John Innes Centre,
Norwich Research Park, Norwich NR4 7UH, UK
e-mail: ravindrapal.1441@gmail.com; Ravindra.Pal-Singh@jic.ac.uk

R. Kothari
Department of Biosciences, Saurashtra University, Rajkot 360005, Gujarat, India

R.P. Singh et al. (eds.), *Understanding Host-Microbiome Interactions - An Omics Approach*, DOI 10.1007/978-981-10-5050-3_1

been unveiled so far and what still needs to be done in order to better understand human and plant microbiomes.

Keywords
Human microbiome • Plant microbiome • Short chain fatty acid • Omics • Biocontrol agent

1.1 Introduction

Advancement in 'omics' technologies permits the quantitative monitoring of the plethora of biological molecules in natural systems in a high-throughput manner. Such technologies allow determination of the variation between different biological states (RNA, DNA and protein) on a genomic scale. From a fundamental knowledge perspective, no single omics technique can completely disentangle the complexities of host microbiomes. As a result, several 'omics' platforms have been developed in order to better understand the systems biology of host-microbiome interactions (Zhang et al. 2010). For instance, metagenomics, meta-transcriptomics, meta-proteomics and metabolomics methods provide information on the meta-genome. Overall changes in the mRNA or proteins levels of the host microbiome can be determined dynamic changes of all classes of molecules within a microbiome over a given time period. Integration of different layers of information obtained from multi-omics approaches are required in order to paint a meaningful canvas of functional and dynamic interactions of host-microbiome communities. To date, the application of more than one -omics technology have been applied to different host microbiomes, including in plant and humans (Tlaskalova-Hogenova et al. 2011; Fernandez et al. 2013; Tkacz and Poole 2015; Addis et al. 2016; Lareen et al. 2016). Application of multi-omics approaches to the microbiome of these hosts unravels their essential functions, which are key throughout the host's life cycle; thus, associated microbial communities are sometimes referred to as the 'secondary genome' of the host (Siboni et al. 2008; Singh and Reddy 2015). This perspective chapter briefly describes what has been unveiled so far and what still needs to be done in order to better understand human and plant microbiomes.

1.2 The Human Microbiome

Comparative functional annotation of genomic sequences and metabolome profiles of the human gut microbiome have provided information regarding genome structures, gene functions, metabolic and regulatory networks, as well as the evolution of associated microbial genomes (Lin and Qian 2007; Chu and Mazmanian 2013; Nakayama et al. 2015; Reck et al. 2015; Addis et al. 2016; Despres et al. 2016). Revolutionary improvements in high-throughput DNA sequencing technology (metagenomics) identified more than 1000 phylogenetically distinct microbial genomes from the human gut, of which the major phylogenetic lineages have been fully sequenced (Medini et al. 2008; Kyrpides 2009). Gut microbiota are reported to have a positive impact on key host functions related to the immune and nervous systems, in the prevention of disease and also in development and behaviour (Tlaskalova-Hogenova et al. 2011; Luczynski et al. 2016; Mu et al. 2016; Schuijt et al. 2016). The structures of gut microbial populations are shaped by food stuffs, in particular (Graf et al. 2015; Nakayama et al. 2015). Moreover, modulating the gut microbiome highlights the importance of common microbial communities which perform several functions, such as maintenance of structural integrity of the gut mucosal barrier, metabolising dietary substances and immunomodulation, providing protection against different pathogens (Zhang et al. 2015). A balanced equilibrium of symbiotic, common and pathogenic microbial species is necessary for correct functioning of the intestine, for instance (Round and Mazmanian 2009).

Cell surface appendages (microbial-associated molecular patterns, MAMPs) of bacteria are key factors for host-microorganism crosstalk through host pattern recognition receptors (PRRs), such as lectins (Martens et al. 2009; Chu and Mazmanian 2013). These MAMPs of probiotic bacteria, such as *Lactobacillus*, *Bacteroides* and *Bifidobacteria* species, are associated which modulate the function of phagocytic cells and could be beneficial for fighting cancerous growths in the host or mitigating pathogen infection (Lebeer et al. 2010). On the basis of clinical, epidemiological and immunological studies, we understand that disruption of these communities can lead to the development of gastrointestinal (GI) tract diseases, including inflammatory bowel disease (IBD) (De Wouters et al. 2012; Greenblum et al. 2012; Forbes et al. 2016; Schulberg and De Cruz 2016). Thus, mapping out MAMP-PRR interactions in the gut would allow for the development of selective probiotic treatments on the GI tract dysfunction. Understanding the molecular interactions concerned will not be straightforward; however, integration of whole genome sequencing through metagenomics and metatranscriptomics as well as metabolomics analyses will certainly add further information to the field.

Gut microbial communities play a vital role in the digestion of food fibre and subsequently converts it into short-chain fatty acids (SCFA) (Rogowski et al. 2015). SCFA (acetate, butyrate and propionate) are important anions and energy sources in the colonic lumen, where they also modulate colonocyte morphology and function (Den Besten et al. 2013). Butyrate produced by *Clostridium butyricum* has been reported to exert anti-tumorigenic and anti-inflammatory effects (Nakanishi et al. 2003) as well as growth inhibition of other enteric pathogens (Zhang et al. 2016). Acetate and propionate are incorporated into lipid and glucose metabolism, respectively, in the liver (Rombeau and Kripke 1990). Supplementing the diet or altering dietary polysaccharide can influence abundance of specific microbial populations. For instance, arabinoxylan can modulate populations of *Bifidobacteria*, *Bacteroides* and *Lactobacillus* (Sanchez et al. 2009; Riviere et al. 2014). This specificity indicates that particular carbohydrates can select for specific microbial populations, as occurs in dysbiosis conditions of the gut. So far, starch utilisation system (sus)-like strategy has been reported for the gut commensal *Bacteroides thetaiotaomicron* for starch degradation (Martens et al. 2009); it remains unclear whether similar examples occur in other probiotics. However, much work is still required to determine precise carbohydrate-microbe associations and how gut microbial communities degrade diverse food fibres. It is expected that the application of multi-omics tools will provide such information. For instance, applying metatranscriptomics approach will provide insight how different carbohydrates alter expressions of genes of a particular bacterium, and then proteomics analysis will give information about enzymes of the bacterium involved in that carbohydrate degradation. Applying different analytical tools will help to figure out what is the structure and linkage pattern in the particular carbohydrate.

Studies have suggested that some bacteria present in the mother's gut can access the mammary glands during late pregnancy and lactation (Fernandez et al. 2013; Rodriguez 2014). Bacteria are transferred from GI to breast milk via entero-mammary pathway—involving mononuclear immune cells (Rodriguez 2014). Accumulating microbial communities in the mammary glands subsequently become part of the human milk that sets up the primary line of protection for infants and contributes to the maturation of their immune system and preventing expression of immune-mediated diseases (Morrow and Rangel 2004; Addis et al. 2016). Milk components, such as galactooligosaccharides (GOSs), act as a prebiotics, modulating the infant gut microbiome and in particular promoting populations of the probiotics, such as *Bifidobacteria* and *Lactobacillus* (Rautava et al. 2012). Little is known about the composition and function of milk microbiota; thus improved understanding is likely to improve our view of the aetiology and dynamics of sub-clinical and culture-negative mastitis as well as informing the management of the mammary gland and offspring health. In addition, it will likely lead to the development of novel strategies for preventing mastitis and improving its' management.

1.3 The Plant Microbiomes

The microbiomes of plants have been categorised into the rhizospheric, the endorhizospheric and the phyllospheric (Berg et al. 2016). The distinctness of microbial communities in these ecological niches can be shaped by aerial surfaces, tissue types and surrounding environmental factors (Gottel et al. 2011; Vorholt 2012). The phyllosphere niche is nutrient poor and regularly exposed to variation of temperature, radiation and moisture (Vorholt 2012). In contrast, the rhizospheric niche is nearest to the plant root system and is relatively rich in nutrients that are derived from and influenced by deposition of mucilage and root exudates (Koranda et al. 2011). It had been established that such host exudates play an important role in shaping specific microbiomes that are influenced by both chemo-attractants as well as repellents (Bais et al. 2006; Wu et al. 2015). The endorhizospheric microbiome is a subgroup of the rhizobacterial communities, which have the capability to enter into root hair (Prieto et al. 2011). Microbial communities of these niches promote growth of the host through suppressing growth of plant pathogens and by producing plant growth hormones (Prieto et al. 2011; Tkacz and Poole 2015; Lareen et al. 2016). Moreover, the rhizospheric microbiome contributes to biogeochemical cycles that produces several chemical elements required for normal plant growth (Rousk and Bengtson 2014). Much of the taxonomical and functional complexities of these communities were not available until recently and have become available due to the development of high-throughput genome sequencing methods (Berendsen et al. 2012; Bakker et al. 2013; Turner et al. 2013).

Multi-omics technologies have been applied to these niches, and it is evident that the associated microbial communities are taxonomically and functionally well separated among different areas of the plant host (Gottel et al. 2011; Vorholt 2012; Bulgarelli et al. 2013; Knief 2014; Mendes et al. 2014). Metagenomics has been applied to the rhizospheric microbiome of a variety of host plants (such as *Arabidopsis*, rice) that had been subjected to different environmental factors in order to understand how microbial communities are shaped by external stimuli, as summarised by Turner et al. (2013) and Mendes et al. (2013). Most of the rhizospheric microbial communities comprise Proteobacteria, Firmicutes and Actinobacteria taxa, though their composition is constantly changing according to the different developmental stages of the host (Lundberg et al. 2012), the cultivars and the types of soil (Mendes et al. 2011; Weinert et al. 2011). Microbial communities of the phyllospheric environment are much more dynamic than those from rhizospheric regions but are commonly composed of similar taxa to rhizospheric zones (Bodenhausen et al. 2013). Knief et al. (2012) applied meta-proteogenomics to a study on the microbiome present in both the phyllosphere and rhizosphere of the rice and concluded that phyllospheric microbial communities are shaped by potentially assimilating plant-derived simple carbohydrates, ammonium and amino acids. Additionally, high expression of microbial stress and transport response proteins (porins, components of ATP binding cassette transporters and TonB-dependent receptors) indicates the nutrient poor environment of the phyllosphere. Moreover, some of the genes involved in the nitrogen fixation (such as nifH) are exclusively expressed in the rhizospheric as compared to phyllospheric microbiome region. Metabolic footprint studies of Ryffel et al. (2016) identified plant-derived simple carbohydrates such as sucrose, fructose and glucose at the phyllospheric zone. Alteration in these sugar in leaves of *Arabidopsis thaliana* was observed only after colonisation by the organoheterotroph *Sphingomonas melonis* or the phytopathogen *Pseudomonas syringae* pv. *tomato*, indicating these are probably the primary carbon sources in the phyllosphere. Metabolomic studies also determined that the phyllospheric microbial communities influence host-derived arginine metabolites and phytoalexin biosynthesis after occupying their surface. Thus, high-throughput proteomics and metabolomics studies can dramatically enhance our perception of molecular basis of plant-microbe association.

Prior to development of high-throughput methods, endophytic communities were considered as contaminants, though some were suggested as symbionts of the host (Ryan et al. 2008). Endophytes of asymptomatic and symptomatic anthracnose of *Paullinia cupana* was assessed by culture-dependent and 16S rRNA clone libraries, where clone libraries revealed more different phylogenetic microbial communities compared to culture-dependent methods (Bogas et al. 2015). However, Lundberg et al. (2012) studied pyrosequencing of the bacterial 16S rRNA gene of the surrounding root rhizosphere and endophytic compartment of more than 600 *A. thaliana* plants, those were grown on different soil types and belonging to different developmental stages. The study identified a core endophytic microbiome, related to *Proteobacteria* and *Actinobacteria*, which is consistent between different soil types and development stages. Such types of microbial community are localised and probably functionally well connected with host metabolic processes; they are therefore sometimes referred to as 'core communities' (Turner et al. 2013).

A vast range of studies have been carried out to understand phylogenetic nature of microbial communities associated with diverse hosts and during their different life stages as mentioned above. Though, there is a strong need to understand which microorganisms are active in different parts of plant and what they are doing during different development life stages of the host. Until now, the majority of studies include biomarkers for explaining the function of genes using different molecular tools, as Yang et al. (2011) used biomarker to determine nitrogen status in maize. Reporter genes enable the assessment of how specific microbial members of the rhizosphere perceive their environment in terms of chemical, physical and biological stimuli (Deangelis et al. 2005; Steindler and Venturi 2007; De Werra et al. 2008). There is no clear picture of the overall functional role of the plant microbiome, though some studies have used -omics tools to understand aspects of their function (Delmotte et al. 2009; Wang et al. 2011; Knief et al. 2012). These studies have identified some of the proteins that could not previously be evaluated using biomarker assays. Identifying functions of the plant microbiome could enhance sustainable agriculture (Berg 2009; Lundberg et al. 2012). These advancements could be achieved by using beneficial microorganisms as biofertilisers, as biocontrol agents or as stress protection products. A better functional understanding of the plant microbiome might be vital to prevent the outbreak of rhizospheric plant diseases. For instance, a succession of plant disease managements has been achieved via transferring active beneficial microbiome by mixing diseased suppressive soils with the infected conducive soils (Mazzola 2007). Given these achievements, we suggest that in future multi-omics tools should be applied to plant-microbial interaction studies for better understanding their functional complexities. Results of those studies will help to improve crop management through active beneficial microbiome modulation.

As mentioned above, much work still needs to be done in different host microbiomes in order to improve experimental protocols and computational methodologies. These improvements will help us to integrate analysis of multiple 'omics' datasets in order to generate new knowledge with which to decipher complex biological systems.

Acknowledgement The authors are thankful to John Innes Centre and Saurashtra University for support and encouragement. Authors would like to thank Prof. Rob Field for insightful comments.

References

Addis MF, Tanca A, Uzzau S, Oikonomou G, Bicalho RC, Moroni P (2016) The bovine milk microbiota: insights and perspectives from -omics studies. Mol Biosyst 12:2359–2372. doi:10.1039/c6mb00217j

Bais HP, Weir TL, Perry LG, Gilroy S, Vivanco JM (2006) The role of root exudates in rhizosphere interactions with plants and other organisms. Annu Rev Plant Biol 57:233–266. doi:10.1146/annurev.arplant.57.032905.105159

Bakker PA, Berendsen RL, Doornbos RF, Wintermans PC, Pieterse CM (2013) The rhizosphere revisited: root microbiomics. Front Plant Sci 4:165. doi:10.3389/fpls.2013.00165

Berendsen RL, Pieterse CM, Bakker PA (2012) The rhizosphere microbiome and plant health. Trends Plant Sci 17:478–486. doi:10.1016/j.tplants.2012.04.001

Berg G (2009) Plant-microbe interactions promoting plant growth and health: perspectives for controlled use of microorganisms in agriculture. Appl Microbiol Biotechnol 84:11–18. doi:10.1007/s00253-009-2092-7

Berg G, Rybakova D, Grube M, Koberl M (2016) The plant microbiome explored: implications for experimental botany. J Exp Bot 67:995–1002. doi:10.1093/jxb/erv466

Bodenhausen N, Horton MW, Bergelson J (2013) Bacterial communities associated with the leaves and the roots of Arabidopsis thaliana. PLoS One 8:e56329. doi:10.1371/journal.pone.0056329

Bogas AC, Ferreira AJ, Araujo WL, Astolfi-Filho S, Kitajima EW, Lacava PT, Azevedo JL (2015) Endophytic bacterial diversity in the phyllosphere of Amazon Paullinia cupana associated with asymptomatic and symptomatic anthracnose. Springerplus 4:258. doi:10.1186/s40064-015-1037-0

Bulgarelli D, Schlaeppi K, Spaepen S, Ver Loren Van Themaat E, Schulze-Lefert P (2013) Structure and functions of the bacterial microbiota of plants. Annu Rev Plant Biol 64:807–838. doi:10.1146/annurev-arplant-050312-120106

Chu H, Mazmanian SK (2013) Innate immune recognition of the microbiota promotes host-microbial symbiosis. Nat Immunol 14:668–675. doi:10.1038/ni.2635

De Werra P, Baehler E, Huser A, Keel C, Maurhofer M (2008) Detection of plant-modulated alterations in antifungal gene expression in Pseudomonas fluorescens CHA0 on roots by flow cytometry. Appl Environ Microbiol 74:1339–1349. doi:10.1128/AEM.02126-07

De Wouters T, Dore J, Lepage P (2012) Does our food (environment) change our gut microbiome ('in-vironment'): a potential role for inflammatory bowel disease? Dig Dis 30(Suppl 3):33–39. doi:10.1159/000342595

Deangelis KM, Ji P, Firestone MK, Lindow SE (2005) Two novel bacterial biosensors for detection of nitrate availability in the rhizosphere. Appl Environ Microbiol 71:8537–8547. doi:10.1128/AEM.71.12.8537-8547.2005

Delmotte N, Knief C, Chaffron S, Innerebner G, Roschitzki B, Schlapbach R, Von Mering C, Vorholt JA (2009) Community proteogenomics reveals insights into the physiology of phyllosphere bacteria. Proc Natl Acad Sci U S A 106:16428–16433. doi:10.1073/pnas.0905240106

Den Besten G, Van Eunen K, Groen AK, Venema K, Reijngoud DJ, Bakker BM (2013) The role of short-chain fatty acids in the interplay between diet, gut microbiota, and host energy metabolism. J Lipid Res 54:2325–2340. doi:10.1194/jlr.R036012

Despres J, Forano E, Lepercq P, Comtet-Marre S, Jubelin G, Yeoman CJ, Miller ME, Fields CJ, Terrapon N, Le Bourvellec C, Renard CM, Henrissat B, White BA, Mosoni P (2016) Unraveling the pectinolytic function of Bacteroides xylanisolvens using a RNA-seq approach and mutagenesis. BMC Genomics 17:147. doi:10.1186/s12864-016-2472-1

Fernandez L, Langa S, Martin V, Maldonado A, Jimenez E, Martin R, Rodriguez JM (2013) The human milk microbiota: origin and potential roles in health and disease. Pharmacol Res 69:1–10. doi:10.1016/j.phrs.2012.09.001

Forbes JD, Van Domselaar G, Bernstein CN (2016) Microbiome survey of the inflamed and non-inflamed gut at different compartments within the gastrointestinal tract of inflammatory bowel disease patients. Inflamm Bowel Dis 22:817–825. doi:10.1097/MIB.0000000000000684

Gottel NR, Castro HF, Kerley M, Yang Z, Pelletier DA, Podar M, Karpinets T, Uberbacher E, Tuskan GA, Vilgalys R, Doktycz MJ, Schadt CW (2011) Distinct microbial communities within the endosphere and rhizosphere of Populus deltoides roots across contrasting soil types. Appl Environ Microbiol 77:5934–5944. doi:10.1128/AEM.05255-11

Graf D, Di Cagno R, Fak F, Flint HJ, Nyman M, Saarela M, Watzl B (2015) Contribution of diet to the composition of the human gut microbiota. Microb Ecol Health Dis 26:26164. doi:10.3402/mehd.v26.26164

Greenblum S, Turnbaugh PJ, Borenstein E (2012) Metagenomic systems biology of the human gut microbiome reveals topological shifts associated with obesity and inflammatory bowel disease. Proc Natl Acad Sci U S A 109:594–599. doi:10.1073/pnas.1116053109

Knief C (2014) Analysis of plant microbe interactions in the era of next generation sequencing technologies. Front Plant Sci 5:216. doi:10.3389/fpls.2014.00216

Knief C, Delmotte N, Chaffron S, Stark M, Innerebner G, Wassmann R, Von Mering C, Vorholt JA (2012) Metaproteogenomic analysis of microbial communities in the phyllosphere and rhizosphere of rice. ISME J 6:1378–1390. doi:10.1038/ismej.2011.192

Koranda M, Schnecker J, Kaiser C, Fuchslueger L, Kitzler B, Stange CF, Sessitsch A, Zechmeister-Boltenstern S, Richter A (2011) Microbial processes and community composition in the rhizosphere of European beech—the influence of plant C exudates. Soil Biol Biochem 43:551–558. doi:10.1016/j.soilbio.2010.11.022

Kyrpides NC (2009) Fifteen years of microbial genomics: meeting the challenges and fulfilling the dream. Nat Biotechnol 27:627–632. doi:10.1038/nbt.1552

Lareen A, Burton F, Schafer P (2016) Plant root-microbe communication in shaping root microbiomes. Plant Mol Biol 90:575–587. doi:10.1007/s11103-015-0417-8

Lebeer S, Vanderleyden J, De Keersmaecker SC (2010) Host interactions of probiotic bacterial surface molecules: comparison with commensals and pathogens. Nat Rev Microbiol 8:171–184. doi:10.1038/nrmicro2297

Lin J, Qian J (2007) Systems biology approach to integrative comparative genomics. Expert Rev Proteomics 4:107–119. doi:10.1586/14789450.4.1.107

Luczynski P, Mcvey Neufeld KA, Oriach CS, Clarke G, Dinan TG, Cryan JF (2016) Growing up in a bubble: using germ-free animals to assess the influence of the gut microbiota on brain and behavior. Int J Neuropsychopharmacol 19. doi:10.1093/ijnp/pyw020

Lundberg DS, Lebeis SL, Paredes SH, Yourstone S, Gehring J, Malfatti S, Tremblay J, Engelbrektson A, Kunin V, Del Rio TG, Edgar RC, Eickhorst T, Ley RE, Hugenholtz P, Tringe SG, Dangl JL (2012) Defining the core Arabidopsis thaliana root microbiome. Nature 488:86–90. doi:10.1038/nature11237

Martens EC, Koropatkin NM, Smith TJ, Gordon JI (2009) Complex glycan catabolism by the human gut microbiota: the Bacteroidetes Sus-like paradigm. J Biol Chem 284:24673–24677. doi:10.1074/jbc.R109.022848

Mazzola M (2007) Manipulation of rhizosphere bacterial communities to induce suppressive soils. J Nematol 39:213–220

Medini D, Serruto D, Parkhill J, Relman DA, Donati C, Moxon R, Falkow S, Rappuoli R (2008) Microbiology in the post-genomic era. Nat Rev Microbiol 6:419–430. doi:10.1038/nrmicro1901

Mendes LW, Kuramae EE, Navarrete AA, Van Veen JA, Tsai SM (2014) Taxonomical and functional microbial community selection in soybean rhizosphere. ISME J 8:1577–1587. doi:10.1038/ismej.2014.17

Mendes R, Garbeva P, Raaijmakers JM (2013) The rhizosphere microbiome: significance of plant beneficial, plant pathogenic, and human pathogenic microorganisms. FEMS Microbiol Rev 37:634–663. doi:10.1111/1574-6976.12028

Mendes R, Kruijt M, De Bruijn I, Dekkers E, Van Der Voort M, Schneider JH, Piceno YM, Desantis TZ, Andersen GL, Bakker PA, Raaijmakers JM (2011) Deciphering the rhizosphere microbiome for disease-suppressive bacteria. Science 332:1097–1100. doi:10.1126/science.1203980

Morrow AL, Rangel JM (2004) Human milk protection against infectious diarrhea: implications for prevention and clinical care. Semin Pediatr Infect Dis 15:221–228

Mu C, Yang Y, Zhu W (2016) Gut microbiota: the brain peacekeeper. Front Microbiol 7:345. doi:10.3389/fmicb.2016.00345

Nakanishi S, Kataoka K, Kuwahara T, Ohnishi Y (2003) Effects of high amylose maize starch and Clostridium butyricum on metabolism in colonic microbiota and formation of azoxymethane-induced aberrant crypt foci in the rat colon. Microbiol Immunol 47:951–958

Nakayama J, Watanabe K, Jiang J, Matsuda K, Chao SH, Haryono P, La-Ongkham O, Sarwoko MA, Sujaya IN, Zhao L, Chen KT, Chen YP, Chiu HH, Hidaka T, Huang NX, Kiyohara C, Kurakawa T, Sakamoto N, Sonomoto K, Tashiro K, Tsuji H, Chen MJ, Leelavatcharamas V, Liao CC, Nitisinprasert S, Rahayu ES, Ren FZ, Tsai YC, Lee YK (2015) Diversity in gut bacterial community of school-age children in Asia. Sci Rep 5:8397. doi:10.1038/srep08397

Prieto P, Schiliro E, Maldonado-Gonzalez MM, Valderrama R, Barroso-Albarracin JB, Mercado-Blanco J (2011) Root hairs play a key role in the endophytic colonization of olive roots by Pseudomonas spp. with biocontrol activity. Microb Ecol 62:435–445. doi:10.1007/s00248-011-9827-6

Rautava S, Luoto R, Salminen S, Isolauri E (2012) Microbial contact during pregnancy, intestinal colonization and human disease. Nat Rev Gastroenterol Hepatol 9:565–576. doi:10.1038/nrgastro.2012.144

Reck M, Tomasch J, Deng Z, Jarek M, Husemann P, Wagner-Dobler I (2015) Stool metatranscriptomics: a technical guideline for mRNA stabilisation and isolation. BMC Genomics 16:494. doi:10.1186/s12864-015-1694-y

Riviere A, Moens F, Selak M, Maes D, Weckx S, De Vuyst L (2014) The ability of bifidobacteria to degrade arabinoxylan oligosaccharide constituents and derived oligosaccharides is strain dependent. Appl Environ Microbiol 80:204–217. doi:10.1128/AEM.02853-13

Rodriguez JM (2014) The origin of human milk bacteria: is there a bacterial entero-mammary pathway during late pregnancy and lactation? Adv Nutr 5:779–784. doi:10.3945/an.114.007229

Rogowski A, Briggs JA, Mortimer JC, Tryfona T, Terrapon N, Lowe EC, Basle A, Morland C, Day AM, Zheng H, Rogers TE, Thompson P, Hawkins AR, Yadav MP, Henrissat B, Martens EC, Dupree P, Gilbert HJ, Bolam DN (2015) Glycan complexity dictates microbial resource allocation in the large intestine. Nat Commun 6:7481. doi:10.1038/ncomms8481

Rombeau JL, Kripke SA (1990) Metabolic and intestinal effects of short-chain fatty acids. JPEN J Parenter Enteral Nutr 14:181S–185S

Round JL, Mazmanian SK (2009) The gut microbiota shapes intestinal immune responses during health and disease. Nat Rev Immunol 9:313–323. doi:10.1038/nri2515

Rousk J, Bengtson P (2014) Microbial regulation of global biogeochemical cycles. Front Microbiol 5:103. doi:10.3389/fmicb.2014.00103

Ryan RP, Germaine K, Franks A, Ryan DJ, Dowling DN (2008) Bacterial endophytes: recent developments and applications. FEMS Microbiol Lett 278:1–9. doi:10.1111/j.1574-6968.2007.00918.x

Ryffel F, Helfrich EJ, Kiefer P, Peyriga L, Portais JC, Piel J, Vorholt JA (2016) Metabolic footprint of epiphytic bacteria on Arabidopsis thaliana leaves. ISME J 10:632–643. doi:10.1038/ismej.2015.141

Sanchez JI, Marzorati M, Grootaert C, Baran M, Van Craeyveld V, Courtin CM, Broekaert WF, Delcour JA, Verstraete W, Van De Wiele T (2009) Arabinoxylan-oligosaccharides (AXOS) affect the protein/carbohydrate fermentation balance and microbial population dynamics of the simulator of human intestinal microbial ecosystem. J Microbial Biotechnol 2:101–113. doi:10.1111/j.1751-7915.2008.00064.x

Schuijt TJ, Lankelma JM, Scicluna BP, De Sousa E, Melo F, Roelofs JJ, De Boer JD, Hoogendijk AJ, De Beer R, De Vos A, Belzer C, De Vos WM, Van Der Poll T, Wiersinga WJ (2016) The gut microbiota plays a protective role in the host defence against pneumococcal pneumonia. Gut 65:575–583. doi:10.1136/gutjnl-2015-309728

Schulberg J, De Cruz P (2016) Characterisation and therapeutic manipulation of the gut microbiome in inflammatory bowel disease. Intern Med J 46:266–273. doi:10.1111/imj.13003

Siboni N, Ben-Dov E, Sivan A, Kushmaro A (2008) Global distribution and diversity of coral-associated Archaea and their possible role in the coral holobiont nitrogen cycle. Environ Microbiol 10:2979–2990. doi:10.1111/j.1462-2920.2008.01718.x
Singh RP, Reddy CRK (2015) Unraveling the functions of the macroalgal microbiome. Front Microbiol 6:1488. doi:10.3389/fmicb.2015.01488
Steindler L, Venturi V (2007) Detection of quorum-sensing N-acyl homoserine lactone signal molecules by bacterial biosensors. FEMS Microbiol Lett 266:1–9. doi:10.1111/j.1574-6968.2006.00501.x
Tkacz A, Poole P (2015) Role of root microbiota in plant productivity. J Exp Bot 66:2167–2175. doi:10.1093/jxb/erv157
Tlaskalova-Hogenova H, Stepankova R, Kozakova H, Hudcovic T, Vannucci L, Tuckova L, Rossmann P, Hrncir T, Kverka M, Zakostelska Z, Klimesova K, Pribylova J, Bartova J, Sanchez D, Fundova P, Borovska D, Srutkova D, Zidek Z, Schwarzer M, Drastich P, Funda DP (2011) The role of gut microbiota (commensal bacteria) and the mucosal barrier in the pathogenesis of inflammatory and autoimmune diseases and cancer: contribution of germ-free and gnotobiotic animal models of human diseases. Cell Mol Immunol 8:110–120. doi:10.1038/cmi.2010.67
Turner TR, James EK, Poole PS (2013) The plant microbiome. Genome Biol 14:209. doi:10.1186/gb-2013-14-6-209
Vorholt JA (2012) Microbial life in the phyllosphere. Nat Rev Microbiol 10:828–840. doi:10.1038/nrmicro2910
Wang HB, Zhang ZX, Li H, He HB, Fang CX, Zhang AJ, Li QS, Chen RS, Guo XK, Lin HF, Wu LK, Lin S, Chen T, Lin RY, Peng XX, Lin WX (2011) Characterization of metaproteomics in crop rhizospheric soil. J Proteome Res 10:932–940. doi:10.1021/pr100981r
Weinert N, Piceno Y, Ding GC, Meincke R, Heuer H, Berg G, Schloter M, Andersen G, Smalla K (2011) PhyloChip hybridization uncovered an enormous bacterial diversity in the rhizosphere of different potato cultivars: many common and few cultivar-dependent taxa. FEMS Microbiol Ecol 75:497–506. doi:10.1111/j.1574-6941.2010.01025.x
Wu L, Wang J, Huang W, Wu H, Chen J, Yang Y, Zhang Z, Lin W (2015) Plant-microbe rhizosphere interactions mediated by Rehmannia glutinosa root exudates under consecutive monoculture. Sci Rep 5:15871. doi:10.1038/srep15871
Yang XFS, Wu JR, Ziegler TE, Yang X, Zayed A, Rajani MS, Zhou DF, Basra AS, Schachtman DP, Peng MS, Armstrong CL, Caldo RA, Morrell JA, Lacy M, Staub JM (2011) Gene expression biomarkers provide sensitive indicators of in planta nitrogen status in maize. Plant Physiology 157(4):1841–1852
Zhang L, Zhang L, Zhan X, Zeng X, Zhou L, Cao G, Chen A, Yang C (2016) Effects of dietary supplementation of probiotic, Clostridium butyricum, on growth performance, immune response, intestinal barrier function, and digestive enzyme activity in broiler chickens challenged with Escherichia coli K88. J Anim Sci Biotechnol 7:3. doi:10.1186/s40104-016-0061-4
Zhang W, Li F, Nie L (2010) Integrating multiple 'omics' analysis for microbial biology: application and methodologies. Microbiology 156:287–301. doi:10.1099/mic.0.034793-0
Zhang YJ, Li S, Gan RY, Zhou T, Xu DP, Li HB (2015) Impacts of gut bacteria on human health and diseases. Int J Mol Sci 16:7493–7519. doi:10.3390/ijms16047493

Uncultivated Lineages and Host–Microbe Interaction in Saline Environment

2

Kruti G. Dangar, Nirali M. Raiyani, Rupal D. Pandya, and Satya P. Singh

Abstract

The culture-independent approaches can contribute to untold properties of microorganisms. The host and microbe interactions explored through the metagenomics, metatranscriptomics, and metaproteomics approaches reveal the function of the ecosystem. The extremophilic communities can be detected by screening of genes, proteins, and enzymes directly from the environmental samples based on the marker genes and reference species. Evaluation of the host–microbe interaction based on models and libraries generates hidden metabolic pathways to explore the types of interactions. The identification of the unfamiliar microbial species based on functions and sequences of the host–microbe interaction opens new arena of the adaptation in extreme ecosystems, including saline habitats.

Keywords

Saline habitats • Metagenomics • Metatranscriptomics • Metaproteomics • Metagenomic libraries • Host–microbe interaction

2.1 Introduction

The host–microbe interactions are based on inter-intra networking (Cordero and Datta 2016) and the new dimensions of host–microbe interactions are identified on the basis of culture-independent approaches (Doolittle and Zhaxybayeva 2010; Franzosa et al. 2015). Culture- independent methods allow the analysis of the ecosystem on the basis of microbial taxonomic group with functional traits and their interactions with host (Herrmann and Shann 1997; Carpenter-Boggs

K.G. Dangar • N.M. Raiyani • R.D. Pandya • S.P. Singh (✉)
UGC-CAS Department of Biosciences, Saurashtra University, Rajkot 360 005, India
e-mail: satypasingh@yahoo.com; satyapsingh125@gmail.com

R.P. Singh et al. (eds.), *Understanding Host-Microbiome Interactions - An Omics Approach*, DOI 10.1007/978-981-10-5050-3_2

et al. 1998; Klamer and Bååth 1998; Riddech et al. 2002; Tiquia and Michel 2002; Behzad et al. 2016). Extreme environment is identified with extreme traits (Gohel and Singh 2015; Thumar and Singh 2009). High-throughput sequencing methods generate information regarding structure and function of genes, metabolic pathways, and evolution of microbial genomes (Zhang et al. 2010). The sequencing of the whole microbial community DNA by shotgun metagenomics provides information on the genes present in the ecosystems. Metagenomic approaches now a days are being used in marine symbiotic systems to search information regarding the microbes and their associated host (Woyke et al. 2006). Metatranscriptomics relates to the expression of the mRNA within microbial community toward the identification of the genes or genetic pathways, while transcriptomics is the analysis of the mRNA molecules, produced in one cell or a population of cells (Zhang et al. 2010). Proteomics is the analysis of the expressed proteins by a cell, tissue, or organism under a specific physiological condition (Woyke et al. 2006). In this chapter, we describe host–microbe interactions in their habitats. Applications of various omics approaches are highlighted with respect to occurrence, adaptation, and detection of interaction in saline habitats.

2.2 Culture-Independent (Meta-analysis) Approaches

2.2.1 Metagenomics

It is well established that only less than 1% of the microorganisms are cultivable under the laboratory conditions, while the remaining majorities are viable in environment and remain non-cultivable in laboratory (Amann 2000; Ward et al. 1990; Øvreås 2000; Floyd et al. 2005; Handelsman 2004). The term metagenomics, first introduced by Handelsman (Handelsman et al. 1998), relates to the environmental genomics and is based on the isolation of total nucleic acids from environmental samples, further subjected to sequencing and construction of libraries. This helps to explore uncultivated microbial communities of the ecosystem (Zeyaullah et al. 2009; Streit and Schmitz 2004; Cowan et al. 2015).

It is difficult to isolate good quality of total genomic DNA in sufficient amount from extreme environments. Several methods have been reported to extract the metagenomic DNA from different extreme environments (Stein et al. 1996; Venter et al. 2004; Purohit and Singh 2009; Daniel 2005; Abulencia et al. 2006; Wu and Sun 2009; Siddhapura et al. 2010; Behzad et al. 2016). High-molecular-weight metagenomic DNA have been analyzed by sequences and function-based approaches by constructing libraries followed by the expression of genes in suitable cloned vectors (Bertrand et al. 2005; Martin-Laurent et al. 2001; Warnecke and Hess 2009). Functional screening of the total genomic

DNA is carried out by constructing the libraries followed by the screening for novel traits (Henne et al. 1999; Rondon et al. 2000; Handelsman et al. 1998; Krsek and Wellington 1999). However, metagenomic libraries generate large number of hypothetical proteins as indicated in the databases (Handelsman 2004).

After successful amplification and sequencing whole genome of single cells, sequencing methods for metagenomics have been developed (Marcy et al. 2007; Mußmann et al. 2007; Bentley 2006). The pyrosequencing based on GS-FLX platform and Illumina sequencing based on bridge clonal amplification by reversible dye terminator generated highest average length 750 and 300 bp, respectively (Edwards et al. 2006; Turnbaugh et al. 2009; Qin et al. 2010). The advance versions of sequencing by NeSSM: a Next-generation Sequencing Stimulator for Metagenomics and Shotgun metagenomics (Jia et al. 2013; Schmidt et al. 2014), added to the knowledge of the unexplored ecosystems (Singh and Campbell 2009; Segata et al. 2013; Warnecke and Hess 2009). The metagenomic DNA from seawater of Northwest Atlantic to the Eastern Tropical Pacific contained 6.3 billion bp of sequences and 1.6 Gbp of unique metagenomic DNA sequences obtained from Sargasso Sea (Rusch et al. 2007; Yooseph et al. 2007; Venter et al. 2004). These generated informations about more than ~264,949 organisms, ~123,516 sequencing projects, ~18,188 biosamples, and ~100,334 analysis, in the database (http://gold.jgi.doe.gov, Genome online Database GOLD: December 2016).

2.2.1.1 Holobionts and Hologenomes

The major challenge is posed in the exploration of the host–microbe interaction using culture-independent techniques. In 2011, new terms, holobiont or metaorganism, were suggested for the host-associated microbial community (Bosch and McFall-Ngai 2011). The analysis of the metagenomic DNA sequences and functions highlighted on the genomes of uncultivated microbes to understand interaction based on metabolic network in ecosystem (Vieites et al. 2009). A marine worm *Olavius algarvensis* was studied by constructing metagenomic library of pMCL200 and pCC1FOS host. It identified the role of symbionts sculpture-oxidizing and sulfate-reducing bacteria for providing nutrients to host (Woyke et al. 2006). The phylogenetic analyses of the host-associated microbial sequences indicated the wide phenomenon of the interaction occurring in the environment (Vieites et al. 2009). However, the analysis of the host–microbe interaction based on the culture-independent approaches faces difficulties in identification of dissimilar sequences. Based on the analysis of large number of samples collected from the surface marine, freshwater, and hypersaline of the Sorcerer II Global Ocean, it was revealed that most of the viral sequences were dissimilar and hard to establish the host-associated viral genome (Williamson et al. 2008; Tettelin et al. 2005).

2.2.2 Metatranscriptomics

Metagenomics and metatranscriptomics relate to the genes and their expression by extracting the total DNA and RNA, respectively, in an ecosystem (Doolittle and Zhaxybayeva 2010; Franzosa et al. 2015). Metatranscriptomics is carried out by reverse transcription to generate cDNA followed by the sequencing using metagenomic platforms (Warnecke and Hess 2009). Metatranscriptomic analysis relates to SSU rRNA database on the basis of the coding and non-coding RNAs (Caporaso et al. 2010; Schloss et al. 2009; Preheim et al. 2013; Gottesman 2002; Bejerano-Sagie and Xavier 2007). The transcripts of marine and freshwater bacterioplankton generated insight into the microbial communities (Poretsky et al. 2005). On a similar account, the interaction of free-living and particle-associated microorganisms with symbiont phytoplankton was explored by metatranscriptomics of carbon, nitrogen, phosphate, and sulfur cycles (Satinsky et al. 2014).

2.2.3 Metaproteomics

Metaproteomics relates to the analysis of the entire protein directly from the environmental samples using 2-dimentional gel electrophoresis or high-performance liquid chromatography and high-throughput mass spectrometry (Wilke et al. 2003; Nesatyy and Suter 2007; Domon and Aebersold 2006; Wilmes and Bond 2004, 2006; Benndorf et al. 2007). Energy and nutrient limitation pathways of free-living and symbiotic bacteria were analyzed by metaproteomics in marine *Olavius algarvensis* (Kleiner et al. 2012).

2.3 Habitats and Host–Microbe Interaction

Various microorganisms are affected by the dissolved metals in different concentrations (Ventosa et al., 2015; Behzad et al. 2016). Culture-independent approaches have identified the distribution of prokaryotes, eukaryotes, archaea, and viruses in saline habitats (Purohit and Singh 2009; Oren 2011; Benlloch et al. 2002; Santos et al. 2012; Luk et al. 2014; Narasingarao et al. 2012; Dillon et al. 2013). *Haloquadratum* was abundantly present in 19% NaCl saltern and being less abundant in reduced NaCl saltern (Ghai et al. 2011; Fernández et al. 2014a, b).

Biological interactions are studied by different length scales, such as 1–10 μm for single cell interaction (Cordero and Datta 2016). The macro-environment is generated by the meta-population at scale of centimeters to meter, further expanded from meters to kilometers (Cordero and Datta 2016). Kirchman et al. (2010)

explored a scale of 1–1000 μm to distinguish the ocean community. The pink berries at 500 μm to 1 cm scale were used for the sulfur cycling consortia of the Sippewissett Salt Marsh (Wilbanks et al. 2014). The microbial communities of the photic and aphotic zones of oceans have been analyzed by metagenomic method (Ghiglione et al. 2012).

The community dynamics and interaction are governed by time and interval of sampling. Dakos et al. (2008) investigated species composition affected by seasonal changes in ocean. The omics data examined on time series techniques identified periodical patterns (Fuhrman et al. 2006; Gilbert et al. 2012; Hekstra et al. 2012, Benincà E et al. 2008). The pH and temperature affect the habitats without the interaction of species (Faust et al. 2015). Population structure of *Prochlorococcus* was studied by cell-cell comparison in large number of subpopulations (Kashtan et al. 2014). Temperature-induced variability in the population of *Prochlorococcus* was observed in Atlantic Ocean (Johnson et al. 2006). Biosynthesis of indole-3-acetic acid (IAA) through the mutualistic interactions of *Pseudo-nitzschia multiseries* with *Sulfitobacter* in costal water was studied by the transcriptomic analysis (Amin et al. 2015).

2.4 Analysis of the Host–Microbe Interaction by Metagenomics, Metatranscriptomics, and Metaproteomics

Marine microorganisms are capable to live under the extremities of high salinity, pressure, and temperature. Therefore, they have developed ways to protect themselves against these challenges. The analysis of the host and microbes interaction can lead to the better understanding of the marine ecosystem. The host and its microbial community are referred as holobiont or metaorganism (Bosch and McFall-Ngai 2011). To access the information on the host–microbe interaction metagenomics, metatranscriptomics and metaproteomics approaches are quite useful. With the advancements in nucleic acid sequencing, high-throughput sequencing platforms are developed. During the last several years, many biomarkers have been developed to characterize microbial population. It includes DNA/RNA sequencing, PhyloChip, GeoChip, and mass spectroscopic-based proteomics for the analysis of the community and their metabolite pathways (Zhou et al. 2015). GS 20 pyrosequencing system analyzed the interaction between Kingman coral reef and Proteobacteria, Actinobacteria, Bacteroidetes, Cyanobacteria, Firmicutes, and Planctomycetes (Krause et al. 2008). Similarly, *Porites astreoides*, another coral, and Proteobacteria, Firmicutes, Cyanobacteria, and Actinobacteria were also analyzed (Dinsdale et al. 2008). Another example of the metagenomic study relates to the coral *Porites astreoides*, where Proteobacteria, Firmicutes, Cyanobacteria, and Actinobacteria were identified (Wegley et al. 2007).

Metatranscriptomics provides information related to gene expression in the entire microbial community. mRNA transcripts code for housekeeping genes, such as elongation factors, DNA gyrase and sigma factors, ATP binding cassette (ABC) and tripartite ATP-independent (TRAP) transporter, extracellular solute binding proteins (SBP), and TonB-dependent transport systems (TBDT) (Kopf et al. 2015). Total RNA extracted from gill of *Solemya velum* from Rhode Island was linearly amplified and converted into cDNA. The metatranscriptome sequenced in Roche Genome Sequencer FLX instrument suggested that bacterial transcripts coded for genes related to sulfur oxidation pathway. The relative abundance of Gammaproteobacteria transcripts clearly suggested its dominant role in sulfide metabolism under the symbiotic conditions (Stewart et al. 2011).

Metaproteomics identifies the expressed proteins in an ecosystem to determine the function of the microbial community (Wang et al. 2014). Metagenomics and metaproteomics in combination explored metabolic interaction of the host and symbiont of Sargasso Sea. The assimilation pathway under the symbiotic conditions of the host has been studied and elucidated up to certain extent. A total of 2215 peptides mapped to 236 SAR11 proteins; 1911 peptides mapped to 402 *Prochlorococcus* proteins and 2407 peptides mapped to 404 *Synechococcus* proteins identified by capillary liquid chromatography (LC)-tandem mass spectrometry in Sargasso Sea to detect microbial proteins (Sowell et al. 2009). Metaproteomics analysis of South Atlantic surface seawater revealed the presence of two different proteins by the mass spectrometry. One of these two proteins, TonB-dependent transporters (TBDT), is a membrane protein involved in nutrients transport across the outer membrane of bacteria. While the other protein identified as rhodopsin closely relates to four bacterial phyla *Chloroflexi*, *Proteobacteria*, Cyanobacteria, and Actinobacteria. Both these proteins are thus involved in energy generation in phototrophic bacterioplankton (Morris et al. 2010). PAGE and liquid chromatography in combination aided proteomics analysis of the chemosynthetic association of a gutless marine worm and its bacterial symbionts, δ-Proteobacteria and γ-Proteobacteria (Kleiner et al. 2012). The various omics approaches help to understand the interactions of sponge and cultivable and uncultivable microorganism.

The metagenomic DNA sequenced on Roche 454 Titanium platform generate data sets comparable to metaproteome analysis. The metabolic interactions are involved with carbohydrate transport, posttranslational modification, protein turnover, chaperone functions, and signal transduction. It's further revealed that molecular chaperonin GroEL (HSP60 family) and outer membrane receptor proteins CirA and OmpA are involved in COG analysis. Therefore, metaproteomics data can be correlated with the physiological properties and activities of the sponge-microbial interaction (Liu et al. 2012). Cyanophycin synthetase (cphA) is responsible for the synthesis of cyanophycin (multi-L-arginyl-poly-L-aspartate), a storage polymer, detected in the deep-sea symbionts (Gardebrecht et al. 2012). Metagenomic DNA of two sponges *Stylissa carteri* (SC) and *Xestospongia testudinaria* (XT) from the Red Sea coast of Saudi Arabia was extracted followed by the PCR amplification with Bac27F and Bac1492R bacterial primer. Genomic DNA and RNA libraries prepared by TruSeq kit (Illumina, USA) were used to analyze the metatranscriptomes of the microbial consortia in SC and XT (Ryu et al. 2016).

2.5 Detection of Phylogenetic Marker

Microbial communities associated with host are studied by using phylogenetic marker genes. Microbial diversity of an environment can be analyzed on the basis of conserved marker genes, such as 16S rRNA genes, *recA* or *radA* and genes encoding heat shock protein 70, elongation factor Tu, or elongation factor G and rpoB (Simon and Daniel 2011; Case et al. 2007). With the advances in sequencing platforms, comprehensive phylogenetic analysis of the microbial community has become possible (Metzker 2010). The 454 pyrosequencing technique allows high-throughput sequencing of hyper variable regions of 16S rRNA genes providing higher coverage of the microbial diversity as compared to the Sanger sequencing technique. Pyrosequencing technique provides sufficient phylogenetic information with the possibility of using multiple environmental samples in a single run.

The marine sponges from Queensland in Australia were studied for "sponge-specific" 16S rRNA PCR amplicons. The study revealed the presence of Poribacteria; Cyanobacteria; Chloroflexi; Acidobacteria; Alpha-, Gamma-, and Deltaproteobacteria; Actinobacteria; Spirochaetes; and Nitrospirae (Webster et al. 2010). Novel 16S rRNA genomic markers were identified from the microbial community in association with the sponge may suggest specific mechanisms for the interaction (Thomas et al. 2010). The proteomics study indicated the presence of the uncultured *Actinobacteria* clusters (Morris et al. 2010). Based on 16S rRNA gene pyrosequencing data, Gammaproteobacteria, Uncultured Deltaproteobacteria, Roseomonas, unclassified Clostridia, Capnocytophaga, Crenarchaeota, Nitrospira, Thermoplasmata, and Acidobacteria are recently detected and reported from the Red Sea (Cao et al. 2015). Similarly, the analysis of the metatranscriptomes of vent snail *Alviniconcha* collected from Kilo Moana, Tow Cam, ABE, and Tui Malila has been carried out using 454 pyrosequencing/quantitative PCR suggested the presence of γ- and ε-proteobacteria (Sanders et al. 2013). Bacteria and Archaea, such as Cyanobacteria, Crenarchaeota, Euryarchaeota, Thaumarchaeota, Actinobacteria, Firmicutes,Alphaproteobacteria, Betaproteobacteria, Deltaproteobacteria, Gammaproteobacteria, and Bacteroidetes, were identified from deep-sea Northeast Pacific Ocean using metagenomic and metatranscriptomic analysis (Wu et al. 2013).

2.5.1 Reference Species for Host–Microbe Interaction

The identification of the reference species specific to host–microbe interactions in saline habitat is difficult. The draft genome of symbiotic cyanobacteria *Candidatus Synechococcus spongiarum* in sponge *Carteriospongia foliascens* from Red Sea indicated the absence of essential genes in cyanobacteria for photosynthesis and DNA repair mechanisms (Gao et al. 2014). Wagner-Döbler et al. (2010) reported complete genome sequence *of Dinoroseobacter shibae* DFL12^T coding for 4198 protein of vitamin B_1 and B_{12} synthesis. This can be beneficially related to its host marine microalgae *Isochrysis galbana* and dinoflagellates.

The symbiotic relationship of cyanobacteria and haptophytes in North Atlantic Ocean was recognized by metagenomic analysis of the phylogenetic markers (Krupke et al. 2014). On a similar note, sulfur oxidizing and reducing bacteria were identified by functional molecular markers *cbb*, *soxB* genes, and *nif*H gene in a hypersaline environment (Turova et al. 2013; Kovaleva et al. 2011; Tourova et al. 2013). The nitrogen fixation and carbon exchange were studied in uncultured cyanobacterium *Candidatus atelocyanobacterium thalassa* under the symbiotic relationship with its host algae (Thompson et al. 2012). Thus, unique interactive features of the host and microbes interaction can be identified on the basis of the molecular signatures.

The advancement in sequencing technology coupled with the computational tools has greatly helped the microbiome analyses to investigate the habitat (Segata et al. 2013; Qin et al. 2010; Chaffron et al. 2010; Larsen et al. 2012). More recently, the omics data were subjected to the Lotka–Volterra model to identify mutualism and antagonism (Cordero and Datta 2016). New tool such as metaSHARK (metabolic search and reconstruction kit) web server available at http://bioinformatics.leeds.ac.uk/shark/ is helpful to visualize KEGG metabolic network (Hyland et al. 2006; Segata et al. 2013). Similarly, SPIEC-EASI, freely accessible at https://github.com/zdk123/SpiecEasi, can be used to elucidate the sequencing based network (Kurtz et al. 2015).

2.6 Detection of Host–Microbe Interaction by Various Techniques Is Displayed in Table 2.1

Conclusion

The culture-independent methods are employed to explore the microbial diversity with respect to host. Advanced computational tools clearly identify the functionally contributing traits to elucidate the host–microbe interaction in saline ecosystem. Uncultivated lineages are unique and can be extremely useful in biotechnology.

Table 2.1 Summary of various techniques to classify the host-microbe interactions in various saline environments

Sr. no.	Technique	Host–microbe interaction	Habitat	Outcome	Reference(s)
1	Environmental genomic tags or EGTs	Microbial communities associated with whale carcasses	Deep sea	Analyze and monitor the microbial communities metabolic framework	Tringe et al. (2005)
2	Roche 454 GS-FLX system	Halovirus and *Haloquadratum walsbyi* and *Salinibacter ruber*	Santa Pola salterns	Many environmental viruses belong to unknown groups or prey on uncultured and little known cellular lineages	Garcia-Heredia et al. (2012)
3	Flow cytometer, Virochip, Nanohaloarchaeal chip	*Haloquadratum walsbyi*, *S. ruber M8*, *S. ruber M31*, *Candidatus Nanosalinarum J07AB56*, *Candidatus Nanosalina sp.* J07AB43, host SAG D14) along with a total of 43 viruses	Santa Pola salterns	Viruses infecting the ubiquitous hyperhalophilic Nanohaloarchaeota (microbial dark matter, the uncultured fraction of the microbial world)	Martínez-García et al. (2014)
4	Combination of FISH and mass spectrometry-based techniques, NanoSIMS	Uncultured methane-oxidizing archaea and sulfate-reducing bacteria	Anoxic marine sediments	Provide evidence for syntrophic coupling through direct electron transfer	McGlynn et al. (2015)
5	Illumina HiSeq2000, 2D-LC-MS/MS	Marine oligochaete worm *Olavius algarvensis* and sulfur-oxidizing and sulfate-reducing bacteria	Bay of Sant' Andrea, Elba, Italy	Bacteria provide nutrition to worm	Wippler et al. (2016)
6	Proteomic approach based on the metagenome sequence	*R. pachyptila*	Hydrothermal vents	Provide benefits in energy metabolism to worm	Markert et al. (2007)

(continued)

Table 2.1 (continued)

Sr. no.	Technique	Host–microbe interaction	Habitat	Outcome	Reference(s)
7	RNA-SIP	Marine sediments with acetate utilizing manganese-reducing bacteria	Manganese oxide-rich sediment	Identified *Arcobacter Colwellia*- and *Oceanospirillaceae*-affiliated bacteria as the main acetate-oxidizing manganese reducers	Vandieken et al. (2012)
8	Combination of SIP and metatranscriptomics	Role of *methanotrophs* in lake sediment	Lake sediment	Transcripts of the methane monooxygenase genes *pmoCAB*	Dumont et al. (2013)
9	Combination of rRNA-SIP, DNA-SIP, and protein-SIP	Microbiota at marine seeps	Marine sediments	Sulfate-reducing alkane degraders	Kleindienst et al. (2014)

Acknowledgment Our work cited in this chapter from our research group has been supported under various programs of UGC, including the current CAS Programme, DST-FIST, and DST-Women Scientist Programme (Ms. Kruti Dangar). The DST-INSPIRE Fellowship and UGC-BSR Meritorious Fellowship to Ms. Nirali Raiyani and Ms. Rupal Pandya are duly acknowledged.

References

Abulencia CB, Wyborski DL, Garcia JA et al (2006) Environmental whole-genome amplification to access microbial populations in contaminated sediments. Appl Environ Microbiol 72(5):3291–3301

Amann R (2000) Who is out there? Microbial aspects of biodiversity. Syst Appl Microbiol 23(1):1–8

Amin SA, Hmelo LR, Van Tol HM et al (2015) Interaction and signalling between a cosmopolitan phytoplankton and associated bacteria. Nature 522(7554):98–101

Behzad H, Ibarra MA, Mineta K et al (2016) Metagenomic studies of the Red Sea. Gene 576(2):717–723

Bejerano-Sagie M, Xavier KB (2007) The role of small RNAs in quorum sensing. Curr Opin Microbiol 10(2):189–198

Beninca E, Huisman J, Heerkloss R et al (2008) Chaos in a long-term experiment with a plankton community. Nature 451(7180):822–825

Benndorf D, Balcke GU, Harms H et al (2007) Functional metaproteome analysis of protein extracts from contaminated soil and groundwater. The ISME journal 1(3):224–234

Benlloch S, López-López A, Casamayor EO et al (2002) Prokaryotic genetic diversity throughout the salinity gradient of a coastal solar saltern. Environ Microbiol 4(6):349–360

Bentley DR (2006) Whole-genome re-sequencing. Curr Opin Genet Dev 16(6):545–552

Bertrand H, Poly F, Lombard N et al (2005) High molecular weight DNA recovery from soils prerequisite for biotechnological metagenomic library construction. J Microbiol Methods 62(1):1–11

Bosch TC, McFall-Ngai MJ (2011) Metaorganisms as the new frontier. Zoology 114(4):185–190

Cao H, Zhang W, Wang Y et al (2015) Microbial community changes along the active seepage site of one cold seep in the Red Sea. Front Microbiol 6:739

Caporaso JG, Kuczynski J, Stombaugh J et al (2010) QIIME allows analysis of high-throughput community sequencing data. Nat Methods 7(5):335–336

Carpenter-Boggs L, Kennedy AC, Reganold JP (1998) Use of phospholipid fatty acids and carbon source utilization patterns to track microbial community succession in developing compost. Appl Environ Microbiol 64(10):4062–4064

Case RJ, Boucher Y, Dahllöf I et al (2007) Use of 16S rRNA and rpoB genes as molecular markers for microbial ecology studies. Appl Environ Microbiol 73(1):278–288

Chaffron S, Rehrauer H, Pernthaler J et al (2010) A global network of coexisting microbes from environmental and whole-genome sequence data. Genome Res 20(7):947–959

Cordero OX, Datta MS (2016) Microbial interactions and community assembly at microscales. Curr Opin Microbiol 31:227–234

Cowan DA, Ramond JB, Makhalanyane TP et al (2015) Metagenomics of extreme environments. Curr Opin Microbiol 25:97–102

Dakos V, Scheffer M, Van Nes EH et al (2008) Slowing down as an early warning signal for abrupt climate change. Proc Natl Acad Sci U S A 105:14308–14312

Daniel R (2005) The metagenomics of soil. Nat Rev Microbiol 3(6):470–478

Dillon JG, Carlin M, Gutierrez A, Nguyen V et al (2013) Patterns of microbial diversity along a salinity gradient in the Guerrero Negro solar saltern, Baja CA Sur, Mexico. Front Microbiol p. 4

Dinsdale EA, Edwards RA, Hall D et al (2008) Functional metagenomic profiling of nine biomes. Nature 452:629–632

Domon B, Aebersold R (2006) Mass spectrometry and protein analysis. Science 312(5771):212–217
Doolittle WF, Zhaxybayeva O (2010) Metagenomics and the units of biological organization. BioScience 60(2):102–112
Dumont MG, Pommerenke B, Casper P (2013) Using stable isotope probing to obtain a targeted metatranscriptome of aerobic methanotrophs in lake sediment. Environ Microbiol Rep 5(5):757–764
Edwards RA, Rodriguez-Brito B, Wegley L et al (2006) Using pyrosequencing to shed light on deep mine microbial ecology. BMC Genomics 7(1):57
Faust K, Lahti L, Gonze D et al (2015) Metagenomics meets time series analysis: unraveling microbial community dynamics. Curr Opin Microbiol 25:56–66
Fernández AB, Vera-Gargallo B, Sánchez-Porro C (2014a) Comparison of prokaryotic community structure from Mediterranean and Atlantic saltern concentrator ponds by a metagenomic approach. Front Microbiol 5:196
Fernández AB, Ghai R, Martin-Cuadrado AB et al (2014b) Prokaryotic taxonomic and metabolic diversity of an intermediate salinity hypersaline habitat assessed by metagenomics. FEMS Microbiol Ecol 88(3):623–635
Floyd MM, Tang J, Kane M, Emerson D (2005) Captured diversity in a culture collection: case study of the geographic and habitat distributions of environmental isolates held at the American Type Culture Collection. Appl Environ Microbiol 71(6):2813–2823
Franzosa EA, Hsu T, Sirota-Madi A et al (2015) Sequencing and beyond: integrating molecular 'omics' for microbial community profiling. Nat Rev Microbiol 13(6):360–372
Gao ZM, Wang Y, Tian RM et al (2014) Symbiotic adaptation drives genome streamlining of the cyanobacterial sponge symbiont "Candidatus Synechococcus spongiarum". MBio 5(2):e00079-14
Garcia-Heredia I, Martin-Cuadrado AB, Mojica FJ, Santos F et al (2012) Reconstructing viral genomes from the environment using fosmid clones: the case of haloviruses. PLoS One 7(3):e33802
Gardebrecht A, Markert S, Sievert SM et al (2012) Physiological homogeneity among theendosymbionts of *Riftiapachyptila* and *Tevniajerichonana* revealed by proteogenomics. ISME J 6:766–776
Gilbert JA, Steele JA, Caporaso JG et al (2012) Defining seasonal marine microbial community dynamics. The ISME journal 6(2):298–308
Ghai R, Pašić L, Fernández AB et al (2011) New abundant microbial groups in aquatic hypersaline environments. Sci Rep 1:135
Ghiglione JF, Murray AE (2012) Pronounced summer to winter differences and higher wintertime richness in coastal Antarctic marine bacterioplankton. Environ Microbiol 14(3):617–29
Gohel SD, Singh SP (2015) Thermodynamics of a Ca^{2+}-dependent highly thermostable alkaline protease from a haloalkliphilic actinomycete. Int J Biol Macromol 72:421–429
Gottesman S (2002) Stealth regulation biological circuits with small RNA switches. Genes Dev 16(22):2829–2842
Fuhrman JA, Hewson I, Schwalbach MS et al (2006) Annually reoccurring bacterial communities are predictable from ocean conditions. Proc Nat Aca Sci 103(35):13104–13109
Hekstra DR, Leibler S (2012) Contingency and statistical laws in replicate microbial closed ecosystems. Cell 149(5):1164–1173
Handelsman J (2004) Metagenomics: application of genomics to uncultured microorganisms. Microbiol Mol Biol Rev 68(4):669–685
Handelsman J, Rondon MR, Brady SF et al (1998) Molecular biological access to the chemistry of unknown soil microbes: a new frontier for natural products. Chem Biol 5(10):R245–R249
Henne A, Daniel R, Schmitz RA et al (1999) Construction of Environmental DNA Libraries in *Escherichia coli* and Screening for the Presence of Genes Conferring Utilization of 4-Hydroxybutyrate Appl. Environ Microbiol 65(9):3901–3907
Herrmann RF, Shann JF (1997) Microbial community changes during the composting of municipal solid waste. Microb Ecol 33(1):78–85
Hyland C, Pinney JW, McConkey GA et al (2006) metaSHARK: a WWW platform for interactive exploration of metabolic networks. Nucleic Acids Res 34(Suppl 2):W725–W728
Jia B, Xuan L, Cai K et al (2013) NeSSM: a next-generation sequencing simulator for metagenomics. PLoS One 8(10):e75448

Johnson ZI, Zinser ER, Coe A et al (2006) Niche partitioning among Prochlorococcus ecotypes along ocean-scale environmental gradients. Science 311(5768):1737–1740

Kashtan N, Roggensack SE, Rodrigue S et al (2014) Single-cell genomics reveals hundreds of coexisting subpopulations in wild Prochlorococcus. Science 344(6182):416–420

Kirchman DL, Cottrell MT, Lovejoy C (2010) The structure of bacterial communities in the western Arctic Ocean as revealed by pyrosequencing of 16S rRNA genes. Environ Microbiol 1:12(5):1132–1134

Klamer M, Bååth E (1998) Microbial community dynamics during composting of straw material studied using phospholipid fatty acid analysis. FEMS Microbiol Ecol 27(1):9–20

Kleindienst S, Herbst FA, Stagars M et al (2014) Diverse sulfate-reducing bacteria of the Desulfosarcina/Desulfococcus clade are the key alkane degraders at marine seeps. ISME J 8(10):2029–2044

Kleiner M, Wentrup C, Lott C et al (2012) Metaproteomics of a gutless marine worm and its symbiotic microbial community reveal unusual pathways for carbon and energy use. Proc Natl Acad Sci U S A 109(19):E1173–E1182

Kopf A, Kostadinov I, Wichels A et al (2015) Metatranscriptome of marine bacterioplankton during winter time in the North Sea assessed by total RNA sequencing. Mar Genomics 19:45–46

Kovaleva OL, Tourova TP, Muyzer G et al (2011) Diversity of RuBisCO and ATP citrate lyase genes in soda lake sediments. FEMS Microbiol Ecol 75(1):37–47

Krause L, Diaz NN, Goesmann A et al (2008) Phylogenetic classification of short environmental DNA fragments. Nucleic Acids Res 36(7):2230–2239

Krsek M, Wellington EM (1999) Comparison of different methods for the isolation and purification of total community DNA from soil. J Microbiol Methods 39(1):1–16

Krupke A, Lavik G, Halm H et al (2014) Distribution of a consortium between unicellular algae and the N2 fixing cyanobacterium UCYN-A in the North Atlantic Ocean. Environ Microbiol 16(10):3153–3167

Kurtz ZD, Müller CL, Miraldi ER et al (2015) Sparse and compositionally robust inference of microbial ecological networks. PLoS Comput Biol 11(5):e1004226

Larsen PE, Field D, Gilbert JA (2012) Predicting bacterial community assemblages using an artificial neural network approach. Nat Methods 9(6):621–625

Liu M, Fan L, Zhong L et al (2012) Metaproteogenomic analysis of a community of sponge symbionts. ISME J 6(8):1515–1525

Luk AW, Williams TJ, Erdmann S et al (2014) Viruses of haloarchaea. Life 4(4):681–715

Marcy Y, Ouverney C, Bik EM et al (2007) Dissecting biological "dark matter" with single-cell genetic analysis of rare and uncultivated TM7 microbes from the human mouth. Proc Natl Acad Sci U S A 104(29):11889–11894

Markert S, Arndt C, Felbeck H (2007) Physiological proteomics of the uncultured endosymbiont of Riftiapachyptila. Science 315(5809):247–250

Martínez-García M, Santos F, Moreno-Paz M (2014) Unveiling viral–host interactions within the 'microbial dark matter'. Nat Commun 5:4542

Martin-Laurent F, Philippot L, Hallet S et al (2001) DNA extraction from soils: old bias for new microbial diversity analysis methods. Appl Environ Microbiol 67(5):2354–2359

McGlynn SE, Chadwick GL, Kempes CP et al (2015) Single cell activity reveals direct electron transfer in methanotrophic consortia. Nature 526(7574):531–535

Metzker ML (2010) Sequencing technologies—the next generation. Nat Rev Genet 11(1):31–46

Morris RM, Nunn BL, Frazar C et al (2010) Comparative metaproteomics reveals ocean-scale shifts in microbial nutrient utilization and energy transduction. ISME J 4(5):673–685

Mußmann M, Hu FZ, Richter M et al (2007) Insights into the genome of large sulfur bacteria revealed by analysis of single filaments. PLoS Biol 5(9):e230

Narasingarao P, Podell S, Ugalde JA et al (2012) De novo metagenomic assembly reveals abundant novel major lineage of Archaea in hypersaline microbial communities. ISME J 6(1):81–93

Nesatyy VJ, Suter MJ (2007) Proteomics for the analysis of environmental stress responses in organisms. Environ Sci Technol 41(20):6891–6900

Oren A (2011) Thermodynamic limits to microbial life at high salt concentrations. Environ Microbiol 13(8):1908–1923

Øvreås L (2000) Population and community level approaches for analysing microbial diversity in natural environments. Ecol Lett 3(3):236–251

Poretsky RS, Bano N, Buchan A et al (2005) Analysis of microbial gene transcripts in environmental samples. Appl Environ Microbiol 71(7):4121–4126
Preheim SP, Perrotta AR, Friedman J et al (2013) Computational methods for high-throughput comparative analyses of natural microbial communities. Methods Enzymol 531:353–370
Purohit MK, Singh SP (2009) Assessment of various methods for extraction of metagenomic DNA from saline habitats of coastal Gujarat (India) to explore molecular diversity. Lett Appl Microbiol 49(3):338–344
Qin J, Li R, Raes J et al (2010) A human gut microbial gene catalogue established by metagenomic sequencing. Nature 464(7285):59–65
Riddech N, Klammer S, Insam H (2002) Characterisation of microbial communities during composting of organic wastes. In: Insam H, Riddech N, Klammer S (eds) Microbiology of composting. Springer, Berlin, pp 43–51
Rondon MR, August PR, Bettermann AD et al (2000) Cloning the soil metagenome: a strategy for accessing the genetic and functional diversity of uncultured microorganisms Appl. Environ Microbiol 66(6):2541–2547
Rusch DB, Halpern AL, Sutton G et al (2007) The Sorcerer II global ocean sampling expedition: northwest Atlantic through eastern tropical Pacific. PLoS Biol 5(3):e77
Ryu T, Seridi L, Moitinho-Silva L (2016) Hologenome analysis of two marine sponges with different microbiomes. BMC Genomics 17(1):158
Sanders JG, Beinart RA, Stewart FJ et al (2013) Metatranscriptomics reveal differences in in situ energy and nitrogen metabolism among hydrothermal vent snail symbionts. ISME J 7(8):1556–1567
Santos F, Yarza P, Parro V, Meseguer I et al (2012) Culture-independent approaches for studying viruses from hypersaline environments. Appl Environ Microbiol 78(6):1635–1643
Satinsky BM, Crump BC, Smith CB et al (2014) Microspatial gene expression patterns in the Amazon River Plume. Proc Natl Acad Sci U S A 111(30):11085–11090
Schloss PD, Westcott SL, Ryabin T et al (2009) Introducing mothur: open-source, platform-independent, community-supported software for describing and comparing microbial communities. Appl Environ Microbiol 75(23):7537–7541
Schmidt HF, Sakowski EG, Williamson SJ et al (2014) Shotgun metagenomics indicates novel family A DNA polymerases predominate within marine virioplankton. ISME J 8(1):103–114
Segata N, Boernigen D, Tickle TL et al (2013) Computational meta'omics for microbial community studies. Mol Syst Biol 9(1):666
Siddhapura PK, Vanparia S, Purohit MK et al (2010) Comparative studies on the extraction of metagenomic DNA from the saline habitats of Coastal Gujarat and Sambhar Lake, Rajasthan (India) in prospect of molecular diversity and search for novel biocatalysts. Int J Biol Macromol 47(3):375–379
Simon C, Daniel R (2011) Metagenomic analyses: past and future trends. Appl Environ Microbiol 77(4):1153–1161
Singh BK, Campbell CD (2009) Soil genomics. Nat Rev Microbiol 7(10):756–756
Sowell SM, Wilhelm LJ, Norbeck AD et al (2009) Transport functions dominate the SAR11 metaproteome at low-nutrient extremes in the Sargasso Sea. ISME J 3(1):93–105
Stein JL, Marsh TL, Wu KY et al (1996) Characterization of uncultivated prokaryotes: isolation and analysis of a 40-kilobase-pair genome fragment from a planktonic marine archaeon. J Bacteriol 178(3):591–599
Stewart F, Dmytrenko O, DeLong E et al (2011) Metatranscriptomic analysis of sulfur oxidation genes in the endosymbion to Solemya velum. Front Microbiol 2:134
Streit WR, Schmitz RA (2004) Metagenomics—the key to the uncultured microbes. Curr Opin Microbiol 7(5):492–498
Tettelin H, Masignani V, Cieslewicz MJ et al (2005) Genome analysis of multiple pathogenic isolates of Streptococcus agalactiae: implications for the microbial "pan-genome". Proc Natl Acad Sci U S A 102(39):13950–13955
Thomas T, Rusch D, DeMaere MZ et al (2010) Functional genomic signatures of sponge bacteria reveal unique and shared features of symbiosis. ISME J 4(12):1557–1567

Thompson AW, Foster RA, Krupke A, Carter BJ, Musat N et al (2012) Unicellular cyanobacterium symbiotic with a single-celled eukaryotic alga. Science 337(6101):1546–1550
Thumar JT, Singh SP (2009) Organic solvent tolerance of an alkaline protease from salt-tolerant alkaliphilic Streptomyces clavuligerus strain Mit-1. J Ind Microbiol Biotechnol 36(2):211
Tiquia SM, Michel FC Jr (2002) Bacterial diversity in livestock manure composts as characterized by terminal restriction fragment length polymorphisms (T-RFLP) of PCR-amplified 16S rRNA gene sequences. In: Insam H, Riddech N, Klammer S (eds) Microbiology of composting. Springer, Berlin, pp 65–82
Tourova TP, Slobodova NV, Bumazhkin BK et al (2013) Analysis of community composition of sulfur-oxidizing bacteria in hypersaline and soda lakes using soxB as a functional molecular marker. FEMS Microbiol Ecol 84(2):280–289
Tringe SG, Von Mering C, Kobayashi A et al (2005) Comparative metagenomics of microbial communities. Science 308(5721):554–557
Turnbaugh PJ, Hamady M, Yatsunenko T et al (2009) A core gut microbiome in obese and lean twins. Nature 457(7228):480–484
Turova TP, Slobodova NV, Bumazhkin BK et al (2013) Diversity of diazotrophs in the sediments of hypersaline salt and soda lakes analyzed with the use of the nifH gene as a molecular marker. Mikrobiologiia 83(5):583–598
Vandieken V, Pester M, Finke N et al (2012) Three manganese oxide-rich marine sediments harbor similar communities of acetate-oxidizing manganese-reducing bacteria. The ISME journal 6(11):2078–2090
Venter JC, Remington K, Heidelberg JF et al (2004) Environmental genome shotgun sequencing of the Sargasso Sea. Science 304(5667):66–74
Ventosa A, de la Haba RR, Sánchez-Porro C et al (2015) Microbial diversity of hypersaline environments: a metagenomic approach. Curr Opin Microbiol 25:80–87
Vieites JM, Guazzaroni ME, Beloqui A et al (2009) Metagenomics approaches in systems microbiology. FEMS Microbiol Rev 33(1):236–255
Wagner-Döbler I, Ballhausen B, Berger M et al (2010) The complete genome sequence of the algal symbiont Dinoroseobactershibae: a hitchhiker's guide to life in the sea. ISME J 4(1):61–77
Wang DZ, Xie ZX, Zhang SF (2014) Marine metaproteomics: current status and future directions. J Proteomics 97:27–35
Ward DM, Weller R, Bateson MM (1990) 16S rRNA sequences reveal numerous uncultured microorganisms in a natural community. Nature 345(6270):63
Warnecke F, Hess M (2009) A perspective: metatranscriptomics as a tool for the discovery of novel biocatalysts. J Biotechnol 142(1):91–95
Webster NS, Taylor MW, Behnam F et al (2010) Deep sequencing reveals exceptional diversity and modes of transmission for bacterial sponge symbionts. Environ Microbiol 12(8):2070–2082
Wegley L, Edwards R, Rodriguez-Brito B et al (2007) Metagenomic analysis of the microbial community associated with the coral Poritesastreoides. Environ Microbiol 9(11):2707–2719
Wilke A, Rückert C, Bartels D et al (2003) Bioinformatics support for high-throughput proteomics. J Biotechnol 106(2):147–156
Wilbanks EG, Jaekel U, Salman V (2014) Microscale sulfur cycling in the phototrophic pink berry consortia of the Sippewissett Salt Marsh. Environ Microbiol 16(11):3398–415
Williamson SJ, Rusch DB, Yooseph S et al (2008) The Sorcerer II Global Ocean Sampling Expedition: metagenomic characterization of viruses within aquatic microbial samples. PLoS One 3(1):e1456
Wilmes P, Bond PL (2004) The application of two-dimensional polyacrylamide gel electrophoresis and downstream analyses to a mixed community of prokaryotic microorganisms. Environ Microbiol 6(9):911–920
Wilmes P, Bond PL (2006) Towards exposure of elusive metabolic mixed-culture processes: the application of metaproteomic analyses to activated sludge. Water Sci Technol 54(1):217–226
Wippler J, Kleiner M, Lott C et al (2016) Transcriptomic and proteomic insights into innate immunity and adaptations to a symbiotic lifestyle in the gutless marine worm Olaviusalgarvensis. BMC Genomics 17(1):942

Woyke T, Teeling H, Ivanova NN et al (2006) Symbiosis insights through metagenomic analysis of a microbial consortium. Nature 443(7114):950–955

Wu C, Sun B (2009) Identification of novel esterase from metagenomic library of Yangtze river. J Microbiol Biotechnol 19(2):187–193

Wu J, Gao W, Johnson RH et al (2013) Integrated metagenomic and metatranscriptomic analyses of microbial communities in the meso-and bathypelagic realm of North Pacific Ocean. Mar Drugs 11(10):3777–3801

Yooseph S, Sutton G, Rusch DB et al (2007) The Sorcerer II Global Ocean Sampling expedition: expanding the universe of protein families. PLoS Biol 5(3):e16

Zeyaullah M, Kamli MR, Islam B et al (2009) Metagenomics—an advanced approach for noncultivable micro-organisms. Biotechnol Mol Biol Rev 4(3):49–54

Zhang W, Li F, Nie L (2010) Integrating multiple 'omics' analysis for microbial biology: application and methodologies. Microbiology 156(2):287–301

Zhou J, He Z, Yang Y et al (2015) High throughput metagenomic technologies for complex microbial community analysis: open and closed formats. MBio 6(1):e02288–e02214

Exploring Metagenomes Using Next-Generation Sequencing

3

Jalpa R. Thakkar, Pritesh H. Sabara, and Prakash G. Koringa

Abstract

Metagenomics or community genomics refers to the study of genomic DNA of any culture-independent analysis of microbial communities. Non-culturable microbial groups represent the huge majority of global microorganisms. Microbial populations present in every biological niche even humans body carry 10 times more bacterial cells and 100 times more bacterial genes than its own cells and genes. Microbes also hold the secret key for generating renewable biofuels and bioremediation. The next-generation sequencing (NGS) technology provides advantage of parallel sequencing of thousands of sequence from any samples including environmental and clinical without cultivation of it. High-throughput data generated by NGS provides information about vibrant nature of microbial populations and its effect on the atmosphere and health. So, advantages of next-generation sequencing (NGS) technology make metagenomics among the fastest growing research field. In this chapter, we tried to explain the advancement in NGS technology as well as its suitability and approaches to explore metagenomics.

Keywords

NGS • Metagenomics • Illumina • PacBio

J.R. Thakkar • P.H. Sabara • P.G. Koringa (✉)
Department of Animal Biotechnology, College of Veterinary Science and Animal Husbandry, Anand Agricultural University, Anand 388001, Gujarat, India
e-mail: jalpatkr@gmail.com; priteshrhsabara@gmail.com; pgkoringa@aau.in

R.P. Singh et al. (eds.), *Understanding Host-Microbiome Interactions - An Omics Approach*, DOI 10.1007/978-981-10-5050-3_3

3.1 Metagenomics: A Step Ahead in Microbial Ecology

The first large-scale metagenomic sequencing study was performed from Sargasso Sea surface water, an environmental sample to examine microbial populations through shotgun sequencing using Sanger platform which generated 1.045 Gb data and found two viral community among which 65% were novel viral sequence with only 2–3% sequences were from dominating population (Venter et al. 2004). They estimated more than 1800 genomic species in the sample, including 148 novel bacterial phylotypes, by assembling only 25% of the reads. They discovered that poor assembly from huge data without getting saturated level of read depth in fairly diverse community can help in data analysis, and partial and near-complete genome assembly could be recovered from relatively simple communities. Gene-centric analysis can yield valuable biological findings in case of adequate sequencing depth for metagenomic assembly. Examples include formation of sequence cluster as compared to assembly due to data complexity to identify novel biomass degrading genes from cow rumen metagenomics (Iwai et al. 2010; Tun et al. 2012).

The advantages of next-generation sequencing have promoted a considerable manyfold increase in metagenomic studies undertaken and submitted at GOLD. These studies are related to wide environmental spectrum, including clinical, engineered and natural communities (Mardis 2008; Roossinck 2012). There was initially a clear partiality towards the long reads produced by Roche 454 pyrosequencing. However, substantial improvements in Illumina read length and data output have showed increase in popularity of it for this kind of studies. Though NGS has high throughput, it should be used with suitable sequencing strategy for environmental samples. The sequencing strategy, such as single or multiple platforms fragment or paired libraries, should be based on the research focus and the composition of the target society. If it is not taken into account, it will severely hamper downstream processes to get fruitful outcome from the experiment.

Various factors which include copy number of genomes, number of species in given environmental sample, genome size of different members of community, method of DNA extraction and purification, relative loads of the species in community, as well as method of sequencing used will decide the number of reads that are to be derived from a given environmental community. For example, shotgun sequencing of the community typically results in more sequencing depth for numerically dominant species, with very few reads from the rarer community members (Brenig et al. 2010; Shah et al. 2011). Thus, the amount of sequencing data/read required to cover the given community is positively correlated with the community diversity and population complexity. The use of approaches complimentary to metagenomics, such as directed sequencing of the 16S rRNA gene amplicons, can help to determine the choice of sequencing platform(s), library types and amount of sequencing required to obtain the throughput necessary to achieve research goals (Shah et al. 2011).

3.2 Approaches to Study Metagenomes

Originally, the term "metagenomics" was only used for taxonomic and functional-based genome analysis of microbial community present in an environmental sample (Handelsman 2004), but at this time, it is also broadly applied to studies performing PCR amplification of certain targeted genes of attention. The former can be referred to as "shotgun metagenomics" and the latter as "marker gene amplification metagenomics/amplicon-based metagenomics/targeted metagenomics/meta-genetics" (e.g. 16S ribosomal RNA gene).

3.2.1 Shotgun Metagenomics

Shotgun sequencing has the advantage of identifying broadly present microbes in ecological sample. This creates a regional biodiversity profile that can be linked with functional concerto analysis of recognized and mysterious organism lineages (i.e. genera or taxa). Shotgun metagenomics has evolved to deal with the questions of who is nearby in an ecological community, what they are performing (function-wise) and how these germs interact to uphold a balanced ecological niche (Brenig et al. 2010).

Since there are no universal markers for all forms of life (including viruses), shotgun metagenomics approach is the lone method to cop up with comparative abundances and functional potential of different organisms in whole microbial communities with the help of databases, such as KEGG (Kyoto Encyclopedia of Genes and Genomes) pathway database and/or COG (Clusters of Orthologous Groups of proteins) database. For example, metagenomic shotgun sequencing provides breakthrough in a study of the gut microbiome and its association with obesity.

Metagenomic shotgun sequence analysis involves comparison of different genomes through different parts of sequence to make accurate classification as well as cost involved in data generation to identify the rare species of the community that may be a limiting factor in many study involving this methodology (Shah et al. 2011).

3.2.2 Targeted Metagenomics

Targeted metagenomics is a cheaper and faster way to obtain a community profile using PCR amplification and sequencing of evolutionarily conserved marker genes, such as the 16S rRNA gene which provides probable taxonomic distribution of different microbes related with ecological data (metadata) derived from the case site under investigation (Shah et al. 2011).

3.2.2.1 Bacterial Community Profiling Through 16S Amplicon Sequencing

Majority of study involves exploring microbiota of any niche interested in the bacterial portion. By far the most popular genomic region for identifying bacterial diversity is the 16S rRNA gene encoding for the ribosomal, known as small subunit (SSU) and found in all living organisms, with the notable exception of viruses and accounts for more than 80% of total bacterial RNA. This gene can be amplified through universally annealed primers of interspersed conserved region to amplified variable region of it, and sequencing of these fragments through amplicon helps to identify microbial community profile of given DNA sample which also lower down sequencing cost as well as monitoring fluctuation in microbial populations.

3.2.2.2 Fungal Community Profiling Through ITS, LSU and SSU Amplicon Sequencing

The three most commonly used ribosomal loci for fungal genome diversity study are known as the internal transcribed spacer (ITS), large subunit (LSU) and SSU regions. ITS is the most resourceful locus for providing species-level identifications, so it is mostly used for fungal "species barcode", while LSU and SSU loci are more preserved which are quite effectively utilized for phylogenetically based microbiome analyses (Table 3.1).

Table 3.1 Pros and cons of amplicon vs. shotgun sequencing

	Amplicon sequencing	Shotgun sequencing
Type of information produced	The taxonomic concerto and phylogenetic organization of a microbial neighbourhood expressed as OTUs	Functional and process-level categorization of microbial niche as a whole and the reform of draft genome sequences for individual community members
Application	Scrutinize population	Detect new members, new genes and determine compound taxonomies
Ability to spot rare members of the community (sensitivity)	Highly sensitive. rRNA makes up to 80% of total bacterial RNA	Requires much deeper sequencing to achieve the same level of sensitivity
Bias	Bias produced by the probes and the PCR itself. The amplified region may not precisely represent the whole genome due to horizontal transfer or mutations	Sequence content bias
Gene content	The gene catalogue and encoded functionality of most microbial species are still mysterious and may also vary considerably among strains	Generate widespread gene inventories and draft genomes. Discovers new genes and biological pathways

3.3 Metatranscriptomics (RNA Transcript Sequencing)

Metatrasncriptomic approach provides full profile of total actively transcribed genes from members of the microbiome which helps to examine the host molecular behaviour and interaction between host-microbiome in particular environment for host survival and specific function. Earlier metagenomic categorizations of germs in database create problems for metatrasncriptomic assessment of diverse assembly of the microbiome (Vikram et al. 2016; Weckx et al. 2011).

3.4 Next-Generation Sequencing Technologies for Metagenomics

Since the initial launch of the NGS platform in 2005, newer sequencing technologies have been developed rapidly and continuously. The recent development of sequencing technologies has revolutionized the much deeper layers of microbial ecology by generating tons of in-depth nucleotide sequences at lower cost with accelerated speed. The various applications of NGS platforms ranged from amplicon sequencing to whole-genome sequencing and shotgun metagenomic sequencing (Mardis 2008). An advantage of next-generation sequencing over traditional Sanger's sequencing method includes cost-effectiveness, higher throughput, no cloning step and less technical knowhow involved. However, the biggest challenge is to draw scientifically and statistically meaningful conclusion by analyzing NGS data analysis. Experimental design and sequence analysis should be considered to resolve different error model and biases in each NGS technology.

3.4.1 Roche 454 Sequencing (GS20, GS-FLX, GS-FLX Titanium, GS-FLX Titanium Plus)

The first NGS platform released in 2005, GS20, used a sequencing by synthesis on microbeads in picotitre plate (PTP) generated just 20 Mb per run with an average read length of 100 bp. The platform is based on pyrosequencing principles, i.e. nucleotide triphosphates are flowed across the plate in a specific sequence and each base incorporation is marked by the release of pyrophosphate (MacLean et al. 2009). The longer read length with greater accuracy at that time was advantageous over other NGS systems. The most upgraded version, i.e. GS-FLX Titanium Plus, has an advantage as it produces around ~850 Mb per run, especially 16S rRNA gene-based surveys, with a read length of 700–750 bp, which is satisfactory to cover partial hypervariable regions of 16S rRNA. The cost per base in recent time, error rates in homopolymeric regions and initial amount of DNA requirement became drawbacks in current time as compared to other systems due to which technical support and supply of reagents for all GS-FLX systems were withdrawn from Dec 2016. Still there are a large number of datasets that have been generated through this technology which are unpublished.

3.4.2 Illumina Sequencing (GA I, GA II, HiSeq, MiSeq, NextSeq 500, HiSeq 2500, HiSeq X Ten)

Illumina came in 2006, having a portfolio of the most widely used family of platforms and quickly accepted by many researchers as it produced cost-effective high-throughput data, but it had short read length drawback in earlier version which was improved in subsequently developed platforms (i.e. 2 × 300 bp in MiSeq) which attracted research community to shift from 454 to Illumina technology (Quail et al. 2012). Due to these many group of researchers shifted to Illumina from 454 looking to the cost-effectiveness of the technology. It uses sequencing by synthesis approach using reversible dye termination method. Sequencing preparation begins with short stretch of DNA that have specific adapters on either end that hybridized to the ends of the specific oligonucleotide fragments in flow cell during washing so it can generate a cluster of identical fragments after replication. Reversible dye terminator nucleotides are passed through the flow cell and within the given time to incorporate in growing chain; the remaining excess nucleotides are washed away so the flow cell image can be captured and the terminators are reversed so that the process of nucleotide addition can be repeated in subsequent cycles (MacLean et al. 2009).

Currently, MiSeq is the longest 300 bp read length producer in all Illumina products with paired-end read method, while HiSeq 2500 gives the highest four billion fragments of 125 bases for each read in a paired-end fashion in a single run (Caporaso et al. 2012). Recently, Illumina has released an arrangement of ten HiSeq machines as a unit for higher-throughput known as HiSeq X Ten. NextSeq 500 is the first high-throughput desktop sequencer, introduced by Illumina. Multiplexing of Illumina samples is handled differently than the barcoding approach pioneered by 454, it involves a separate indexing of each fragment through separate indexing primer, and dual indexing can be performed by utilizing second index from the adaptors of the flaw cell lawn which can help in greater degree for sample multiplexing.

3.4.3 Applied Biosystem's SOLiD

In 2006, SOLiD (Sequencing by Oligonucleotide Ligation and Detection) was introduced by Applied Biosystem under Life Technology which used chemistry of "sequencing by ligation" with the help of DNA ligase and dibase (Hedges et al. 2011). The SOLiD 5500xl produces up to 300 Gb data with three billion reads per run with 75 bp long reads. Chemistry is now no longer available in market though it gave huge amount of data with reduced rate of $0.0001/base but high cost per run, and low read length was the biggest drawback of this platform.

3.4.4 Ion Torrent Sequencing (PGM, Proton, S5)

Ion Torrent is the first company which introduced feasible and affordable personal genome machine (PGM) in 2010 which is very helpful to small and low-budget researcher groups; as a result, it quickly gets mass response from them (Merriman et al. 2012). The cost of the machine as well as per base generation was considered to be the lowest at the time of invention in NGS market. The Ion Torrent system contains micro-well plate for beads to which DNA fragments are attached, as found in GS-FLX. It has unique systems as base incorporation in growing DNA strand releases a proton ion which alters the pH of a well and micro-detector at the bottom of chip which itself is a semiconductor chip recorder (Merriman et al. 2012). When the individual base is added to a growing DNA strand, a proton is released, which alters the pH of individual well. As the different nucleotide bases sequentially flush through the chip and its incorporation is recorded from each well that helps to infer sequences of individual well.

The Ion Proton platform produces up to 10 Gb data, with a maximum of 50 million reads per run having read lengths of 200 bases, while the PGM has the longest reads at *400 bases but gives a maximum 2 Gb data and 5.5 million reads with Ion 318 V2 chip. One interesting feature of this system is that sequencing of longer fragments is omitted by a size-selection step. Bidirectional sequencing is not feasible with this technology as "pairing" the reads themselves does not seem to be reliable at its current state but multiplexing is possible by availability of the standard in-line molecular barcode sequences (Quail et al. 2012). Like GS-FLX, this system is also susceptible to homopolymer-related errors. Recently they introduced Ion S5 which produced up to 15 Gb data and 60–80 million reads with read length of 200 bp by using 540 chip. This approach can be effective for generating microbiome data, even though strict size selection and lack of bidirectional sequencing get in the way to this technology from accepted by microbiome researchers.

3.4.5 Pacific Biosciences (PacBio RS II)

Pacific Biosciences use sequencing by synthesis for single-molecule, real-time sequencing approach (SMRT sequencing) and introduced it in 2012 (Bachall 2009). Although Helicos BioSciences was first in single-molecule sequencing but PacBio, became a successful market leader as the first single-molecule sequencing platform. The circular consensus sequencing (CCS) used to sequence a fragment repeatedly to decrease the error rate (Bachall 2009; Quail et al. 2012).

The PacBio sequencing system does not involve amplification step as used in most NGS platforms. Instead, they utilizes the zero-mode waveguide (ZMW) in which a single DNA polymerase enzyme is fixed to the bottom of a well with a

Table 3.2 Comparison of next-generation sequencing platforms

Platform family	Clonal amplification	Chemistry	Highest average read length (bp)	Highest output with any machine of family
GS-FLX	Emulsion PCR	Pyrosequencing	750	850 Mb
Illumina	Bridge Amplification	Reversible dye terminator	300	1.6–1.8 Tb
SOLiD	Emulsion PCR	Oligonucleotide 8-mer chained ligation	75	25 Gb
Ion Torrent	Emulsion PCR	Proton detection	400	30 Gb
PacBio	Not required	Phospholinked fluorescent nucleotide	8500	8 Gb
Oxford Nanopore	Not required	Proprietary nanopore sensing technology	250,000	12 Tb

single molecule of DNA as a template whose illumination after addition of single nucleotide in growing sequencing natural DNA strand is captured. All four phospholinked nucleotides are individually labelled with four different florescent dyes which release and give illumination after incorporation of nucleotides by DNA polymerase that is detected in recorded by detector, and the base call is made according it. PacBio gives much longer read lengths (10,000–60,000 bp) with 99.999% accuracy compared to the other technologies, thus having obvious advantages over annotation and assembly for shotgun metagenomics (Bachall 2009; Quail et al. 2012). For PacBio, they use a process called strobing for paired-end read sequencing, and standard in-line barcoding is also available for multiplexing (Table 3.2).

3.4.6 Oxford Nanopore Sequencing

Oxford Nanopore technologies introduce an innovative "strand sequencing" that could sequence completely intact DNA strands passed through a small-sized protein nanopore which records change in the ionic current when any biomolecule passes through it (Branton et al. 2008). MinION mk1B is a pocket-sized device which contains 512 nanopore channels that can be linked directly to any computer for real-time data collection. The chip is recently launched in May 2016. PromethION is under development, it's a benchtop having up to 48 flow cells of 3000 nanopore channels each which combinedly makes total 144,000 channels which can be used through PromethION Early Access Programme (PEAP). They are also developing SmidgION which can be operated through smartphone at any place for any type of samples. This is even smaller than MinION and PromethION. Oxford Technology have smart and rapid sample preparation advantage within 10 min which is going to be improved after development of VolTRAX that can be operated through USB, and when a sample is loaded in cartridge, it passes through a path controlled by software. The advantage of long reads eliminates the need for shotgun sequencing and revolutionizes the sequencing industry in the future. This technology has advantages, as it will eliminate erroneous shotgun sequencing and exclude the error-prone assembly step during data analysis. However, nanopore sequencing is at the moment non-commercialized (available only through the MinION™ Access Program) which is still being optimized using specific template and sequencing needs (Goodwin et al. 2015).

3.4.7 Irys Technology of BioNano Genomics

It is another innovative as well as promising technology, which uses micro- and nanostructures for constructing de novo genome maps. Unamplified, native-state DNA molecules, each up to a megabase long, can be loaded into the IrysChip's NanoChannels. The input is DNA labelled at specific sequence motifs, utilized for imaging and identification through IrysChips. These labelling steps result in a uniquely identifiable, sequence-specific pattern which can be used for de novo map assembly or anchoring sequencing contigs (Xiao et al. 2015).

3.5 Basic Steps in Metagenomic Data Analysis

Tools and databases utilized for metagenomic data analysis are becoming more and more efficient and elaborate, but no standardized analysis pipelines are available for metagenomic experiments till date. Nanopore sequencing technology offers new horizon for the analysis pipelines as well as new options for the assembly and concurrent annotation process. Basic metagenomic steps and tools widely used in practice nowadays are summarized in Fig. 3.1. Some popular tools that have been used extensively by the metagenomics community are shown for every step as well as the databases and algorithms in common practice.

3.6 Future of Metagenomics

Metagenomic studies were performed on simple ecosystem with precise target due to costing and lack of proper knowledge of sequencing depth, unavailability of resources for computing facilities and trained bioinformaticians, confusion in statistical design and analysis strategies that's why previous metagenomic studies cannot

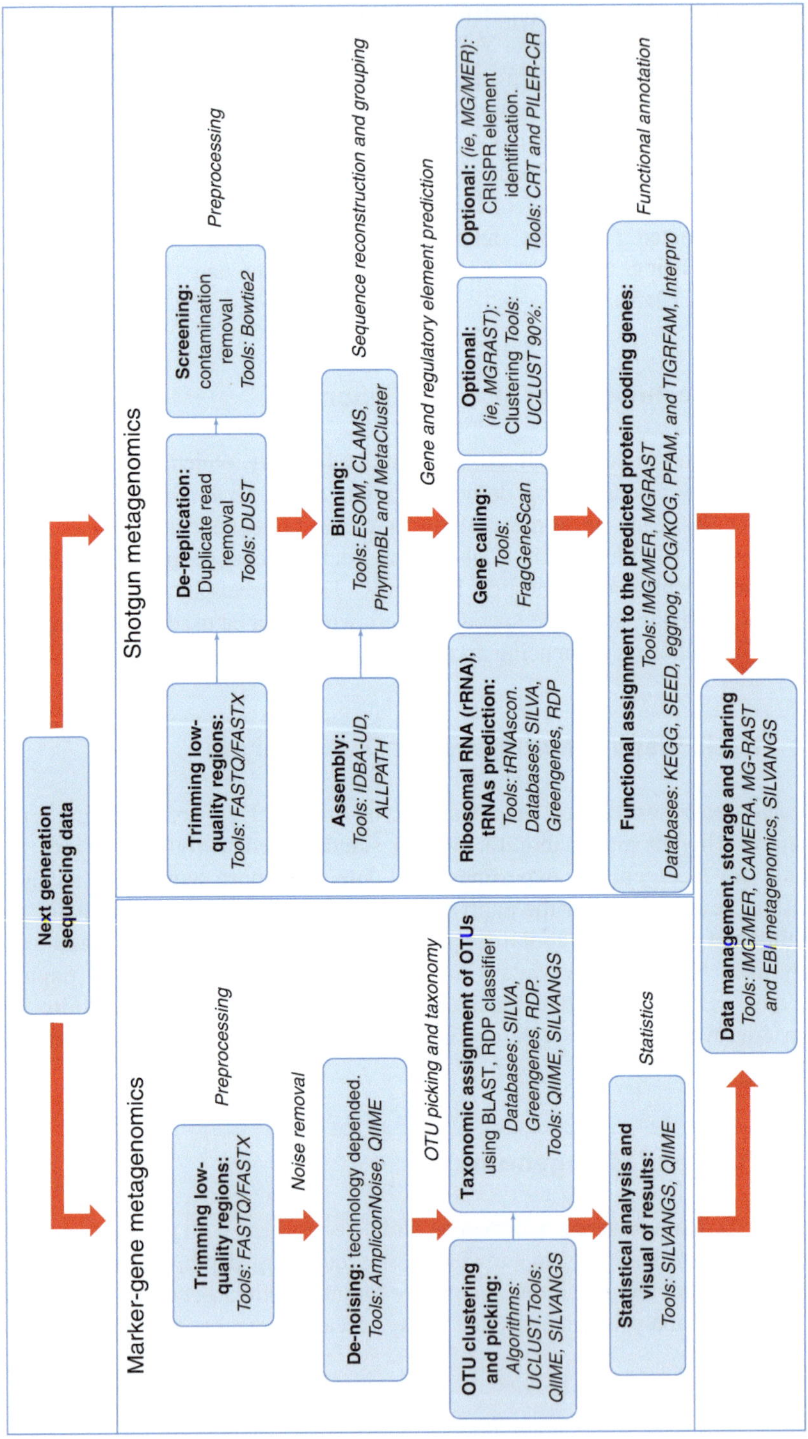

Fig. 3.1 Flowchart of basic metagenomics steps and tools currently in practice. Modified from Oulas et al. (2015)

be replicated but after reduction in sequencing cost, proper knowledge of ecosystem, statistical design for sampling and data analysis with addition of metadata, availability of efficient new enzymes and molecular tools, accessibility of centralized computing and bioinformatic facilities developed by funding agencies. Nowadays, researchers are using large number of ecological samples with huge number of biological replication and provide marker gene, genome and metadata information during submission in curetted database which become standard practice and provide meaningful information of particular study, but still improvement is going on in collecting metadata and improving statistical analysis that can help to give broad picture of particular ecosystem which can be compared with other similar kinds of study or other ecosystems.

Metagenomic data analyses have other limitations related to short read and low coverage for genome that creates problem during assembly and annotation of sequences, but researchers are using co-assembly from single cell genomics as well as collaborative analysis of their data to fill their analysis gaps by considering common species that must exist in their system. The metagenomic studies are targeted to provide a broad understanding of our different ecosystem. In the near future, metagenomics will be one of the key parts of screening different ecosystems through combination with other '-omics' tools such as metatranscriptomics and metaproteomics.

References

Bachall O (2009) Pac Bio sequencing. Nat Genet 41(2):147–148

Branton D, Deamer DW, Marziali A, Bayley H, Benner SA, Butler T, Di Ventra M, Garaj S, Hibbs A, Huang X (2008) The potential and challenges of nanopore sequencing. Nat Biotechnol 26(10):1146–1153

Brenig B, Beck J, Schutz E (2010) Shotgun metagenomics of biological stains using ultra-deep DNA sequencing. Forensic Sci Int Genet 4(4):228–231. doi:10.1016/j.fsigen.2009.10.001

Caporaso JG, Lauber CL, Walters WA, Berg-Lyons D, Huntley J, Fierer N, Owens SM, Betley J, Fraser L, Bauer M (2012) Ultra-high-throughput microbial community analysis on the Illumina HiSeq and MiSeq platforms. ISME J 6(8):1621–1624

Goodwin S, Gurtowski J, Ethe-Sayers S, Deshpande P, Schatz MC, McCombie WR (2015) Oxford Nanopore sequencing, hybrid error correction, and de novo assembly of a eukaryotic genome. Genome Res 25(11):1750–1756

Handelsman J (2004) Metagenomics: application of genomics to uncultured microorganisms. Microbiol Mol Biol Rev 68(4):669–685. doi:10.1128/MMBR.68.4.669-685.2004

Hedges DJ, Guettouche T, Yang S, Bademci G, Diaz A, Andersen A, Hulme WF, Linker S, Mehta A, Edwards YJ (2011) Comparison of three targeted enrichment strategies on the SOLiD sequencing platform. PLoS One 6(4):e18595

Iwai S, Chai B, Sul WJ, Cole JR, Hashsham SA, Tiedje JM (2010) Gene-targeted-metagenomics reveals extensive diversity of aromatic dioxygenase genes in the environment. ISME J 4(2):279–285. doi:10.1038/ismej.2009.104

MacLean D, Jones JD, Studholme DJ (2009) Application of 'next-generation' sequencing technologies to microbial genetics. Nat Rev Microbiol 7(4):287–296

Mardis ER (2008) Next-generation DNA sequencing methods. Annu Rev Genomics Hum Genet 9:387–402. doi:10.1146/annurev.genom.9.081307.164359

Merriman B, Ion Torrent R&D Team, Rothberg JM (2012) Progress in ion torrent semiconductor chip based sequencing. Electrophoresis 33(23):3397–3417

Oulas A, Pavloudi C, Polymenakou P, Pavlopoulos GA, Papanikolaou N, Kotoulas G, Arvanitidis C, Iliopoulos I (2015) Metagenomics: tools and insights for analyzing next-generation sequencing data derived from biodiversity studies. Bioinform Biol Insights 9:75–88. doi:10.4137/BBI.S12462

Quail MA, Smith M, Coupland P, Otto TD, Harris SR, Connor TR, Bertoni A, Swerdlow HP, Gu Y (2012) A tale of three next generation sequencing platforms: comparison of Ion Torrent, Pacific Biosciences and Illumina MiSeq sequencers. BMC Genomics 13(1):1

Roossinck MJ (2012) Plant virus metagenomics: biodiversity and ecology. Annu Rev Genet 46:359–369. doi:10.1146/annurev-genet-110711-155600

Shah N, Tang H, Doak TG, Ye Y (2011) Comparing bacterial communities inferred from 16S rRNA gene sequencing and shotgun metagenomics. Pac Symp Biocomput:165–176

Tun HM, Brar MS, Khin N, Jun L, Hui RK, Dowd SE, Leung FC (2012) Gene-centric metagenomics analysis of feline intestinal microbiome using 454 junior pyrosequencing. J Microbiol Methods 88(3):369–376. doi:10.1016/j.mimet.2012.01.001

Venter JC, Remington K, Heidelberg JF, Halpern AL, Rusch D, Eisen JA, Wu D, Paulsen I, Nelson KE, Nelson W (2004) Environmental genome shotgun sequencing of the Sargasso Sea. Science 304(5667):66–74

Vikram A, Lipus D, Bibby K (2016) Metatranscriptome analysis of active microbial communities in produced water samples from the Marcellus Shale. Microb Ecol 72(3):571–581. doi:10.1007/s00248-016-0811-z

Weckx S, Allemeersch J, Van der Meulen R, Vrancken G, Huys G, Vandamme P, Van Hummelen P, De Vuyst L (2011) Metatranscriptome analysis for insight into whole-ecosystem gene expression during spontaneous wheat and spelt sourdough fermentations. Appl Environ Microbiol 77(2):618–626. doi:10.1128/AEM.02028-10

Xiao S, Li J, Ma F, Fang L, Xu S, Chen W, Wang ZY (2015) Rapid construction of genome map for large yellow croaker (Larimichthys crocea) by the whole-genome mapping in BioNano Genomics Irys system. BMC Genomics 16(1):670

4 Metagenomics: An Era of Throughput Gene Mining

Bhupendra Singh Panwar and Ruchi Trivedi

Abstract

As far as our understanding of the microbial world is considered, until the present only a small fraction of microbial communities from an ecosystem can be characterized, hence having enormous gene bioprospecting potential. Gene mining/prospecting is referring to the process of identifying gene(s) or their allele(s) involved in the economically important biological process of industrial importance. Industrial biotechnology and its vigorous growth over the decade increase the demand of novel genes with increased efficiency for industrial application. The metagenomic studies are one of the ways to explore the hidden potential of cultured and unculturable microbial communities of the ecosystem. Metagenome from the complex microbial community is a good bioresource for bioprospecting of industrial important biocatalyst. Despite having a remarkable contribution of metagenomics in gene prospecting from prokaryotes, it is still in its formative years. In the present chapter, we have tried to provide am efficacy of soil and marine metagenomic in gene mining/prospecting and describe various methods available to narrow down to the specific gene in the metagenome. Accomplishments made thus far, limitations, and future prospects of this new science are discussed.

Keyword

Gene mining • Metagenomics • Function-based screening • Structure-based screening • Soil metagenomics • Marine metagenomics

---

B.S. Panwar • R. Trivedi (✉)
Department of Animal Biotechnology, College of Veterinary Science and Animal Husbandry, Anand Agricultural University, Anand, Gujarat, India
e-mail: bhupendrasinh170886@gmail.com; ruchitrivedi84@gmail.com

R.P. Singh et al. (eds.), *Understanding Host-Microbiome Interactions - An Omics Approach*, DOI 10.1007/978-981-10-5050-3_4

4.1 Introduction

Microbes are one of the most indispensable constituents of the Earth's ecosystems. Microbes in the ecosystem act as a small organic factory which processes elements (such as carbon, hydrogen, nitrogen, oxygen, sulfur, phosphorous, etc.) into complex biologically active organic molecules accessible to the higher organism including humans (Adkins et al. 2012). Since microorganism is a critical source of bioresource, they are even hosted by higher life forms that include plant and animals. They help to make vital biomolecules (e.g. vitamins) and other nutraceuticals; moreover, they also help to process and digest food. In addition, microbes are also beneficial to the environment. Microbes act as tiny disasters which have the capability to remove the human introduced contaminants and pollutants from air, soil, and water. Besides having these much functional superiority, the understanding of this microscopic word remains limited.

In the absence of DNA structure and function, all organisms were characterized on the basis of their phenotypes. Therefore, to characterized microbes, culturing was the only way, according to the Bergey's Manual (1923) (Handelsman 2005). Cumulative research of the last three decades help us to realize that there are many bacteria which are not amenable to culturing, and sub-culturing may be due to lack of nutrient required or distortion in the amount of nutrient required, absence of symbionent or presence of inhibitory compound, etc. (Zengler et al. 2002). The idea of characterizing microbes thriving an ecosystem without cultivating them was given by Torsvik and coworker (Torsvik et al. 1990). Furthermore, Lane and coworker performed phylogenetic analysis based on 16s ribosomal rRNA gene sequence directly retrieved from the environment. After gaining widespread acceptance of genomic technology, microbial genomic (cultivated and uncultivated) has started materializing. Before the introduction of new term "metagenomics," the science existed in literature and was described by different terms that include microbial population genomics, soil DNA library, environmental genomics, ecogenomics, community genomics, recombinant environmental libraries, etc. (Vakhlu et al. 2008). The term "metagenomic" was given by Handelsman and coworker (Handelsman et al. 1998) and defined as the study of the genetic material directly recovered from the environment. Although Healy and coworker first reported the application of metagenomic for gene mining/bioprospecting for cellulase enzyme (Healy et al. 1995).

Gene mining or gene bioprospecting refers to the process of identifying genes or their alleles involved in the economically important biological process. Technically, gene mining is used to identify and isolate genes that are previously characterized to conferring economically important traits (e.g., amylase, esterase, lipase, lactonase, racemase, etc.) or identification of gene controlling particular biological process of industrially importance. For the purpose of gene mining from microorganism, two methods can apply: (1) method involving screening of cultivated microorganism and (2) method involving screening of metagenome. The first approach allowed us to screen microorganism based on targeted enzyme activity, antimicrobial activities, stress resistance activities, etc. However, with the luxury of prior phenotypic screening, this approach also has some limitation of incapability of culturing wide group of microorganism using standard techniques (Amann et al. 1995). In contrast, second approach allows us to go deep insight into the hidden world of uncultivable microorganism in search of new enzymes and functional proteins with limitation of prior screening.

Presently, metagenomic is an established science which has been deciphering the secrets of earth microbial communities from time to time. Metagenomic predominantly exploited in the studies related to taxonomic diversity, ecology, and evolution of microbes. In addition, they also have huge potential in biomedicine, renewable energy resources, agriculture, etc. (National Research Council Committee on Metagenomics and Functional 2007). The evolution of metagenomic is in synchronization with the evolution of genomic technology that includes microarray technology, high-throughput DNA sequencing, and bioinformatics. Introduction of this advance genomic methodology further increases the resolution and speed of gene mining from new and available metagenomic database. Some of the advances are discussed in succeeding sections.

4.2 Gene Mining from Soil Metagenome

Recently, several attempts were made to identify and characterize novel enzyme and bioactive compounds from soil. Esterase/lipases are one of the most prevailing novel biocatalysts found from the soil metagenome. Esterase/lipase is the lipolytic enzymes belonging to hydrolase group, which involved in breaking of chemical bonds in the presence of water. The attractive feature of lipolytic enzymes includes stability in organic solvent, no requirement of cofactors, stereoselectivity, broad substrate specificity, and positional selectivity (Lee et al. 2004). Isolating enzyme through gene mining from metagenome with specific chiral resolution is an attractive feature, especially for the processes involving organic synthesis. Henceforth, the identification and isolation of novel esterase/lipase from soil metagenome acquire much attention as compared to other industrially important enzymes. Although lipolytic enzyme is considered to be the most studied and characterized among biocatalyst, discovery of novel enzyme is quite surprising. To illustrate the application of metagenomic in lipolytic gene mining, the study conducted by Lee and coworker (Lee et al. 2010) was discussed in this section. In order to identify novel lipolytic enzyme, Lee and coworker used functional metagenomic approach (describe below) in which they prepare metagenomic library and screen clones for lipolytic activity. Screening followed by identification of target gene revealed the presence of the new lipolytic enzyme (EstD2) having a low degree of sequence homology to the well-characterized lipolytic enzyme gene family member. Further sequence analysis showed positive hit with hypothetical protein from *Phenylobacterium zucineum* HLK1 indicating presence of novel lipolytic enzyme which was further confirmed by lipolytic assay. Similar studies conducted by different groups of researchers using metagenome lead into the discovery of novel lipolytic enzymes (Elend et al. 2006; Hong et al. 2007; Lee et al. 2010; Tao et al. 2012; Yu et al. 2011).

Cell wall-degrading enzymes (CWDEs)/cellulolytic enzymes are the second group of enzymes after lipolytic enzymes having industrial as well as biotechnological interest specifically in biomass degradation for bioenergy generation. Cell wall-degrading enzymes (CWDEs)/cellulolytic enzymes are the biocatalyst involved in cellulose degradation. Cellulose is a polysaccharide and when degraded by cellulase produces two molecule of glucose which is easily fermentable biomolecule and yield ethanol (Polizeli et al. 2005). Furthermore, cellulose degradation using biocatalyst has many industrial applications like preparation of foods, kraft pulp bleaching, textile processing, etc. (Ando et al. 2002; Hu et al. 2008; Lynd et al. 2002). Identification and cloning of cellulase gene from metagenome have been documented in several reports. Healy and coworker (Healy et al. 1995) initially before the introduction of term "metagenomic" by using metagenome identified endo-β-1,4-glucanase gene. In order to identify cellulase enzyme, Healy and coworker prepared metagenomic library from the DNA isolated from thermophilic, anaerobic digester on lignocellulosic feedstocks. Function screening of metagenomic libraries on CMC (carboxymethylcellulose) and MUC (4-methylumbelliferyl β-D-cellobiopyranoside) revealed the presence of endo-β-1,4-glucanase gene in metagenomic library. A similar type of study was performed by Voget and coworker (Voget et al. 2003); they used microbial-enriched culture and identified one cellulase gene. By using same approach, Voget and coworker (Voget et al. 2006) isolated and characterized halotolerant endoglucosidase from soil metagenome. A similar kind of report was also documented by Liu and coworker (Liu et al. 2011) for the identification of novel endo-β-1,4-glucanase (Cel5G) which showed a low degree of homology (<39%) with endoglucanases deposited in GenBank. Moreover, Cel5G hydrolyzed wide range of complex carbohydrates that includes β-1,4-, β-1,3/β-1,4-, or β-1,3/β-1,6-linked polysaccharides, microcrystalline cellulose, filter paper, and amorphous cellulose.

Recently, by using metagenomic library prepared from slug DNA, Matsuzawaand and Yaoi identified novel saccharide-stimulated glycosidase (MeBglD2) which has β-glucosidase, β-galactosidase, and β-fucosidase activities. A similar study conducted for other industrially important genes from soil metagenome suggested a plethora of microbial diversity bearing novel microbial enzyme (Matsuzawa and Yaoi 2016).

4.3 Gene Mining from Marine Metagenome

Seventy percent of the Earth's surface comprise of water; hence covered with marine environment, enclosing extreme habitats includes freezing Arctic and Antarctica to the warm tropics. Microorganism inhabited complete oceans despite of extreme environment includes depth of 11,000 m with 100 Mpa pressure, low temperature, high salinity, or hydrothermal vents with temperature higher than 100 °C. Their evolution over the millennia introduces a high level of diversity both in genotypes and phenotypes. Diversity in genotypes is generally reflected into the enzymes that potentially may endow unique properties. Therefore, marine metagenomic is an excellent tool to perform gene mining to unlock the novel superior extremozymes of industrial and biotechnological importance. Marine metagenomics is found to be very effective in the discovery and isolation of extremozymes from extreme habitats. For instance, bacteria thriving in marine snow have an intrinsic property to produce enzymes which have potential to work efficiently within the snow (Azam and Long 2001).

As described in the above section, lipase/esterase is a predominant enzymes used in industrial application and includes foods, laundry, textile, pulp and paper industry, bioenergy, and organic chemical synthesis. A screening of a metagenomic library prepared from the metagenome recovered from the Arctic intertidal zone allowed to identify and isolate novel cold-tolerant esterase (Est97) (Fu et al. 2013). Another example depicting the potential of marine metagenomic to identify a novel class of enzyme comes from a study conducted by Lee and coworker (Lee et al. 2012). They prepared metagenomic library from the DNA recovered from the tidal flat sediment from the Korean west coast. Sequence analysis showed gene to be phospholipase A1. Phospholipases are the enzymes which catalyze lysis of phospholipid into fatty acid and lipophilic substrate; they used phospholipid as a substrate and hence didn't act on another form of lipid-like lipases. Interestingly, the enzyme isolated by Lee and coworker has both phospholipase and lipase activity. Marine metagenomic also has application in the dairy industry. In dairy industry they can be exploited for identification of the novel β-galactosidase enzyme (enzyme involved in the hydrolysis of lactose into galactose and glucose). Wierzbicka-Wos and coworker prepared metagenomic library from metagenome isolated from the Baltic Sea water sample. By using this library, they identified a monomeric cold-tolerant glycoside hydrolase (BglMkg). With cold tolerance, these enzymes also have β-galactosidase, β-fucosidase, and β-glucosidase activity in a pH range from 6.0 to 8.0 (Wierzbicka-Wos et al. 2011).

Besides cold, ocean also host other ecological niches thrives by microorganisms called thermophiles, hyperthermophiles, halophiles, and barophiles possessing biocatalyst of economical value. By using similar metagenomic approaches, several biocatalysts include lipase/esterase (Tirawongsaroj et al. 2008; Zhu et al. 2013), β-glucosidase (Schroder et al. 2014), glycoside hydrolases (Wang et al. 2011), and fumarase (Jiang et al. 2010) that were identified and characterized from thermophilic marine ecosystem. Marine environment also includes halophilic condition therefore providing opportunities to identify salt-tolerant biocatalyst using metagenomics approaches. This opportunity was very well used by Fang and coworker and Jeon and coworker; they identified bacterial laccase showing activity in alkaline condition and salt-tolerant esterase, respectively, by using metagenomic approach (Fang et al. 2012; Jeon et al. 2012).

Furthermore, gene bioprospecting from polluted seas using metagenomic approaches also facilitates identification of biocatalyst tolerance to heavy metals. For example, Mohamed and coworker, by using a metagenomic library from metagenome isolated from the Red sea brine pool, identified novel esterase with thermophilic as well as heavy metal tolerance activity (Mohamed et al. 2013). Similarly, Sayed and coworker identified novel mercuric reductase, which shows activity at high temperature and concentration of heavy metals and salts. Henceforth, the above study indicated the vast potential of marine metagenomics in gene mining and bioprospecting (Sayed et al. 2014).

4.4 Metagenomic Approaches for Gene Mining

4.4.1 Sequence-Based Approach

A sequence-based approach for gene mining/bioprospecting was further divided into three strategies. The first strategy involves the use of metagenomic libraries to screen genes, either through hybridization probes (Pham et al. 2007) or DNA microarray (Park et al. 2010) or by using PCR-based scheme (Fig. 4.1) (Israel 1993). Gene mining based on hybridization involves probe designing from homologous sequences available in online databases (e.g., National Centre for Biotechnological Information). Typically, this probe is designed from partial targeted gene encoding sequences such as catalase, nitrite reductase, glycerol dehydratase, dioxygenase, xenobiotic degradation gene, etc.

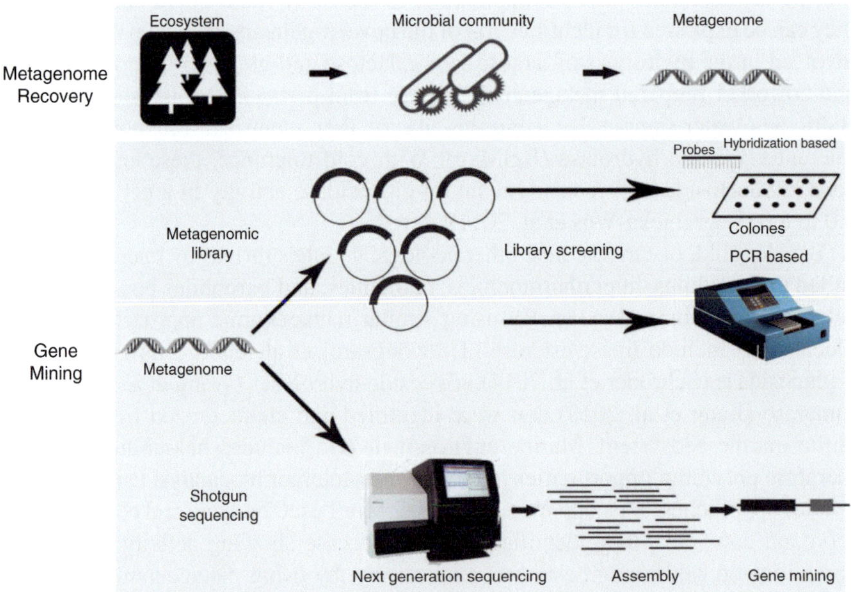

Fig. 4.1 Schematic of acquisition and processing of metagenomic information via sequencing-based approach. Sequence-based metagenomic approach having three strategies of which two main strategies are describe above. All strategies have same metagenome recovery step (i.e., DNA isolation from ecosystem sample) followed by gene mining either via metagenomic library screening or by shotgun sequencing approach. There are two schemes available for metagenomic library screening; one is based on hybridization and the other on PCR

The finest example of exploitation of heterologous probe for gene mining was present in the study conducted by Jacquiod and coworker (Jacquiod et al. 2013). In their study, they use metagenomic library prepared from soil sample of Park Grass (Rothamsted, England) to screen economically important genes that includes glycoside hydrolases GH18 (role in chitin degradation), dehalogenases (bioremediation), and bacterial laccases (bioremediation). Overall, two million clones were generated out of which 405,504 clones were used for screening purposes. Screening of clones using 33p-dCTP-labeled probes identified 88 positive clones. Further sequencing followed by analysis of DNA from positive clones using 454 pyrosequencing techniques generated 42 metagenomic contigs (of length >5 kb) encompassing 94 putative sequences. One of the main advantages of this technique is that large amount of metagenomic materials can be simultaneously screened and analyzed. Furthermore, by controlling stringency of hybridization, one can control the degree of homology during hybridization in order to get the most similar or divergent sequence. In order to further improve the throughput of gene mining using hybridization method, microarray has been widely used as it allows simultaneously screening of a library against a wide array of different target genes.

The next effective method for metagenomic library screening for target gene is through PCR. For example, Courtois and coworker used PCR to identify polyketide synthase gene from metagenomic library (Courtois et al. 2003). They prepared metagenomic library from the metagenome recovered from the field in La Cite Saint Andre (Isere, France). In order to identify target gene from metagenomic library, it was screened by using PCR. For the screening purpose, degenerative primers were designed from the conserve region of type I polyketide synthase genes (PKS), and primer efficacy was further tested on PKS-producing strains (e.g., *Streptomyces coelicolor* ATCC 101478). Once the efficacy of degenerative primers was checked, 96 clones were screened. Out of 96 clones, eight clones were found to be PKS positive.

Despite having a remarkably high efficiency and probability of getting novel gene, these methods have their own limitation of having low throughput and being labor intensive and time consuming. The second strategy involved direct screening of DNA sample recovered from the environment using PCR by designing degenerative primers (Itoh et al. 2014). The approach used by Itoh and coworker also known as screening of gene-specific amplicons from metagenome (S-GAM) allows successful identification of two superior alcohol dehydrogenase (*adh*) (HLADH-012 and HLADH-021) for the production of anti-prelog chiral alcohols. In their investigation, they selected metagenome recovered from 20 different environments to generate metagenomic library. For the identification of superior *adh* gene, they designed degenerative primer from the conserved region of the gene which have the potential to amplify almost full-length gene. Cloning of amplified fragments in *E. coli* plasmid library recovered overall 2000 clones out of which 1200 clones were found to be *adh* positive. Therefore, this strategy proves itself efficient in terms of identifying genes' functionally as well as structurally related family. Although this strategy has great potential in metagenome gene mining but has some limitation of requiring prior sequence information for degenerative primer designing, moreover, only gene family member with conserved domain gets identified (Maurer 2011).

In contrast, the third strategy based on high-throughput sequencing (NGS) has great potential in gene mining/prospecting. The standard protocol for gene mining using next-generation sequencing (NGS) technology is divided into four main steps: (1) high-quality metagenomic DNA isolation followed by next-generation sequencing, (2) assembly of NGS reads into contigs and scaffolds, (3) analysis of the sequencing data in search of target genes, and (4) functional complementation assay in heterologous system (Vester et al. 2015) (Fig. 4.1). When the objective of metagenomic investigation is gene mining, hybrid sequencing strategy (strategy in which more than one sequencing platform was used) is useful to fish full-length gene. For this purpose, Illumina shotgun sequencing strategy coupled with 454 pyrosequencing/PacBio sequencing technology is good enough. Illumina shotgun reads increase the coverage and hence the confidence on the correctness of discovered gene(s), whereas long reads from 454 pyrosequencer/PacBio increase the probabilities of getting full-length gene(s). All sequencing platforms have their advantages or disadvantages; therefore, sequencing platform should be selected carefully. Details describing sequencing platforms and their sequence handling were very well reviewed by Loman and coworker and Kim and coworker, respectively (Kim et al. 2013; Loman et al. 2012). After getting cleaned and trimmed with high-quality sequence, the next critical and challenging task is to perform assembly with minimum of chimeras. Recent development in bioinformatics tools dedicated especially for the metagenomic analysis makes it easy to perform accurate assembly of short reads. Once the good quality of assembly was obtained, gene mining task was initiated either by using local BLAST approaches or by creating custom algorithms. The study conducted by Zhao and coworker illustrated the above approach very well (Zhao et al. 2016). In the study conducted by Zhao and coworker, they identify novel cellulase from the cellulose-degrading microbial consortium. The metagenome isolated from microbial consortium was sequenced using 454 pyrosequencing platform followed by assembly and annotation. They identified overall 28 genes including 15 glycoside hydrolase families. Further confirmation of identified cellulase was performed using functional screening.

4.4.2 Functional-Based Approach

The sequence-based approach exploits sequence information based on homology for gene mining, which often revealed the new members of the same family. However, it doesn't provide information related to the functionality of biomolecules. Therefore, despite generating large amount of metagenomic DNA sequencing data, library construction followed by functional screening for biocatalyst molecule is more interesting (Schoenfeld et al. 2010). Furthermore, functional-based screening has the potential to identify all isoforms of proteins available in metagenome for particular enzymatic processes or function. Functional

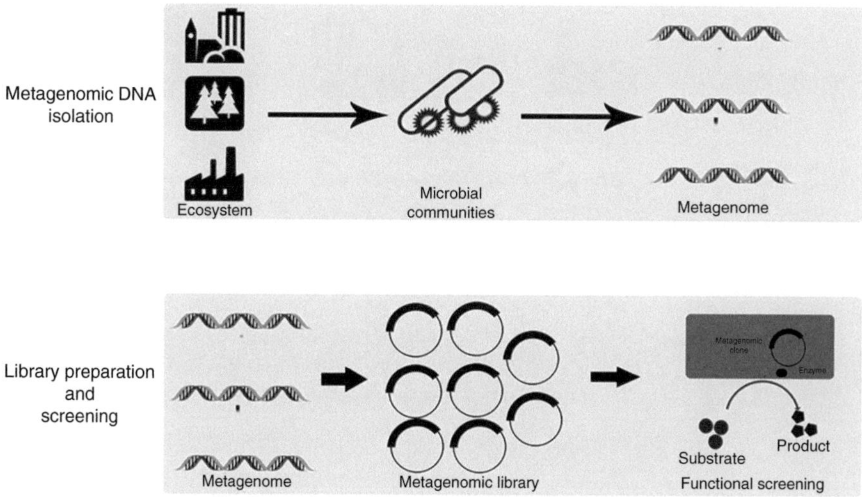

Fig. 4.2 Schematic of acquisition and processing of metagenomic information via function-based approach. Likewise sequence-based approach, function-based approach also involved initial step of metagenome isolation followed by library construction in expression vector. In order to identify functionally relevant clone, protein assay was performed. After the identification of positive clone, sequencing was performed to identify gene sequence

activity-driven screening detects function of interest and hence doesn't require any prior sequence information (Felczykowska et al. 2012) (Fig. 4.2). Function-based metagenomic for gene mining utilizes four different approaches as follows: (1) direct identification of gene product in clones (this type of screening may use fluorescent catabolite to evaluate enzyme activity), (2) complementation assay, (3) induce gene expression, and (4) enzymatic assay.

There is a plethora of studies available that uses functional screening to identify new genes. Same approach was used by Suenaga et al. (2007), in which they used fosmid metagenomic library to identify genes for extradiol dioxygenase enzymes (EDOs). EDO gene is involved in the catabolism of aromatic substance. Analysis of 91 positive clones through sequencing revealed 25 genes belonging to the new subfamily. Likewise, Tannieres et al. (2013) obtained new NAHLase, a xenobiotic degradation enzyme from a metagenomic library. Current metagenomic function screening-based approaches are not able to identify and isolated catabolic genes induced by certain stimuli or against certain chemicals. To overcome this limitation to high-throughput screening methods, viz., substrate induces gene expression screening (SIGEX) (Uchiyama et al. 2005), and intercellular biosensor method (Williamson et al. 2005) was used (Fig. 4.3). SIGEX is a high-throughput screening approach that consists of four steps including (1) constructing metagenomic libraries in liquid growth medium, (2) removing green fluorescent proteins (GFP) constitutively expressing empty clones, (3) providing stimuli/induction/substrate for the co-expression of target protein and GFP, and (4) selecting clones using fluorescence-activated cell sorting (FACS) (Uchiyama et al. 2005). By using SIGEX, Uchiyama

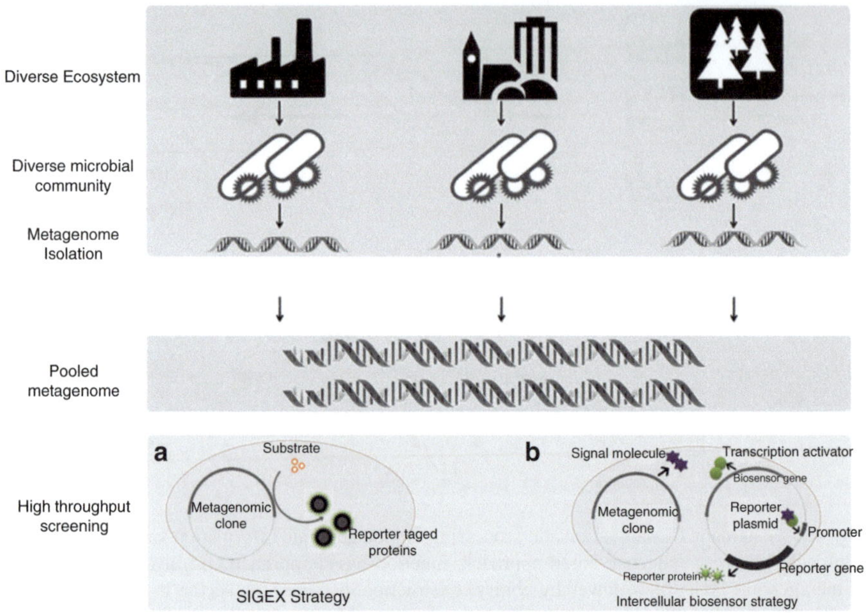

Fig. 4.3 Systematic workflow for high-throughput screening for function-based metagenome. (**a**) SIGEX strategy for metagenomic library screening. In SIGEX strategy metagenomic clones s fused with reporter gene. The expression of reporter gene in the presence of certain stimuli/substance or condition revealed the presence of fragment of interest, hence, helping to identify clones on the basis of function. (**b**) Intercellular biosensor strategy is designed specifically for the identification of biologically active small biomolecules. In this strategy expression of reporter gene depends upon the presence of target molecule. In the presence of target molecule transcription factor get activated and in response activate the transcription followed by translation of reporter gene. In both the strategy once the reporter gene get activated, positive bacterial clones can be separated by using FACS

et al. (2005) identified catabolic genes induced by aromatic compound from metagenomic library prepared from ground water.

The other high-throughput strategy for metagenomic library screening is intercellular biosensor also called METREX and was successfully used by Willamson and coworkers, to identify metabolites/compound induce quorum sensing in bacteria. They design a system in which, if metagenomic clone produces inducer of quorum sensing, the presence of inducer resulted into the production of GFP and can be easily sorted by FACS. By using METREX, overall, 11 clones were identified which induce quorum sensing and two clones that inhibit. Both of the above described high-throughput screening methodologies can act as a powerful tool for identification of biologically active small molecules and functionally active clone in metagenomic library.

The detection of new genes through function-based screening is only useful if the target gene gets expressed in heterologous host (Schoenfeld et al. 2010). Therefore, the main challenge of the function-based screening is the expression of target gene, often impeded by various factors that include truncated gene clone, incompatibility between the regulatory machinery of gene and host, codon bias, improper protein folding, etc. (Felczykowska et al. 2012; Schoenfeld et al. 2010). To overcome some of the difficulties, the use of a different expression host is recommended. For example, mostly exploited prokaryotic laboratory host *E. coli* can express only 40% of the metagenomic gene pool. To avoid such adversity, other organisms need to be tested that include genera *Bacillus*, *Pseudomonas*, and *Streptomyces*. Recently, Iqbal and coworker used alternate host strategy to identify six enzymes involved in antibiosis activity using *Ralstonia metallidurans* as a host (Iqbal et al. 2014). This same fragment was unable to express in *E. coli* which demonstrate the perspective of alternative host in metagenomic library screening. In summary, wide acceptance of function-based screening for gene mining required the development of a versatile method for library preparation and screening in high-throughput manner, e.g., in nano-scale by microfluidic technology (Arnold et al. 2016).

Conclusion

This chapter is an attempt to summarize the ongoing research and development in the area of metagenomic to study microbe community in gene mining/bioprospecting point of view. Metagenomic studies help us to understand the functioning mechanism of microbial communities and ecosystem inhabited by them. In addition, they also improve our perception and shed light on the biology of microbial individual constituting the communities which could provide census of genes and protein of economical importance. Identification of novel protein or biocatalyst with no sequence homology to any previously known proteins through activity-based discovery, highlighting the importance of metagenomic in gene mining. The enormous diversity of microbes present in metagenome can be made accessible to its full potential by using recently developed genomic tools. In order to increase the efficiency of gene mining using function-based approach, development of new lab-on-a-chip tools was expected.

References

Adkins J, Pugh S, McKenna R, Nielsen DR (2012) Engineering microbial chemical factories to produce renewable "biomonomers" Frontiers in microbiology 3:313. doi:10.3389/fmicb.2012.00313

Amann RI, Ludwig W, Schleifer KH (1995) Phylogenetic identification and in situ detection of individual microbial cells without cultivation. Microbiol Rev 59(1):143–169

Ando S, Ishida H, Kosugi Y, Ishikawa K (2002) Hyperthermostable endoglucanase from Pyrococcus horikoshii. Appl Environ Microbiol 68(1):430–433

Arnold JW, Roach J, Azcarate-Peril MA (2016) Emerging technologies for gut microbiome research. Trends Microbiol 24(11):887–901. doi:10.1016/j.tim.2016.06.008

Azam F, Long RA (2001) Sea snow microcosms. Nature 414(6863):495, 497–495, 498. doi:10.1038/35107174

Bergey DH, American Society for M (1923) Bergey's manual of determinative bacteriology. Williams & Wilkins Co., Baltimore

Courtois S, Cappellano CM, Ball M, Francou FX, Normand P, Helynck G, Martinez A, Kolvek SJ, Hopke J, Osburne MS, August PR, Nalin R, Guerineau M, Jeannin P, Simonet P, Pernodet JL (2003) Recombinant environmental libraries provide access to microbial diversity for drug discovery from natural products. Appl Environ Microbiol 69(1):49–55

Elend C, Schmeisser C, Leggewie C, Babiak P, Carballeira JD, Steele HL, Reymond JL, Jaeger KE, Streit WR (2006) Isolation and biochemical characterization of two novel metagenome-derived esterases. Appl Environ Microbiol 72(5):3637–3645. doi:10.1128/aem.72.5.3637-3645.2006

Fang ZM, Li TL, Chang F, Zhou P, Fang W, Hong YZ, Zhang XC, Peng H, Xiao YZ (2012) A new marine bacterial laccase with chloride-enhancing, alkaline-dependent activity and dye decolorization ability. Bioresour Technol 111:36–41. doi:10.1016/j.biortech.2012.01.172

Felczykowska A, Bloch SK, Nejman-Falenczyk B, Baranska S (2012) Metagenomic approach in the investigation of new bioactive compounds in the marine environment. Acta Biochim Pol 59(4):501–505

Fu J, Leiros HK, de Pascale D, Johnson KA, Blencke HM, Landfald B (2013) Functional and structural studies of a novel cold-adapted esterase from an Arctic intertidal metagenomic library. Appl Microbiol Biotechnol 97(9):3965–3978. doi:10.1007/s00253-012-4276-9

Handelsman J (2005) Sorting out metagenomes. Nat Biotechnol 23(1):38–39. doi:10.1038/nbt0105-38

Handelsman J, Rondon MR, Brady SF, Clardy J, Goodman RM (1998) Molecular biological access to the chemistry of unknown soil microbes: a new frontier for natural products. Chem Biol 5(10):R245–R249

Healy FG, Ray RM, Aldrich HC, Wilkie AC, Ingram LO, Shanmugam KT (1995) Direct isolation of functional genes encoding cellulases from the microbial consortia in a thermophilic, anaerobic digester maintained on lignocellulose. Appl Microbiol Biotechnol 43(4):667–674

Hong KS, Lim HK, Chung EJ, Park EJ, Lee MH, Kim JC, Choi GJ, Cho KY, Lee SW (2007) Selection and characterization of forest soil metagenome genes encoding lipolytic enzymes. J Microbiol Biotechnol 17(10):1655–1660

Hu Y, Zhang G, Li A, Chen J, Ma L (2008) Cloning and enzymatic characterization of a xylanase gene from a soil-derived metagenomic library with an efficient approach. Appl Microbiol Biotechnol 80(5):823–830. doi:10.1007/s00253-008-1636-6

Iqbal HA, Craig JW, Brady SF (2014) Antibacterial enzymes from the functional screening of metagenomic libraries hosted in Ralstonia metallidurans. FEMS Microbiol Lett 354(1):19–26. doi:10.1111/1574-6968.12431

Israel DI (1993) A PCR-based method for high stringency screening of DNA libraries nucleic acids research 21:2627–2631

Itoh N, Kariya S, Kurokawa J (2014) Efficient PCR-based amplification of diverse alcohol dehydrogenase genes from metagenomes for improving biocatalysis: screening of gene-specific amplicons from metagenomes. Appl Environ Microbiol 80(20):6280–6289. doi:10.1128/aem.01529-14

Jacquiod S, Franqueville L, Cecillon S, Vogel TM, Simonet P (2013) Soil bacterial community shifts after chitin enrichment: an integrative metagenomic approach. PLoS One 8(11):e79699. doi:10.1371/journal.pone.0079699

Jeon JH, Lee HS, Kim JT, Kim SJ, Choi SH, Kang SG, Lee JH (2012) Identification of a new subfamily of salt-tolerant esterases from a metagenomic library of tidal flat sediment. Appl Microbiol Biotechnol 93(2):623–631. doi:10.1007/s00253-011-3433-x

Jiang C, LL W, Zhao GC, Shen PH, Jin K, Hao ZY, Li SX, Ma GF, Luo FF, GQ H, Kang WL, Qin XM, Bi YL, Tang XL, Wu B (2010) Identification and characterization of a novel fumarase gene by metagenome expression cloning from marine microorganisms. Microb Cell Fact 9:91. doi:10.1186/1475-2859-9-91

Kim M, Lee KH, Yoon SW, Kim BS, Chun J, Yi H (2013) Analytical tools and databases for metagenomics in the next-generation sequencing era. Genomics Inform 11(3):102–113. doi:10.5808/gi.2013.11.3.102

Lee MH, Hong KS, Malhotra S, Park JH, Hwang EC, Choi HK, Kim YS, Tao W, Lee SW (2010) A new esterase EstD2 isolated from plant rhizosphere soil metagenome. Appl Microbiol Biotechnol 88(5):1125–1134. doi:10.1007/s00253-010-2729-6

Lee MH, KH O, Kang CH, Kim JH, TK O, Ryu CM, Yoon JH (2012) Novel metagenome-derived, cold-adapted alkaline phospholipase with superior lipase activity as an intermediate between phospholipase and lipase. Appl Environ Microbiol 78(14):4959–4966. doi:10.1128/aem.00260-12

Lee SW, Won K, Lim HK, Kim JC, Choi GJ, Cho KY (2004) Screening for novel lipolytic enzymes from uncultured soil microorganisms. Appl Microbiol Biotechnol 65(6):720–726. doi:10.1007/s00253-004-1722-3

Liu J, Liu WD, Zhao XL, Shen WJ, Cao H, Cui ZL (2011) Cloning and functional characterization of a novel endo-beta-1,4-glucanase gene from a soil-derived metagenomic library. Appl Microbiol Biotechnol 89(4):1083–1092. doi:10.1007/s00253-010-2828-4

Loman NJ, Misra RV, Dallman TJ, Constantinidou C, Gharbia SE, Wain J, Pallen MJ (2012) Performance comparison of benchtop high-throughput sequencing platforms. Nat Biotechnol 30(5):434–439. doi:10.1038/nbt.2198

Lynd LR, Weimer PJ, van Zyl WH, Pretorius IS (2002) Microbial cellulose utilization: fundamentals and biotechnology. Microbiol Mol Biol Rev 66(3):506–577, table of contents

Matsuzawa T, Yaoi K (2017) Screening, identification, and characterization of a novel saccharide-stimulated beta-glycosidase from a soil metagenomic library. Appl Microbiol Biotechnol 101(2):633–646. doi:10.1007/s00253-016-7803-2

Maurer JJ (2011) Rapid detection and limitations of molecular techniques. Annu Rev Food Sci Technol 2:259–279. doi:10.1146/annurev.food.080708.100730

Mohamed YM, Ghazy MA, Sayed A, Ouf A, El-Dorry H, Siam R (2013) Isolation and characterization of a heavy metal-resistant, thermophilic esterase from a Red Sea brine pool. Sci Rep 3:3358. doi:10.1038/srep03358

National Research Council Committee on Metagenomics C, Functional A (2007) The National Academies Collection: reports funded by National Institutes of Health. In: The new science of metagenomics: revealing the secrets of our microbial planet. National Academies Press, Washington, DC. doi:10.17226/11902

Park SJ, Chae JC, Rhee SK (2010) Application of DNA microarray for screening metagenome library clones Methods in molecular biology (Clifton, NJ) 668:313–324. doi:10.1007/978-1-60761-823-2_22

Pham VD, Palden T, DeLong EF (2007) Large-Scale Screens of Metagenomic Libraries Journal of Visualized Experiments: JoVE:201 doi:10.3791/201

Polizeli ML, Rizzatti AC, Monti R, Terenzi HF, Jorge JA, Amorim DS (2005) Xylanases from fungi: properties and industrial applications. Appl Microbiol Biotechnol 67(5):577–591. doi:10.1007/s00253-005-1904-7

Sayed A, Ghazy MA, Ferreira AJ, Setubal JC, Chambergo FS, Ouf A, Adel M, Dawe AS, Archer JA, Bajic VB, Siam R, El-Dorry H (2014) A novel mercuric reductase from the unique deep brine environment of Atlantis II in the Red Sea. J Biol Chem 289(3):1675–1687. doi:10.1074/jbc.M113.493429

Schoenfeld T, Liles M, Wommack KE, Polson SW, Godiska R, Mead D (2010) Functional viral metagenomics and the next generation of molecular tools. Trends Microbiol 18(1):20–29. doi:10.1016/j.tim.2009.10.001

Schroder C, Elleuche S, Blank S, Antranikian G (2014) Characterization of a heat-active archaeal beta-glucosidase from a hydrothermal spring metagenome. Enzyme Microb Technol 57:48–54. doi:10.1016/j.enzmictec.2014.01.010

Suenaga H, Ohnuki T, Miyazaki K (2007) Functional screening of a metagenomic library for genes involved in microbial degradation of aromatic compounds. Environ Microbiol 9:2289–2297. doi:10.1111/j.1462-2920.2007.01342.x

Tao W, Lee MH, Wu J, Kim NH, Kim JC, Chung E, Hwang EC, Lee SW (2012) Inactivation of chloramphenicol and florfenicol by a novel chloramphenicol hydrolase. Appl Environ Microbiol 78(17):6295–6301. doi:10.1128/aem.01154-12

Tannieres M, Beury-Cirou A, Vigouroux A, Mondy S, Pellissier F, Dessaux Y, Faure D (2013) A metagenomic study highlights phylogenetic proximity of quorum-quenching and xenobiotic-degrading amidases of the AS-family PloS one 8:e65473. doi:10.1371/journal.pone.0065473

Tirawongsaroj P, Sriprang R, Harnpicharnchai P, Thongaram T, Champreda V, Tanapongpipat S, Pootanakit K, Eurwilaichitr L (2008) Novel thermophilic and thermostable lipolytic enzymes from a Thailand hot spring metagenomic library. J Biotechnol 133(1):42–49. doi:10.1016/j.jbiotec.2007.08.046

Torsvik V, Goksoyr J, Daae FL (1990) High diversity in DNA of soil bacteria. Appl Environ Microbiol 56(3):782–787

Uchiyama T, Abe T, Ikemura T, Watanabe K (2005) Substrate-induced gene-expression screening of environmental metagenome libraries for isolation of catabolic genes Nature biotechnology 23:88–93. doi:10.1038/nbt1048

Vakhlu J, Sudan AK, Johri BN (2008) Metagenomics: future of microbial gene mining. Indian J Microbiol 48(2):202–215. doi:10.1007/s12088-008-0033-2

Vester JK, Glaring MA, Stougaard P (2015) Improved cultivation and metagenomics as new tools for bioprospecting in cold environments. Extremophiles 19(1):17–29. doi:10.1007/s00792-014-0704-3

Voget S, Leggewie C, Uesbeck A, Raasch C, Jaeger KE, Streit WR (2003) Prospecting for novel biocatalysts in a soil metagenome. Appl Environ Microbiol 69(10):6235–6242

Voget S, Steele HL, Streit WR (2006) Characterization of a metagenome-derived halotolerant cellulase. J Biotechnol 126(1):26–36. doi:10.1016/j.jbiotec.2006.02.011

Wang H, Gong Y, Xie W, Xiao W, Wang J, Zheng Y, Hu J, Liu Z (2011) Identification and characterization of a novel thermostable gh-57 gene from metagenomic fosmid library of the Juan de Fuca Ridge hydrothemal vent. Appl Biochem Biotechnol 164(8):1323–1338. doi:10.1007/s12010-011-9215-1

Wierzbicka-Wos A, Cieslinski H, Wanarska M, Kozlowska-Tylingo K, Hildebrandt P, Kur J (2011) A novel cold-active beta-D-galactosidase from the Paracoccus sp. 32d--gene cloning, purification and characterization. Microb Cell Fact 10:108. doi:10.1186/1475-2859-10-108

Williamson LL, Borlee BR, Schloss PD, Guan C, Allen HK, Handelsman J (2005) Intracellular screen to identify metagenomic clones that induce or inhibit a quorum-sensing biosensor Applied and environmental microbiology 71:6335–6344. doi:10.1128/aem.71.10.6335-6344.2005

Yu EY, Kwon MA, Lee M, JY O, Choi JE, Lee JY, Song BK, Hahm DH, Song JK (2011) Isolation and characterization of cold-active family VIII esterases from an arctic soil metagenome. Appl Microbiol Biotechnol 90(2):573–581. doi:10.1007/s00253-011-3132-7

Zengler K, Toledo G, Rappe M, Elkins J, Mathur EJ, Short JM, Keller M (2002) Cultivating the uncultured. Proc Natl Acad Sci U S A 99(24):15681–15686. doi:10.1073/pnas.252630999

Zhao C, Chu Y, Li Y, Yang C, Chen Y, Wang X, Liu B (2017) High-throughput pyrosequencing used for the discovery of a novel cellulase from a thermophilic cellulose-degrading microbial consortium. Biotechnol Lett 39(1):123–131. doi:10.1007/s10529-016-2224-y

Zhu Y, Li J, Cai H, Ni H, Xiao A, Hou L (2013) Characterization of a new and thermostable esterase from a metagenomic library. Microbiol Res 168(9):589–597. doi:10.1016/j.micres.2013.04.004

Prospects and Progress in Extreme Biosphere Microbiome

5

Pravin Dudhagara, Ramesh Kothari, Anjana Ghelani, Jalpa Rank, and Rajesh Patel

Abstract

There is a considerable market potential of extremophiles and their biomolecules; however, their high evolutionary rates and culturability limitations restrict their industrial exploration. The impressive development of "omics" technologies has allowed the culture-free technique, i.e., metagenomics as a valuable tool for mining the hidden information of the extremophiles and their biospheres. At present, environmental microbiomes are being studied using functional-based and sequence-based approaches. The growth of the extreme environmental metagenomics literature and projects is increased nearly tenfold in the last decade due to the advancement in the sequencing technologies. Here, in the context, we summarize all the aspects of extreme biosphere metagenomics with recent prospects and progress.

Keywords

Extreme biosphere • Microbiome • Extremophiles • Metgenomics • Environmental

P. Dudhagara (✉)
Department of Biosciences (UGC-SAP), Veer Narmad South Gujarat University, Surat 395007, India
e-mail: dudhagarapr@gmail.com

R. Kothari • J. Rank
Department of Biosciences (UGC-CAS), Saurashtra University, Rajkot 360005, India

A. Ghelani • R. Patel
Department of Life Sciences, Hemchandracharya North Gujarat University, Patan 384265, India

R.P. Singh et al. (eds.), *Understanding Host-Microbiome Interactions - An Omics Approach*, DOI 10.1007/978-981-10-5050-3_5

5.1 Extremophiles and Extreme Environments

These are various extreme habitats situated across the globe since the genesis of the earth. These extreme habitats harbor a rich microbial diversity. Certain old evidence of life like microfossils, stromatolites, microfibrous sedimentary rocks, and sedimentary carbon pool suggests that the microorganisms inhabit the earth since the archaean period, the time before 2.5 billion years (Stanley 2005). Such ancient microbial life had developed robust metabolic functions similar to many present-day living extremophiles thriving into extreme environments. The extremophiles hold secret survival "kits" to shelter at either single or multiple extreme conditions. Microbiologists are exploring their cellular properties like new gene pools, robust biomolecules, and metabolic uniqueness through the culturing methods since the discovery of extremophiles. However, despite the technological advancement for the investigation of extremophiles and their habitats, we have decoded the very limited information from the extreme biosphere (Rampelotto 2013).

Extreme territories support all the three taxonomic forms of life to flourish. However, the largest membership is represented by the Archaea followed by Bacteria and Eukaryotes. The ability of these extremophilic microorganisms to proliferate under extreme conditions is of immense importance for understanding microbial physiology and evolution. Extremophiles are best characterized according to their growth profiles, using marginal data, under certain culture conditions including salt concentration, temperature profile, pH scale, and growth under hydrostatic pressure (Mesbah and Wiegel 2008). The representative examples of extremophiles thriving in different extremities are thermophiles (45–60 °C temperature), hyperthermophiles (60–120 °C temperature), psychrophiles (below 0 °C temperature), acidophiles (below 4.0 pH), alkaliphiles (over 9.0 pH), piezophiles or barophiles (>0.5 MPa pressure), halophiles (>1 M NaCl concentration), and xerophiles (<0.85 water activity) (Horikoshi et al. 2010).

Culturability of the extreme habitats is very less due to the persistence of abiotic stresses and differences between natural environments and laboratory conditions. Additionally, culturability is a very complex physiological process that depends on various phenomena (Barer and Harwood 1999). Due to these limitations, the majority of such environment has remained unexplored. However, during the recent years, development and application of molecular techniques, such as PCR, cloning, and next-generation high-throughput sequencing, have proved quite valuable in judging the distribution and diversity of extreme habitats. Hence, due to the limitation of available culturing methods for the extremophiles, it is now being studied by the uncultivable approach referred as a metagenomics to translate the potentials of various extremophilic microorganisms. Furthermore, the holistic community can be deciphered through the metagenomics approach, whereas the traditional microbiology relies upon the cultivation of few clones or colonies. So, the metagenomics application provides the profiling of the microbial diversity of any extreme environment to analyze the entire community and can be delineated broadly as an environmental genomics, ecogenomics, or community genomics (Hugenholtz et al. 1998).

5.2 Metagenomics and Microbial Diversity

Development and discovery of various molecular biology techniques after the 1980s extend the genomic discipline toward its associated "omics" technologies. Metagenomics emerges in the ending of last century, which eliminates the culturability for mining the microbial information and revolutionizing microbial ecology. Metagenomics is the study of the collective forms of genomes directly isolated from the environmental sample for the comprehensive analysis of microbial diversity and ecology of a specific environment. Metagenomics studies provide the mechanism for analyzing previously unknown organisms, and at the same time, one can examine the diversity of organisms present in specific environments as well as analyze the complex interactions between members of a specific environment (Handelsman 2004). Metagenomics studies are conducted by two different approaches. One is function-based analysis, which deals with the total DNA extraction from environmental sample followed by cloning into suitable host and detection of expressed phenotypes in the host cells, whereas sequence-based analysis is mainly concerned with decoding of the extracted DNA and/or RNA using various sequencing platforms followed by assessment of taxonomic diversity. Both approaches are applicable to decipher hidden microbial gene pools and profiling of the microorganisms (Fig. 5.1). Early environmental gene sequencing is dealt with cloning of 16S rRNA gene to analyze the microbial taxonomic profile. Such work revealed the vast majority of microbial biodiversity that had been explored by cultivation-based methods (Hugenholtz et al. 1998). Slowly the shotgun Sanger sequencing or massively parallel pyrosequencing is applied to get largely unbiased samples of all genes from all the members of the

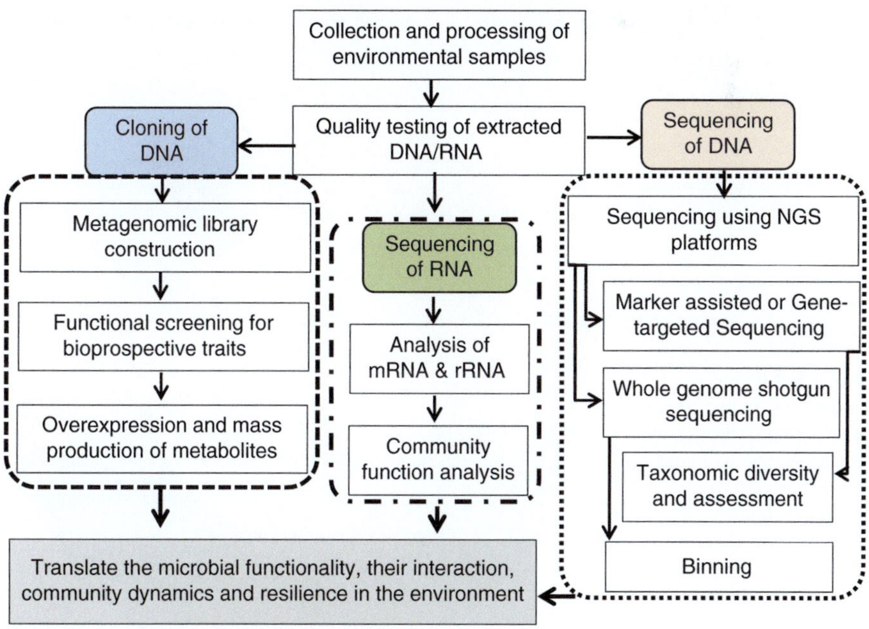

Fig. 5.1 Standard metagenomics pipeline for environmental microbiomes research

sampled communities (Eisen 2007). The first metagenomics studies conducted using high-throughput sequencing by massively parallel 454 pyrosequencing transformed the studies of the microbial universe (Poinar et al. 2006; Edwards et al. 2006). Nowadays the various next-generation sequencing (NGS) platforms are utilized for the metagenomics studies, and continuous improvements in the existing sequencing technologies are often done by the original developers.

5.3 Environmental Metagenomics

According to the meeting report of Earth microbiome project, one quintillion (10^{30}) microbial cells are present on the earth. Theoretical estimation of the average quantity of DNA in the microbial cell is 10 million base pairs. As yet, we have investigated hardly 1% of the total environmental DNA by global environmental DNA sequencing efforts (Gilbert et al. 2010). This statistics may also be far greater than actual analysis; so, the massive information of microbial life on the Earth is yet unknown and/or under-sampled. Hence, we are in the beginning stage in the study of the extreme environmental metagenomics.

Early environmental metagenomics projects are considered a key trigger to drive the field of extreme microbiomes. Environmental genomic studies of the Sargasso Sea (Venter et al. 2004) are a major breakthrough in the environmental metagenomics, which leads to developing the interest among the scientific communities to initiate and explore the microbial diversity of extreme habitats using metagenomics approach. The scientific literature on "extreme environmental metagenomics" available in public domains increased quickly in the last decade indicating the development of the field (Fig. 5.2). The rapid escalation of metagenomics projects in the

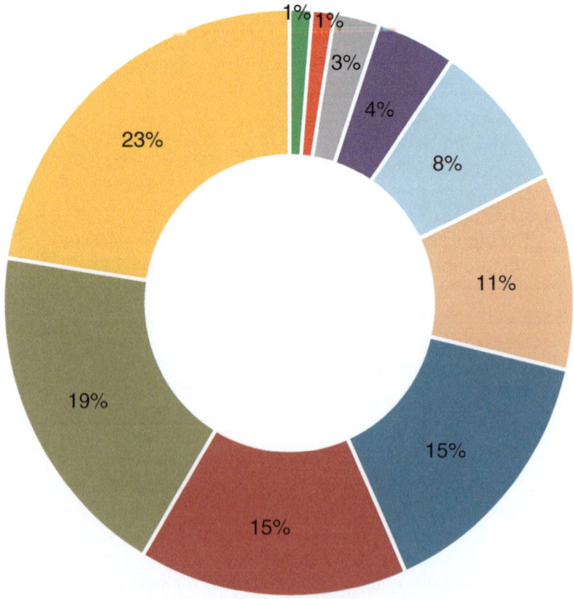

Fig. 5.2 Literature available in public domain on extreme environmental metagenomics assessed on 8 February 2017 on Google Scholar ($n = 15,700$). Data presented in the graph for the last 10 years are clockwise from the year 2007 (*green* series 1%) to 2016 (*yellow* series 23%)

various online databases indicated the quick growth of metagenomics field. Currently, more than 20% metagenomes submitted into public domains are derived from the various extreme biosphere including marine/ocean, abyssal plain, desert, hydrothermal vents, permafrost, glacier, salt marsh, thermal hot springs, geyser, soda lake, hypersaline lake, submarine volcano, black smoker, acid mine drainage, etc. (Table 5.1). Undoubtedly, it is due to the recent advances in high-throughput sequencing technologies with more sophisticated bioinformatics analysis pipeline making the metagenomics study very easy and rapid.

The recent identification of new gene pools and species of extremophiles from the extreme habitats geared up the exploration of microbial species for the industrial and biomedical potentials. At the beginning of metagenomics era, the giant vector, i.e., BAC, was used to construct the metagenomics library of the environmental

Table 5.1 Metagenomes of extreme environments available in public domains (Assessed on 8 February 2017)

Public database/web portal with URL	Total submitted metagenomes	Extreme environment-related metagenomes (%)	Extreme biome examples
MGRAST (http://metagenomics.anl.gov/)	276,607	20	Marine sample, thermal hot springs, geyser, salt lake, desert, hypersaline lake, soda lake, glacier, submarine volcano, black smoker, acid mine drainage, contaminated site
EBI metagenomics (https://www.ebi.ac.uk/metagenomics/)	12,412	25	Oceanic/marine, abyssal plane, desert, hydrothermal vents, permafrost, glacier, salt marsh, and contaminated habitats
NCBI metagenome (https://www.ncbi.nlm.nih.gov/genbank/metagenome)	11,929	18	Marine samples, hypersaline lake, mine drainage, marine sediment
IMG/M (https://img.jgi.doe.gov/cgi-bin/m/main.cgi)	7341	20	Oceanic and marine sample, hot springs, saline and hypersaline lake
GOLD (Genomes Online Database) (https://gold.jgi.doe.gov/)	18,408	18	Hot springs, oceanic/marine water, saline lake, geyser, geothermal water, marine deep subsurface hydrothermal vent, subseafloor sediment

DNA (Rondon et al. 2000) and function, and the sequence-based analysis was performed from each clone. Modern metagenomics approach makes it possible to know the physiology of the extremophiles, their role in the habitats and adaption to environmental pressures. So, the microbiome of extreme biosphere may help to establish the microbial community network structure, which is very useful to decode the microbial functionality, interaction, and community dynamics (Cowan et al. 2015). However, the various experimental challenges from sampling to sequencing should be addressed before conducting environmental metagenomics projects.

5.4 Challenges in Environmental DNA Extraction

The key challenges of conducting metagenomics studies include the sampling and transporting of the adequate intact environmental sample and extraction of the high-quality nucleic acid from the environmental sample. The stresses in the extreme site are the key hurdles for the extraction and the purification of the high-quality nucleic acids; so, the sample processing is prerequisite for environmental metagenomics project (Thomas et al. 2012). Isolation of poor-quality DNA may hamper the subsequent analysis, i.e., cloning and sequencing; so, the specific methods and protocol are needed to extract the high-quality and high molecular weight (HMW) community genomic DNA from the environmental sample. Various direct environmental DNA extraction methods including freezing-thawing, bead beating, and ultrasonication along with indirect extraction methods like PEG-NaCl-based and enzyme lysis with hot detergent treatment are being used for the extraction of environmental DNA and viable for the functional and structural profiling of the microorganisms (Delmont et al. 2011; Narayan et al. 2016). Based on direct and indirect extraction, nowadays commercial kits are also developed by many manufacturers to extract the nucleic acid in good quality and quantity from soil and water samples. However, the success of the DNA extraction depends on the microbial population and the physiological status of the cells.

Heterogeneous microbial communities exist in the environment, and all the microbial cells have the substantial structural variation in the cell wall and cell membrane. So the cellular dissimilarity in the microbial species restricts the selection of single universal cell lysis method to extract the nucleic acid. However, harsh treatment can be used but such conditions cause damage to DNA and mild treatment leads to less recovery. So, the combination of chemical, physical/mechanical, and biological cell lysis is best suitable for the extraction of high-quality environmental DNA (Bag et al. 2016). The environmental stresses have also increased the difficulties in extraction procedure as the extremities may give the adaptive and protective mechanisms to the microbial cell to survive in the extreme conditions, which may make the cell very resilient to lysis and consequently inadequate nucleic acid will be taken out that missing genomic contents of rare species. Hence, before starting on the extreme environmental metagenomics project, one should thoroughly study all the geological, biological, and cellular features to extract the high-quality HMW community genomic DNA (Table 5.2).

Table 5.2 Challenges in DNA extraction from the various extreme biosphere and methodologies required for the success of extraction (Note: *CW* cell wall and *CM* cell membrane)

Extreme biosphere (stresses)	Microorganisms and its adaptive and protective cellular mechanisms	Challenges in DNA extraction	Methodologies and modification required for DNA extraction	References
Hot spring Volcanoes (high temperature) 	Thermophiles and hyperthermophiles: CW: very thick with pseudomurein and rigid CM: long, branched, and cyclic fatty acid chain, ether lipid in membrane Extra cell envelope called S-layer	High G + C % content, high content of histones like protein chaperonin protein, archaeal cell resistant to lysozyme	Enzymatic lysis is required for removal of proteins	Koga (2012), Klingl (2014), Ramakrishnan and Adams (1995)
Glacier Cold arid desert (low temperature) 	Psychrophiles and psychrotrophs CW: fragile to high temperature CM: membrane lipids have high unsaturation and average chain length with methyl branching High content of cold-shock proteins	Cell lysis during the sampling and transportation due to the temperature changes	No special modification required except the protein removal and physical lysis of cells	Goordial et al. (2016), De Maayer et al. (2014)

(continued)

Table 5.2 (continued)

Extreme biosphere (stresses)	Microorganisms and its adaptive and protective cellular mechanisms	Challenges in DNA extraction	Methodologies and modification required for DNA extraction	References
Desert Arid soil (desiccation, high radiation, and high temperature) 	Xerophiles or xerotolerant CW: desiccation-resistant proteins CM: membranes contain glycerol The cell contains various compatible solutes, i.e., trehalose Sporulation state	Cells are hardy due to sporulation or viable-but-not-culturable state	Cell lysis difficulties due to the accumulation of compatible solute and dormancy. Pretreatment and bead beating are necessary	Kumar Gothwal et al. (2007), Kuske et al. (1997)
Salt lake Saline pond Marine environment (high salt and high pH) 	Halophiles Halotolerant Haloversatile CW: negative-charge amino acid in cellular proteins, ether in membrane lipid of Archaea CM: salt-stable proteins and more hydrophilic proteins S-layer and biopolymers on cell surface	High accumulation of K+ and compatible solute, the high salt stability of cells	Salt hinders the cell lysis Excesses salt and protein removal is required	Purohit and Singh (2009), Narayan et al. (2016), Horikoshi and Grant (1998)

Alkaline lake Soda lake (high pH) 	Alkaliphiles Haloalkaliphiles CW: excess of hexosamines and amino acid, high cross-linking rate in peptidoglycan, acidic polymers CM: membrane proteins are alkali stable Cells internalize high Na+	High Na+ concentration in cell, alkali, and detergent-tolerant proteins	High Mg concentration will facilitate the cell lysis and purification agents are required	Horikoshi and Grant (1998), Verma and Satyanarayana (2011)
Acid mine drainage Fumarole (low pH and high temperature) 	Acidophiles Thermoacidophiles CW: hydrophobic proteins CM: very less positively charged residues Exopolymer and rusticyanin in cell	High metal tolerance via complexation of free metals, in few cases high G + C% content Overall negative charges on protein	$CaCO_3$ pretreatment necessary	Sagova-Mareckova et al. (2008)
Submarine volcano (hydrostatic pressure, salt, and high temperature) 	Piezothermophiles Acidothermophiles Polyextremophiles CW: very thick and resistant to pressure CM: monounsaturated fatty acids Compact macromolecular structure	Helix-coil transition of DNA and histone-like Isopiestically and isothermally sample collection is required to avoid the cell lysis	Pretreatment with lytic enzymes required followed by heating with chemical lysis for high-quality DNA extraction	Michels and Clark (1997), Morono et al. (2014)

5.5 Sequencing Platforms

Metagenomics analysis using DNA sequencing technique is performed either through gene-targeted metagenomics (i.e., 16s rRNA or 18s rRNA) for taxonomic assessment or whole genome shotgun sequencing for structural and functional analysis (Ghelani et al. 2015; Dudhagara et al. 2015; Patel et al. 2015). Presently, various NGS platforms are effectively used for the metagenomics pipeline including the AB SOLiD System, 454 GS FLX, Illumina MiSeq, Roche-454, and Ion Torrent (Liu et al. 2012; Mardis 2013). All these sequencing techniques are not feasible in off-grid analysis and offer short read length, creating the difficulties in assembly process which consequently affects the downstream analysis including the taxonomic and functional profiling. Two modern sequencing platforms (1) PacBio sequencing and (2) Oxford nanopore sequencing recently emerge, which offer the advantages over the limitations of the above-discussed NGS techniques. Both provide longer reads mainly useful for the analysis of a diverse pool of microorganisms.

5.5.1 PacBio Sequencing

PacBio sequencing is a real-time sequencing developed by Pacific Biosciences, California, USA. The single-molecule real-time (SMRT) sequencing of single-stranded circular DNA is based on a template called SMRTbell, which is loaded into a chip referred as an SMRT cell (Travers et al. 2010). The key features of this sequencing method are the long read length up to 10^4 bp which makes it suitable for microbiome analysis by the full-length sequencing of target genes, i.e., 16S rRNA (~1500 bp) and 18S rRNA (~1800 bp) (Schadt et al. 2010). Longer read output is important to improve the contiguity in the assembly process. However, the higher error rate, high cost, and lower sequencing depth are major demerits of the technique (Rhoads and Au 2015). Recently, large contigs and minimizing the errors with >99% Q20 accuracy can be achieved using long read circular consensus sequencing (CCS) and place it comparatively affordable for the metagenomics analysis pipelines (Frank et al. 2016). So the aim of the metagenomics projects can be easily achieved by obtaining long contig sizes with negligible possibilities of misassemblies. PacBio sequencing is suggested for the analysis of microbial abundance and taxonomic assessment. Furthermore, the fusion assemblies using PacBio CCS and Illumina HiSeq contigs improve statistics of assembly, overall contig length and number.

5.5.2 Oxford Nanopore Sequencing

It is a very impressive fourth-generation sequencing method. It is based on the nanopore embedded in the membrane, which is kept at a certain voltage. When the ssDNA or ssRNA passes through the nanopore, the current level variation is detected resulting into decoding of nucleotide order (Ashkenasy et al. 2005). Oxford nanopore technologies have devised the portable MinION sequencer. This is very fast, is small in size, and produces the 200 kb long reads with high accuracy. Ultra-portability offers the in-field metagenomics analysis and hence overcomes the difficulties associated with the preservation and transportation of extreme environmental sample to the laboratory. Environmental samples from the glacier and hot springs are easily getting degraded in transit and biases acquired by taphonomic degradation during storage and subsequent extraction. The in situ microbial community analysis using portable nanopore sequencing methods improved agility to analyze environmental microbiomes (Edwards et al. 2016). However, the off-grid metagenomics analysis should be cross-validate before applying to the search the microbial life and extraterrestrial life in their habitation.

5.6 Future Prospects

Environmental microbiomics will bring new insights very shortly in the comprehensive determination of the microbial composition. The newest subdisciplines within the metagenomics field referred as metatranscriptomics and metaproteomics are also new hopes to offer the more resolution in structure and function of the microbial community. In the future, the single-step DNA extraction, rapid library preparation, and fast real-time in situ DNA and RNA sequencing will uplift the extreme biosphere microbiomics. The recent emergence of progressive miniaturization in the sophisticated tools and techniques will move the laboratory-dependent analysis toward in-field study to capture the more real microbial profile. Fusions of the sequencing and Raman spectroscopy, as well as mineralization of bioinformatics tools, are also the good hope in the future for search and analysis of microbial life. However, the universal standards of procedures and protocols should be established for uniformity research on environmental metagenomics, like human microbiome.

References

Ashkenasy N, Sánchez-Quesada J, Bayley H, Ghadiri MR (2005) Recognizing a single base in an individual DNA strand: a step toward DNA sequencing in nanopores. Angew Chem Int Ed 117(9):1425–1428

Bag S, Saha B, Ojasvi Mehta DA, Kumar N, Dayal M, Pant A et al (2016) An improved method for high quality metagenomics DNA extraction from human and environmental samples. Sci Rep 6:26775

Barer MR, Harwood CR (1999) Bacterial viability and culturability. Adv Microb Physiol 41:93–137

Cowan DA, Ramond JB, Makhalanyane TP, De Maayer P (2015) Metagenomics of extreme environments. Curr Opin Microbiol 25:97–102

De Maayer P, Anderson D, Cary C, Cowan DA (2014) Some like it cold: understanding the survival strategies of psychrophiles. EMBO Rep 15(3):e201338170

Delmont TO, Robe P, Clark I, Simonet P, Vogel TM (2011) Metagenomic comparison of direct and indirect soil DNA extraction approaches. J Microbiol Methods 86(3):397–400

Dudhagara P, Ghelani A, Bhavsar S, Bhatt S (2015) Metagenomic data of fungal internal transcribed Spacer and 18S rRNA gene sequences from Lonar lake sediment, India. Data Brief 4:266–268

Edwards A, Debbonaire AR, Sattler B, Mur LA, Hodson AJ (2016) Extreme metagenomics using nanopore DNA sequencing: a field report from Svalbard, 78 N. bioRxiv 2016:073965
Edwards RA, Rodriguez-Brito B, Wegley L, Haynes M, Breitbart M, Peterson DM et al (2006) Using pyrosequencing to shed light on deep mine microbial ecology. BMC Genomics 7(1):57
Eisen JA (2007) Environmental shotgun sequencing: its potential and challenges for studying the hidden world of microbes. PLoS Biol 5(3):e82
Frank JA, Pan Y, Tooming-Klunderud A, Eijsink VG, McHardy AC, Nederbragt AJ, Pope PB (2016) Improved metagenome assemblies and taxonomic binning using long-read circular consensus sequence data. Sci Rep 6:25373
Ghelani A, Patel R, Mangrola A, Dudhagara P (2015) Cultivation-independent comprehensive survey of bacterial diversity in Tulsi Shyam Hot Springs, India. Genomics Data 4:54–56
Gilbert JA, Meyer F, Antonopoulos D, Balaji P, Brown CT, Brown CT et al (2010) Meeting report: the terabase metagenomics workshop and the vision of an Earth microbiome project. Stand Genomic Sci 3(3):243
Goordial J, Davila A, Lacelle D, Pollard W, Marinova MM, Greer CW et al (2016) Nearing the cold-arid limits of microbial life in permafrost of an upper dry valley, Antarctica. ISME J 10(7):1613–1624
Handelsman J (2004) Metagenomics: application of genomics to uncultured microorganisms. Microbiol Mol Biol Rev 68(4):669–685
Horikoshi K, Grant WD (eds) (1998) Extremophiles: microbial life in extreme environments, vol 20. Wiley-Liss, Chichester
Horikoshi K, Antranikian G, Bull AT, Robb FT, Stetter KO (eds) (2010) Extremophiles handbook. Springer Science & Business Media, Berlin
Hugenholtz P, Goebel BM, Pace NR (1998) Impact of culture-independent studies on the emerging phylogenetic view of bacterial diversity. J Bacteriol 180(18):4765–4774
Klingl A (2014) S-layer and cytoplasmic membrane–exceptions from the typical archaeal cell wall with a focus on double membranes. Front Microbiol 5:624
Koga Y (2012) Thermal adaptation of the archaeal and bacterial lipid membranes. Archaea 2012:789652
Kumar Gothwal R, Kumar Nigam V, Mohan MK, Sasmal D, Ghosh P (2007) Extraction of bulk DNA from Thar Desert soils for optimization of PCR-DGGE based microbial community analysis. Electron J Biotechnol 10(3):400–408
Kuske CR, Barns SM, Busch JD (1997) Diverse uncultivated bacterial groups from soils of the arid southwestern United States that are present in many geographic regions. Appl Environ Microbiol 63(9):3614–3621
Liu L, Li Y, Li S, Hu N, He Y, Pong R et al (2012) Comparison of next-generation sequencing systems. Biomed Res Int 2012:251364
Mardis ER (2013) Next-generation sequencing platforms. Annu Rev Anal Chem 6:287–303
Mesbah NM, Wiegel J (2008) Life at extreme limits. Ann N Y Acad Sci 1125(1):44–57
Michels PC, Clark DS (1997) Pressure-enhanced activity and stability of a hyperthermophilic protease from a deep-sea methanogen. Appl Environ Microbiol 63(10):3985–3991
Morono Y, Terada T, Hoshino T, Inagaki F (2014) Hot-alkaline DNA extraction method for deep-subseafloor archaeal communities. Appl Environ Microbiol 80(6):1985–1994
Narayan A, Jain K, Shah AR, Madamwar D (2016) An efficient and cost-effective method for DNA extraction from athalassohaline soil using a newly formulated cell extraction buffer. 3 Biotech 6(1):1–7
Patel R, Mevada V, Prajapati D, Dudhagara P, Koringa P, Joshi CG (2015) Metagenomic sequence of saline desert microbiota from wild ass sanctuary, Little Rann of Kutch, Gujarat, India. Genomics Data 3:137–139
Poinar HN, Schwarz C, Qi J, Shapiro B, MacPhee RD, Buigues B et al (2006) Metagenomics to paleogenomics: large-scale sequencing of mammoth DNA. Science 311(5759):392–394

Purohit MK, Singh SP (2009) Assessment of various methods for extraction of metagenomic DNA from saline habitats of coastal Gujarat (India) to explore molecular diversity. Lett Appl Microbiol 49(3):338–344

Ramakrishnan V, Adams MWW (1995) Preparation of genomic DNA from sulfur-dependent hyperthermophilic Archaea. Archaea: a laboratory manual—thermophiles. Cold Spring Harbor Laboratory Press, Cold Spring Harbor, NY, pp 95–96

Rampelotto PH (2013) Extremophiles and extreme environments. Life (Basel) 3(3):482–485

Rhoads A, Au KF (2015) PacBio sequencing and its applications. Genomics Proteomics Bioinformatics 13(5):278–289

Rondon MR, August PR, Bettermann AD, Brady SF, Grossman TH, Liles MR et al (2000) Cloning the soil metagenome: a strategy for accessing the genetic and functional diversity of uncultured microorganisms. Appl Environ Microbiol 66(6):2541–2547

Sagova-Mareckova M, Cermak L, Novotna J, Plhackova K, Forstova J, Kopecky J (2008) Innovative methods for soil DNA purification tested in soils with widely differing characteristics. Appl Environ Microbiol 74(9):2902–2907

Schadt EE, Turner S, Kasarskis A (2010) A window into third-generation sequencing. Hum Mol Genet 19(R2):R227–R240

Stanley SM (2005) Earth system history. Macmillan, New York, NY, pp 302–323

Thomas T, Gilbert J, Meyer F (2012) Metagenomics-a guide from sampling to data analysis. Microb Inform Exp 2(1):3

Travers KJ, Chin CS, Rank DR, Eid JS, Turner SW (2010) A flexible and efficient template format for circular consensus sequencing and SNP detection. Nucleic Acids Res 38(15):e159

Venter JC, Remington K, Heidelberg JF, Halpern AL, Rusch D, Eisen JA et al (2004) Environmental genome shotgun sequencing of the Sargasso Sea. Science 304(5667):66–74

Verma D, Satyanarayana T (2011) An improved protocol for DNA extraction from alkaline soil and sediment samples for constructing metagenomic libraries. Appl Biochem Biotechnol 165(2):454–464

Part II

Human Microbiome

16S rRNA Metagenomics of Asian Gut Microbiota

6

Juma Kisuse and Jiro Nakayama

Abstract

To characterize the diversity of gut microbial community structures of Asian people, Asian Microbiome Project (AMP) has been established. AMP notably aims to understand the linkage of their gut microbiota with diets and its impact on their health. To that end, AMP began with phase I which focused on the gut microbiota of school-age children who must follow the regional dietary habit. Stool samples were collected from 303 school-age children living in urban or rural regions in five countries spanning temperate and tropical areas of Asia. Bacterial compositions of those samples were determined by using the hypervariable sequences of 16S rRNA V6–V8 region analyzed by 454 pyrosequencing platform. Their community profiles were characterized into two enterotype-like clusters, each driven by *Prevotella* (P-type) or *Bifidobacterium/Bacteroides* (BB-type), respectively. Moreover, random forest analysis marked the participant country of residence through fecal species analysis by demonstrating accumulating gut microbiota. The predicted metagenomics using PICRUSt program has suggested overrepresentation of certain enzymes which may reflect their intestinal environment, such as amylase for nondigestible starch in P-type subjects and choloylglycine hydrolase for bile acid metabolism in BB-type subjects. Following this pilot study using 454 sequencing platform, MiSeq pair-end sequencing platform has been introduced into AMP. The MiSeq platform covered more than 99% of gut microbial community profile. Enterotyping was reproduced regardless of the read regions and taxonomy levels. Further study using the MiSeq 16S rRNA metagenomics is promising to gain deep insight of gut microbial community of Asian people.

J. Kisuse • J. Nakayama (✉)
Department of Bioscience and Biotechnology, Faculty of Agriculture, Kyushu University, 6-10-1 Hakozaki, Higashi-ku, Fukuoka 812-8581, Japan
e-mail: nakayama@agr.kyushu-u.ac.jp

R.P. Singh et al. (eds.), *Understanding Host-Microbiome Interactions - An Omics Approach*, DOI 10.1007/978-981-10-5050-3_6

Keywords
Asian gut microbiome • Enterotype • 16S rRNA • Metagenome

6.1 Introduction of Asian Microbiome Project (AMP)

Human gastrointestinal tract is the home of thousands of normal flora microbial species which are influenced by several intrinsic and extrinsic factors (Simon and Gorbach 1984; Lozupone et al. 2012). Among many extrinsic factors, diet is a driving key as on top of nutrition provision while it also contribute to the diversity of microbial composition (De Filippo et al. 2010; Lee 2012; Wu et al. 2012; Zhang et al. 2014). Hence, Asian Microbiome Project (AMP) was established to characterize the diversity of gut microbial community structures of Asian people, with notable aim to understand the linkage of their gut microbiota with diets and its impact on their health and diseases. Hence of all information, it will enable to create a microbiota database of Asian society.

AMP involves different phases, whereas **Phase I** involved the profiling of gut microbiota of school-age children whose microbiota expected to follow the regional dietary habit (Nakayama et al. 2015). Stool samples were collected from 303 school-age children living in urban or rural regions in five countries spanning temperate and tropical areas of Asia. Fecal microbiomes of Asian children showed a local variation and geographical locations which reflect to their country of residence as it represents the dietary habits and lifestyle.

Phase II aims to build a comprehensive gut microbiome database covering whole age samples from newborns to elderlies. Notably, we will gain an insight in age-related change in gut microbial community structure which may associate with decay of host physiology and immunology.

Phase III studied the highlighted important differences in microbial composition from phase I and made cohort studies for deducing the cause of differences, i.e., the place of origin, dietary habit, lifestyle, and the cause of enterotype differences among the age groups.

6.2 Phase I Pilot Study (Nakayama et al. 2015)

6.2.1 Bacterial Composition and Enterotype Clustering of 303 Children Samples

The human gut microbiota varies between and within individuals, due to influence of different internal and external factors. Fecal bacterial compositions of 303 children from ten cities of five countries were profiled by using the hypervariable sequences of 16S rRNA V6–V8 region analyzed by 454 pyrosequencing platform. With the advantage of the sequence-based community data, the bacterial compositions were summarized at all taxonomic level from phylum to OTU (operational taxonomic unit) and analyzed comprehensively in depth. In particular, family-level structure clearly

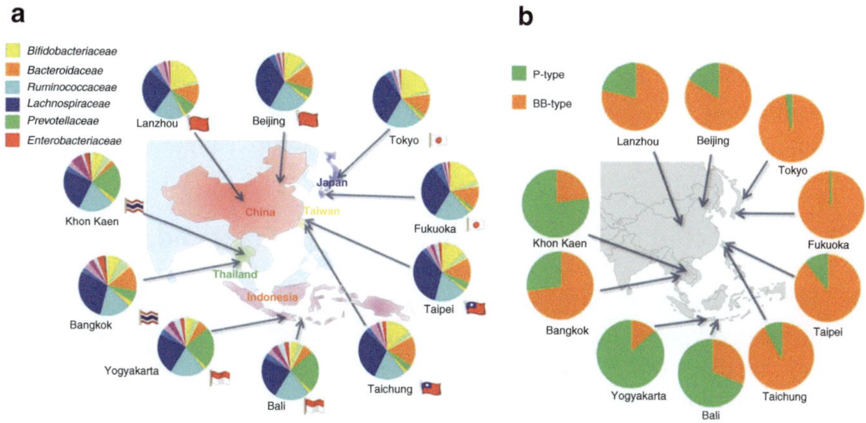

Fig. 6.1 Comparison of fecal microbiota of school-age children living in ten cities of five countries in Asia. (**a**) Fecal microbial composition by city. Pie chart represents relative abundance of dominant bacterial families. The data are averaged from 25 to 43 subjects in each city. (**b**) Distribution of enterotypes in each city

represents the characteristics of gut microbiota of each city (Fig. 6.1a). It is mainly represented by five dominant families, *Lachnospiraceae*, *Ruminococcaceae*, *Prevotellaceae*, *Bacteroidaceae*, and *Bifidobacteriaceae*. *Lachnospiraceae* and *Ruminococcaceae* are dominant members of order Clostridiales in phylum Firmicutes. But they consist of a number of species including unknown ones. In most Asian children, these two Clostridiales families account for about 50% of total community. The other three families mainly consist of one or a few genera; *Prevotellaceae* and *Bifidobacteriaceae* mostly consist of single genus, *Prevotella* or *Bifidobacterium*. Abundance of these three families is characteristic; children in East Asia are more colonized by *Bifidobacteriaceae* and *Bacteroidaceae*, while those in Southeast Asia are highly colonized by *Prevotellaceae*. These features are further highlighted by cluster analysis on a principal component analysis plot, suggesting samples in each cluster share the same type of microbial community and samples from each cluster have distinct community with each other. Eventually, these two microbiota communities were termed *Prevotella*-type (P-type) and *Bacteroides*/*Bifidobacterium*-type (BB-type).

These two types probably correspond to enterotypes observed in a European consortium study named MetaHIT, although the numbers of enterotypes were originally reported to be three (Arumugam et al. 2011). This discrepancy might be due to the differences in subject ethnicity and age. AMP phase I samples were from school-age preadolescent children known to be more colonized by *Bifidobacterium* than adult GI tract and also Western population. Indeed, *Bifidobacterium* did not appear as a main loading factor in the Western enterotyping graph. Instead of *Bifidobacterium* loading, *Ruminococcus*-driven clusters were observed in the Western enterotyping. Perhaps, due to the higher number of analyzed samples, the border between *Bacteroides*-type and *Ruminococcus*-type became unclear. This less-clear cut was observed in the study of human microbiome project in the United

States with larger number of samples (Koren et al. 2013). In AMP dataset, BB-type cluster of Asian population appears to include the two enterotype clusters of *Ruminococcus*-type and *Bacteroides*-type, while P-type cluster is not completely but significantly segregated from BB-type cluster. It suggests that these two types of microbiota consist of independent stable community, respectively.

6.2.2 Predicted Function of Each Enterotype Community

To get hints to understand the functional difference between two enterotypes, we performed the "phylogenetic investigation of communities by reconstruction of unobserved states (PICRUSt)." In this simulation, abundance of each functional gene in each sample was estimated based on phylotype composition in each sample deduced from the 16S rRNA count data and the gene count and annotation matrix for each phylotype. The PICRUSt simulation of 303 Asian children's samples showed overrepresentations of some genes in either enterotype group, which may, respectively, reflect their intestinal environment. P-type samples are enriched in genes for α-amylase and pectinase, suggesting that P-type microbial community favors plant oligosaccharides. Indeed, it is known that vegetarians commonly harbor P-enterotype (Ruengsomwong et al. 2016). On the other hand, choloylglycine hydrolase, which is a key enzyme for deconjugation of bile acid, is overrepresented in BB-type samples, suggesting that BB-type microbial community is exposed to high fat environment in the intestine. Indeed, it is known that BB-type associates with high fat and high animal protein diet (Wu et al. 2012) and is common in developed countries where people favor high fat and high animal protein diet. In the case of Asian population, distribution of two enterotypes nicely correlates with cultivar of rice daily consumed as a staple food. In Southeast Asia where P-type is dominant, Indica rice is a major cultivar, while B-type is dominant in East Asia where Japonica rice is mainly eaten. Nondigestible starch highly present in Indica rice may deliver amylose to the large intestine and promote the growth of *Prevotella* with amylase activity. In BB-type subjects, high concentration of bile acids secreted in response to high fat intake may hamper the colonization of bile-sensitive species, like *Prevotella*. Indeed, BB-type microbiota is less diverse within an individual (alpha diversity) compared with P-enterotype.

6.2.3 Local Variation Associated with Country

Following enterotype analysis, local variation of gut microbiota among Asian children was investigated. For this purpose, random forest analysis, which is a machine-learning classifier, was employed. The bacterial composition data of the 303 children was subjected to the random forest algorithm to simulate origin of samples. Identification was attempted for city or country of residence of sample providers, by using the 16S rRNA-based bacterial composition data at different taxonomic levels from phylotypes to genus. Classification was achieved at high probability according to country (74.3%) but not city (56.1%), suggesting the local variation in microbiota more associates with the country of residence. Particularly, children of Japan and Indonesia could be identified at 97% and 86% probability, respectively. It is interesting to see a multidimensional scaling (MDS) plot based on the calculated proximity, which mirrors geographical location of countries. Indeed, Chinese samples are localized in the root, Japanese and Indonesian are localized at a tip of two branches, and Taiwanese and Thai samples are localized in the middle between Chinese and Japanese and between Chinese and Indonesian, respectively. These associations between gut microbiota and geography may be linked with the migration of agricultural products which could be one of major determinants for gut microbial composition.

Furthermore, we performed hierarchical clustering using top 30 bacterial species with the highest Gini score for the identification of countries in the random forest classification analysis (Fig. 5b in Nakayama et al. 2015). As a result, the species were clustered into four groups, each associated with country of residence. Cluster I mainly comprises the BB-type bacteria such as *Bifidobacterium* and *Bacteroides* species, which are abundant in China, Japan, and Taiwan. Cluster I includes *Phascolarctobacterium faecium* which is particularly abundant in China and Taiwan. Cluster III comprises P-type species such as *Prevotella copri* and *Desulfovibrio piger*. Clusters II and IV display a unique distribution profile independent of the two enterotypes. For example, two *Dorea* species in cluster II were abundant in China in addition to the P-type countries, and *Dialister invisus* in cluster IV was detected from 67% of children in Japan but only 18% from other cities. The next question should address the factors recruiting these local variations and then the effect of these characteristic microbiota on the host health.

6.3 Comparison of 16S rRNA Metagenomic Data from MiSeq Platform to that from 454 Platform

6.3.1 Reanalysis of AMP Phase I Samples Using Illumina MiSeq Platform

With the new establishment of Illumina MiSeq pair-end sequencing platform, lower-cost 16S rRNA metagenomics has been realized. Indeed, sequencing cost per reads has been reduced to lower than one tenth. AMP has also introduced the MiSeq pair-end sequencing system for the metagenomic 16S rRNA analysis. To evaluate the data quality for the gut microbiota analysis, we herewith carefully compare the results from 454 MiSeq platforms. An output detail in each step of sequence data processing is summarized in Table 6.1.

Table 6.1 Summary of 16S rRNA sequencing for AMP phase I by MiSeq and 454 Titanium platforms (data overview of different NGS platforms)

	MiSeq		454 Titanium
	V1V2	V3V4	V6V8
# Samples	300	300	303
# Raw seqs.	7,939,563	8,119,460	2,296,414
# High quality seqs.	5,275,081	3,870,108	1,866,525
# Seqs. per sample	20,523 ± 13,564	20,846 ± 13,713	5623 ± 2038
# OUTs	1752	1032	3003
# OUTs per sample	238 ± 69	193 ± 63	418 ± 135
# P-type / BB-type	104/196	97/203	88/215
Good's coverage	0.992	0.996	0.937

From 303 samples of phase I, 300 samples were reanalyzed by the MiSeq pair-end sequencing platform. Since the available read length is somewhat shorter than that obtained by 454 Titanium platform, V1–V2 and V3–V4 region was analyzed instead of V6–V8 region analyzed by 454. Also in MiSeq, multiplex barcode sequencing system was employed to analyze up to 392 samples at a run.

The total number of raw reads obtained by the MiSeq platform was approximately four times higher than those obtained by the 454 Titanium platform. Yield of high-quality sequences from raw sequences was significantly lower in the MiSeq platform compared to the 454 platform. This should be due to the difference in the sequencing systems; pair-end sequencing was performed in the MiSeq platform, while single-pass sequencing was done in the 454.The paired reads obtained in the MiSeq system should be merged and this extra step reduced the number of sequences used for the following step. Notably in the case of V3–V4, yield of high-quality merged sequences was significantly lower. This is because the V3–V4 region was significantly longer than the V1–V2 region and the lower-quality regions were obliged to be used for merging. The merged sequences were then clustered into operational taxonomic units (OTUs) by using UPARSE program which is evaluated as one of the best algorithms for OTU clustering in terms of the number of obtained OTUs comparable with real number of species (Edgar 2013). Indeed, the number of resulted OTUs was 1752 (238 per sample) in V1–V2 and 1032 (193 per sample) in V3–V4, while that in V6–V8 by the 454 platform using UCLUST algorithm was 3003 (418 per sample). A shotgun metagenomics study, which provided much finer taxonomic information using up to full-length 16S rRNA sequences than 16S rRNA metagenomics using its partial sequences, has indicated that the entire cohort of 124 Europeans harbors between 1000 and 1150 prevalent bacterial species and each individual at least 160 such species (Qin, J. et al. 2010). Referring to this study, the data obtained by MiSeq platform is considered to be closer to real picture of human gut microbiota. Good's coverage estimated by the OTU composition of each sample, the so-called OTU table, showed more than 99% in both V1–V2 and V3–V4, suggesting adequate depth of obtained data for gut bacterial community structure.

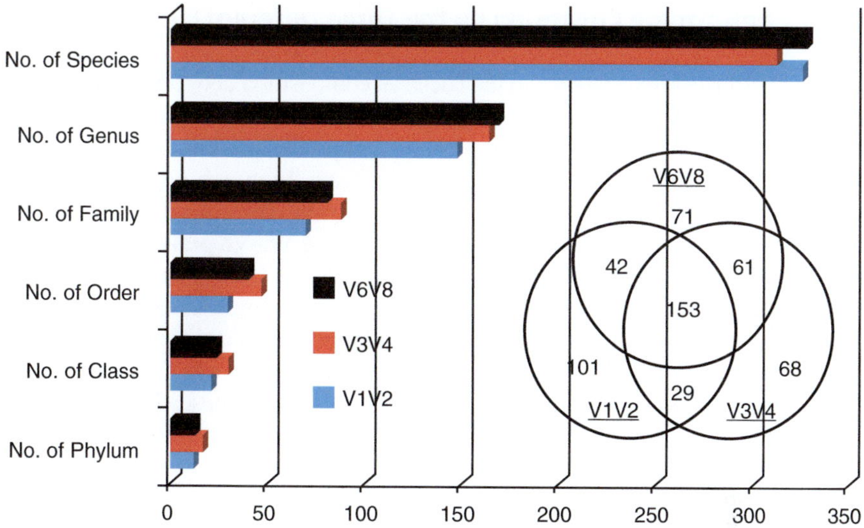

Fig. 6.2 Comparison of the number of classified taxa observed by the different NGS platforms. 300 samples of AMP phase I study were analyzed by sequencing different 16S rRNA regions in each platform

6.3.2 Comparison of Bacterial Composition Data at Each Taxonomic Level Obtained by Different NGS Platforms

Figure 6.2 shows the number of classified taxonomic groups at different taxonomy levels from phylum to species. They are totally comparable at all taxonomic level. Species identification by using short 16S rRNA amplicon sequences is generally difficult. Here we use SeqMatchQ400 algorithm (Nakayama 2010) to find the most probable species from the Ribosomal Database Project II (RDP-II) database. The number of species identified from each amplicon data (S_ab score higher than 0.84) is shown in Venn diagram in Fig. 6.2. This data indicates that approximately 50% corresponding to 153 species are commonly identified among the three different sequenced regions (Fig. 6.2).

6.3.3 Comparison of Family-Level Composition Profiles Determined by Different NGS Platforms

Figure 6.3 showed family-level gut bacterial compositions of phase I subjects, each determined by MiSeq V1–V2, MiSeq V3–V4, and 454 V6–V8. Similarly to Fig. 6.1a, five dominant families were commonly observed in all analyses. Although the relative abundance of each family differs remarkably depending on the read regions, substantial difference was observed in the total profile between three cities in right side and the rest in left side. This corresponds to enterotypes and will be addressed in more detail in the next subchapter.

6.3.4 Enterotype Clustering on the Three Different Datasets

Figure 6.3 shows big difference in the bacterial composition between left seven cities from East Asia and Bangkok and right three cities from Southeast Asia, which corresponds to enterotype variation. This distinct profile was commonly observed independently of the 16S rRNA regions sequenced. Then, we further confirmed enterotype clustering using the three different datasets. Enterotype clustering was performed according to the same methods used for the V6–V8 pilot study shown in Fig. 6.1b (Arumugam et al. 2011; Nakayama et al. 2015). Here, composition data at genus level and OTU level were applied in addition to family-level data. Jensen-Shannon distance was calculated for genus and family levels and weighted UniFrac distance was calculated for OTU level. Then, PAM clustering was performed based on the distance matrix. The optimal number of clusters was chosen by maximizing the Calinski–Harabasz index, and the resultant clusters were validated based on the prediction strength (PS) (Tibshirani and Walther 2005) and average silhouette width (SW) (Rousseeuw 1987). Independently of the read regions and taxonomy levels, two significant clusters were obtained, although validation scores differ in a certain extent (Fig. 6.4). Regarding taxonomy, family level shows stability with highest scores of PS and SI, while V6–V8 marked the highest stability among the three regions. The enterotypes identified to the 300 samples were mostly consistent; more than 91.3% samples were identified to same enterotype among the three analyses. Taken together, enterotypes significantly represent global types of gut microbiota of

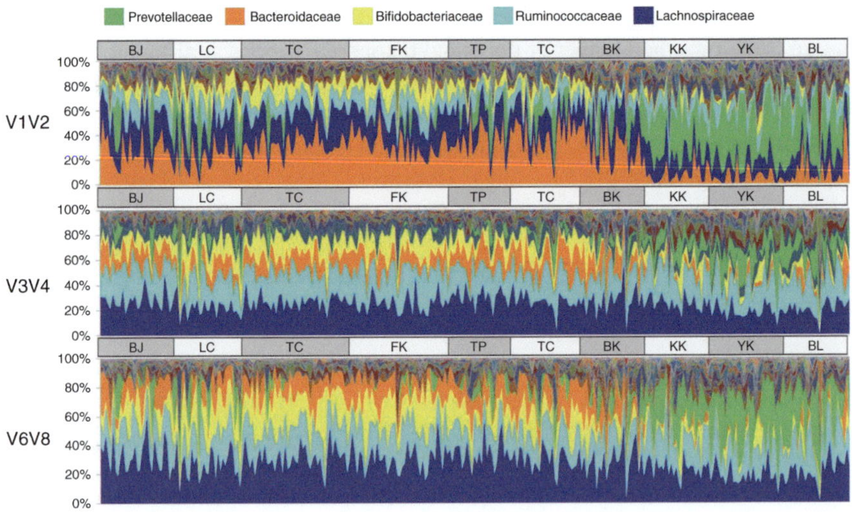

Fig. 6.3 Comparison of family-level composition profiles determined by different NGS platforms. 300 samples of AMP phase I study were analyzed by sequencing different 16S rRNA regions in each platform, and the relative abundances of five dominant families in each sample were graphed. The 300 bar graphs were arranged horizontally by city. *BJ* Beijing, *LC* Lanzhou, *TC* Tokyo, *FK* Fukuoka, *TP* Taipei, *TC* Taichung, *BK* Bangkok, *KK* Khon Kaen, *YK* Yogyakarta, *BL* Bali

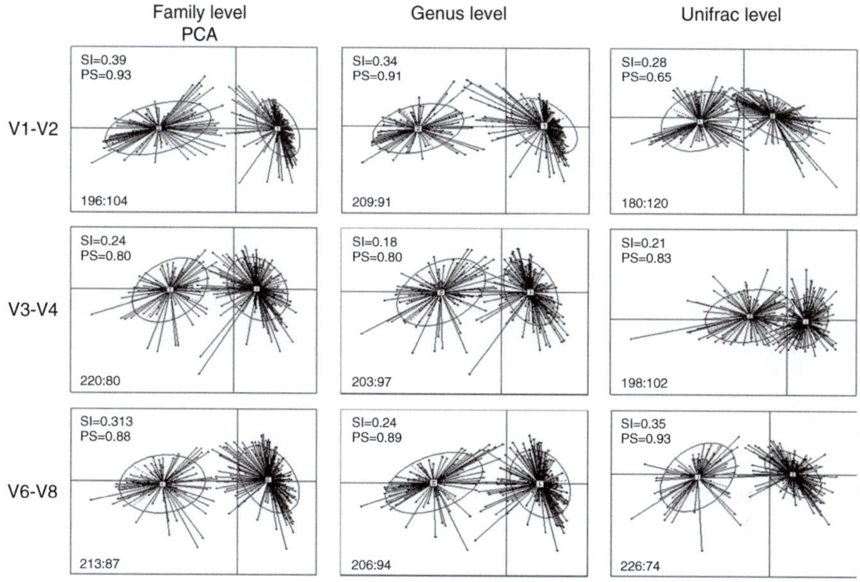

Fig. 6.4 Enterotype clustering on different NGS platforms and different taxonomic levels. SI and PS represent the silhouette width and prediction strength, respectively. Ratio values at the bottom of each analysis represent the number of samples identified BB- and P-types, respectively

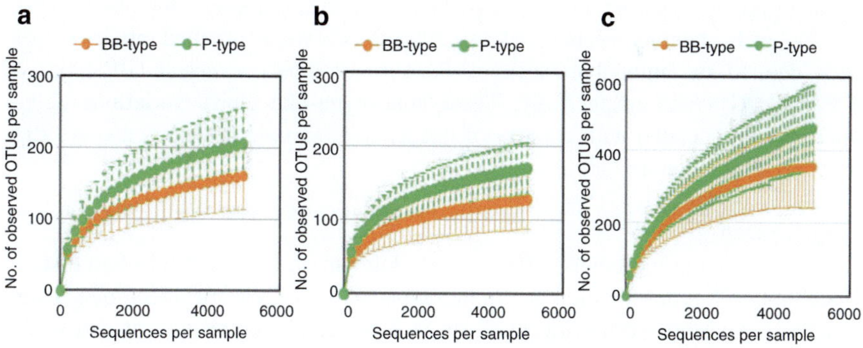

Fig. 6.5 Rarefaction curve for the number of OTUs observed in each enterotype. The values are simulated based on the sample OTU matrix obtained from V1–V2 (**a**), V3–V4 (**b**), and V6–V8 (**c**) datasets

Asian children, although enterotype concept is controversial in intestinal microbiology at present in terms of whether the number of clusters is two or three, and the difference in the community profile is distinctive or continuous (Koren et al. 2013).

As shown in Fig. 6.5, P-type samples have distinctive bacterial composition in which *Prevotellaceae* is highly abundant, while the rest of other four dominant families, notably *Bacteroidaceae*, are reduced compared to BB-type. In addition to this bacterial composition difference, alpha diversity corresponding to species richness within individuals also differs between the two enterotypes. Regardless of the regions sequenced, the number of OTUs observed in P-type samples was significantly higher than those in BB-type. As indicated by PICRUSt analysis, BB-type community may be exposed to high level bile acids with antimicrobial toxicity, which may cause the less variety of inhabiting bacteria.

According to the result of phase I study, gut microbiota of Asians differs largely but can be classified into a few community types defined as "enterotype." To gain deeper insight of gut microbiota of Asian people and understand the structure and function as an interface between foods and host health, we definitely need more and larger-scale study and AMP has started to phase II study to collect the 16S rRNA data of all age groups of Asians. The MiSeq platform will be a powerful tool to achieve our objective.

6.4 Summary

6.4.1 Summary of the Chapter

By using Illumina MiSeq and Roche 454 sequencing platforms, we analyzed 16S rRNA metagenomics for 300 stool samples of school-age children living in east and southeast Asia. The MiSeq platform has realized low-cost and high-performance 16S rRNA metagenomics to cover more than 99% portion of microbial community in each sample. Regardless of the sequencing systems, sequenced regions, and taxonomy levels, the microbiota profiles of 300 subjects were classified into two types, each defined by high abundance of *Bacteroides/Bifidobacterium* (BB-type) or *Prevotella* (P-type), respectively. These enterotype-like global variations of gut microbiota associated with country of residence of sample providers and are supposed to reflect dietary habit.

Conclusion

Gut microbiota of Asians differs largely but can be classified into two distinct community types defined as "enterotype." 16S rRNA metagenomics using MiSeq pair-end sequencing platform is a powerful tool at present and in the future to understand the structure and function of gut microbial community of Asians as an interface between varied foods and host physiology and health.

References

Arumugam M et al (2011) Enterotypes of the human gut microbiome. Nature 473:174–180. doi:10.1038/nature10187

De Filippo C et al (2010) Impact of diet in shaping gut microbiota revealed by a comparative study in children from Europe and rural Africa. Proc Natl Acad Sci 107:14691–14696. doi:10.1073/pnas.1005963107

Edgar RC (2013) UPARSE: highly accurate OTU sequences from microbial amplicon reads. Nat Methods 10:996–998. doi:10.1038/nmeth.2604

Koren O et al (2013) A guide to enterotypes across the human body: meta-analysis of microbial community structures in human microbiome datasets. PLoS Comput Biol 9:e1002863. doi:10.1371/journal.pcbi.1002863

Lee Y-K (2012) Effects of diet on gut microbiota profile and the implications for health and disease. Biosci Microbiota Food Health 32:1–12. doi:10.12938/bmfh.32.1

Lozupone CA et al (2012) Diversity, stability and resilience of the human gut microbiota. Nature 489:220–230. doi:10.1038/nature11550

Nakayama J (2010) Pyrosequence-based 16S rRNA profiling of gastro-intestinal microbiota. Biosci Microflora 29:83–96. doi:10.12938/bifidus.29.83

Nakayama J et al (2015) Diversity in gut bacterial community of school-age children in Asia. Sci Rep 5:8397. doi:10.1038/srep08397

Qin J et al (2010) A human gut microbial gene catalogue established by metagenomic sequencing. Nature 464(7285):59–65

Rousseeuw PJ (1987) Silhouettes-a graphical aid to the interpretation and validation of cluster analysis. J Comput Appl Math 20:53–65

Ruengsomwong S et al (2016) Microbial community of healthy Thai vegetarians and non-vegetarians, their core gut microbiota and pathogens risk. J Microbiol Biotechnol 26:1723–1735. doi:10.4014/jmb.1603.03057

Simon GL, Gorbach SL (1984) Intestinal flora in health and disease. Gastroenterology 86:174–193

Tibshirani R, Walther G (2005) Cluster validation by prediction strength. J Comput Graph Stat 14:511–528. doi:10.1198/106186005X59243

Wu GD et al (2012) NIH public access. Science 334:105–108. doi:10.1126/science.1208344. Linking

Zhang J et al (2014) Mongolians core gut microbiota and its correlation with seasonal dietary changes. Sci Rep 4:5001. doi:10.1038/srep05001

Human Milk Microbiome: A Perspective to Healthy and Infected Individuals

7

Chaitanya Joshi and Anju Kunjadiya

Abstract

Human milk is a vital source of nutrient as well as a continuous source of bacteria to newborn. Microbes are present in milk aid to initiation and development of infant gut microflora. These bacteria play a vital role in lessening of incidences and severity of infection to the child. Breast milk protects the newborn against infectious diseases, as it consists of different antimicrobial compounds, immunoglobulin, immune component cells, and bacteriocins secreted by probiotic bacteria, which all together provoke the growth of the helpful bacteria in neonate gut. However, breastfeeding mothers may also experience a condition called mastitis. Mastitis, one of the most common conditions experienced by breastfeeding mother, is an inflammation of connective tissue within the mammary gland. It is caused by a mixture of pathogenic bacteria and often treated with antimicrobials. The recent advances in metagenomic sequencing and amplicon sequencing technologies, which try to capture all the DNA information from the biological sample, have been widely used for the characterization of microbial community present within a sample and identification of unknown etiological agents involved in diseased condition. In the present review, effort has been made to understand the development of milk microflora and also the microbial diversity in healthy and infected breast. The present article reveals that breast milk is a source of more life than we envision.

Keywords

Human milk • Microbiota • Metagenomics • Mastitis

C. Joshi
Department of Animal Biotechnology, College of Veterinary Science and A.H., Anand Agricultural University, Anand 388 001, Gujarat, India

A. Kunjadiya (✉)
Center for Interdisciplinary Studies in Science and Technology (CISST), Sardar Patel University, V.V. Nagar, Anand, Gujarat, India
e-mail: anjukunjadia@gmail.com

R.P. Singh et al. (eds.), *Understanding Host-Microbiome Interactions - An Omics Approach*, DOI 10.1007/978-981-10-5050-3_7

7.1 Introduction

The human microbiome is defined as collection of microbial species that colonize many body sites, including human milk. The human microbiome project was undertaken by the National Institutes of Health with a goal to conduct survey of microbes present within the body and those resting on human body and the potential impact these communities have on health. However, one of the key systems was ignored, human milk. Human milk was conventionally considered as sterile; however, recent examination discovered a constant foundation of commensal, mutualistic, and probiotic bacteria in human milk.

7.2 Human Milk

Human milk is an intricate biological fluid which fulfills the nutritional supplies of newborn baby, helps in the development of infant immune system, and provides defense against pathogens (Morrow and Rangel 2004). Bioactive molecules like polyamines, oligosaccharides, fatty acids, lactoferrin, lysozyme, immunoglobulin, immune-competent cells, and antimicrobial peptides present in colostrum and milk (Newburg 2005) are the main constituents involved in providing defense. Recent studies articulated the presence of not only environmental bacteria but also the symbiotic and probiotic bacteria in the milk which are transmitted through milk to the infant and hence contribute in initial colonization of gut microflora of the infant (Martín et al. 2009). Daily consumption of human milk by an infant is 800 ml/day; this in fact contributes to transport of 1×10^5 to 1×10^7 bacteria each day leading to their colonization in gut microflora (Heikkilä and Saris 2003). Human milk protects against gastrointestinal infections (Duijts et al. 2010), respiratory infections (Nishimura et al. 2009), and allergic diseases (Greer et al. 2008; Ip et al. 2008). According to the American Academy of Pediatrics (AAP), it also trims down the possibility of diseases like inflammatory bowel disease (IBD), obesity, or diabetes

As neonates are born with immature immune system, they are more prone to get infected. At that time breastfeeding helps in building up the immune system by providing fatty acids, α-lactalbumin, sIgA, oligosaccharides, lactoferrin, lysozyme, antioxidants, and cytokine molecules bearing immune-protective role (Chirico et al. 2008; Goldman 2007). Human milk proteome consists of 976 proteins, out of which plentiful possess immunogenic property (Molinari et al. 2013; Gao et al. 2012). In addition to immune molecules, human milk also consists of blood-derived leukocytes which get transported to milk via the paracellular pathway. Bacteria present in human milk play numerous roles in the infant gut; they reduce the occurrence and severity of infections, produce antimicrobial compounds, or improve intestinal barrier function by enhancing mucin production and dropping intestinal permeability (Olivares et al. 2014). Studies have shown that accumulation of *Lactobacillus* strain, isolated from human milk, reduces the incidence of gastrointestinal infection, upper respiratory tract infections, and total number of infections to 46%, 27%, and 30% (Maldonado et al. 2012). These microorganisms also contribute in digestion by breaking down sugars and proteins and also participate in the right maturation of the infant immune system.

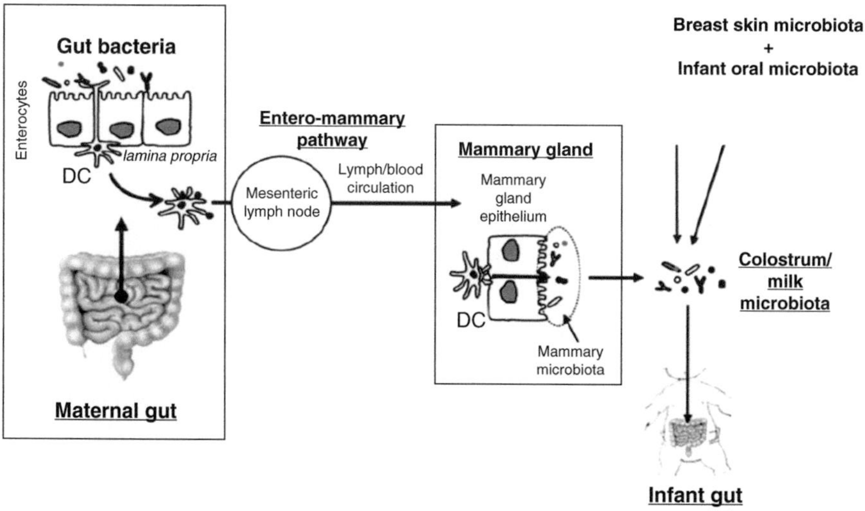

Fig. 7.1 Origin of microflora in human breast milk. Source: Fernández et al. (2013)

7.3 Origin of Microflora in the Human Milk

Physiological and hormonal alteration occurring during and after pregnancy increased gut permeability which in turn helps in the transfer of gut microflora to the mammary gland. Dendritic cells and macrophages also play an important role in the migration of microbes to the mammary gland (Fernández et al. 2013). These bacteria are transferred from maternal community to breast milk via the entero-mammary pathway (Fig. 7.1). Along with above apparent mechanisms, the retrograde flux between the mother's skin microbes and infant's oral microbes may also help in the development of the human milk microbiome (Makino et al. 2011; Albesharat et al. 2011).

7.4 Mechanism of Health-Promoting Probiotic Bacteria

The milk microbiota plays a significant role in decreasing the frequency of infection to the newborn babies due to their probiotic properties (Fig. 7.2). Probiotics have a potential to produce antimicrobial substance like bacteriocins which work as antagonists to the pathogenic bacteria and their efficient antagonistic activity is by alone or synergistically. These antimicrobial substances can be protein and bioactive peptides. Bacteriocins are important antimicrobial peptides which have therapeutic activity against intestinal pathogenic microbes (Thirabunyanon et al. 2009; Verdenelli et al. 2009; Gaudana et al. 2010). They also produce metabolites, i.e., acetic and lactic acids, which reduce the pH in the intestine and generate adverse environment for pathogen to survive (Ridwan et al. 2008). Probiotics can remove pathogens using competitive exclusion and/or blocking their attachment at the intestinal epithelium cells by competing for the glycoconjugate receptors (Vanderpool et al. 2008).

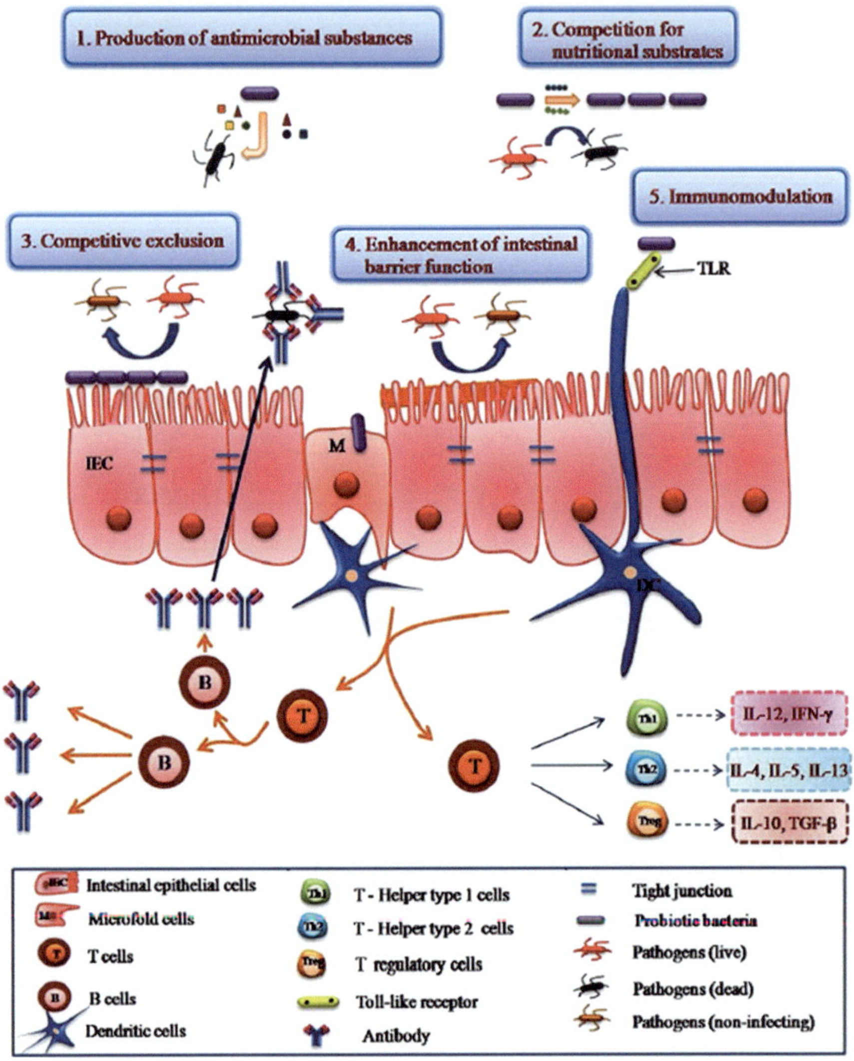

Fig. 7.2 Mechanisms of action of probiotic bacteria

7.5 Cell of Human Milk

Human milk alters in composition since colostrum to late lactation and varies within feeds and between mothers. Human milk consists of 75% leukocytes, i.e., neutrophils, erythrocytes, macrophages, and lymphocytes, and 25% epithelial cells (Paape and Weinland 1988). The epithelial cells of the glands are normally shed and get renewed, but at the time of infection, the number increases. The white blood cells work as a defense mechanism which fight against the infection and help in the repair of damaged tissue. During inflammation, it was observed that the level of neutrophils increases by 90% in human milk to fight against infection (Miller et al. 1985; Cooey and Harmon 1994). Moreover, composition of somatic cells in human milk changes with respect to lactation cycle and type of secretion (Table 7.1). Generally, the number of SCC in human milk from healthy mammary gland is approximately 1×10^5 cells/ml, while challenge with bacterial infection causes it to increase above 1×10^6 cells/ml (Bytyqi et al. 2010). Of the somatic cells, leukocyte is the most studied cell type in human milk, and depending on stage of lactation and health status of breastfeeding dyad, it may account for considerable portion of human milk (Boutinaud and Jammes 2002; Hassiotou et al. 2012; Cregan 2002; Ho et al. 1979). Many of these leukocytes are activated, motile, and interactive (Smith and Goldman 1970). This suggests that they confer active immunity to the infant (Wirt et al. 1991).

Table 7.1 Microbial diversity of human milk studied by culture-dependent and culture-independent methods

Sr. no	Author's name	Country and sample size	Experimental techniques	Identified microbial profiles
1	Martín et al. (2003)	*Spain* No. of samples 8 4 days postpartum	Culturing and identification of lactic acid bacteria using RAPD analysis	Lactic acid bacteria, specifically *Lactobacillus gasseri* and *Enterococcus faecium*, were present in all the milk samples
2	Grönlund et al. (2007)	*Finland* No. of samples 61 mothers and infant pairs	Real-time PCR	Bifidobacteria were noticed in all milk samples with the *Bifidobacterium longum* being most abundant
3	Collado et al. (2009)	*Spain* No. of samples 50	qPCR	*Lactobacillus*, *Bifidobacterium*, *Staphylococcus*, *Streptococcus*, *Enterococcus*, and *Clostridium* clusters XIVa–XIVb were the mainly abundant
4	Solís et al. (2010)	*Spain* No. of samples 20 mothers and infants. At day 1, 10 days, 1 month, and 3 months postpartum	Culturing and identification of lactic acid bacteria and bifidobacteria using 16S rRNA sequencing and RAPD	*Streptococcus*, i.e., *Streptococcus salivarius*, was predominant followed by *Lactobacillus* and *Bifidobacterium*
5	Albesharat et al. (2011)	*Syria* No. of samples 30 mothers and infant pain (1 month to 2 years postpartum). Human milk, maternal/infant feces, and fermented foods were collected	Culturing and identification of lactic acid bacteria using RAPD, 16S rRNA sequencing, and matrix-assisted laser desorption/ionization (MALDI)	Lactic acid bacteria like *Lactobacillus*, *Enterococcus*, *Pediococcus*, *Streptococcus*, *Staphylococcus*, and *Weissella* were isolated

(continued)

Table 7.1 (continued)

Sr. no	Author's name	Country and sample size	Experimental techniques	Identified microbial profiles
6	Hunt et al. (2011)	*United States* No. of samples 16 22–26 weeks postpartum three samples collected from each subjects	Pyrosequencing approach	Most abundant genera were *Streptococcus*, *Staphylococcus*, *Serratia*, and *Corynebacterium*. "Core" microbiome includes *Staphylococcus*, *Streptococcus*, *Serratia*, *Pseudomonas*, *Corynebacterium*, *Ralstonia*, *Propionibacterium*, *Sphingomonas*, *Bradyrhizobium*
7	Collado et al. (2012)	*Finland* No. of samples 56 mothers (22 overweight and 34 normal weight) and their infants. 1–2 days (colostrum), 1 month, and 6 months postpartum	qPCR	Most abundant genera were *Lactobacillus*, *Bifidobacterium*, and *Staphylococcus* *Staphylococcus* occurred in higher abundance, and *Bifidobacterium* and *Lactobacillus* were observed in lower abundance in overweight mother
8	Cabrera-Rubio et al. (2012)	*Finland* No. of samples 18 0–2 days, 1 month, and 6 months postpartum	Pyrosequencing, qPCR	*Weissella*, *Leuconostoc*, *Staphylococcus*, *Streptococcus*, and *Lactococcus* dominant in colostrum, whereas *Leuconostoc*, *Weissella*, *Lactococcus*, and *Staphylococcus* in mature milk
9	Bhatt et al. (2012)	*India* No. of samples 7 Randomly milk samples collected	Cultured probiotic bacteria	*Lactobacillus fermentum*, *Enterococcus mundtii*, *Enterococcus faecium*, *Lactobacillus reuteri, and* *Bacillus subtilis* were identified by 16S approach
10	Gonzalez et al. (2013)	*Mozambique* No. of samples 55 (29 of whom tested positive for HIV) 14 days, 15–90 days, 91–180 days, and 181–360 days postpartum	Culturing of nonfastidious bacteria, yeasts, molds, qPCR	44 genera and 124 species were identified; commonly cultured isolates belonged to *Staphylococci*, *Streptococci*, and *Lactobacilli*

Table 7.1 (continued)

Sr. no	Author's name	Country and sample size	Experimental techniques	Identified microbial profiles
11	Jost et al. (2013)	*Switzerland* No. of samples 7 3–6 days, 9–14 days, and 25–30 days postpartum	Pyrosequencing, RAPD, Sanger sequencing	Firmicutes and Proteobacteria dominated. *Staphylococcus*, *Streptococcus*, *Pseudomonas*, and *Ralstonia* were the most abundant genera
12	Ward et al. (2013) and Khodayar-Pardo et al. (2014)	*Canada* No. of samples 1 (ten milk samples pooled) 9–30 days postpartum	Metagenomic sequencing on Illumina	360 genera were identified. Proteobacteria (65%) and Firmicutes (34%) dominated; *Pseudomonas* and *Staphylococcus* were the most abundant genera
13	Khodayar-Pardo et al. (2014)	*Spain* No. of samples 32 1–5 days, 6–15 days, and 17–18 days postpartum	qPCR	*Lactobacillus*, *Streptococcus*, and *Enterococcus* spp. were most prevalent
14	Olivares et al. (2014)	*Spain* No. of samples 24 (half with celiac disease) 1 month postpartum	qPCR	*Bifidobacterium* spp. were observed in all milk samples. *Bifidobacterium bifidum* and *Bifidobacterium breve* were the most abundant
15	Urbaniak et al. (2014)	*Canada* No. of samples 9 (one undergoing chemotherapy related to Hodgkin's lymphoma)	Ion Torrent sequencing	Chemotherapy was associated with lower microbial diversity and altered bacterial profiles: decreased percentage abundances of *Acinetobacter* and Xanthomonadaceae with chemotherapy
16	Soto et al. (2014)	*Germany and Austria* No. of samples 160 Mainly 1–4 weeks postpartum	Culturing of *Lactobacilli* and bifidobacteria Its identification by 16S sequencing	*Lactobacilli* and bifidobacteria were isolated and identified

(continued)

Table 7.1 (continued)

Sr. no	Author's name	Country and sample size	Experimental techniques	Identified microbial profiles
17	Vaidya et al. (2015)	*India* No. of samples 32	Culture-dependent method Sanger sequencing	At species level, *Enterococcus faecalis*, *Lactococcus lactis*, *Bacillus litoralis*, *Bacillus licheniformis*, *Bacillus safensis*, *Lactobacillus oris*, *Pseudomonas aeruginosa*, *Staphylococcus aureus*, *Staphylococcus epidermis*, *Lysinibacillus* spp. were identified
18	Cabrera-Rubio et al. (2015)	*Spain* No. of samples 10 (six vaginally and four cesarean delivered mothers)	Pyrosequencing qPCR	Alteration in microbiome of human milk based on mode of delivery. *Streptococcus*, *Staphylococcus*, *Enterobacter*, and *Pseudomonas* were the most abundant genera
19	Urbaniak et al. (2016)	*Canada* No. of samples 39	Illumina sequencing	No statistical difference was observed in human milk microbiome based on birthing method, gestation time, and infant gender. *Staphylococcus*, *Pseudomonas*, *Streptococcus*, and *Lactobacillus* were the most abundant genera
20	Sakwinska et al. (2016)	*China* No. of samples 90 (30 samples without aseptic cleansing and 60 samples collected aseptically)	Illumina 16S sequencing qPCR	*Streptococci* and *Staphylococci* dominated in both collection procedures. *Acinetobacter* was predominant in milk collected without aseptic cleansing

This was further supported by in vivo studies in animal models showing active transfer of milk leukocyte through the intestinal epithelium into the blood circulation, and movement to and engraftment in different organs, including the mesenteric nodes, liver, and spleen (Weiler et al. 1983; Zhou et al. 2000; Michie et al. 1998; Schnorr and Pearson 1984).

7.6 Microbial Profiling of Human Milk

During the last decades, microbiological studies that focused on human milk were restricted to the identification of potential pathogenic bacteria in stored milk or milk retrieved from maternal infected human milk, but microbes present in healthy human milk are unexplored (El-Mohandes et al. 1993; Wright et al. 1998). Standard microbiological based culturing methods can only detect small proportion of bacteria because the great majority of bacteria on earth are not culturable in laboratory condition. To identify these unculturables and estimate real bacterial diversity, culture-independent method is required. Sequence-based identification of microbial species through sequencing has overcome the limitation. The nine hypervariable regions of 16S rRNA can be used for identification of bacterial species. Amplification of 16S rRNA region using universal primer is useful for estimation of bacterial diversity.

7.7 Culture-Dependent Assessment of Human Milk Microbial Diversity

Initial report of culture-dependent methods for studying human milk microbial diversity came in 2003 by Dr. Juan Rodriguez with his associate researcher R. Martin. They isolated a total of 178 isolates from each mother and infant pair (human milk, nipple areola, infant's mouth and feces) and subjected it to randomly amplified polymorphic DNA (RAPD) analysis and identified by 16S rDNA sequencing. Bacteria having identical profiles in mother and child pair were identified as *Lactobacillus gasseri* and *Enterococcus faecium.* Surprisingly, none of the lactic acid bacteria isolated from breast skin shared RAPD profiles into other sources (Martín et al. 2003).

After that Grönlund et al. (2007) studied the association of maternal fecal and breast milk bifidobacteria and infant fecal bifidobacteria using real-time PCR from 61 mother-infant pairs. They found that *Bifidobacterium longum* was the most abundant species isolated from breast milk. Moreover, they concluded that *Bifidobacterium adolescentis* and *Bifidobacterium bifidum* colonization frequency and count correlated significantly among mother and infant pairs (Grönlund et al. 2007).

Collado et al. (2009) in their study examined 50 breast milk samples for the presence of differential bacterial genera by using qPCR technique. They found that *Staphylococcus*, *Streptococcus*, *Bifidobacterium*, and *Lactobacillus* were the most abundant genera in all the samples. In addition, Collado et al. (2012) studied the effect of maternal weight and weight gain during pregnancy on milk microbiota (56 mothers, 22 overweight and 34 normal weight) using qPCR. *Staphylococcus* group bacteria were observed in higher number, whereas *Bifidobacterium* group was in lower level, in overweight mother compared to normal-weight mother. Moreover, they found higher prevalence of *Akkermansia muciniphila* in higher number in breast milk of overweight mothers (Collado et al. 2009).

Solís et al. (2010) studied the development of lactic acid bacteria and bifidobacteria during the first 3 months of life in 20 vaginally delivered breastfed infants and mothers. *Streptococcus*, *Lactobacillus*, and *Bifidobacterium* were the most dominant genera in breast milk contributing to the initial establishment of microbiota in newborn (Solís et al. 2010).

Albesharat et al. (2011) isolated a total 700 isolates of LAB from fecal sample of breastfeeding mother, feces of their infant, from breast milk, and from fermented food that is normally consumed in Syria, and characterized it by RAPD and matrix-assisted laser desorption/ionization time-of-flight mass spectrometry (MALDI-TOF MS). Their study demonstrates occurrence of 36 different species of *Lactobacillus*, *Enterococcus*, *Streptococcus*, *Weissella*, and *Pediococcus*. Interestingly, they found identical RAPD genotype of *L. plantarum*, *L. fermentum*, *L. brevis*, *Enterococcus faecium*, *Enterococcus faecalis*, and *P. pentosaceus* in feces of mother, in breast milk of mother, and in feces of her babies (Albesharat et al. 2011).

In 2014, Khodayar-Pardo et al. studied the bacterial population present in 32 Spanish breastfeeding women using quantitative PCR and determine the influence of lactational stage, gestational age, and delivery mode on milk microbiota. They identified *Enterococcus*, *Lactobacillus*, and *Streptococcus* spp. as the dominant bacterial group. They also concluded that *Bifidobacterium* is found more commonly in vaginal than cesarean deliveries (Khodayar-Pardo et al. 2014).

Afterward, Soto et al. (2014) isolated *Bifidobacterium*, *Lactobacillus*, *Enterococcus*, and *Staphylococcus* species from breast milk of 47 Slovenian lactating mother. Moreover, Gonzalez et al. (2013) also found *Staphylococcus*, *Streptococcus*, and *Lactobacillus* genera in breast milk collected from 121 Mozambique women (Albesharat et al. 2011).

7.8 Culture-Independent Assessment of Human Milk Microbial Diversity

In 2011, Hunt et al. used a new approach (454 pyrosequencing), which utilized specific primer targeting the V1–V2 hypervariable region of 16S rRNA gene of bacteria. They characterized microbial diversity and temporal stability of bacterial profiles in healthy human milk collected from 16 US women over a 4-week period (Hunt et al. 2011). Half of the bacterial sequences were contributed by nine "core" OTUs which include *Pseudomonas*, *Staphylococcus*, *Serratia*, *Corynebacterium*, *Ralstonia*, *Streptococcus*, *Sphingomonas*, *Bradyrhizobium*, and *Propionibacterium*. Moreover, the proportion of these core OTUs varied greatly between subjects.

Similarly, Cabrera-Rubio et al. (2012) studied bacterial diversity in human milk over three different time points (colostrum 1 and 6 months postpartum) in 18 Finnish women (Cabrera-Rubio et al. 2012). They found that human milk microbiome changes over lactation period. Bacteria belonging to *Weissella*, *Leuconostoc*, *Staphylococcus*, *Streptococcus*, and *Lactococcus* were more abundant in colostrum. While in 1- and 6-month milk samples, *Veillonella*, *Leptotrichia*, and *Prevotella*, typical inhabitants of oral cavity, increased significantly. Moreover, they concluded that milk from obese mother tends to be altered and less diverse than normal-weight mothers.

Jost et al. (2013) examined bacterial diversity in breast milk of seven mothers at three different sampling points (days 3–6, 9–14, and 25–30 postpartum) using culture-dependent and culture-independent techniques. They found that Firmicutes, Proteobacteria, Bacteroidetes, and Actinobacteria were the most abundant phyla, including representatives from the genera *Pseudomonas*, *Staphylococcus*, *Ralstonia*, *Streptococcus*, *Bacteroides*, *Blautia*, and *Bifidobacterium*. Moreover, they also found, for the first time, bacteria belonging to *Faecalibacterium* and *Roseburia*, which are butyrate producers and important for colonic health (Jost et al. 2013).

After that, Ward et al. (2013) performed metagenomic functional analysis of a pooled milk samples from ten donor mothers using Illumina sequencing. Over 360 bacterial genera were identified with predominance of sequences belonging to Proteobacteria and Firmicutes. In addition, they also concluded that human milk is less diverse than the feces of infant and mother at the phylum level. Human milk contained prominent amounts of genetic component which link with nitrogen membrane transport, stress response, metabolism, and immunomodulatory functions (Ward et al. 2013).

In addition, Urbaniak et al. (2014) studied bacterial diversity in human milk collected every 2 weeks over a 4-month period from lactating mothers undergoing chemotherapy of Hodgkin's lymphoma. They found that chemotherapy causes substantial alteration in microbiome from healthy controls, with reduction in genera such as *Bifidobacterium*, *Eubacterium*, *Staphylococcus*, and *Cloacibacterium* (Urbaniak et al. 2014).

Two recent independent studies by Cabrera-Rubio et al. (2015) and Urbaniak et al. (2016) studied the milk microbiota composition of healthy women and correlated it to birthing method. In addition to birthing method, Urbaniak et al. (2016) also studied alteration of milk microbiota with gestation time and infant gender. Urbaniak et al. (2016) in their study collected human milk from 39 Canadian mothers and analyzed microbial profiles by 16S rRNA sequencing using Illumina platform. They found Proteobacteria and Firmicutes as most dominant phyla and *Staphylococcus*, *Pseudomonas*, *Streptococcus*, and *Lactobacillus* as most abundant genera. However, comparison of bacterial profile between term and preterm infants, vaginal and C-section deliveries, and male and female showed no statistical significant difference (Urbaniak et al. 2016). In contrast, Cabrera-Rubio et al. (2015), in their study, compared milk microbiome of six vaginally delivered mothers and four cesarean delivered mothers and found significant separation of milk microbiome based on mode of delivery (Cabrera-Rubio et al. 2015).

The microbiota of breast milk from 90 Chinese lactating women was analyzed with two different collection procedures (without aseptic cleansing and after aseptic cleaning) by Olga Sakwinska et al. (2016). They found that *Streptococci* and *Staphylococci* were the most abundant in both the group and results were consistent with that of previous study. However, they revealed that breast milk collected without aseptic cleansing and rejection of foremilk had higher abundance of *Acinetobacter* sp. Moreover, bifidobacteria and *Lactobacilli* were present in few samples but with low abundance (Sakwinska et al. 2016).

7.9 Overview of Mastitis

Mastitis is an inflammation of connective tissue within the mammary gland (Gianneechini et al. 2002; Zhao and Lacasse 2008). The term comes from the Greek word masto-referring to the mammary gland and its meaning inflammation. It is characterized by physical, chemical, and bacteriological changes in the breast milk.

It is the most common condition experienced by lactating mothers. Incidence of occurrence of mastitis varies extensively because of difference in breastfeeding method. As per the data of WHO (World Health Organization), overall 2–33% of breastfeeding mothers are thought to be infected with mastitis (WHO 2000). Studies conducted in the USA, New Zealand, Finland, and Australia suggest that 20–25% of breastfeeding women have chances of developing mastitis (Kinlay et al. 1998; Fetherston 1997; Foxman et al. 2002). Although mastitis is a very common condition, very few studies are conducted on it till date (Foxman et al. 2002). Mastitis usually affects lactating women; hence, it is known as lactational mastitis.

Mastitis is a deliberately painful condition experienced by breastfeeding mothers. It is mainly found to be prevalent during second and third week postpartum. Mastitis can be caused by an infection or excess of milk remaining in the milk tissue (milk stasis). Mastitis is usually the result of a blocked milk duct that hasn't cleared. Milk banked up behind the blocked duct can be forced into nearby breast tissue, causing the tissue to become inflamed. Sometimes it may occur due to sudden stop of breastfeeding. Infectious mastitis develops when bacteria commonly found on skin enter the mammary gland through cracked nipples and multiply in the fatty tissue of mammary gland resulting in infection.

7.9.1 Mastitis: A Dysbiosis of Breast Milk Bacteria

Breast milk has got vibrant bacterial diversity mainly that related with skin and non-skin. These bacteria are transported from maternal community to breast milk via the entero-mammary pathway. Pathogenesis of mastitis could have resulted from enrichment of pathogenic bacteria in milk and mutual healthy milk microflora killed due to toxins released by pathogenic bacteria.

7.9.1.1 Types of Mastitis

Scandinavian researchers suggested in the 1980s that mastitis should be classified into two classes: clinical mastitis and subclinical mastitis.

Clinical mastitis: It is characterized by the presence of gross inflammatory signs and symptoms.

Clinical mastitis can be divided into three types:

1. *Preacute mastitis*: Inflammation and changes in milk composition. Systemic signs like fever, depression, shivering, loss of appetite, and loss of weight.
2. *Acute mastitis*: Similar to preacute mastitis, but with mild signs like fever and mild depression.
3. *Subacute mastitis*: In this type of mastitis, signs of inflammation are minimal and no visible systemic signs.

Subclinical mastitis: This form of mastitis is characterized by change in milk composition with no signs of gross inflammation or milk abnormalities.

Mastitis is associated with increased somatic cells, free fatty acids, and interleukin-8 concentrations (Hunt et al. 2013). However, fresh milk produced by a mastitis gland has free fatty acids (FFAs) and when stored at 4 °C exhibits greater rates if lipolysis occurs (Randolph and Erwin 1974; Murphy et al. 1989).

7.9.2 Mastitis and Somatic Cell Count

Somatic cells are white blood cells; their number increases during bacterial infection in order to fight against pathogenic bacteria (Sharma et al. 2011). Thus somatic cells can be a better indicator of infectious condition in mammary gland. Somatic cell count in women with mastitis usually has an elevated count compared to healthy women (Hunt et al. 2013; Hassiotou et al. 2013). Intramammary infection results in a significant increase in the somatic cell count level in the breast milk. In response to invasion of mammary gland by bacteria, leukocytes are released into the milk to kill the bacteria, which results in increases in somatic cell numbers and ultimately leads to inflammation and blocked milk ducts. Moreover somatic cells contain lipolytic and proteolytic enzymes, which degrade fats and proteins, respectively. Upon challenge by bacterial infection, the amount of destructive enzymes carried out by increased somatic cells results into deterioration of milk fat and protein. Somatic cell count is often used for diagnosis of mastitis in case of bovine animals.

7.9.3 Etiology of Mastitis

Etiological agents of mastitis can be infectious or noninfectious. Organisms which may cause mastitis are bacteria, viruses, mycoplasma, yeasts, and algae. Gram-positive, catalase-positive bacteria are mostly isolated from mastitis-infected milk. It can be caused by microbes, such as *Staphylococcus aureus*, *Streptococcus dysgalactiae*, *Streptococcus uberis*, *Streptococcus agalactiae*, *Staphylococcus epidermidis*, *Corynebacterium bovis*, *Corynebacterium pyogenes*, *Klebsiella* sp., and *Candida albicans*. Among all of these microorganisms, the most important are *Staphylococcus aureus* and *Staphylococcus epidermidis*, which is a common cause of mastitis, and it is commonly isolated from mastitis-infected milk. In Brazil, studies reported the predominance of *Staphylococcus aureus* over other disease-causing agents in all regions of the country (Rodrigues et al. 2015). Other than this, coagulase-negative *Staphylococci* are considered as minor mastitis-causing pathogens. Nineteen distinct species of coagulase-negative *Staphylococci* have been revealed to date. Members of the *Staphylococcus epidermidis* subgroup include *S. hominis*, *S. warneri*, *S. capitis*, and *S. haemolyticus*. Variety of pathogenic organism causing mastitis can be divided into two groups: contagious mastitis pathogens and environmental mastitis pathogens.

Contagious mastitis pathogens: These are commonly found on the skin and enter into the mammary gland through cracked or sore nipples. The major contagious pathogens are *Staphylococcus aureus* and *Streptococcus agalactiae*.

Environmental mastitis pathogens: Environmental mastitis pathogens are also known as opportunistic mastitis pathogens because they will take the opportunity to cause mastitis and cause intramammary infections sporadically. The most common environmental mastitis pathogens are *Staphylococcus chromogenes*, *Staphylococcus simulans*, *Staphylococcus xylosus*, *Staphylococcus epidermidis*, *Staphylococcus hyicus*, and *Staphylococcus haemolyticus*.

7.9.3.1 *Staphylococcus aureus* and Its Virulence Gene

Staphylococcus aureus is a gram-positive bacteria associated with many serious diseases in humans as well as animals, and it is found to be the most predominant bacteria causing human mastitis with relevant losses in the dairy industry (Bjork et al. 2014; Li et al. 2009). *S. aureus* is the most common species of *Staphylococci* to cause *Staphylococcus* infections. It is frequently found in the human respiratory tract and on the skin. Although *S. aureus* is not always pathogenic, it is a common cause of skin infections (e.g., boils), respiratory disease (e.g., sinusitis), and food poisoning. Other than this *S. aureus* can cause a range of illnesses, from minor skin infections, such as pimples, impetigo, boils (furuncles), cellulitis folliculitis, carbuncles, scalded skin syndrome, and abscesses, to life-threatening disease such as pneumonia, meningitis, osteomyelitis, endocarditis, toxic shock syndrome, bacteremia, and sepsis. Virulence factors, such as surface proteins that promote colonization of host tissues, surface factors that inhibit phagocytic engulfment (protein A), biochemical properties that enhance their survival in phagocytes (catalase production), immunological disguises (protein A, coagulase clotting factor), and acquired resistance to microbial agents, are often expressed by *S. aureus*. Clumping factor is the surface agent that acts as adhesions. Coagulase is tightly bound to the surface of *S. aureus* and coats its surface with fibrin upon contact with blood. This fibrin-coated *S. aureus* resists phagocytosis. Protein A binds to IgG in wrong orientation in serum, thus preventing opsonization and phagocytosis.

The role of bacteria in lactational mastitis is still questionable. Although it is most common among lactating women, there is lack of scientific analysis on it. Culture-dependent and culture-independent assessment of mastitis-infected breast milk can provide in-depth analysis of microflora involved in diseased condition. Culture-dependent studies involve classical microbiological techniques and it has several drawbacks, while culture-independent studies involve assessment of microflora by metagenomic approach with the recent next-generation sequencing technology.

7.9.4 Culture-Dependent Assessment of Mastitis

There are many conventional techniques used for isolation and identification of pathogenic bacteria. Isolation of pathogenic bacteria on the sheep blood agar is widely used in many laboratories because pathogenic bacteria grow via engulfing the red blood cell and appear as greenish colony. Otherwise if pathogens are not present in breast milk, they cannot grow on blood agar. Molecular typing (molecular markers) techniques such as polymerase chain reaction (PCR) technology provided additional approaches that have been reported and is considered as the most powerful technique for the control and investigation of pathogens. But culture-based approach to isolate microorganism from any environment does not provide comprehensive information on composition of bacterial communities. This technique also failed to determine microorganism which cannot grow in laboratory condition. Most of the studies performed till date on mastitis involve classical microbiological techniques to identify etiology of mastitis.

Kvist et al. (2008) compare bacterial composition in milk samples collected from 192 women with a clinical mastitis and 466 healthy donors. They found that *S. aureus* was present in 45% of women with mastitis and 31% of healthy donors. In both the group, mean colony counts were identical and no correlation was observed between colony counts and symptom severity. Finding hints that the presence of *S. aureus* in breast milk does not always result in clinical mastitis and it is always present in healthy human milk.

Delgado et al. (2008) recognized the role of coagulase-negative *Staphylococcus* spp. in human mastitis. Employing pulsed-field gel electrophoresis, they found that *S. epidermidis* was present in 85% (17/20) of samples collected, while *S. aureus* in 40% (8/20) of samples. After that they compared strains of *S. epidermidis* present in women with mastitis and women with healthy human milk. They found that women with clinical signs of infection were more likely to harbor strains of *S. epidermidis* with the *icaD* (33 vs. 11%, *p* ¼ 0.03), which was correlated with biofilm production. Thus, virulence factors of *S. epidermidis* strains found in breast milk may play a vital role in pathogenesis.

Using 16S-specific PCR primers, Shriram et al. (2015) identified bacteria belonging to *Staphylococcus* and *Pseudomonas* genera from human milk of 32 mastitis women. Moreover, the authors found 17 genera and 30 different species from mastitis milk suggesting diverse community in diseased condition (Patel et al. 2016).

7.9.5 Culture-Independent Assessment of Mastitis

Traditionally microbial genome sequencing has been restricted to only a small number of organisms which can be grown in pure culture in laboratory. Progressive development of culture-independent methods has allowed researchers to sequence microbial communities directly from environmental samples. Culture-independent techniques deal with the isolation of total DNA from the environmental sample. Culture-independent approach is commonly referred to as "metagenomic" or "community genomics." Metagenomics is applied literally to describe any culture-independent analysis of microbial communities. With the recent development in more advances sequencing techniques, which try to capture all the DNA information from the biological sample have been widely used for the characterization of microbial community present within a sample and identification of unknown etiological agents involved in diseased condition. Moreover, this type of technology provides identification of thousands of sequences per sample, which increases the possibility to observe less frequent phylotypes that may have significant importance in disease condition. Metagenomics can also be applied to solve practical challenges in the field of medicine, agriculture, sustainability, and ecology. Numerous microbiome studies have been carried out to assess the composition of the bacterial communities inhabiting a variety of human body locations, including the gut (Zhao and Lacasse 2008), oral cavity (Nasidze et al. 2009; Belda-Ferre et al. 2012), vagina (Ravel et al. 2011), skin (Costello et al. 2009), and human milk (Jost et al. 2013; Belda-Ferre et al. 2012; Ward et al. 2013). All of these studies were focused on the bacterial component of the microbiome.

So far only one study has been reported discussing metagenome of breast milk from mastitis-infected women. Jimenez et al. (2015) performed shotgun sequencing of ten healthy and ten mastitis-infected breast milk samples. They found that *Staphylococcus aureus* clearly dominated the microbiome in the samples from the women with acute mastitis, whereas high abundance of *Staphylococcus epidermidis*-related reads was observed in the milk of those suffering from subacute mastitis (Jimenez et al. 2015).

7.9.6 Prevention and Control

Antibiotics are regularly used to treat mastitis. But nowadays development of multiple resistances by different bacteria has led to failure of treatment. It is due to indiscriminate use of antimicrobials without checking its in vitro sensitivity to the causing bacteria (Oliver and Murinda 2012). In addition to antibiotic resistance, formation of biofilm is also an important virulence factor implicated by mastitis-causing pathogens, which allow survival of bacteria at high antimicrobial concentration (Hoiby et al. 2010). Alternative treatment for the antibiotics can be probiotic therapy and herbal therapy.

7.9.7 Probiotic Therapy

Development of new strategies based on probiotics is an alternative or complement to antibiotic therapy for the management of mastitis and is particularly appealing. Use of lactic acid bacteria as oral administration of lactobacilli isolated from breast milk for the treatment of mastitis has been used by researchers (Jimenez et al. 2008). Human milk consists of bacterial species like *Lactobacillus gasseri*, *Lactobacillus reuteri*, *Lactobacillus salivarius*, *Lactobacillus fermentum*, or *Bifidobacterium breve* with probiotic properties. These bacteria have shown promising results as probiotic agents that might be useful in treating mastitis.

7.9.8 Herbal Therapy

There has been not a single study published till date indicating use of herbal therapy on human mastitis. But in veterinary field, there are some studies focused on the use of natural herbal plant as a remedy for mastitis. It has been reported that garlic tincture or aloe gel can be used as a fast remedy from mastitis (Pol and Ruegg 2007). In literature antimicrobial properties of garlic extracts and *Aloe vera* gels have already been reported (Ross et al. 2001; Agarry et al. 2005). But the use of these compounds to successfully treat mastitis has not been described. In one clinical trial, they have specifically evaluated the clinical efficacy of a botanical treatment to treat subclinical mastitis (Abaineh and Sintayehu 2001). Two different doses of a dried leaf powder of an African perennial herb (*Persicaria senegalense*) were fed for 3–5 days to cows infected with subclinical mastitis. Results of this trial indicated positive effect of herbal medicines in eradication of mastitis but conceded that more research is necessary.

7.10 Conclusion and Future Aspects

In conclusion, there are now convincing proofs that human milk consists of diverse and feasible microbial population, which initially colonize the infant gut. Somehow, variations in microbial profiling in different studies were due to behavioral, environmental, and genetic differences or a consequence of methodological variation. As such, the era has moved away from the past belief that breast milk is sterile and acknowledged the rich microbial community present in human milk.

However, dysbiosis of breast milk microbial community results in a development of mastitis. Monitoring changes in mastitis-causing microflora with metagenomic platforms might be helpful in building a strategy to overcome this problem.

References

Abaineh D, Sintayehu A (2001) Treatment trial of subclinical mastitis with the herb Persicaria senegalense (Polygonaceae). Trop Anim Health Prod 33(6):511–519

Agarry OO, Olaleye MT, Bello-Michael CO (2005) Comparative antimicrobial activities of aloe vera gel and leaf. Afr J Biotechnol 4(12):1413–1414

Albesharat R, Ehrmann MA, Korakli M, Yazaji S, Vogel RF (2011) Phenotypic and genotypic analyses of lactic acid bacteria in local fermented food, breast milk and faeces of mothers and their babies. Syst Appl Microbiol 34(2):148–155

Belda-Ferre P, Alcaraz LD, Cabrera-Rubio R, Romero H, Simon-Soro A, Pignatelli M, Mira A (2012) The oral metagenome in health and disease. ISME J 6(1):46–56. doi:10.1038/ismej.2011.85

Bhatt VD, Vaidya YH, Kunjadia PD, Kunjadia AP, Patel R (2012) Isolation and characterization of probiotic bacteria from human milk. Int J Pharm Sci Health Care 3(2):62–70

Bjork S, Bage R, Kanyima BM, Andre S, Nassuna-Musoke MG, Owiny DO, Persson Y (2014) Characterization of coagulase negative staphylococci from cases of subclinical mastitis in dairy cattle in Kampala, Uganda. Ir Vet J 67(1):12. doi:10.1186/2046-0481-67-12

Boutinaud M, Jammes H (2002) Potential uses of milk epithelial cells: a review. Reprod Nutr Dev 42(2):133–147
Bytyqi H, Rrustemi M, Mehmeti H, Kryeziu A, Gjinovci V, Gjonbalaj M (2010) Milk production in commercial cattle dairy farms in Kosova. Stočarstvo 63(4):275–285
Cabrera-Rubio R, Carmen Collado M, Laitinen K, Salminen S, Isolauri E, Mira A (2012) The human milk microbiome changes over lactation and is shaped by maternal weight and mode of delivery. Am J Clin Nutr 96(3):544–551
Cabrera-Rubio R, Mira-Pascual L, Mira A, Collado MC (2015) Impact of mode of delivery on the milk microbiota composition of healthy women. J Dev Orig Health Dis 7(1):54–60. doi:10.1017/S2040174415001397
Chirico G, Marzollo R, Cortinovis S, Fonte C, Gasparoni A (2008) Antiinfective properties of human milk. J Nutr 138(9):1801S–1806S
Collado MC, Delgado S, Maldonado A, Rodríguez JM (2009) Assessment of the bacterial diversity of breast milk of healthy women by quantitative real-time PCR. Lett Appl Microbiol 48(5):523–528
Collado MC, Laitinen K, Salminen S, Isolauri E (2012) Maternal weight and excessive weight gain during pregnancy modify the immunomodulatory potential of breast milk. Pediatr Res 72(1):77–85
Cooey PM, Harmon MW (1994) Religious imagination and the body: a feminist analysis: a feminist analysis. Oxford University Press, Oxford
Costello EK, Lauber CL, Hamady M, Fierer N, Gordon JI, Knight R (2009) Bacterial community variation in human body habitats across space and time. Science 326(5960):1694–1697
Cregan MD (2002) The paracellular pathway and the lactating human breast. University of Western Australia, Australia
Delgado S, Arroyo R, Martin R, Rodriguez JM (2008) PCR-DGGE assessment of the bacterial diversity of breast milk in women with lactational infectious mastitis. BMC Infect Dis 8:51. doi:10.1186/1471-2334-8-51
Duijts L, Jaddoe VWV, Hofman A, Moll HA (2010) Prolonged and exclusive breastfeeding reduces the risk of infectious diseases in infancy. Pediatrics 126(1):e18–e25
El-Mohandes AE, Schatz V, Keiser JF, Jackson BJ (1993) Bacterial contaminants of collected and frozen human milk used in an intensive care nursery. Am J Infect Control 21(5): 226–230
Fernández L, Langa S, Martín V, Maldonado A, Jiménez E, Martín R, Rodríguez JM (2013) The human milk microbiota: origin and potential roles in health and disease. Pharmacol Res 69(1):1–10
Fetherston C (1997) Characteristics of lactation mastitis in a Western Australian cohort. Breastfeed Rev 5(2):5–11
Foxman B, D'Arcy H, Gillespie B, Bobo JK, Schwartz K (2002) Lactation mastitis: occurrence and medical management among 946 breastfeeding women in the United States. Am J Epidemiol 155(2):103–114
Gao X, Zhang Q, McMahon RJ, Woo JG, Davidson BS, Morrow AL (2012) Semi-quantitative analysis of milk proteomes reveals new evolving activities in carbohydrate metabolism in breastfeeding women. FASEB J 26(1_MeetingAbstracts):lb287
Gaudana SB, Dhanani AS, Bagchi T (2010) Probiotic attributes of Lactobacillus strains isolated from food and of human origin. Br J Nutr 103(11):1620–1628
Gianneechini R, Concha C, Rivero R, Delucci I, Moreno Lopez J (2002) Occurrence of clinical and sub-clinical mastitis in dairy herds in the West Littoral Region in Uruguay. Acta Vet Scand 43(4):221–230
Goldman AS (2007) The immune system in human milk and the developing infant. Breastfeed Med 2(4):195–204
Gonzalez R, Maldonado A, Martin V, Mandomando I, Fumado V, Metzner KJ, Sacoor C et al (2013) Breast milk and gut microbiota in African mothers and infants from an area of high HIV prevalence. PLoS One 8(11):e80299. doi:10.1371/journal.Pone.0080299

Greer FR, Sicherer SH, Wesley Burks A (2008) Effects of early nutritional interventions on the development of atopic disease in infants and children: the role of maternal dietary restriction, breastfeeding, timing of introduction of complementary foods, and hydrolyzed formulas. Pediatrics 121(1):183–191

Grönlund M-M, Gueimonde M, Laitinen K, Kociubinski G, Grönroos T, Salminen S, Isolauri E (2007) Maternal breast-milk and intestinal bifidobacteria guide the compositional development of the Bifidobacterium microbiota in infants at risk of allergic disease. Clin Exp Allergy 37(12):1764–1772

Hassiotou F, Trengove N, Tat Lai C, Filgueira L, Blancafort P, Hartmann PE (2012) Breastmilk stem cells: an overview of the current knowledge. In: Breastfeeding and lactation symposium, Vienna, Austria

Hassiotou F, Hepworth AR, Metzger P, Tat Lai C, Trengove N, Hartmann PE, Filgueira L (2013) Maternal and infant infections stimulate a rapid leukocyte response in breastmilk. Clin Transl Immunol 2(4):e3. doi:10.1038/cti.2013.1

Heikkilä MP, Saris PEJ (2003) Inhibition of *Staphylococcus aureus* by the commensal bacteria of human milk. J Appl Microbiol 95(3):471–478

Ho FC, Wong RL, Lawton JW (1979) Human colostral and breast milk cells: a light and electron microscopic study. Acta Paediatr 68(4):389–396

Hoiby N, Bjarnsholt T, Givskov M, Molin S, Ciofu O (2010) Antibiotic resistance of bacterial biofilms. Int J Antimicrob Agents 35(4):322–332. doi:10.1016/j.Ijantimicag.2009.12.011

Hunt KM, Foster JA, Forney LJ, Schutte UM, Beck DL, Abdo Z, Fox LK, Williams JE, McGuire MK, McGuire MA (2011) Characterization of the diversity and temporal stability of bacterial communities in human milk. PLoS One 6(6):e21313

Hunt KM, Williams JE, Shafii B, Hunt MK, Behre R, Ting R, McGuire MK, McGuire MA (2013) Mastitis is associated with increased free fatty acids, somatic cell count, and interleukin-8 concentrations in human milk. Breastfeed Med 8(1):105–110. doi:10.1089/bfm.2011.0141

Ip S, Chung M, Raman G, Chew P, Magula N, DeVine D, Trikalinos T, Lau J (2008) Breastfeeding and maternal and infant health outcomes in developed countries. US Department of Health and Human Services, Agency for Healthcare Research and Quality, Rockville, MD. Evidence Report/Technology Assessment (153)

Jimenez E, Delgado S, Maldonado A, Arroyo R, Albujar M, Garcia N, Jariod M, Fernandez L, Gomez A, Rodriguez JM (2008) *Staphylococcus epidermidis*: a differential trait of the fecal microbiota of breast-fed infants. BMC Microbiol 8:143. doi:10.1186/1471-2180-8-143

Jimenez E, de Andres J, Manrique M, Pareja-Tobes P, Tobes R, Martinez-Blanch JF, Codoner FM, Ramon D, Fernandez L, Rodriguez JM (2015) Metagenomic analysis of milk of healthy and mastitis-suffering women. J Hum Lact. doi:10.1177/0890334415585078

Jost T, Lacroix C, Braegger C, Chassard C (2013) Assessment of bacterial diversity in breast milk using culture-dependent and culture-independent approaches. Br J Nutr 110(7):1253–1262. doi:10.1017/S0007114513000597

Khodayar-Pardo P, Mira-Pascual L, Collado MC, Martinez-Costa C (2014) Impact of lactation stage, gestational age and mode of delivery on breast milk microbiota. J Perinatol 34(8):599–605

Kinlay JR, O'Connell DL, Kinlay S (1998) Incidence of mastitis in breastfeeding women during the six months after delivery: a prospective cohort study. Med J Aust 169(6):310–312

Kvist LJ, Larsson BW, Hall-Lord ML, Steen A, Schalen C (2008) The role of bacteria in lactational mastitis and some considerations of the use of antibiotic treatment. Int Breastfeed J 3:6. doi:10.1186/1746-4358-3-6

Li JP, Zhou HJ, Yuan L, He T, Hu SH (2009) Prevalence, genetic diversity, and antimicrobial susceptibility profiles of *Staphylococcus aureus* isolated from bovine mastitis in Zhejiang Province, China. J Zhejiang Univ Sci B 10(10):753–760. doi:10.1631/jzus.B0920072

Makino H, Kushiro A, Ishikawa E, Muylaert D, Kubota H, Sakai T, Oishi K, Martin R, Ben Amor K, Oozeer R (2011) Transmission of intestinal *Bifidobacterium longum* subsp. longum strains from mother to infant, determined by multilocus sequencing typing and amplified fragment length polymorphism. Appl Environ Microbiol 77(19):6788–6793

Maldonado J, Cañabate F, Sempere L, Vela F, Sanchez AR, Narbona E, López-Huertas E, Geerlings A, Valero AD, Olivares M (2012) Human milk probiotic *Lactobacillus fermentum* CECT5716 reduces the incidence of gastrointestinal and upper respiratory tract infections in infants. J Pediatr Gastroenterol Nutr 54(1):55–61
Martín R, Langa S, Reviriego C, Jimínez E, Marín ML, Xaus J, Fernández L, Rodríguez JM (2003) Human milk is a source of lactic acid bacteria for the infant gut. J Pediatr 143(6):754–758
Martín R, Jiménez E, Heilig H, Fernández L, Marín ML, Zoetendal EG, Rodríguez JM (2009) Isolation of bifidobacteria from breast milk and assessment of the bifidobacterial population by PCR-denaturing gradient gel electrophoresis and quantitative real-time PCR. Appl Environ Microbiol 75(4):965–969
Michie CA, Tantscher E, Rot A (1998) The long term effects of breastfeeding: a role for the cells in breast milk? [Editorial]. J Trop Pediatr 44(1):2–3
Miller WR, Scott WN, Morris R, Fraser HM, Sharpe RM (1985) Growth of human breast cancer cells inhibited by a luteinizing hormone-releasing hormone agonist. Nature 313(5999):231–233
Molinari CE, Casadio YS, Hartmann BT, Arthur PG, Hartmann PE (2013) Longitudinal analysis of protein glycosylation and β-casein phosphorylation in term and preterm human milk during the first 2 months of lactation. Br J Nutr 110(01):105–115
Morrow AL, Rangel JM (2004) Human milk protection against infectious diarrhea: implications for prevention and clinical care. Semin Pediatr Infect Dis 15(4):221–228
Murphy SC, Cranker K, Senyk GF, Barbano DM, Saeman AI, Galton DM (1989) Influence of bovine mastitis on lipolysis and proteolysis in milk. J Dairy Sci 72(3):620–626
Nasidze I, Li J, Quinque D, Tang K, Stoneking M (2009) Global diversity in the human salivary microbiome. Genome Res 19(4):636–643. doi:10.1101/gr.084616.108
Newburg DS (2005) Innate immunity and human milk. J Nutr 135(5):1308–1312
Nishimura T, Suzue J, Kaji H (2009) Breastfeeding reduces the severity of respiratory syncytial virus infection among young infants: a multi-center prospective study. Pediatr Int 51(6):812–816
Olivares M, Albrecht S, De Palma G, Ferrer MD, Castillejo G, Schols HA, Sanz Y (2014) Human milk composition differs in healthy mothers and mothers with celiac disease. Eur J Nutr 54(1):119–128
Oliver SP, Murinda SE (2012) Antimicrobial resistance of mastitis pathogens. Vet Clin North Am Food Anim Pract 28(2):165–185. doi:10.1016/j.Cvfa.2012.03.005
Paape MJ, Weinland BT (1988) Effect of abraded intramammary device on milk yield, tissue damage, and cellular composition. J Dairy Sci 71(1):250–256
Patel SH, Vaidya YH, Joshi CG, Kunjadia AP (2016) Culture-dependent assessment of bacterial diversity from human milk with lactational mastitis. Comp Clin Pathol 25(2):437–443
Pol M, Ruegg PL (2007) Treatment practices and quantification of antimicrobial drug usage in conventional and organic dairy farms in Wisconsin. J Dairy Sci 90(1):249–261. doi:10.3168/jds.S0022-0302(07)72626-7
Randolph HE, Erwin RE (1974) Influence of mastitis on properties of milk: X. Fatty acid composition. J Dairy Sci 57(8):865–868. doi:10.3168/jds.S0022-0302(74)84978-7
Ravel J, Gajer P, Abdo Z, Schneider GM, Koenig SS, McCulle SL, Karlebach S et al (2011) Vaginal microbiome of reproductive-age women. Proc Natl Acad Sci U S A 108(Suppl 1):4680–4687. doi:10.1073/pnas.1002611107
Ridwan BU, Koning CJM, Besselink MGH, Timmerman HM, Brouwer EC, Verhoef J, Gooszen HG, Akkermans LMA (2008) Antimicrobial activity of a multispecies probiotic (Ecologic 641) against pathogens isolated from infected pancreatic necrosis. Lett Appl Microbiol 46(1):61–67
Rodrigues MDA, Gindri L, Silva ADD, Guex CG, Santos SOD, Hörner R (2015) Prevalence of methicillin-resistant Staphylococcus aureus in a University Hospital in the South of Brazil. Braz J Pharm Sci 51(1):35–41
Ross ZM, O'Gara EA, Hill DJ, Sleightholme HV, Maslin DJ (2001) Antimicrobial properties of garlic oil against human enteric bacteria: evaluation of methodologies and comparisons with garlic oil sulfides and garlic powder. Appl Environ Microbiol 67(1):475–480. doi:10.1128/AEM.67.1.475-480.2001

Sakwinska O, Moine D, Delley M, Combremont S, Rezzonico E, Descombes P, Vinyes-Pares G, Zhang Y, Wang P, Thakkar SK (2016) Microbiota in breast milk of Chinese lactating mothers. PLoS One 11(8):e0160856. doi:10.1371/journal.Pone.0160856

Schnorr KL, Pearson LD (1984) Intestinal absorption of maternal leucocytes by newborn lambs. J Reprod Immunol 6(5):329–337

Sharma N, Singh NK, Bhadwal MS (2011) Relationship of somatic cell count and mastitis: an overview. Asian Australas J Anim Sci 24(3):429–438

Shriram HP, Vaidya YH, Joshi CG, Kunjadia AP (2015) Culture-dependent assessment of bacterial diversity from human milk with lactational mastitis. Comparative Clinical Pathology 25(2);437–443

Smith CW, Goldman AS (1970) Interactions of lymphocytes and macrophages from human colostrum: characteristics of the interacting lymphocyte. J Reticuloendothel Soc 8(1):91–104

Solís G, de Los Reyes-Gavilan CG, Fernández N, Margolles A, Gueimonde M (2010) Establishment and development of lactic acid bacteria and bifidobacteria microbiota in breast-milk and the infant gut. Anaerobe 16(3):307–310

Soto A, Martín V, Jiménez E, Mader I, Rodríguez JM, Fernández L (2014) Lactobacilli and bifidobacteria in human breast milk: influence of antibiotherapy and other host and clinical factors. J Pediatr Gastroenterol Nutr 59(1):78

Thirabunyanon M, Boonprasom P, Niamsup P (2009) Probiotic potential of lactic acid bacteria isolated from fermented dairy milks on antiproliferation of colon cancer cells. Biotechnol Lett 31(4):571–576

Urbaniak C, McMillan A, Angelini M, Gloor GB, Sumarah M, Burton JP, Reid G (2014) Effect of chemotherapy on the microbiota and metabolome of human milk, a case report. Microbiome 2:24. doi:10.1186/2049-2618-2-24

Urbaniak C, Angelini M, Gloor GB, Reid G (2016) Human milk microbiota profiles in relation to birthing method, gestation and infant gender. Microbiome 4(1):1. doi:10.1186/s40168-015-0145-y

Vaidya Y, Patel S, Patel R, Joshi C, Kunjadia A (2015) Exploring the microbiota of human milk using the culture-dependent method. Int J 3(5):462–471

Vanderpool C, Yan F, Brent Polk D (2008) Mechanisms of probiotic action: implications for therapeutic applications in inflammatory bowel diseases. Inflamm Bowel Dis 14(11):1585–1596

Verdenelli MC, Ghelfi F, Silvi S, Orpianesi C, Cecchini C, Cresci A (2009) Probiotic properties of *Lactobacillus rhamnosus* and *Lactobacillus paracasei* isolated from human faeces. Eur J Nutr 48(6):355–363

Ward TL, Hosid S, Ioshikhes I, Altosaar I (2013) Human milk metagenome: a functional capacity analysis. BMC Microbiol 13:116. doi:10.1186/1471-2180-13-116

Weiler IJ, Hickler W, Sprenger R (1983) Demonstration that milk cells invade the suckling neonatal mouse. Am J Reprod Immunol 4(2):95–98

WHO (2000) Mastitis causes and management

Wirt DP, Adakins LT, Palkowetz KH, Schmalsteig FC, Goldman AS (1991) Activated-memory T-cells in human-milk (HM). In: Pediatric research. Williams & Wilkins, Baltimore, MD

Wright KC, Feeney AM (1998) The bacteriological screening of donated human milk: laboratory experience of British Paediatric Association's published guidelines. J Infect 36(1):23–27

Zhao X, Lacasse P (2008) Mammary tissue damage during bovine mastitis: causes and control. J Anim Sci 86(13 Suppl):57–65. doi:10.2527/jas.2007-0302

Zhou L, Yoshimura Y, Huang Y-Y, Suzuki R, Yokoyama M, Okabe M, Shimamura M (2000) Two independent pathways of maternal cell transmission to offspring: through placenta during pregnancy and by breast-feeding after birth. Immunology 101(4):570–580

Role of Gut Microbiome in Neuromodulation

8

Suganya Kannan, Govindan Krishnamoorthy,
Prabha Palanichamy, and Murugan Marudhamuthu

Abstract

Gut microbial populations plays a significant role in human health and maintains bidirectional signalling between the brain and gut. The bacterial population endures a dynamic developmental progression all over the existence and begins its associated relationship with the host early in life. Latest investigational evidences propose that the microbial population in gut has a significant influence on the brain–gut axis, and it plays a major role in epithelial cell function, gastrointestinal motility, visceral sensitivity, perception and behaviour. Results from physiological experiments propose the communication of gut bacteria not lone with the enteric nervous system but also with the central nervous system via neural, neuroendocrine, neuroimmune and humoral links. This chapter describes how the microbiota-derived products and variations of microbiome modulates the conditions of neuroimmune and neuropsychiatric disorders and the role of gut microbiome as a neuromodulator.

Keywords

Gut microbiome • Central nervous system • GABA • Gut–brain axis • Neuroimmune disorders

8.1 Introduction

The hypothesis of the human microbiome was initially introduced to mainstream researchers by Joshua Lederberg, who characterized it as 'the natural group of commensal, symbiotic, and pathogenic microorganisms that truly occupy human

S. Kannan • G. Krishnamoorthy • P. Palanichamy • M. Marudhamuthu (✉)
Department of Microbial Technology, School of Biological Sciences, Madurai Kamaraj University, Madurai 625 021, Tamil Nadu, India
e-mail: murubio2001@yahoo.com

R.P. Singh et al. (eds.), *Understanding Host-Microbiome Interactions - An Omics Approach*, DOI 10.1007/978-981-10-5050-3_8

body space and have been everything except disregarded as determinants of well-being and illness' (Lederberg and McCray 2001). Human body outsides are resident to microbial groups whose aggregate participation exceeds our human substantial and germ cells by no less than a request of extent (Gill et al. 2006). The microbial 'factory' adds to a comprehensive assortment of biological tasks that the human body couldn't generally perform. Late reports recommend that the human microbiota influence nutrient obtaining, energy acquisition and a group of host metabolic activities. The human gut region has the significant quantities of microscopic organisms and the outrageous quantities of species distinguished with different regions of the body, and it appears to play an essential part in human digestion and illness (Turnbaugh et al. 2009). Previous investigations gathered an extensive collection of proofs concerning modifications in the gut microbial alignment to numerous diseases including inflammatory bowel disease, asthma, arthritis, obesity and cardio vascular disease. In addition, the regular abdominal microbiota also influences several functional characteristics of the normal body such as morphogenesis of organs, development of immunity and maturation of gastrointestinal tract (Neish 2009; Sommer and Bäckhed 2013). Gastric microbial communities can effectively influences the gene expression in the gut mucosal area, eventually influencing the capability of the gastro-intestinal tract. A report with germ-free and normal mice discovered that the gut microbial population can alter or influence the expression of numerous genes in the gastro intestinal tract of human and mouse. The genes which are involved in the development of immunity, absorption of nutrients, energy harvesting and intestinal barrier development have been greatly influenced by the action of colonic microbiome (Boulangé et al. 2016).

Recently an integrative exploration has found strong confirmation for the presence of signalling between intestine and the brain from both the directions, which is extensively called as 'brain–gut axis' (Burokas et al. 2015). This bidirectional signalling system coordinates immunological, hormonal and neural signalling between the gut and the brain and is vital to retain homeostasis (Grenham et al. 2011). This gut–brain hypothesis was extended to the 'microbiota–gut–brain axis', which comprises 100 trillion microbial populaces and effects the functioning of central nervous system (CNS) with the effect of mood and behaviour .The brain communicates or transmits the signals with the intestinal microbiota by secreting the signalling molecules into the lumen of gut and ultimately modifying gastric motility (Rhee et al. 2009). Receptor-mediated signalling and intestinal epithelial cells are the noticeable mediators which connect the intestinal microbiome with the host (O'Mahony et al. 2011). Nowadays, the microbial endocrinology-based studies claimed that microorganisms can act as medication conveyance vehicles because of their capacity to create neuroactive metabolites. In this chapter, we discuss the role of gut microbiome, in relation to central nervous system (CNS) and the effect of neuroactives synthesized by gut microbes.

8.2 Diversity of Gut Flora

After 1 or 2 years of birth, the gut microbial flora became matured, and at the same time the intestinal epithelial cells and the mucosal barrier begin discharging which becomes tolerant towards adverse conditions and fight against the pathogen invasion (Cho and Blaser 2012; Gill et al. 2006). Based on the gut microbial studies, abundant population of microbial communities were recorded in gut than any other region of the body. The healthy human gut possesses bacteria, archaea and eukarya which are the three major domains of life, and it has the most astounding cell population than of any other biological community (Whitman et al. 1998). Among them 55 phyla belong to bacteria and 13 of archaea have been assigned existing on earth till date. But just nine of the bacterial phyla have been characterized in the digestive system, out of which five are uncommon (Ley et al. 2006). Bacterial taxa such as Bacteroidetes, Firmicutes, Proteobacteria, Actinobacteria, Fusobacteria, Verrucomicrobia and Cyanobacteria are the major part of gut, among them Bacteroidetes and Firmicutes are the dominant, which comprises 85% of the sequences in the intestinal tract (Bäckhed et al. 2005; Durbán et al. 2011). Actinobacteria and Proteobacteria are the two most ample bacterial phyla in the intestinal microbiota in addition to Bacteroidetes and Firmicutes (Arumugam et al. 2011). Based on the phylotype census, results more than 395 bacterial phylotypes were described in the gut, among them, 244 (62%) phylotypes were novel and 80% groupings from species that have not been well studied and cultured. The members of the phyla Firmicutes and Bacteroidetes occupy the majority. The Firmicutes phylum consisted of 301 phylotypes, 191 of which were novel; most (95%) of the Firmicutes sequences were members of the Clostridia class (Barcenilla et al. 2000; Pryde et al. 2002). Major intestinal species are *Bacteroides* spp., Bifidobacterium spp., and Clostridia (Delgado et al. 2004; Moore and Holdeman, 1974). *Bacteroides* spp. alone makes up approximately 30% of all intestinal bacteria (Salyers 1984). *Bifidobacterium* spp. account for 1–5%, and *Lactobacillus* spp. comprise less than 1–2% of bacteria (Franks et al. 1998; Sghir et al. 2000). Other bacteria, such as *Enterobacteria*, exist in slight numbers (Delgado et al. 2004). The total of Enterobacteriaceae species is about 0.1% of entire intestinal bacteria (Eckburg et al. 2005).

8.3 Gut Microbiota and Brain

Gut microbiota, the inhabitant bacterial communities resides in the healthy gastrointestinal tract (GIT), is representing that microbial actions taking place in the gut have a great impact on the growth and role of the CNS. Remarkable studies evidenced the overall framework connection between the enteric nervous system (ENS), the neuroendocrine and neuroimmune systems and the sympathetic vs. parasympathetic arms of the autonomic nervous system (ANS) and finally the gut microbiota (Borre et al. 2014a, b). This complete structure popularly depicted as gut–brain axis and brain–gut–enteric microbiota axis. With the effect of bidirectional signalling network, indications from the brain can influence the motor, sensory and secretory modalities of the gastrointestinal tract, and similarly, instinctual communications from the intestinal microbiome can stimulus brain function (Thakur et al. 2014; Furness et al. 1999). Investigations with gnotobiotic mice exhibited a minor constraint stress induced a blown-up discharge of corticosterone and adrenocorticotrophic hormone (ACTH) compared to the control animals (Sudo et al. 2004). Treatment of gnotobiotic mice with the faecal matter of normal healthy mice induces the stress response in the germ-free animal in a time-dependent manner. These microbiota-dependent stress alterations clearly support that the gastrointestinal microbial load is an essential factor for the stress response and development of the hypothalamic–pituitary–adrenal axis is greatly influenced by the gut microbial colonization. (Neufeld and Foster 2009).

8.4 Neuroactive Compounds Produced by Gut Microbiome

Various neuroactive molecules, for instance, gamma-aminobutyric acid (GABA), serotonin, catecholamine and acetylcholine, have been stated to be microbial origin, which have been obtained from the gastrointestinal microbiota of human (Barrett et al. 2012; Özogul 2011; Roshchina 2010). Dopamine has been produced by *Bacillus* spp., *Escherichia* spp. and *Enterococcus* spp. Norepinephrine or noradrenaline, a catecholamine family neurotransmitter, is secreted by *Escherichia* spp. and *Bacillus* spp. GABA production has been reported in certain probiotics such as *Lactobacillus* spp. and *Bifidobacterium* spp. The presence of serotonin a monoamine neurotransmitter has been reported in *Streptococcus* spp.; acetyl choline which plays a significant role in neuromuscular junction has been reported from certain *Lactobacillus* spp. Short-chain fatty acids (SCFAs) and long-chain fatty acids such as conjugated linoleic acid (CLA) are also proven to have an effective role as a neuroprotective agent (Thomas et al. 2012; Kawashima et al. 2007; Tsavkelova et al. 1999). These neurochemicals which are secreted from gastrointestinal bacteria greatly influence the brain functioning; further the probiotics which releases these neurochemicals has been recommended as an innovative treatment for neuropsychiatric conditions. (Galland 2014). Intestinal bacteria possess and transfer the genomic characteristics among the bacterial communities, which are in control for the secretion of neuroprotective chemicals by horizontal gene transfer. They encode certain enzymes involved in the synthesis of vital neuroactive molecules such as catecholamine, acetylcholine and GABA. For example, *Escherichia coli* O157:H7 possesses receptor for host-derived epinephrine/norepinephrine which activates the transcription of virulence genes in bacteria (Clarke et al. 2006) (Fig. 8.1).

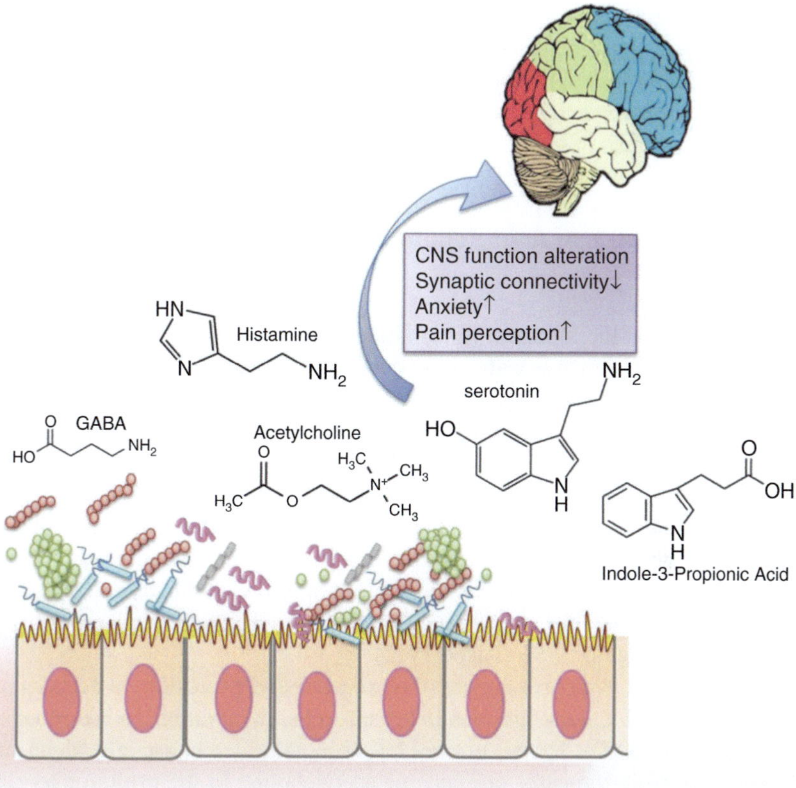

Fig. 8.1 Gut microbial metabolic by-products and their effect on behavioural function of brain

8.4.1 Influence of Gamma-Aminobutyric Acid in Central Nervous System Activity

Gamma-aminobutyric acid, a nonprotein amino acid, owns distinguished biological functions such as neurotransmission, stimulus of hypotension and diuretic and anaesthetic effects. Synthesis of GABA was influenced by the enzyme glutamate decarboxylase (GAD) (EC 4.1.1.15), which depends on pyridoxal 5-phosphate that catalyzes the irreparable α-decarboxylation of L-glutamate to GABA (Wang et al. 2011). A number of reports (Komatsuzaki et al. 2005; Park and Oh 2004; Park and Oh 2007) have shown the existence of the enzyme GAD from the healthy individuals intestinal microbial residents such as lactobacilli and also from Bifidobacteria.

Both indigenous bacteria and neuronal tissue are adapting the similar biosynthetic pathway for biological synthesis of GABA in which GAD acts on the transformation of glutamate with the help of the cofactor pyridoxal phosphate (De Biase et al. 1999). In intestinal bacteria, the GABA synthesis encompasses the uptake of glutamate with the proton exchange, and it leads to the protection of the bacteria from the acidic nature of the digestive tract. In the brain, GABA binds with the plasma membrane-transmembrane receptors and acts as inhibitory synapses. Moreover, both presynaptic and postsynaptic neuronal progressions were influenced by the effect of GABA.

8.4.2 Influence of Acetylcholine in Motor and Memory Functions

Acetylcholine is the first neurotransmitter discovered in the central and peripheral nervous systems which is critical in conduction of pain, the regulation of neuroendocrine function, REM sleep cycles and the process of learning and memory formation (López-Muñoz and Alamo 2009). Acetylcholine is produced by a single stride response catalyzed by the enzyme choline acetyltransferase. But in the case of bacterial acetylcholine synthesis, carnitine acetyltransferase plays a major role (Horiuchi et al. 2003). Acetylcholine is present within the central nervous system and used by cholinergic cells (neurons that use ACh as a neurotransmitter) which are found in several different locations of the brain, including the striatal complex, the basal forebrain, the diencephalon, pontomesencephalic cell groups, and the medulla (Rylett et al. 1993; Tuček 1982).

Experimental evidences suggest the existence and expression of acetylcholine in prokaryotic and eukaryotic non-neuronal cells in addition to neuronal matter (Wessler et al. 2003). The occurrence has been identified in bacteria, algae and protozoa suggesting the early presence of acetylcholine in the evolutionary process. Acetylcholine and cell-free enzymes responsible for acetylcholine biosynthesis have been reported in a strain of *L. plantarum* which is an important probiont (Girvin and Stevenson 1954; Rowatt 1948). Bacterial spp. such as *E. coli*, *Staphylococcus aureus* and *Bacillus subtilis* also possesses trace amount of acetyl choline.

8.4.3 Histamine as a Modulatory Neurotransmitter

The name histamine for imidazole ethylamine designates an amine occurring in tissues mainly in the grey matter and was first shown by Kwiatkowski in 1943 and later White in 1959 which demonstrated its formation and catabolism in the brain (Kahlson 1962). Histamine involved in local immune responses and biological function regulation in the gut also acts as a neurotransmitter. Histamine plays a significant role in the maintenance of wakefulness during the dysfunction of histaminergic system and acts as a modulatory neurotransmitter in the mammalian brain (Panula and Nuutinen 2013). Certain fermentative bacterial strains such as *Lactobacillus*, *Lactococcus*, *Streptococcus*, *Pediococcus* and *Enterococcus* have been testified to possess the histidine decarboxylase gene and to synthesis histamine at altered levels. Nervous communication in the enteric nervous has been influenced by the luminal expression of histamine by *L. reuteri* (Thomas et al. 2012).

8.4.4 Role of 3-Indolepropionic Acid (IPA) as a Neuroprotective Antioxidant

3-Indolepropionic acid (IPA) is a potent antioxidant and acts as an inhibitor of beta-amyloid fibril formation and to be an effective neuro-protectant against a variety of oxidotoxins (Bendheim et al. 2002). The occurrence of IPA was observed in plasma and cerebrospinal fluid (CSF) of human, and it has been synthesized by the deamination of tryptophan by intestinal bacteria (Young et al. 1980). IPA entirely safeguards primary rat hippocampal neurons and human neuroblastoma cells against oxidative damage and death. Moreover, IPA was shown to protect the death of neuroblastoma cells exposed to H_2O_2 or diethyl-dithiocarbamate (DDTC), an inhibitor of superoxide dismutase (Chyan et al. 1999). Studies with germ-free mice colonized with the bacterium *Clostridium sporogenes* demonstrated the production of IPA (Jellet et al. 1980).

8.5 Approaches Used to Examine Gut–Brain Axis

Involvement of the gut microbiota in the human health via the gut–brain axis has been recently highlighted by an increasing number of studies using germ-free (GF) animals, bacterial infections, probiotic and prebiotic administration and antibiotic administration (Grenham et al. 2011).

8.5.1 Germ-Free Animals as a Tool to Study Gut–Brain Relationship

Germ-free animals, free from any microbiota throughout its organismal build-up, is a well-established and valuable tool to determine the involvement of microbiota towards regulating the development and expression of physiological and behavioural parameters in the host (Bercik et al. 2011a, b; Clarke et al. 2013; Hsiao et al. 2013). They were maintained in the gnotobiotic units to eliminate the chance for the postnatal colonization in their GI tract. Thus, becoming a 'microbiota-free' control group against the conventionally colonized gut of their counterparts. The GF animals have higher susceptibility towards infection (O′Hara and Shanahan 2006) but have decreased digestive enzyme activities and intestinal muscle wall thickness (Round and Mazmanian 2009). By colonizing GF animals at a particular age, the continuous effects of the functional microbiota can be distinguished from the developmental effects. So, GF animals are considered as a powerful tool to examine the relationship between the gut microbiota and brain function.

Adult GF mice exhibit altered brain physiology, which exhilarated the levels of corticotropin-releasing hormone gene expression in the hypothalamus and also elevated the secretion of ACTH and corticosterone after acute restraint stress (Sudo et al. 2004). The absence of gut microbiota in a stress-induced rat strain aggravates the neuroendocrine and behavioural response to an acute stress and deeply alters the turnover rate of dopamine and its metabolites in brain upper structures involved in the regulation of stress and anxiety (Crumeyrolle-Arias et al. 2014). Behavioural studies suggested that GF mice show weakened anxiety-like behaviour, but increase in the motor activity, when compared to the conventional mice (Heijtz et al. 2011). These behavioural changes were reversed by colonization with probiotics like *B. infantis* but not with the enteropathogenic *E. coli* (Sudo et al. 2004). The behavioural parameter alterations of the gut microbiota in specific pathogen-free mice were correlated by antibiotic treatment which results in increased motor activities when compared to the conventional mice (Bercik et al. 2011a, b) (Fig. 8.2).

At the molecular level, GF mice possess decreased levels of *N*-methyl-D-aspartate receptors (NMDARs), specially the NR1 and NR2A subunits, in the hippocampus (Sudo et al. 2004), or NR2B subunits in the amygdala (Neufeld et al. 2011), which shown to play a key role in neuropsychiatric disorders (Lakhan et al. 2013). GF animals have reduced levels of brain-derived neurotrophic factor (BDNF), a key neurotrophin involved in neuronal growth and survival (Sudo et al. 2004). Moreover, genetic background appears to play an extensive role in inflecting the microbiome-brain-gut axis. In summary, the studies involving GF animals demonstrate utility in evaluating the underlying mechanisms of microbiota–gut–brain axis communication apposite to brain function.

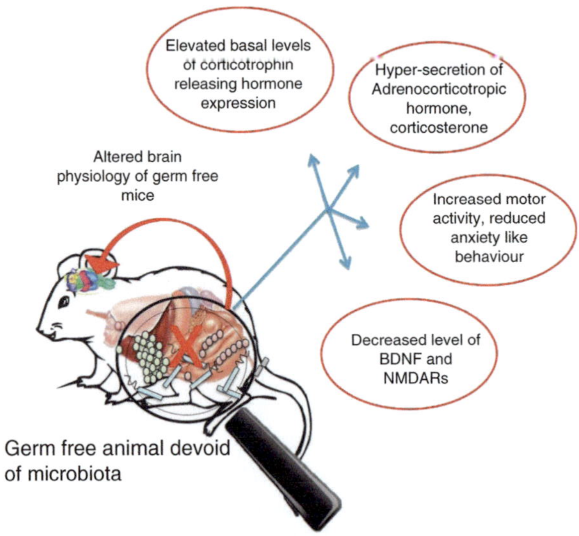

Fig. 8.2 Role of GF animals as a powerful tool to examine brain alterations

8.5.2 Role of Bacterial Infections in Gut–Brain Relationship

Considering the impact of infections by enteric pathogens on brain and behaviour has been a decisive strategy to investigate the function of the microbiota–gut–brain axis. Experiments were needed to explore how the chronic inflammation of the gut influences behavioural changes. For example, *Trichuris muris* infection, a very closely associated species to *Trichuris trichiura*, decreased hippocampal levels of *Bdnf* mRNA, inflated the concentration of kynurenine–tryptophan ratio in plasma (an indication of alterations in tryptophan metabolism) and also increased the anxiety-like behaviour in mice. Vagotomy in the infected mice did not avert the anxiety-like behaviour, which indicates that the behavioural effects caused by *T. muris* infection are not mediated by vagus nerve. Treatment with the probiotic *B. longum* regulated the behaviour of the infected mice and improved the *Bdnf* mRNA level in the hippocampal region, but kynurenine–tryptophan ratio was not restored. It establishes that the mechanism of action of probiotic treatment inversed the behavioural changes, demonstrating that the gut microbiota influences the brain signal via multiple routes (Bercik et al. 2010).

Number of studies has been carried out using *Citrobacter rodentium* to explore gut–brain axis function. Infection to the animal caused an increase in anxiety-like behaviour (Lyte et al. 2006) and cognitive dysfunction. This effect was averted by pretreatment with a combination of probiotics begin 7 days before the infection (Gareau et al. 2010). This pretreatment regimen decreased the serum corticosterone levels and averted the alterations in hippocampal BDNF mRNA and central FOS gene expression (an indicator for neuronal activity) induced by the bacterial infection. An infection with *Campylobacter jejuni* was resulted in increased anxiety-like behaviour, while treatment with a probiotic *Bifidobacterium* strain reduced the anxiety and depressive behaviour (Bercik et al. 2012).

Infection with *C. rodentium* validated that the vagus nerve is one of the most aided route for gut–brain signalling (Lyte et al. 2006). The animals showed increased anxiety-like behaviour, and the level of anxiety was assessed by evaluating the number of cells expressing FOS proteins in the bed nucleus of stria terminalis and was correlated with vagotomy (Goehler et al. 2008). Vagotomy studies in the *Salmonella enterica* infected rats confirmed the involvement of vagus nerve in the transmission of signals from the GIT to the brain (Wang et al. 2002). Even though the studies involving pathogens precisely do not address the ability of gut microbiome to signal the CNS, they bid key observations towards elucidating the pathways through which microbiota can signal the CNS and affect the behaviour of the infected animals. Concisely, it propose that the infection and stress can combine and act together to alter cognitive function of the CNS and behaviour of the animals (Kennedy et al. 2012).

8.5.3 Probiotic and Prebiotic Administration

Probiotics are living microorganisms, when consume in sufficient quantities, and induce a health aid to the host. Studies involving probiotics demonstrate that they have wide range of influence towards both human and animal health (Clarke et al. 2012). Furthermore, there are clinical affirmations to evident the effect of probiotic administration towards reducing the anxiety-like behaviour and stress responses in irritable bowel syndrome (IBS) patients with chronic fatigue (Rao et al. 2009). Recently, probiotic cocktail treatment of *L. helveticus* and *B. longum* decreased anxiety-like behaviour in animals and exerted beneficial cognitive effects and also reduced serum cortisol levels in humans (Messaoudi et al. 2011). It also conversed the depression-related behavioural effects observed in rats after myocardial infarction (Arseneault-Bréard et al. 2012).

The probiotic *B. longum* NCC3001 aversed anxiety-like behaviour and alterations in hippocampal *Bdnf* mRNA levels in mice with colitis, without reducing the gut inflammation (Bercik et al. 2011a, b). Ingestion of *L. rhamnosus* NCC4007 reduced anxiety and anguish-like behaviour and decreased the concentration of plasma corticosterone levels in stress-induced mice and modified the expression of both GABAA and GABAB receptors in the brain (Bravo et al. 2011). These alterations were reported to be associated with depression and anxiety-like behaviours in animal models. As vagotomy influenced the antidepressant effects, anxiety-like behaviour and GABA receptor expression in CNS induced by *L. rhamnosus* NCC4007 administration suggest that the alterations are via vagus nerve signalling. It demonstrates that parasympathetic neuro signals are necessary for *L. rhamnosus* to aid in the interaction with the brain (Bercik et al. 2010; Wang and Kasper 2014).

The reversal of anxiety like behaviour by *B. longum* was not observed in vagotomized mice and proposes that a neural mechanism underlies this effect. Neuronal route of action was confirmed by treating the myenteric neurons with *B. longum*-leavened medium to evaluate the effect of bacterial products in altering the excitatory properties of enteric nerves. The fermented medium decreased the variation in action potentials in response to the artificial stimulation and indicates that probiotic-leavened products directly modulate the neural signals (Bercik et al. 2011a, b). It was also confirmed by administration of other probiotic fermented medium such as *L. rhamnosus* and *L. johnsonii*. An intra-duodenal administration of *L. johnsonii* aids the gastric vagal nerve activity via histaminergic pathways (Tanida et al. 2005).

It shows that probiotic strains and prebiotics can regulate diverse aspects of brain function and behaviour and some functions are vagus nerve dependent. While generalizing the influence of one bacterial strain to another, with or without prebiotics, serious attention should be exercised, and efforts need to be destined at identifying the mechanism (Paulus and Stein 2006).

8.5.4 Antibiotic Administration

The use of antibiotics is one of the most frequently used methods to cause intestinal dysbiosis in experimental animals. The distress of the microbiota by oral administration of the antibiotics like neomycin, bacitracin and natamycin to the adult mice inflated the visceral hypersensitivity induced by colorectal distension. This effect was altered by treating with *L. paracasei* (Verdu et al. 2006). Antimicrobial regimen increases exploratory behaviour and varied the BDNF expression in hippocampal and amygdala regions in the brain (Bercik et al. 2011a, b). These medications failed to influence the behaviour in GF mice demonstrate that the effects were not due to any off-target or intrinsic effects of these medications. Both vagotomy and sympathectomy did not influence the role of antibiotics towards modulating the behaviour. It confirms that there are other underlying mechanisms involved in gut–brain axis communication in the dysbiosis-induced animal models (Bercik et al. 2012).

Reducing the level of microbiota by chronic antibiotic administration during adolescence altered the anxiety-related and social behaviours collectively by inducing the disruption in kynurenine–tryptophan ratio and other behaviour-related factors of the endocrine system (Desbonnet et al. 2015). This focuses the usage of antibiotics towards examining the influence of microbiota in gut–brain axis functions. Furthermore, it also demonstrates that evaluating the impact of widespread use of antibiotics in human CNS should be justified. Further studies with antibiotics could examine the role of gut microbiome on brain function and physiology.

8.6 Therapeutic Opportunities Involving Gut–Brain Axis

Targeted therapeutics to shape up the microbiota is still in the primary stage when compared to other pharmaceutical products. Prebiotics and probiotics are the most commonly used ailments for gastrointestinal disorders (Preidis et al. 2012; Preidis and Versalovic 2009). Probiotics are generally a cocktail of bacterial strains which blend with the broader microbiota in the GIT with minimal global effect (Shanahan et al. 2012; Quigley 2011) except the microbiota is interimly weakened by antibiotics (Hickson et al. 2007). Prebiotics are nutrients intended towards stimulating the growth of a specific microbial species, exhibited a greater potential for manipulating the environmental pressures which shape up the microbiome of an individual, specifically in the developmental stages of life (Arslanoglu et al. 2007).

It is widely established that an entire biodiversity is important for a healthy microbiota and single species administration might have slight effect on the long-term host–microbiota interaction (Sekirov et al. 2010). Lactic acid bacteria species that thrive on specific cofactors are the closest probiotic field that has come to targeted microbial therapies (Bailey et al. 2011). Most lactic acid species do not require iron, and they can be out-competed by normal pathogens in high-iron environments such as that accompanied by trauma and internal bleeding prior to surgery. *Streptococcus thermophilus* strains have a positive response to increased iron concentrations; the negative effects of pathogenic bacteria could be potentially counteracted. Other intense therapeutic strategies are intentional infections by parasitic helminths (Weinstock Elliott 2009) and faecal transplants (Landy et al. 2011).

Antimicrobials are the most commonly used treatment against infectious diseases and inflammation. But, current research in the field of microbial contribution towards health and advancements in understanding the complex diseases resulted in evaluation of some antibiotic (Dethlefsen et al. 2008; Blaser 2011) and immunosuppressant ailments (Proal et al. 2011). Antibiotics were considered to be a poor choice for gut microbiota alterations because of the tolerance issues combined with long-term usage and the absence of bacterial species specificity. But, antimicrobials can positively alter chronic disease conditions like diabetes and obesity in rodents (Kootte et al. 2012). A potential gut microbiome modulator should have specificity to the bacterial species and rich bioavailability in the gastrointestinal tract. Interestingly, these are the type of molecules considered as failed candidates in antibiotic drug development (Payne et al. 2007). The fact that more than 80% of the microbial species in the gut cannot be grown in laboratory conditions reduces the usage of high-throughput compound screening attempts to explore the microbiota-modulating compounds (Eckburg et al. 2005). But, in vitro human gut model development (Feria-Gervasio et al. 2011; Kim et al. 2012) and the usage of bacteria phylum-specific antibiotics, for example, against Gram-positive Firmicutes, will pave way to the development of precise microbiota modulators. The unique proteins encoded by gut microbiota can act as a source for potential high-specificity drug targets (Ellrott et al. 2010).

Another potential therapeutic strategy is focussing on host genes involved in the cross talk with microbiome. Current research is focussing on human receptors engaged in maintaining the gut microbiome (Zaneveld et al. 2008). Toll-like receptors (TLRs) are involved in cellular responses towards bacterial infections, initiation of inflammation, production of antimicrobial peptides, maturation of antigen-presenting cells and activation of cellular repair and survival pathways (Saleh and Trinchieri 2011). Specifically, the receptors TLR2 and TLR4 are the primary sensors of pathogenic bacteria and also involved in maintaining bacterial gut flora equilibrium. Microbiota synthesize a wide range of bio-activated signalling low molecular weight molecules and metabolites, which are similar to neurotransmitters (Shenderov 2011; Rajpal and Brown 2013). GABA and serotonin are neurotransmitters that can influence host behaviour and are produced directly or indirectly by certain commensal microbes (Barrett et al. 2012; Reigstad et al. 2015). SCFAs are considered to be neuroactive microbial metabolites which can navigate through the blood–brain barrier and alter the CNS functions, brain development and behaviour. Propionate and butyrate were found to modulate behavioural changes in rodents, while it induces changes similar to autism (MacFabe 2012; MacFabe et al. 2007). Oral capsaicin decreases adiposity via global nutrigenomic changes in different tissues (Baboota et al. 2014). Further, the gut communicates to the brain via hormonal signalling pathways through the release of peptides from enteroendocrine cells in gut, which can directly influence the brain. Gut peptides such as orexin, galanin, ghrelin, gastrin and leptin modulate feeding behaviour, energy homeostasis, circadian rhythm, sexual behaviour, arousal and anxiety (Borre et al. 2014a, b).

Conclusion

Even though we are at the primary phases of analyzing multifaceted signalling coordination among colonic microbiome and brain, we understand that several bacterial communities which reside in the gastrointestinal tract own the capability to secrete neuroactive chemicals with various modulatory effects which directly influences the brain function. The evidences suggest that the adaptations in bacterial colonization in gut can alter the conditions in brain or central nervous system. Deciphering the mechanisms that elicit these corollaries will advance our understanding about the aetiology of neurodevelopmental psychiatric illnesses, assist analysis of biomarkers of dysfunction and allow the identification of new openings of chance for the growth of innovative therapeutic interferences onset later in life. At the end the perception of unique microbiome–neuromodulation, focused therapeutic approaches open new avenues for the treatment of neurodevelopmental deficits and brain disorders.

References

Arseneault-Bréard J, Rondeau I, Gilbert K, Girard SA, Tompkins TA, Godbout R, Rousseau G (2012) Combination of *Lactobacillus helveticus* R0052 and *Bifidobacterium longum* R0175 reduces post-myocardial infarction depression symptoms and restores intestinal permeability in a rat model. Br J Nutr 107(12):1793–1799

Arslanoglu S, Moro GE, Boehm G (2007) Early supplementation of prebiotic oligosaccharides protects formula-fed infants against infections during the first 6 months of life. J Nutr 137(11):2420–2424

Arumugam M, Raes J, Pelletier E, Le Paslier D, Yamada T, Mende DR, Fernandes GR, Tap J, Bruls T, Batto JM, Bertalan M (2011) Enterotypes of the human gut microbiome. Nature 473(7346):174–180

Baboota RK, Murtaza N, Jagtap S, Singh DP, Karmase A, Kaur J, Bhutani KK, Boparai RK, Premkumar LS, Kondepudi KK, Bishnoi M (2014) Capsaicin-induced transcriptional changes in hypothalamus and alterations in gut microbial count in high fat diet fed mice. J Nutr Biochem 25(9):893–902

Bäckhed F, Ley RE, Sonnenburg JL, Peterson DA, Gordon JI (2005) Host-bacterial mutualism in the human intestine. Science 307(5717):1915–1920

Bailey JR, Probert CS, Cogan TA (2011) Identification and characterisation of an iron-responsive candidate probiotic. PLoS One 6(10):e26507

Barcenilla A, Pryde SE, Martin JC, Duncan SH, Stewart CS, Henderson C, Flint HJ (2000) Phylogenetic relationships of butyrate-producing bacteria from the human gut. Appl Environ Microbiol 66(4):1654–1661

Barrett E, Ross RP, O'Toole PW, Fitzgerald GF, Stanton C (2012) γ-Aminobutyric acid production by culturable bacteria from the human intestine. J Appl Microbiol 113(2):411–417

Bendheim PE, Poeggeler B, Neria E, Ziv V, Pappolla MA, Chain DG (2002) Development of indole-3-propionic acid (OXIGON™) for Alzheimer's disease. J Mol Neurosci 19(1–2):213–217

Bercik P, Verdu EF, Foster JA, Macri J, Potter M, Huang X, Malinowski P, Jackson W, Blennerhassett P, Neufeld KA, Lu J (2010) Chronic gastrointestinal inflammation induces anxiety-like behavior and alters central nervous system biochemistry in mice. Gastroenterology 139(6):2102–2112

Bercik P, Park AJ, Sinclair D, Khoshdel A, Lu J, Huang X, Deng Y, Blennerhassett PA, Fahnestock M, Moine D, Berger B (2011a) The anxiolytic effect of *Bifidobacterium longum* NCC3001 involves vagal pathways for gut–brain communication. Neurogastroenterol Motil 23(12):1132–1139

Bercik P, Denou E, Collins J, Jackson W, Lu J, Jury J, Deng Y, Blennerhassett P, Macri J, McCoy KD, Verdu EF (2011b) The intestinal microbiota affect central levels of brain-derived neurotropic factor and behavior in mice. Gastroenterology 141(2):599–609

Bercik P, Collins SM, Verdu EF (2012) Microbes and the gut-brain axis. Neurogastroenterol Motil 24(5):405–413

Blaser M (2011) Antibiotic overuse: stop the killing of beneficial bacteria. Nature 476(7361):393–394

Borre YE, O'Keeffe GW, Clarke G, Stanton C, Dinan TG, Cryan JF (2014a) Microbiota and neurodevelopmental windows: implications for brain disorders. Trends Mol Med 20(9):509–518

Borre YE, Moloney RD, Clarke G, Dinan TG, Cryan JF (2014b) The impact of microbiota on brain and behavior: mechanisms & therapeutic potential. In: Microbial endocrinology: The microbiota-gut-brain axis in health and disease. Springer, New York, pp 373–403

Boulangé CL, Neves AL, Chilloux J, Nicholson JK, Dumas ME (2016) Impact of the gut microbiota on inflammation, obesity, and metabolic disease. Genome Med 8(1):1

Bravo JA, Forsythe P, Chew MV, Escaravage E, Savignac HM, Dinan TG, Bienenstock J, Cryan JF (2011) Ingestion of *Lactobacillus* strain regulates emotional behavior and central GABA receptor expression in a mouse via the vagus nerve. Proc Natl Acad Sci U S A 108(38):16050–16055

Burokas A, Moloney RD, Dinan TG, Cryan JF (2015) Microbiota regulation of the mammalian gut–brain axis. Adv Appl Microbiol 91:1–62

Cho I, Blaser MJ (2012) The human microbiome: at the interface of health and disease. Nat Rev Genet 13(4):260–270

Chyan YJ, Poeggeler B, Omar RA, Chain DG, Frangione B, Ghiso J, Pappolla MA (1999) Potent neuroprotective properties against the Alzheimer β-amyloid by an endogenous melatonin-related indole structure, indole-3-propionic acid. J Biol Chem 274(31):21937–21942

Clarke MB, Hughes DT, Zhu C, Boedeker EC, Sperandio V (2006) The QseC sensor kinase: a bacterial adrenergic receptor. Proc Natl Acad Sci U S A 103(27):10420–10425

Clarke G, Cryan JF, Dinan TG, Quigley EM (2012) Review article: probiotics for the treatment of irritable bowel syndrome–focus on lactic acid bacteria. Aliment Pharmacol Ther 35(4):403–413

Clarke G, Grenham S, Scully P, Fitzgerald P, Moloney RD, Shanahan F, Dinan TG, Cryan JF (2013) The microbiome-gut-brain axis during early life regulates the hippocampal serotonergic system in a sex-dependent manner. Mol Psychiatry 18(6):666–673

Crumeyrolle-Arias M, Jaglin M, Bruneau A, Vancassel S, Cardona A, Daugé V, Naudon L, Rabot S (2014) Absence of the gut microbiota enhances anxiety-like behavior and neuroendocrine response to acute stress in rats. Psychoneuroendocrinology 42:207–217

De Biase D, Tramonti A, Bossa F, Visca P (1999) The response to stationary-phase stress conditions in *Escherichia coli*: role and regulation of the glutamic acid decarboxylase system. Mol Microbiol 32(6):1198–1211

Delgado S, Suárez A, Otero L, Mayo B (2004) Variation of microbiological and biochemical parameters in the faeces of two healthy people over a 15 day period. Eur J Nutr 43(6):375–380

Desbonnet L, Clarke G, Traplin A, O'Sullivan O, Crispie F, Moloney RD, Cotter PD, Dinan TG, Cryan JF (2015) Gut microbiota depletion from early adolescence in mice: implications for brain and behaviour. Brain Behav Immun 48:165–173

Dethlefsen L, Huse S, Sogin ML, Relman DA (2008) The pervasive effects of an antibiotic on the human gut microbiota, as revealed by deep 16S rRNA sequencing. PLoS Biol 6(11):e280

Durbán A, Abellán JJ, Jiménez-Hernández N, Ponce M, Ponce J, Sala T, D'Auria G, Latorre A, Moya A (2011) Assessing gut microbial diversity from feces and rectal mucosa. Microb Ecol 61(1):123–133

Eckburg PB, Bik EM, Bernstein CN, Purdom E, Dethlefsen L, Sargent M, Gill SR, Nelson KE, Relman DA (2005) Diversity of the human intestinal microbial flora. Science 308(5728):1635–1638

Ellrott K, Jaroszewski L, Li W, Wooley JC, Godzik A (2010) Expansion of the protein repertoire in newly explored environments: human gut microbiome specific protein families. PLoS Comput Biol 6(6):e1000798

Feria-Gervasio D, Denis S, Alric M, Brugère JF (2011) In vitro maintenance of a human proximal colon microbiota using the continuous fermentation system P-ECSIM. Appl Microbiol Biotechnol 91(5):1425–1433

Franks AH, Harmsen HJ, Raangs GC, Jansen GJ, Schut F, Welling GW (1998) Variations of bacterial populations in human feces measured by fluorescent in situ hybridization with group-specific 16S rRNA-targeted oligonucleotide probes. Appl Environ Microbiol 64(9):3336–3345

Furness JB, Kunze WA, Clerc N (1999) Nutrient tasting and signaling mechanisms in the gut: II. The intestine as a sensory organ: neural, endocrine, and immune responses. Am J Physiol 277(5):G922–G928

Galland L (2014) The gut microbiome and the brain. J Med Food 17(12):1261–1272

Gareau MG, Wine E, Rodrigues DM, Cho JH, Whary MT, Philpott DJ, MacQueen G, Sherman PM (2010) Bacterial infection causes stress-induced memory dysfunction in mice. Gut 60(3):307–317

Gill SR, Pop M, DeBoy RT, Eckburg PB, Turnbaugh PJ, Samuel BS, Gordon JI, Relman DA, Fraser-Liggett CM, Nelson KE (2006) Metagenomic analysis of the human distal gut microbiome. Science 312(5778):1355–1359

Girvin GT, Stevenson JW (1954) Cell free "choline acetylase" from Lactobacillus plantarum. Can J Biochem Physiol 32(2):131–146

Goehler LE, Park SM, Opitz N, Lyte M, Gaykema RP (2008) Campylobacter jejuni infection increases anxiety-like behavior in the holeboard: possible anatomical substrates for viscerosensory modulation of exploratory behavior. Brain Behav Immun 22(3):354–366

Grenham S, Clarke G, Cryan JF, Dinan TG (2011) Brain–gut–microbe communication in health and disease. Front Physiol 2:94

Heijtz RD, Wang S, Anuar F, Qian Y, Björkholm B, Samuelsson A, Hibberd ML, Forssberg H, Pettersson S (2011) Normal gut microbiota modulates brain development and behavior. Proc Natl Acad Sci U S A 108(7):3047–3052

Hickson M, D'Souza AL, Muthu N, Rogers TR, Want S, Rajkumar C, Bulpitt CJ (2007) Use of probiotic *Lactobacillus* preparation to prevent diarrhoea associated with antibiotics: randomised double blind placebo controlled trial. BMJ 335(7610):80

Horiuchi Y, Kimura R, Kato N, Fujii T, Seki M, Endo T, Kato T, Kawashima K (2003) Evolutional study on acetylcholine expression. Life Sci 72(15):1745–1756

Hsiao EY, McBride SW, Hsien S, Sharon G, Hyde ER, McCue T, Codelli JA, Chow J, Reisman SE, Petrosino JF, Patterson PH (2013) Microbiota modulate behavioral and physiological abnormalities associated with neurodevelopmental disorders. Cell 155(7):1451–1463

Jellet JJ, Forrest TP, Macdonald IA, Marrie TJ, Holdeman LV (1980) Production of indole-3-propanoic acid and 3-(p-hydroxyphenyl) propanoic acid by *Clostridium sporogenes*: a convenient thin-layer chromatography detection system. Can J Microbiol 26(4):448–453

Kahlson G (1962) New approaches to the physiology of histamine. Perspect Biol Med 5(2):179–197

Kawashima K, Misawa H, Moriwaki Y, Fujii YX, Fujii T, Horiuchi Y, Yamada T, Imanaka T, Kamekura M (2007) Ubiquitous expression of acetylcholine and its biological functions in life forms without nervous systems. Life Sci 80(24):2206–2209

Kennedy PJ, Clarke G, Quigley EM, Groeger JA, Dinan TG, Cryan JF (2012) Gut memories: towards a cognitive neurobiology of irritable bowel syndrome. Neurosci Biobehav Rev 36(1):310–340

Kim HJ, Huh D, Hamilton G, Ingber DE (2012) Human gut-on-a-chip inhabited by microbial flora that experiences intestinal peristalsis-like motions and flow. Lab Chip 12(12):2165–2174

Komatsuzaki N, Shima J, Kawamoto S, Momose H, Kimura T (2005) Production of γ-aminobutyric acid (GABA) by *Lactobacillus paracasei* isolated from traditional fermented foods. Food Microbiol 22(6):497–504

Kootte RS, Vrieze A, Holleman F, Dallinga-Thie GM, Zoetendal EG, de Vos WM, Groen AK, Hoekstra JB, Stroes ES, Nieuwdorp M (2012) The therapeutic potential of manipulating gut microbiota in obesity and type 2 diabetes mellitus. Diabetes Obes Metab 14(2):112–120

Kwiatkowski H (1943) Histamine in nervous tissue. J Physiol 102(1):32

Lakhan SE, Caro M, Hadzimichalis N (2013) NMDA receptor activity in neuropsychiatric disorders. Front Psychiatry 4:52

Landy J, Al-Hassi HO, Mclaughlin SD, Walker AW, Ciclitira PJ, Nicholls RJ, Clark SK, Hart AL (2011) Review article: Faecal transplantation therapy for gastrointestinal disease. Aliment Pharmacol Ther 34(4):409–415

Lederberg J, McCray AT (2001) Ome Sweet Omics—a genealogical treasury of words. Scientist 15(7):8–8

Ley RE, Peterson DA, Gordon JI (2006) Ecological and evolutionary forces shaping microbial diversity in the human intestine. Cell 124(4):837–848

López-Muñoz F, Alamo C (2009) Historical evolution of the neurotransmission concept. J Neural Transm 116(5):515–533

Lyte M, Li W, Opitz N, Gaykema RP, Goehler LE (2006) Induction of anxiety-like behavior in mice during the initial stages of infection with the agent of murine colonic hyperplasia *Citrobacter rodentium*. Physiol Behav 89(3):350–357

MacFabe DF (2012) Short-chain fatty acid fermentation products of the gut microbiome: implications in autism spectrum disorders. Microb Ecol Health Dis 23. doi:10.3402/mehd.v23i0.19260

MacFabe DF, Cain DP, Rodriguez-Capote K, Franklin AE, Hoffman JE, Boon F, Taylor AR, Kavaliers M, Ossenkopp KP (2007) Neurobiological effects of intraventricular propionic acid in rats: possible role of short chain fatty acids on the pathogenesis and characteristics of autism spectrum disorders. Behav Brain Res 176(1):149–169

Messaoudi M, Lalonde R, Violle N, Javelot H, Desor D, Nejdi A, Bisson JF, Rougeot C, Pichelin M, Cazaubiel M, Cazaubiel JM (2011) Assessment of psychotropic-like properties of a probiotic formulation (*Lactobacillus helveticus* R0052 and *Bifidobacterium longum* R0175) in rats and human subjects. Br J Nutr 105(05):755–764

Moore WEC, Holdeman LV (1974) Human fecal flora: the normal flora of 20 Japanese-Hawaiians. Appl Microbiol 27(5):961–979

Neish AS (2009) Microbes in gastrointestinal health and disease. Gastroenterology 136(1):65–80

Neufeld KA, Foster JA (2009) Effects of gut microbiota on the brain: implications for psychiatry. J Psychiatry Neurosci 34(3):230–231

Neufeld KM, Kang N, Bienenstock J, Foster JA (2011) Reduced anxiety-like behavior and central neurochemical change in germ-free mice. Neurogastroenterol Motil 23(3):255–e119

O'Mahony SM, Hyland NP, Dinan TG, Cryan JF (2011) Maternal separation as a model of brain–gut axis dysfunction. Psychopharmacology 214(1):71–88

O'Hara AM, Shanahan F (2006) The gut flora as a forgotten organ. EMBO Rep 7(7):688–693

Özogul F (2011) Effects of specific lactic acid bacteria species on biogenic amine production by foodborne pathogen. Int J Food Sci Technol 46(3):478–484

Panula P, Nuutinen S (2013) The histaminergic network in the brain: basic organization and role in disease. Nat Rev Neurosci 14(7):472–487

Park KB, Oh SH (2004) Cloning and expression of a full length glutamate decarboxylase gene from *Lactobacillus plantarum*. J Food Sci Nutr 9(4):324–329

Park KB, Oh SH (2007) Cloning, sequencing and expression of a novel glutamate decarboxylase gene from a newly isolated lactic acid bacterium, *Lactobacillus brevis* OPK-3. Bioresour Technol 98(2):312–319

Paulus MP, Stein MB (2006) An insular view of anxiety. Biol Psychiatry 60(4):383–387

Payne DJ, Gwynn MN, Holmes DJ, Pompliano DL (2007) Drugs for bad bugs: confronting the challenges of antibacterial discovery. Nat Rev Drug Discov 6(1):29–40

Preidis GA, Versalovic J (2009) Targeting the human microbiome with antibiotics, probiotics, and prebiotics: gastroenterology enters the metagenomics era. Gastroenterology 136(6):2015–2031

Preidis GA, Saulnier DM, Blutt SE, Mistretta TA, Riehle KP, Major AM, Venable SF, Finegold MJ, Petrosino JF, Conner ME, Versalovic J (2012) Probiotics stimulate enterocyte migration and microbial diversity in the neonatal mouse intestine. FASEB J 26(5):1960–1969

Proal AD, Albert PJ, Blaney GP, Lindseth IA, Benediktsson C, Marshall TG (2011) Immunostimulation in the era of the metagenome. Cell Mol Immunol 8(3):213–225

Pryde SE, Duncan SH, Hold GL, Stewart CS, Flint HJ (2002) The microbiology of butyrate formation in the human colon. FEMS Microbiol Lett 217(2):133–139

Quigley EM (2011) Gut microbiota and the role of probiotics in therapy. Curr Opin Pharmacol 11(6):593–603

Rajpal DK, Brown JR (2013) Modulating the human gut microbiome as an emerging therapeutic paradigm. Sci Prog 96(3):224–236

Rao S, Srinivasjois R, Patole S (2009) Prebiotic supplementation in full-term neonates: a systematic review of randomized controlled trials. Arch Pediatr Adolesc Med 163(8):755–764

Reigstad CS, Salmonson CE, Rainey JF, Szurszewski JH, Linden DR, Sonnenburg JL, Farrugia G, Kashyap PC (2015) Gut microbes promote colonic serotonin production through an effect of short-chain fatty acids on enterochromaffin cells. FASEB J 29(4):1395–1403

Rhee SH, Pothoulakis C, Mayer EA (2009) Principles and clinical implications of the brain–gut–enteric microbiota axis. Nat Rev Gastroenterol Hepatol 6(5):306–314

Roshchina VV (2010) Evolutionary considerations of neurotransmitters in microbial, plant, and animal cells. In: Microbial endocrinology. Springer, New York, pp 17–52

Round JL, Mazmanian SK (2009) The gut microbiota shapes intestinal immune responses during health and disease. Nat Rev Immunol 9(5):313–323

Rowatt E (1948) The relation of pantothenic acid to acetylcholine formation by a strain of *Lactobacillus plantarum*. Microbiology 2(1):25–30

Rylett RJ, Goddard S, Schmidt BM, Williams LR (1993) Acetylcholine synthesis and release following continuous intracerebral administration of NGF in adult and aged Fischer-344 rats. J Neurosci 13(9):3956–3963

Saleh M, Trinchieri G (2011) Innate immune mechanisms of colitis and colitis-associated colorectal cancer. Nat Rev Immunol 11(1):9–20

Salyers AA (1984) Bacteroides of the human lower intestinal tract. Annu Rev Microbiol 38(1):293–313

Sekirov I, Russell SL, Antunes LCM, Finlay BB (2010) Gut microbiota in health and disease. Physiol Rev 90(3):859–904

Sghir A, Gramet G, Suau A, Rochet V, Pochart P, Dore J (2000) Quantification of bacterial groups within human fecal flora by oligonucleotide probe hybridization. Appl Environ Microbiol 66(5):2263–2266

Shanahan F, Dinan TG, Ross P, Hill C (2012) Probiotics in transition. Clin Gastroenterol Hepatol 10(11):1220–1224

Shenderov BA (2011) Probiotic (symbiotic) bacterial languages. Anaerobe 17(6):490–495

Sommer F, Bäckhed F (2013) The gut microbiota—masters of host development and physiology. Nat Rev Microbiol 11(4):227–238

Sudo N, Chida Y, Aiba Y, Sonoda J, Oyama N, Yu XN, Kubo C, Koga Y (2004) Postnatal microbial colonization programs the hypothalamic–pituitary–adrenal system for stress response in mice. J Physiol 558(1):263–275

Tanida M, Yamano T, Maeda K, Okumura N, Fukushima Y, Nagai K (2005) Effects of intraduodenal injection of *Lactobacillus johnsonii* La1 on renal sympathetic nerve activity and blood pressure in urethane-anesthetized rats. Neurosci Lett 389(2):109–114

Thakur AK, Shakya A, Husain GM, Emerald M, Kumar V (2014) Gut-microbiota and mental health: current and future perspectives. J Pharmacol Clin Toxicol 2:1016–1031

Thomas CM, Hong T, van Pijkeren JP, Hemarajata P, Trinh DV, Hu W, Britton RA, Kalkum M, Versalovic J (2012) Histamine derived from probiotic *Lactobacillus reuteri* suppresses TNF via modulation of PKA and ERK signaling. PLoS One 7(2):e31951

Tsavkelova EA, Botvinko IV, Kudrin VS, Oleskin AV (1999) Detection of neurotransmitter amines in microorganisms with the use of high-performance liquid chromatography. Dokl Biochem 372(1–6):115–117

Tuček S (1982) The synthesis of acetylcholine in skeletal muscles of the rat. J Physiol 322(1):53–69

Turnbaugh PJ, Hamady M, Yatsunenko T, Cantarel BL, Duncan A, Ley RE, Sogin ML, Jones WJ, Roe BA, Affourtit JP, Egholm M (2009) A core gut microbiome in obese and lean twins. Nature 457(7228):480–484

Verdu EF, Bercik P, Verma-Gandhu M, Huang XX, Blennerhassett P, Jackson W, Mao Y, Wang L, Rochat F, Collins SM (2006) Specific probiotic therapy attenuates antibiotic induced visceral hypersensitivity in mice. Gut 55(2):182–190

Wang Y, Kasper LH (2014) The role of microbiome in central nervous system disorders. Brain Behav Immun 38:1–12

Wang X, Wang BR, Zhang XJ, Xu Z, Ding YQ, Ju G (2002) Evidences for vagus nerve in maintenance of immune balance and transmission of immune information from gut to brain in STM-infected rats. World J Gastroenterol 8(3):540–545

Wang Q, Xin Y, Zhang F, Feng Z, Fu J, Luo L, Yin Z (2011) Enhanced γ-aminobutyric acid-forming activity of recombinant glutamate decarboxylase (gadA) from *Escherichia coli*. World J Microbiol Biotechnol 27(3):693–700

Weinstock JV, Elliott DE (2009) Helminths and the IBD hygiene hypothesis. Inflamm Bowel Dis 15(1):128–133

Wessler I, Kilbinger H, Bittinger F, Unger R, Kirkpatrick CJ (2003) The non-neuronal cholinergic system in humans: expression, function and pathophysiology. Life Sci 72(18):2055–2061

White T (1959) Formation and catabolism of histamine in brain tissue in vitro. J Physiol 149(1):34–42

Whitman WB, Coleman DC, Wiebe WJ (1998) Prokaryotes: the unseen majority. Proc Natl Acad Sci U S A 95(12):6578–6583

Young SN, Anderson GM, Gauthier S, Purdy WC (1980) The origin of indole acetic acid and indole propionic acid in rat and human cerebrospinal fluid. J Neurochem 34(5):1087–1092

Zaneveld J, Turnbaugh PJ, Lozupone C, Ley RE, Hamady M, Gordon JI, Knight R (2008) Host-bacterial coevolution and the search for new drug targets. Curr Opin Chem Biol 12(1):109–114

Part III

Plant Microbiome

Metagenome of Rhizosphere and Endophytic Ecosystem

9

P.U. Krishnaraj and Malik Ahmed Pasha

Abstract

Microbes inhabit almost every corner of earth with highest known to be in soil. The diversity and activity of microbes have been found to be high in plants (as endophytes) and the surrounding ecosystem (rhizosphere). Within plant, the roots seem to harbour more diverse microbes than any other part. Among other parameters, plant species and its age drive the define distinctness of microbial communities within plant and in soil. In addition, soil type and its management also influence the microbial diversity. These microbes produce various compounds that influence plant growth, productivity, susceptibility and resistance to biotic and abiotic stresses. Hence, structural and functional analysis of microbial communities associated with plant is very much essential to understand the various processes that define their functions. Culture-based analysis of all existing microbes is difficult largely due to limited knowledge of their culture requirements in laboratory. Molecular finger printing and high throughput sequencing of DNA isolated directly from the niches have proved to be an effective alternative to culture-based analysis. In recent past, much of the metagenomics work has been dedicated to study soil microbes, but limited information is available regarding the endophytic microbes. The information obtained by culture-independent analysis of soil microbes can help understand interaction between plant, soil and resident microbes. This is expected to pave the way for effective modulation of soil biological processes by rhizosphere engineering. Endophytic bacteria have been shown to have several beneficial effects on their host plant. Hence, further improvement in crop protection, production and soil health can be achieved by modulating plant's own processes through amending rhizospheric and endophytic microbes.

P.U. Krishnaraj (✉) • M.A. Pasha
Department of Agricultural Microbiology, College of Agriculture, Vijayapura University of Agricultural Sciences, Dharwad, Karnataka, India
e-mail: krishnarajpu@gmail.com

R.P. Singh et al. (eds.), *Understanding Host-Microbiome Interactions - An Omics Approach*, DOI 10.1007/978-981-10-5050-3_9

Keywords
Rhizosphere • Endosphere • Metagenomics • Sequencing • Plant production • Plant protection

9.1 Introduction

The soil surrounding plant is the most biodiverse environment on the earth followed by internal spaces of a plant. A gram of soil contains as many as 10^{10}–10^{11} bacterial cells (Horner-Devine et al. 2003) belonging to 10^3–10^4 species (Curtis et al. 2002), whereas up to 10^9 bacterial cells are estimated in a gram of plant tissue (Chi et al. 2005). Microbes may exist as free-living organisms in soil or attached to the surface of root or phyllosphere and may establish symbiotic relations with plant (Smith and Goodman 1999). The association of these microbes plays an important role in number of vital ecosystem processes such as nutrient cycling, decomposition of organic waste, detoxification of toxic chemicals and heavy metals and carbon sequestering. The interaction between soil, microbe and plant is highly complex and dynamic. Soil acts as medium and provides nutrients for plants to grow which in turn inhabit diverse microbial species. In turn, the physical and chemical properties of soil are influenced by the microbial activities and plant-root exudates (Taylor et al. 2009). However, plants provide nutrients and protect some of these microbes (endophytes) from biotic and abiotic stresses.

The structure and function of microbial species in the soil and within plant affect the crop productivity (Bever et al. 2013). Microorganisms can provide beneficial effects on plants directly by enhancing crop nutrition or indirectly by reducing damage caused by pathogens or environmental stress. Another advantage of endophytes to the plants is the protection they confer against herbivory by producing toxic alkaloids (Schardl et al. 2004). The phenotype of most plants in nature is now known to be the effect of concerted and highly co regulated expression of both plant and microbial genes. In fact, significant parts of the plant phenotype represent the extended phenotype of one or several microorganisms. A better knowledge on the interplay of soil, plants and microbes is currently needed and requires in-depth studies on the role that plant-associated microbes play in the ecosystems. This chapter recalls the meaning and definition of rhizosphere and endosphere, diversity of microbes in and around the plants, how the microbes enter into the plant system, methods to study this population of microbes, importance of culture-independent analysis of rhizosphere and endosphere, the factors affecting these microbes in their respective niche, role of yet uncultured microbes in plant production and protection and finally the future prospects and applications of these microbes through metagenomic approach.

9.2 Meaning and Definition of Rhizosphere and Endophytic Microbes

The term rhizosphere was first coined by German agronomist and plant physiologist Lorenz Hiltner in 1904 to describe the plant-root interface. The word rhizosphere originates in part from the Greek word 'rhiza', meaning root (Hiltner 1904 and Hartmann et al. 2008). According to Hiltner, rhizosphere is the area around a plant root which is inhabited by a unique population of microbes structured by the compounds secreted from plant roots. Presently, the rhizosphere definition has been refined to include three zones which are defined based on their relative proximity to the root.

Endorhizosphere includes portions of the cortex and endodermis in which microbes and cations can occupy the 'free space' between cells (apoplastic space).

Rhizoplane is the medial zone directly adjacent to the root including the root epidermis and mucilage.

Ectorhizosphere is the outermost zone which extends from the rhizoplane out into the bulk soil.

Because of the complexity and diversity of plant-root system, there is no definite size and shape which can be used to define rhizosphere. But the region can be defined based on gradient in chemical, biological and physical properties which change both radially and longitudinally along the roots.

The interaction between plant and microbe is not only limited to root surface but also to the root tissue. They enter through natural openings, spread throughout the plant system, multiply and colonize. The microbes residing within the plant are called endophytes. Endophytes were first described by the German botanist Heinrich Friedrich Link in 1809 (Link 1809). He used the termed *Entophytae* to describe a distinct group of partly parasitic fungi living in plants. Orlando Petrini in 1991 redefined endophytes as 'all organisms inhabiting plant organs that at some time in their life cycle can colonize internal plant tissues without causing apparent any harm to their host' (Petrini 1991). Since then, endophytes have been defined in various forms (Stone et al. 2000). All the definitions related to the microorganisms which invade tissues of living plants without causing disease remain there for short period or throughout their life.

Currently endophytes are described as any microbe that can be isolated from asymptomatic plant tissue (Hardoim et al. 2015). This definition includes beneficial, commensal, pathogens and neutral microorganisms. While for some others, microbes with generally neutral or positive effects on host fitness represent the 'classical' endophytes (Partida-Martínez and Heil 2011). Endophytic microbes are a class of endosymbiotic microorganisms that live in internal plant tissues of apparently healthy host plants (Schulz and Boyle 2006). Putative endophyte has been recommended for the microbes which are not microscopically validated. True endophytes should reinfect the disinfected seedlings.

9.3 The Process of Colonization

Majority of endophytes are presumed to be facultative endophytes; they remain inside the plant system during some stages of their life cycle and spend the remaining life outside the plant system (Hardoim et al. 2008). The microbes which reside inside the plant are naturally selected or move in from a pool of microbes. The strains of *Rhizobium etli* isolated from inside maize stem (endophytes) are the selected subset from the pool of *Rhizobium etli* found in rhizosphere or roots (Rosenblueth and Martinez-Romero 2004). It was also observed that some amount of competition do exist among the endophytes; aggressive colonizers displace other endophytes (Verma et al. 2004). The sequence of events leading to colonization of microbes in and around the plant roots is similar, at least in the early stages. Bacterial colonization in and around plant is a multistep complex process. This includes (a) migration towards root, (b) attachment on root surface, (c) distribution along root, (d) growth and survival of the population and (e) entry into root and formation of microcolonies (Dudeja and Giri 2014).

The interaction of microbes with plant and their colonization route depends on the strain of microbe. Some microbes make their entry inside the plant through seeds (vertical transmission), while others have a mechanism to infect and colonize the plant system (horizontal transmission). The endophytes enter into plant through root by colonizing lateral root hairs (Mercado-Blanco and Prieto 2012). Some endophytes also penetrate plants through flowers and fruits via colonization of the anthosphere and carposphere (Compant et al. 2010 and Compant et al. 2011). Plants secrete various carbohydrates, amino acids and other compounds through root as root exudates or root mucilage (Philippot et al. 2013). These compounds act as chemoattractants which attract a particular group of microbes and also serve as nutritional source to these microbes. In some cases, the leaf and stem surfaces also secrete exudates which attract selected microbes (Compant et al. 2010). The composition of rhizo-deposits (root exudates and root mucilage) depends on the plant genotype, its age and physiological status and also to some extent agronomic practices (Haichar et al. 2008). The profile of these compounds is unique for a plant type which attracts few selected microbes. Some microbe use organs like flagella for their movement towards their host plant. The attracted microbes use these exudates as nutrients for their growth and multiplication and colonize around the plant tissue (Lutenberg and Kamilova 2009). Due to spatial variation in exudation process, some microbes colonize some areas better than the other, leading to spatial variation in colonization (Gamalero et al. 2004).

Endorhizosphere microbes often colonize and penetrate the epidermis of lateral root emergence, below the root hair zone and in root cracks (Zakria et al. 2007). While most of the endophytes are active colonizer, which penetrate the plant system by infection process, some are passive colonizers and make their entry through wounds or natural openings in plants (Gaiero et al. 2013). Invasion of microbes on root surface is a coordinated process and involves multiple signalling pathways and reciprocal signalling between plants and endophytes and between endophytes

(Rudrappa et al. 2008). The well-studied signalling mechanism between microbes which is required for colonization is cell density-based regulation of microbial behaviour, called quorum sensing (Teplitski 2000). The microbial cells produce and secrete low molecular weight molecules called autoinducer, which helps in coordinated expression of microbial genes to augment colonization process (Elasri et al. 2001). Most commonly described low molecular weight quorum sensing signal molecule in Gram-negative bacteria is N-acyl homoserine lactones (AHLs) (Elasri et al. 2001).

The colonization process is also affected by plant-specific compounds like flavonoids. The flavonoids are unique to a host plant that specifically stimulates or inhibits AHL-dependent quorum sensing responses through interactions with microbial AHL receptors (Gao and Teplitski 2003). Giri and Dudeja (2013) did not find any specificity among 11 endophytic bacteria with four hosts at any stage of root colonization. However, some authors believe that the colonization process is not general; only few selected species of microbes can colonize a particular plant species. They believe that this specificity is regulated by interaction between microbe-specific autoinducer and plant-specific flavonoids. The preferential colonization of a host plant is mediated by expression of genes required for colonization by microbes upon the release of specific flavonoids by plant (Bais et al. 2004).

Upon infection and entry into host root system, the endophytes need to be equipped with internal spreading mechanism for successful establishment and colonization inside the plant. Like infection, spreading of microbes inside the plant system is also mediated by both active and passive mechanism. Some bacteria may spread passively through the disrupted endodermal layer (Gregory 2006). But the active mechanism requires cell wall-degrading enzymes and type IV pili (James et al. 2002). Some endophytes move to aerial parts of plant by entering the vascular tissues through moderate expression of degradative enzymes such as glucanase, pectinases and cellulases and spreading systemically (Johnston-Monje and Raizada 2011). Endoglucanase helps in loosing larger cellulose fibres which helps in entry of microbe into the plant. *Azoarcus* strain lacking endoglucanase was not effective in colonizing the rice plants. Reinhold-Hurek and Hurek (2011) identified an exoglucanase having cellobiohydrolase and β-glucosidase activity, playing a key role in colonization process of *Azoarcus* sp. BH72. It has been reported that *Klebsiella* strains produce pectate lyase to infect and colonize plant tissue (Kovtunovych et al. 1999). *Burkholderia* sp. infects *Vitis vinifera* by producing cell wall-degrading enzymes like endoglucanase and polygalacturonase (Compant et al. 2005). The migration of *Xylella* spp. in the host vascular system is mediated by the use of type IV pili (Meng et al. 2005). *Azoarcus* sp. type IV pili are involved in the adherence to plant surfaces which is an essential step towards endophytic colonization (Dorr et al. 1998). Type IV pili of *Azoarcus* sp. BH72 found to be essential for biofilm formation on the surface of rice roots and subsequently for endophytic spreading into shoots (Reinhold-Hurek et al. 2006).

Successful colonization by endophytes depends on many factors like plant genotype, tissue type, colonizing species and biotic and abiotic conditions.

Endophytes play an active role in colonization. Difference in colonizing capabilities in different strains of *Rhizobium etli* was observed by Rosenblueth and Martinez-Romero (2004). Two strains of *Klebsiella* differ significantly in their colonization capacity in *Medicago sativa*, *Medicago truncatula*, *Arabidopsis thaliana*, *Triticum aestivum* and *Oryza sativa* plant hosts. Bacterial mutants impaired in production of necessary secretary proteins failed to colonize plant host. Environmental factors like lack of nutrients, desiccation and UV generally reduce colonization (Compant et al. 2010). The plant hosts also vary in their ability to be colonized endophytically by the same bacterium. Molecular analysis showed that a plant defence response also limits bacterial populations inside plants (Rosenblueth and Martínez-Romero 2006).

9.4 How to Study These Microbes?

Soil microbes consist of a large portion of the genetic diversity on earth (Whitman et al. 1998). Standard traditional techniques of soil or rhizosphere microbial ecology involve serial dilution of the sample, isolation and purification of microbes by plating on nutrient-rich synthetic media in laboratory condition (Kirk et al. 2004), followed by a series of morphological, physiological and biochemical tests to identify species and their function (Maron et al. 2011). Similarly, the microbes inhabiting the plant system are studied by cultivation-dependent methods. The plant tissue is subjected to a series of processes to remove the surface inhabiting microbes and then culturing the endophytes on nutrient-rich media for further characterization. Cultivation-based techniques generally include (1) thorough washing of the plant tissue to remove adhering soil particles, debris and major epiphytes, (2) surface sterilization of plant tissue to kill microbes present on the host surface, (3) isolation and purification of microbes on nutrient supplemented agar media and (4) structural and functional characterization of the isolated microbes based on morphological, physiological and biochemical characteristics in cultures (Sun et al. 2011). The second step and to some extent first step is required for the isolation of endophytic microbes.

Majority of the microbes in the environment are not readily culturable in laboratory condition. It has been said that only 0.1 to 1% of actual bacteria can be cultured on common media under standard conditions (Torsvik and Ovreas 2002). Specific or complex nutrient requirement, interdependence and requirement of signalling molecules are the reasons for inability to culture larger fraction of microbial species in laboratory condition. Culturing microbes in isolation also did not give information about possible interaction between microorganism and their habitat. The fast growers will exhaust nutrients and occupy the space much ahead of slow grower and are overrepresented. Because of the inherent limitations associated with culture-based analysis, the microbial ecologists are moving to culture-independent analysis of environmental sample.

9.5 Metagenomic Approach

A modern method of culture-independent genomic analysis known as metagenomics (in Greek, meta means transcendent) has been developed to access the information about both the culturable and unculturable microorganisms. Metagenomics can provide a cultivation-independent assessment of the largely untouched genetic reservoir of environmental microbial communities. The term 'metagenomics' was first used by Handelsman et al. (1998) for isolating DNA directly from soil organism, cloning it into culturable organism and screening the resultant clones for the production of new chemicals or compounds. Metagenomics can be defined as 'the application of modern genomics techniques to the study of microbial communities directly in their natural environments, bypassing the need for isolation and laboratory cultivation of individual species' (Chen and Pachter 2005). Metagenomics is the culture-independent analysis of a mixture of microbial genomes (Schloss and Handelsman 2003; Patrick and Handelsman 2005). Metagenomics is also known by other names such as environmental genomics or community genomics or microbial ecogenomics. Nucleic acid (metagenomic)-based analysis of environmental sample helps to identify the phylogenetic relationship and functional potential of microbial community.

The general approach in metagenomic studies relies on the extraction of DNA directly from environmental samples, amplification of phylogenetic marker gene (generally 16S rDNA for bacteria and ITS for fungi) and sequencing or molecular fingerprinting to categorize their phylogenetic relationship. This approach will help in understanding the microbial species content, their relative proportion, and dynamics in a specific environment or niche. The other approach involves cloning of isolated metagenomic DNA into large carrying capacity vectors (Fosmid, BAC and YAC) and its expression in laboratory-culturable species (generally *E. coli*). The metabolic potential embedded in unculturable microbes can be exploited by this approach. The earlier approach is called structural metagenomics and the later one functional metagenomics. There is one more approach wherein the whole metagenomic DNA is sequenced using next-generation sequencing technology to know the species diversity and the protein coding genes to understand the potential functional capability of the niche.

For successful metagenomic studies, good-quality and quantity DNA is prerequisite. Isolation of DNA from environmental sample (especially soil) is very challenging as it harbours diverse microbes varying in their cell wall composition. Soil DNA isolation protocols are broadly classified into direct and indirect lysis. Indirect lysis involves separation of cells from soil matrix followed by cell lysis and DNA extraction (Holben et al. 1988). Cell lysis in the soil matrix followed by separation and purification of DNA from matrix and cell debris forms the direct lysis method (Ogram et al. 1987). Releasing of microbial cells from soil particle, cell lysis and purification of soil DNA from contaminants like humic substances are critical and challenging. There are several methods to extract DNA from bulk or rhizosphere soil, but none of them can be accepted as standard for different soil types and among scientific community (Schneegurt et al. 2003). The soil DNA isolation protocol needs to be optimized at each step. The various steps involved in the isolation of soil DNA may be read in Ogram et al. (1987), Tsai and Olson (1991), Lakay et al. (2007), Mahmoudi and Slate (2011) and Islam et al. (2012).

Similarly, isolation of endophytic microbial community DNA also faces some technical challenges. Separation of endophytic microbial DNA from plant nuclei, plastid and mitochondrial DNA demands sound knowledge and technical skills. It is difficult to isolate and sequence only bacterial community DNA as plant cells are much more abundant than bacterial cells in a target tissue. Enrichment of bacterial DNA prior to DNA amplification should be followed to reduce the amount of interference from host plant DNA (Govindasamy et al. 2014). To exclude interference of host plant DNA, Jiao et al. (2006) enriched bacterial endophytes by hydrolysis of the plant cell walls, followed by differential centrifugation. Ikeda et al. (2009) used series of differential centrifugation steps followed by density gradient centrifugation to enrich the bacterial DNA in stem and leaves of soybean and rice. The technique developed by Nikolic et al. (2011) involved overnight shaking of the small pieces of tubers in sodium chloride solution for extraction of endophytes from internal tissues of potato tubers. Though these methods reduced the contamination of plant DNA from endophytic DNA, the representation of rare endophytes was also reduced.

Once a good-quality and quantity DNA is obtained, there are several techniques to study microbial community structure, composition and their functional capabilities in rhizosphere and endosphere. Detailed discussion about each of these techniques is out of the scope of this chapter. The principle, applicability, advantages and limitations of each method for both structural and functional metagenomics including a range of statistical tools to estimate diversity and dynamics of microbial population have been described (Krishnaraj and Pasha, 2017). Due to improvement in sequencing technology and reduction in cost, next-generation sequencing is most preferred method among molecular ecologists to study environmental samples. This method generates huge data in a shortest possible time and can effectively provide deep insights into complex microbial communities in ecological niches (Fakruddin and Mannan 2012). But it demands technically sound knowledge of bioinformatics to process the sequences data, analyze it and interpret the results in a scientific manner. In functional metagenomics, cloning of metagenomic DNA in vectors and expression in *E. coli* is still the preferable method. But the methods for screening library to isolate the desired clone have been revived. Though there are different methods like stable isotope probing and substrate-induced gene expression, the enrichment of sample before construction of metagenomic library is most preferable one. This method does not require any specialized technique or instruments but increases the chance of isolating desirable clone from complex habitat like soil (Knietsch et al. 2003).

9.6 Diversity of Microbes in and Around the Plants

The number of species and their relative proportion in a sample encompasses microbial diversity. Due to the intimate association of plant, the microbial diversity is higher in the soil close to the plant compare to the bulk soil (Nunes da Rocha et al. 2009). The bacterial diversity in rhizosphere is of 4–7 order magnitude more than in bulk soil (Foster 1988). Next to the soil, plants constitute vast and diverse niches for endophytic organisms. Thinking of microbe-free plant is close to a myth as such plant would hardly be capable of surviving under natural conditions (Partida-Martinez and Heil 2011). In general, the diversity of microbes colonizing inside the plant is lower than they are observed around the plant near root zone (Rosenblueth and Martinez-Romero 2004). The diversity of endophytic bacteria is highly variable in different parts or tissues of plant and is found to be from hundreds to 9×10^9 per gram of plant tissue (Chi et al. 2005). Endophytes colonize almost every part of plant like apoplast intercellular spaces of the cell walls, xylem vessels of plant roots, stems, leaves, flowers, fruits and seeds (Hardoim et al. 2015). Generally higher diversity of endophytes are observed in roots and other below-ground tissues of plant as compared to above-ground tissues indicating endophytes enter the plant system through roots and migrates towards aerial parts (Rosenblueth and Martinez-Romero 2004 and Chi et al. 2005). While most of the endophytes are found in rhizosphere also, some endophytes are very rare or totally absent in soil (Reinhold-Hurek and Hurek 1998). Species belonging to *Clostridia* were detected in surface-sterilized leaves, stem and roots of a grass but not found in surrounding soil (Miyamoto et al. 2004). Janssen (2006) surveyed 32 libraries of 16S rDNA from published articles of different bulk soils to assess the bacterial diversity. He found 32 bacterial phyla of which Proteobacteria, Acidobacteria, Actinobacteria, Verrucomicrobia, Bacteroidetes, Chloroflexi, Planctomycetes, Gemmatimonadetes and Firmicutes are dominant. Members of these nine phyla make up an average of 92% of soil libraries. Culture-independent analysis of bulk soil and soybean rhizosphere soil identified 17 bacterial phyla, but the relative proportion of *Proteobacteria* was increased, while Acidobacteria and Firmicutes were decreased in rhizosphere soil during soybean growth (Sugiyama et al. 2014). At subgroup level, Alphaproteobacteria were the most abundant, followed by *Beta-*, *Gamma-* and Deltaproteobacteria in different soil types. At genus level, the three most abundant are Burkholderia (class Betaproteobacteria), *Pseudomonas* (class Gammaproteobacteria) and *Chitinophaga* (class Sphingobacteria), which constitute 2.7, 1.6 and 1.0% of all the sequences, respectively (Janssen. 2006).

The endophyte communities isolated so far mainly comprised of species belong to phylum *Proteobacteria* followed by Firmicutes, Actinobacteria and Bacteroides (Reinhold-Hurek and Hurek 2011). The presence of cellulose attracts more of *Proteobacteria* in endosphere (Bulgarelli et al. 2012). At subgroup level, most of the endophytic species belongs to the α-, β- and γ-Proteobacteria subgroups (Kuklinsky-Sobral et al. 2004). The γ-Proteobacteria group is the most diverse and dominant. Streptomycetaceae was found to be abundant in the root endosphere among Actinobacteria phyla (Bulgarelli et al. 2012). *Bacillus* and *Pseudomonas* species are frequently observed species in most of the crop plants (Souza et al. 2013). The analysis based on the 16S rDNA sequence data of endophytes deposited in the nucleotide database repository (as of 1 March 2014) revealed presence of 23 recognized and a candidate phyla of prokaryotes (2 from Archaea and 21 from Bacteria) in endophytes (Hardoim et al. 2015). However, more than 96% of the endophytes belong to only four bacterial phyla. The highest number is represented by Proteobacteria (54%) followed by Actinobacteria (20%), Firmicutes (16%) and Bacteroidetes (6%). Other phyla detected were Planctomycetes, Verrucomicrobia and Acidobacteria. Among them, most of the prokaryotic endophytes belong to Gammaproteobacteria, including 56 recognized and seven unidentified genera as well as the 'Candidatus Portiera' genus (Hardoim et al. 2015). No Archaea has been found to associate tightly with plants as endophytes. Although there are metagenomic-based studies (Chelius and Triplett 2001 and Sun et al. 2008) on few Archaea associated with roots, they have found to occur on the surface of older roots and not been detected convincingly in internal plant tissues (Chelius and Triplett 2001).

Fungi are next most dominant microbes in soil after bacteria. Earlier methods of fungal diversity were not sufficient to identify fungi at lower taxonomic level. Classification of fungi based on 18S rDNA or internal transcribed spacer (ITS) sequence has partially solved this problem and allowed for better species classification (Gehlot et al. 2012). Nine fungal species belonging to Dothideomycetes, Eurotiomycetes and Sordariomycetes classes of the phylum Ascomycota have been isolated from the roots of plants in volcanic islands of Korean Peninsula. These species belong to the genera *Alternaria*, *Cladosporium*, *Exserohilum*, *Phoma*, *Pyrenochaeta*, *Aspergillus*, *Neosartorya*, *Penicillium* and *Fusarium* (Nam et al. 2015). Fungal diversity in soil and roots of three different poplar plantations was studied using metagenomic-based 454 pyrosequencing of ITS region (Danielsen et al., 2012). Fungal species richness in the soil was found to be higher than in roots. They also found that soil saprophytic, pathogenic and endophytic fungi are dominant in soil, whereas 87% of root endophytes belong to ectomycorrhizal fungi.

A large diversity of fungal species was also found in plant endophytes. Thirty three, 60, 18 and 58 taxa of endophytic fungi were discovered in leaves of European aspen (*Populus tremula*) (Albrectsen et al., 2010), leaves and twigs of *Quercus ilex* (Fisher et al. 1994), leaves of switch grass (*Panicum virgatum* L. Ghimire et al. 2011) and leaves of wild rubber trees (*Hevea brasiliensis*) (Gazis and Chaverri 2010), respectively. On an average 5.73 ± 1.94 species were found to co-occur in leaf discs of ca. 3cm^2. The diversity of microbes can be different even with different plant parts or tissues of oat (DeAngelis et al. 2009). *Trichoderma*, *Penicillium*, *Paecilomyces*, *Pestalotiopsis* and *Basidiomycota* were found to be dominant fungal groups in wild rubber, whereas *Trichoderma*, *Pestalotiopsis*, *Colletotrichum*, Xylariales and Basidiomycota were most common in planted trees.

Sequence analysis of full-length ITS region of endophytic species deposited in National Center for Biotechnology Information (NCBI) nucleotide database revealed that most of the endophytes belong to Glomeromycota (40%), Ascomycota (31%), Basidiomycota (20%), unidentified phyla (8%) and, to a lesser extent, *Zygomycota* (0.1%) (Hardoim et al. 2015). Most (around 39%) of the eukaryotic endophytes can be assigned to the class Glomeromycetes. Among the Ascomycota, a large number of endophytes are identified in the class Dothideomycetes (15%). Members belonging to the class Agaricomycetes contain a large number of endophytes (18%) among the Basidiomycota (Hardoim et al. 2015).

9.7 Factors Affecting Microbes in and Around Plant

The factors affecting microbial community composition and their diversity in rhizosphere belong to edaphic factors (like mineral nutrients, ionic composition, pH, oxidation-reduction potential, carbon and energy sources), environmental factors (like soil moisture, temperature, pressure, air composition and electromagnetic radiation), vegetation type and land management practices. Like rhizosphere microbes, endophytes are also susceptible to biotic and abiotic factors, but they are more protected as they are not directly exposed to outside environment (Hallmann et al. 1997; Kuklinsky-Sobral et al. 2004). The composition and diversity of endophytic microbial community is affected by plant type, soil type, environmental conditions, interaction between host plant and endophyte, interaction among endophytes and also the agricultural practices (Ryan et al. 2008).

9.7.1 Environment-Related Factors

The metabolism and the survival of microbes in and around the plants are affected largely by environmental conditions and soil parameters. A fraction of the microbial community dies during each drying-and-wetting cycle (Kieft et al. 1987). As a consequence, the composition of soil microbial communities fluctuates. Distribution of microbial cells in soil matrix is not uniform (Gonod et al. 2003) and is dependent on the localized factors like pH, moisture content, nutrient availability, predation, parasitism and competition (Rousk et al. 2010). In upper layer of cropping soil where nutrient distribution is uniform, the distribution of bacterial cell is random. However, in deeper layers where nutrient availability is limiting factor, the bacterial cells are localized to soil pores (Nunan et al. 2003). Changes in environmental conditions, such as temperature, humidity, light intensity, geographic location and vegetation, significantly affect the distribution pattern of endophytic fungi (Suryanarayanan et al. 2005; Song et al. 2007).

The effect of nutrient availability and soil pH on soil microbial community composition and their functional diversity was analyzed by using combination of shotgun metagenom sequencing and biolog (Uroz et al. 2013). Soil resource availability impacts the functional diversity and to a lesser extent the taxonomic diversity of the bacterial communities. The microbial communities inhabiting the organic horizon are well adapted to degrade easily accessible carbon substrates such as soluble carbohydrates or polysaccharides and, on the contrary, those inhabiting the mineral horizon are better adapted to degrade amino acid derivatives and proteins resulting from the leaching of by-products of organic matter and litter decomposition. Xu et al. (2014) used computational tools in bioinformatics to understand the complex relationships between structural diversity of microbes and their habitat. They analyzed 33 metagenome samples from publicly available five soil habitats (i.e. grassland, forest soil, desert, Arctic soil and mangrove sediment). Microbiota from same soil habitat was more similar to each other. Comparison within habitat indicated, grassland soils are more similar, and desert soils are more dissimilar in bacterial community composition. The difference in desert community composition was due to different geographic location; three samples were from hot desert and one from cold desert. Between-habitat comparison of microbial community composition indicates the forest and grassland soils are more similar and the mangroves and desert soil are more dissimilar.

The ecological niche consisting of environmental factors and edaphic factors like soil type, pH, nutrition status and other physical properties determines the production and secretion of type and amount of secondary metabolites by host plant. The plants in the tropical region of high temperature and high mean annual rainfall will produce more nutrients and diverse secondary metabolites which will attract more endophytes for the colonization, reproduction and dissemination (Wu et al. 2013). In contrast, the plants in the temperate region of low light and low rainfall conditions will face inappropriate rates of photosynthesis, respiration, oxygen concentration and pH value and hence attract only those species of microbes which can sustain in such poor conditions (Jiang et al. 2010). Variation in nitrate level also influences the soil and root-associated microbial communities. Soil nitrate is involved in shaping root architecture; the presence of exogenous nitrate induces the plant to produce lateral roots. It has also been proposed that root architecture can affect root microbiota and vice versa (Cheng et al. 2016). A change in soil microbial composition and diversity is strongly correlated with changes in soil pH. The population of bacteria increased with increase in pH, whereas fungal population remains significantly unaltered (Rousk et al. 2010).

The effect of soil type is found to be highest among the environmental factors on the community composition of endophytes. The type of species and their relative distribution were found to be similar in different host in the same locality or region. A high degree of similarity in terms of endophytic species taxonomy and their distribution was observed in different host plants grown in the same regions (D'Amico et al. 2008). Conversely, the endophytic microbial profile was found to be dissimilar even in the same host plant but grown in different regions (Jiang et al. 2010). The soil type determines the endophytic population in wheat to a large extent (Conn and Franco 2004). Rasche et al. (2006) investigated the effect of soil type and plant type on bacterial endophytes of potato. They used different varieties, including genetically modified plants, and were grown in contrasting soil types in green house. The plants were sampled at different growth stages to study the effect of plant growth stage-specific compounds. Molecular community analysis showed that the soil type was the most important driver of bacterial community composition, followed by the plant developmental stage. Similar study was conducted recently by Lundberg et al. (2012). They studied root endophytes of eight *Arabidopsis* accessions at different growth stages cultivated in two different soil types using next-generation sequencing technologies. The results showed that the effect of soil type is more prominent than plant genotype on endophytic microbial structure.

9.7.2 Host-Related Factors

The experiments of Grayston et al. (1998) and Miethling et al. (2000) showed that the plant species have greater effect than soil type on rhizosphere microbial community composition. Microbial activity in soil is greatly influenced by plant roots (Bais et al. 2006). Soil microbes use sugars, amino acids and organic acids excreted by plant root as their substrate and improve their biomass and activity near the root zone. The composition of root exudates depends on plant species and cultivar, developmental stage, plant growth substrate, and stress factors (Uren 2000). Smalla et al. (2001) using cultivation-independent methods showed that roots of different species of plant are colonized by their own bacterial communities. And this difference does not exist only at species level but also at cultivar level (Germida and Siciliano 2001).

Bacterial community composition associated with the roots of *Medicago truncatula* changes with the developmental stages of crop (Mougel et al. 2006). The effect of plant growth stage-specific root exudates on rhizosphere was analyzed by Hussain et al. (2012). They collected the rhizosphere samples at tillering, grain filling and ripening stage of rice and compared it with the bulk soil. The highest diversity of bacteria and fungi was observed at grain filling stage; however, no such difference was observed in bulk soil. Above-ground diversity (plant) influences the below-ground microbial community and structure (He et al. 2010). They used metagenomic-based DNA microarray technology to analyze the rhizosphere sample from monoculture and mixed cropping system. Their result indicated that monocultures of plants lead to a decrease in the below-ground microbial community diversity. Similarly, metagenomic-based pyrosequencing analysis of potato rhizosphere soil indicates that monoculturing of potato for 7 years gradually declined the bacterial riches and diversity (Liu et al., 2014). Moreover, the population and incidence of soil-borne plant pathogen increased gradually due to monoculture.

The genetic background of the host plant, its age and physiological status considerably affects the population structure of endophytic microbes. Bacterial endophytes associated with leaves of distantly related plant species grown under natural conditions were studied by Ding et al. (2013). They found that the plant type is the major factor determining the endophytic community composition followed by sampling dates and sampling locations. Similar results were observed from other related studies suggesting that the distribution of certain endophytic microbial populations is only restricted to particular host plant species (or families) and particular genetic background (genotypes) of a species (D'Amico et al. 2008). Generally, genetically related plants seem to host more similar bacterial endophyte communities (Ferrando et al. 2012). The plants secrete unique compounds in the form of root exudates which are likely to attract distinct microorganisms by each plant type, and hence plant-specific endophytic community composition is observed (Bodenhausen et al. 2013). Some exudates are known to have negative effect on endophytic microbes.

Fitness of the endophytic community largely depends on the fitness of the host plant suggesting the role of plant physiology in the colonization and distribution of endophytic microbes in the host plants (Saikkonen et al. 2004). Hallmann et al. (1997) also suggested that the changes in plant physiology can lead to the development of a distinct endophytic population. The effect of fertilizer on endophytes has been studied in rice wherein rapid change in nitrogen-fixing population was observed within 15 days after nitrogen fertilization (Tan et al. 2003). Increasing the dose of nitrogen fertilizers reduced the sucrose content in sugarcane which in turn affected the colonization capacity of *Gluconacetobacter diazotrophicus* (Fuentes-Ramírez et al. 1999). Furthermore, age and tissues of host plants also influence species composition of the endophytic community (Sieber. 2007). Rodrigues (1994) observed different endophytic species in different tissues of host plant. Such a specific distribution of endophytic species might be related to their ability to utilize specific substrates produced and secreted by specific tissues of plants at different stages (Rodrigues 1994). This indicates that the metabolites or compounds produced by different tissues of plant at different developmental stages determine the type and intensity of plant colonization by endophytic microbes.

9.7.3 Microbe-Related Factors

The process of plant colonization is regulated by certain traits possessed by microbes which are collectively known as colonization traits. Communication between plant and microbe plays a key role during colonization process (Rosenblueth and Martinez-Romero 2006). The process of colonization often starts with recognition of specific compounds in the root exudates by the microbes (De Weert et al. 2002). Specific microbes respond to these unique exudates by chemotactic movement (Lugtenberg et al. 2001). The type of response may differ in different microbes. The organic acids secreted by tomato roots are major chemoattractant for *Pseudomonas fluorescens* (De Weert et al. 2002), whereas in case of rice the carbohydrates and amino acids attract *Corynebacterium flavescens* and *Bacillus pumilus* (Bacilio-Jimenez 2003). A mutant strain of *P. fluorescens* which lacked *che*A gene responsible for chemotaxis showed reduced movement towards tomato root exudates and also decreased colonization on roots (De Weert et al. 2002). Mark et al. (2005) studied the response of *P. aeruginosa* strain PAO1 to the root exudates of two cultivars of sugar beet. Interestingly, different set of genes responded to the exudates of two different cultivars of sugar beet.

In addition to root exudates, some bacteria are known to be attracted by hydrated polysaccharides excreted from the root tip, commonly called as root mucilage (Knee et al. 2001). Root mucilage produced by maize is known to attract *Azospirillum* spp. strains (Mandimba et al. 1986); on the other hand, it prevents colonization by *P. fluorescens* strain SBW25 (Humphris et al. 2005). The establishment of bacteria in root is also affected by its surface characteristics. The composition of lipopolysaccharide, particularly the rhamnose content is very important for attachment to and colonization of maize roots by *H. seropedicae* (Balsanelli et al. 2010). The production and secretion of exopolysaccharide was found to be necessary for colonization of rice plant by the nitrogen-fixing endophyte *Gluconacetobacter diazotrophicus* (Carlos et al. 2011). It is observed that the superoxide dismutase and glutathione reductase produced by this bacteria is also crucial for endophytic colonization of rice roots (Sylvia et al. 2013).

Root exudates and root mucilage attract large number of microbes, but only the competent microbe will colonize and infect the host plant. The competent microbe produces secondary metabolites, antibiotics and siderophores to outcompete with other microbes in the rhizosphere (Lugtenberg and Kamilova 2009). Commonly produced secondary metabolites include 2,4-diacetylphloroglucinol (DAPG), hydrogen cyanide, oomycin A, phenazine, pyoluteorin, pyrrolnitrin, thiotropocin, tropolone as well as many others such as cyclic lipopeptides, rhamnolipids, oligomycin A, kanosamine, zwittermicin A, and xanthobaccin (Nielsen et al. 2002, Raaijmakers et al. 2002; de Souza et al. 2003). Genome analysis of *Bacillus amyloliquefaciens* FZB42 (Chen et al., 2007) and *P. fluorescens* Pf-5 (Paulsen et al. 2005) revealed that these bacteria contain large gene clusters responsible for detoxification and production of antibiotics and siderophores, which makes them efficient colonizer of the host plant. In some cases, the presence of locomotory organs like flagella is an added advantage which allows bacteria to get into contact with exudates and root mucilage components (Turnbull et al. 2001). The polysaccharides and other related compounds present in plant cell wall sometimes became obstacles for the active colonization of host plant. To overcome this problem, the endophytic microbe needs to secrete the relevant enzymes, such as protease, xylanase, cellulases and lactase to decompose these secondary metabolites before they penetrate the host plants (Reinhold-Hurek et al. 2006). The cellulolytic and pectinolytic enzymes produced by endophytes are involved in the infection process (Hallmann et al. 1997). Pectate lyase produced by *Klebsiella* strains participates actively during plant colonization (Kovtunovych et al. 1999). *Burkholderia* sp. infects *Vitis vinifera* by producing cell wall-degrading enzymes like endoglucanase and polygalacturonase (Compant et al. 2005). For successful colonization within plant, the endophytes need to be equipped with a set of cell wall-degrading enzymes with confined and localized activity.

9.7.4 Factors Related to Agricultural Practices

Shift in bacterial community structure due to land management practices has been reported by many (Smit et al. 2001; Buckley and Schmidt 2003). Soil disturbance can cause significant modifications of soil habitat, which affects the microbial community. The soil organic matter and microbial biomass decline due to land disturbance (Sparling 1997). Application of chemical fertilizers and pesticides inhibits the activity and function of rhizosphere microbes and also leads to the changes in microbial community diversity and dynamics (Kaur et al. 2008). Application of organics like farm yard manure and green manure increases soil nutrient availability, microbial biomass and their activity in soil (Widmer et al. 2006; Liu et al. 2009). The effect of organic farming on rhizosphere microbes was analyzed by 16S rDNA sequencing of metagenomic DNA (Aparna et al., 2014). Organic farming improved the bacterial richness and diversity, and the species were more evenly distributed among the phyla. However, no significant difference in the bacterial diversity of 5-year-old organic and organically grown cotton rhizosphere was observed by DGGE-based analysis of metagenomic DNA (Pasha et al., 2015). Stark et al. (2007) and Lazcano et al. (2013) showed significant difference in bacterial community composition between short-term organic and mineral-amended soil. However, no significant difference was observed in diversity, species richness and evenness in bacterial community of 4–5-year-old (Wu et al. 2008), more than 5-year-old (Van Diepeningen et al., 2006) and 20-year-old (Liu et al. 2007) organic and conventional farm soils.

The effect of agricultural land use on community structure, composition and metabolic profiles of soil microbiomes was studied by Carbonetto et al. (2014). They used three different production farms; two plots (samples) were under cultivation for at least one century under conventional tillage systems and one with no agricultural land use (no tillage nor cultivation) was recorded for the last 30 years. Agronomical land use and the type of tillage system induced microbiomes to shift their life-history strategies. Microbiomes of cultivated fertilized soils presented tendencies to copiotrophy, while microbiomes of noncultivated homogenous soils appeared to have a more oligotrophic lifestyle. Application of chemical fertilizers and pesticides reduces the overall microbial diversity compared to organic farming. Long-term application of inorganic fertilizers reduced the dependency of root feeding bacteria on root exudates. Souza et al. (2013) studied the effect of tillage on soil microbial structural and functional diversity using pyrosequencing of soil metagenome. Significantly high diversity of microbes was found in soil with conventional tillage. The microbes in conventional tilled soil (ploughing and disking) were found to be associated with residue decomposition and carbon and nitrogen cycling. However, the microbes known to inhabit nutrient-rich environment were found abundant in no tillage soil.

Information on the effect of agricultural practices on endophyte population is limited to several studies. Application of high level of nitrogenous fertilizers to sugarcane reduced the colonization ability of nitrogen-fixing endophytic bacterium *Acetobacter diazotrophicus* (Fuentes-Ramirez et al. 1999). Specific and distinct group of microbes were observed in the endosphere of maize treated with chemical fertilizers and herbicides compared to organic-based fertilizers and herbicides (Seghers et al. 2004). Application of glyphosate changed the endophytic composition of soybean (Kuklinsky-Sobral et al. 2004). These studies did not reveal that the changes in endophytic population are the direct effect of agronomic practices or it is the result of changes in overall soil microbial population upon the fertilizer. The composition of endophytic population is not directly determined by microbial diversity of the rhizosphere but involves other factors related to plant biochemistry. Application of chitin resulted in changes in the soil and root endophytes of cotton; this could be due to enhanced concentrations of chitinase and peroxidise which might have attracted certain type of microbes in rhizosphere and endosphere (Hallman et al. 1999).

Hallmann et al. (1997) showed that application of fertilizers or pesticides to plants in the form of organic amendments also influences the endophytic populations. Repeated application of manures may pose environmental hazards and have the potential to alter the endogenous microbial structure (Soupir et al. 2006). Even the type and quality of organic inputs affect the microbial community composition (Liu et al., 2009). Application of organic manures contaminated with faecal bacteria leads to the colonization and accumulation of human pathogens like *Escherichia coli*, *Vibrio cholerae* strains and *Pseudomonas aeruginosa* in plants (Deering et al. 2012 and Akhtyamova 2013).

9.8 Role of Microbes Associated with Plants: A Metagenomic Approach

Though the association of microbes with plants is known for many years, their role in plant's survival and its production was dissected very recently. The rhizosphere and endophyte colonizing microbes are considered to be part of the plant's pan-genome (Tkacz and Poole 2015). Some others consider these microbes genetic material as plant's second genome (Berg et al. 2014). Some of the phenotypic appearance and functional capabilities of plants are not coded by the plant's own genome but governed by the microbes it is associated with. There are many reports of successful isolation, characterization and functional validation of microbes associated with plants involved in plant growth promotion, protection and quality improvement. Since the interaction between plant and microorganism is very complex and open system, most of the interactions that happen in rhizosphere or within plant are unknown. The results obtained based on cultured microbes give only partial information about these interactions and identify only few selected microbes which can grow in nutrient-rich media, leaving a large pool unidentified which could be potential contributor towards the observed effect.

The microbes associated with plant modulate almost every aspect of plant's life either directly or indirectly. For simplicity in understanding, the role of these microbes is discussed in three broad categories: plant growth promotion, plant protection and quality improvement. Plant growth promotion includes production or secretion of nutrients, production of hormones and plant growth-promoting substances. Plant-associated microbes are known to protect its host plant from range of pests including bacterial and fungal pathogen, nematodes and herbivores. These microbes are also known to protect the host plant from abiotic stresses. The quality of product produced by plant is also shown to be influenced to some extent by microbes present in and around the plants.

9.8.1 Plant Production

Plant growth is promoted through improved nutrient acquisition, including nitrogen fixation and phosphate solubilization and production of plant growth-enhancing substances such as cytokinins (Garcia de Salamone et al. 2001) and indole acetic acid (IAA) (Naveed et al. 2015). Application of microarray to the metagenomic DNA-detected key genes involved in nitrogen cycle and provided information on the composition and activity of the complex soil microbial community (Wu et al. 2001). Metagenomic analysis of rice root endophytes detected the presence and expression of protein domains involved in nitrogen fixation, denitrification and nitrification (Sessitsch et al. 2012). This result indicates that the endophytes can be involved in the entire nitrogen cycle. The detailed analysis of how plant adopts and survives under poor nutrient conditions was analyzed by Ikeda et al. (2014). They applied metagenomic approach to study and compare the rhizosphere and endophytes of rice under low and standard nitrogen application. The relative abundances of *Burkholderia, Bradyrhizobium* and *Methylosinus* were significantly increased in rice root grown in low nitrogen relative to the standard nitrogen condition. The genes involved in methane oxidation and metabolism of nitrogen, sulphur, iron and aromatic compounds were also more abundant in the low nitrogen root microbiome. This result clearly indicates the importance of microbes associated with plant in sustaining plant growth under low-nutrient regime.

Shotgun metagenome sequencing of *Lotus japonicus* rhizosphere soil revealed that the microbial communities are involved in phytic acid utilization which would allow the plant to survive under low-nutrient availability (Unno and Shinano 2013). Sequence analysis showed improvement in the relative abundance of the classes *Bacteroidetes*, *Betaproteobacteria*, *Chlorobi*, *Dehalococcoidetes* and *Methanobacteria.* These classes are known to contain species that potentially promote plant growth and phytic acid utilization. The gene clusters relating to phytic acid utilization (like alkaline phosphatise, citrate, glutamine, glutamate, aspartate and asparagine biosynthesis) and glyoxylate synthesis were also found abundant. The soil was managed without phosphate fertilizer for over 90 years yet supporting the plant growth with the help of its associated microbes. The bacterial community associated with tap root of sugar beet (*Beta vulgaris* L.) was analyzed by metagenome sequencing (Tsurumaru et al. 2015). Genes involved in plant growth-promoting traits like phosphate solubilization (quinoprotein glucose dehydrogenase), methanol utilization (methanol dehydrogenase), siderophore production (isochorismate pyruvate lyase) and ACC deaminase were found abundant.

Metagenomic analysis of tamarisk (*Tamarix nilotica*), soybean (*Glycine max*), *Arabidopsis* (*Arabidopsis thaliana*), clover (*Trifolium repens*) and rice (*Oryza sativa*) phyllosphere has identified an abundance of various known and novel microbial rhodopsins (Atamna-Ismaeel et al. 2012). These light-sensing proteins and proton pumps showed non-overlapping absorption spectra with their host plant. This result indicates that the energy metabolism in the plant is also dependent on its associated microbes. The plant-microbe interaction and the response of plant to its surrounding microbes are well explained by Costa et al. (2014). They proposed a model based on categorical principal component analysis to explain the expression of bacterial plant growth-promoting traits according to the soil nutritional status. They applied their model to data sets of seven independent but similar studies and found that the plants favour interaction with growth hormone producers under nutrient-rich conditions but favour nutrient solubilizers under nutrient-poor conditions. Analysis of 16Sr DNA sequencing of soil metagenome showed that the bacterial communities of the soybean rhizosphere changed significantly during growth stages. It was observed that the potential plant growth-promoting rhizobacterial community like *Bacillus*, *Bradyrhizobium*, and *Rhizobium* changed in a stage-specific manner (Sugiyama et al. 2014).

9.8.2 Protection from Biotic and Abiotic Stress

In addition to enhance growth of plants by nutrient assimilation and production of plant growth hormones, plant-associated microbes are also known to involve in adaptation of plant to biotic and abiotic stress. Endophytic microbes present a special interest for improved crop adaptation to stress as they have the advantage of being relatively protected from the harsh environment of the soil under draught, high salt or other stress conditions (Sturz et al. 2000).

Rondon et al. (2000) used BAC vector to construct libraries of genomic DNA isolated from soil. Initial screening of libraries in *E. coli* identified several clones that express heterologous gene from the insert. The phenotypes expressed by these clones include antibacterial, lipase, amylase, nuclease and hemolytic activities. Novel bacterial chitinase Chi18H8 was isolated from disease-suppressive soil and cloned in fosmid vector. The expressed protein in *E. coli* showed antifungal activity against *Colletotrichum gloeosporioides*, *Fusarium oxysporum* and *Fusarium graminearum* (Hjort et al. 2014). Tejesvi et al. (2016) constructed the metagenomic library from endosymbiotic fungal microbiome of black crowberry. One of the clone produced folded protein En-AP1, with no similarity to known protein. The tryptic digest of this protein showed antibacterial activity against *Staphylococcus aureus* and *E. coli.* The metagenomic analysis of sugar beet root-associated bacterial microbiome (Tsurumaru et al. 2015) revealed the abundance of β-1,3-glucanase coding genes which play an important role in suppressing plant pathogen and maintaining plant health. Mendes et al. (2011) used the combination of culture-dependent functional analysis and phylo-chip based on metagenome of rhizosphere microbiome. They identified that the microbes belonging to Proteobacteria, Firmicutes and Actinobacteria produce non-ribosomal peptide synthetases which are involved in plant disease suppression. They explained how plant recruits and exploit microbial consortia from soil for protection against fungal root pathogen.

Application of soil slurry prepared from rhizosphere soil of different plants to the *Arabidopsis* rhizosphere reduces the herbivore behaviour of *Trichoplusia ni* insect larvae (Badri et al. 2013). They found the application of microbial consortia in the form of soil slurry prepared from different rhizosphere changed the leaf metabolome, which modulated the insect herbivory. Based on principal component analysis of leaf metabolome and soil metagenome, they targeted the species belonging to *Balneimonas*, *Skermanella* and *Nocardioides* could be responsible for changing leaf metabolome and hence reduced insect feeding.

Though there is no direct report of isolation and characterization of drought-tolerant genes from plant-associated microbial metagenome, but there are few reports which can reveal their potential application in near future. The study of Zolla et al. (2013) unrevealed the importance of soil microbiome in combating the drought stress in *Arabidopsis*. A sympatric microbiome (i.e., having a history of exposure to Arabidopsis at a natural site) significantly increased the *Arabidopsis* biomass under drought condition, but the soil from two different rhizospheres (i.e., pine and maize) when inoculated to *Arabidopsis* did not improve the plant biomass in same conditions. Further, metagenomic analysis of one sympatric and two non-sympatric microbiomes revealed 84 bacterial OTU's from 41 genera which were significantly higher in sympatric microbiome, but there was no up-regulation of *Arabidopsis* drought response marker genes. This study clearly indicates that microbes associated with plant can modify the plant's ability to sense abiotic stress and increase its biomass production. Using culture-independent genomic analysis of farm in desert, Koberl et al. (2011) identified that prolonged farming in desert ecosystem improved the antagonistic microbial population like *Bacillus* and *Paenibacillus.* These species were not applied in the form of bio-inoculants, but they were native to the desert. This study indicates that the indigenous desert microorganisms promoted plant health in desert agroecosystems, paving the way for designing similar strategy to mitigate abiotic stresses.

Metagenomic analysis of grey mangrove (*Avicennia marina*) rhizosphere in Northern Red sea revealed an interesting relationship between host and its associated microbiome to alleviate salt stress (Alzubaidy et al. 2016). The grey mangrove is known to absorb salt from its surrounding which is later deposited in roots and rhizosphere. High-salt concentration in rhizosphere will select and multiply a special group of bacteria which can alleviate this situation. The metagenomic analysis of mangrove rhizosphere microbiome identified 11 pathways that are involved in osmolyte utilization. Mirete et al. (2015) constructed metagenomic library from *Arthrocnemum macrostachyum* rhizosphere grown in moderate salinity. The metagenomic DNA was expressed in osmosensitive strain MKH13 of *E. coli* and screened for salt resistance. Eleven genes that conferred salt resistance were identified, some of which encoding for well-known proteins previously related to osmoadaptation such as a glycerol transporter and a proton pump and some coded for novel genes. Yuan et al. (2016) studied the functional capability of microbes associated with halo-tolerant seepweed *Suaeda salsa* and then inoculated the identified core microbiome to non-host plant to enhance salt tolerance. Pyrosequencing of *S. salsa* associated microbial metagenome identified the core microbes and abundance of genes contributing to salt stress acclimatization, nutrient solubilization and competitive root colonization. The tolerance to salinity increased significantly in non-host plants, cucumber (*Cucumis sativus)* and rice (*Oryza sativa*) upon inoculation of core microbial species to their rhizosphere.

9.8.3 Quality Improvement

We were unable to spot literature highlighting the importance of metagenomics on the quality of products. There are some studies which explained the role of land management and application of pure culture or consortia of isolated species on quality of fruits. The quality of orange was improved by inoculation of *Pseudomonas fluorescence* and *Azospirillum brasilense* to tree rhizosphere (Shamseldin et al. 2010). Inoculation with *P. fluorescence* significantly improved fruit yield, fruit weight, fruit length, TSS and juice volumes, while inoculation with strain *A. brasilense* increased fruit size but did not significantly improved fruit quality.

Amarante et al. (2008) assessed the fruit quality of apple produced with conventional and organic production systems using physical-chemical and organoleptic properties. The apples grown organically had higher flesh firmness, higher soluble solid content, higher density, a more yellowish skin background colour, lower titratable acidity and higher percentage of blush in the fruit skin than fruit from the conventional orchard. Despite differences in terms of physicochemical quality, the untrained sensory panellists detected no differences in terms of taste, flavour and texture between organic and conventional fruits. The apples grown organically had lower average weight, and also there was higher incidence of disease which can significantly reduce yield, fruit quality and market price. Another group studied and compared the spatial and compositional variation of fungal communities in apple fruit grown organically and inorganically using high through put sequencing of apple fruit metagenome (Abdelfattah et al. 2016). Although, no significant diversity was observed, few taxa were exclusively detected in organic apples. Also there was difference in relative abundance of several taxa between organic and conventionally grown apple; Ascomycota were more prevalent in samples obtained from organic apples, while Basidiomycota were more abundant in samples from conventionally grown apples. The abundant taxa in organic apple represented both known plant pathogens also potential antagonists. But they did not analyze and correlate the observed fungal patterns with fruit quality.

Several authors have reported the role of endophytes in production and maintaining quality of secondary metabolites and medicinally important drugs (Chen et al. 2016). The best known example is production of anticancer compound taxol by endophytic fungi *Taxomyces andreanae* isolated from *Taxus brevifolia* (Stierle et al. 1995). There are reports of colonization and production of taxol by such fungi in other non-host species like *Seimatoantlerium tepuiense* and *Seimatoantlerium nepalense* (Bashyal 1999). Thus, there are several reports of culturable endophytic microbes influencing production and accumulation of medicinally important drug. Analysis on the interaction of metagenome on quality enhancement is required.

9.9 Future Prospects and Applications

Although plant-associated microbes are one of the most important elements that have significant influences on the growth and development of host plant, our knowledge about the exact relationships between these microbes and their host plants is far from being complete. The colonization and growth promotion activity of beneficial microbes is expected to perform better when they are applied to the environment similar to that from where they have been isolated (Herrmann and Lesueur 2013). Culture-independent genomic analysis of rhizosphere and endosphere microbes will help in understanding the plant-microbe interaction in true sense. In order to improve our crop management practices, we need to understand the variability of beneficial microbes in different environments. Based on the information generated through metagenomics of respective niche, it might be possible to modulate the composition of root microbiome to improve crop health and growth (Rascovan et al. 2016). In addition to microbes having beneficial role in general, there are certain group of microbes whose potential can be realized only when they interact with their respective host. Based on soil microbial community analysis, we can plan and choose right type of crop to get desirable and sustainable yield with fewer inputs.

The data obtained from culture-independent genomic analysis of plant-associated microbes along with metadata will help in understanding the physiological requirement of some unculturable species. This information could be used to design and develop synthetic media and growth conditions to culture the yet uncultured species. Tyson et al. (2005) successfully applied the metagenomic sequence data to culture previously unculturable species *Leptospirillum ferrodiazotrophum*. Further, this information can also be used to design synthetic microbial consortia (SMC). This SMC can be prepared targeting a specific objective and applied for specific purpose like biotic or abiotic stress tolerance (Qin et al. 2016). The benefits of microbiome inhabiting in a particular host (endophyte) can be transferred to a non-host plant by co-culturing the callus in tissue culture media. Up to 2.5-fold increased biomass production was observed in *Sporobolus cryptandrus* grass when its seedlings were cultured along with endophyte contained calli of *Atriplex canescens* (Lucero et al. 2008).

A large number of novel compounds and secondary metabolites produced by rhizosphere and endorhizosphere microbes posses antibacterial or antifungal properties which can be employed directly in crop improvement. A near-future application may consider the use of genetically engineered rhizosphere and endophytic microbes with biological control potential in agricultural crops. The endophytes *Herbaspirillum seropedicae* and *Clavibacter xyli* have been genetically modified to produce and excrete the δ-endotoxin of *Bacillus thuringensis* to control insect pests (Downing et al. 2000). *Streptomyces* spp. was genetically modified to introduce enhanced green fluorescent protein marker and resistance to apramycin to track the colonization process (Bonaldi et al., 2015). The transformed strain was able to colonize soil, developing roots and rhizosphere. The re-isolation of modified strain from the rhizosphere and the inner tissues of surface-sterilized lettuce roots demonstrated that it is both rhizospheric and endophytic. The transformed strain also inhibited the soil-borne pathogen *Sclerotinia sclerotiorum* under in vitro.

The phenotypic appearance and physiological performance of both host plant and its associated microbes are interconnected with each other. This interaction plays a major role in performance and fitness of both the host and its associated microbes. With this understanding, the plant microbiome should be recognized as an additional factor for breeding strategies in near future. The quality of production (fruits and medicinal compounds) is known to be influenced by the plant-associated microbiome to some extent. Understanding and exploiting such relationships will facilitate the ideal production of better drugs and fruits in future by manipulating the growth conditions to aid the growth of those specific microbes to improve the quality of produce (Firakova et al. 2007). In addition, these compounds can induce the production of a plethora of known and novel biologically active secondary metabolites (Rodriguez et al. 2009) that can be exploited to produce novel drugs to cure diseases.

Several authors have reported that the plants are alternate host for human pathogens. They have ability to attack, infect, colonize, persist, multiply and survive on plants or inside the plants. The ability of these pathogens to survive for prolonged periods on fruits and vegetables has been described by Natvig et al. (2002). An enteric pathogenic strain *Salmonella enterica* have been isolated from barley roots (Kutter et al. 2006). *Campylobacter jejuni* was isolated from radish roots and the spinach rhizosphere. This enteric pathogen also persisted in the rhizosphere of spinach for prolonged periods of time at 16 °C (Brandl et al., 2004). The reports about the outbreak of human disease caused by these contaminated human pathogens are continuously rising (Behravesh et al., 2012). We may in near future need to think before advising the application of farm waste, especially human waste as a source of organic nutrients to crop plants. Or at least we need to advice the precocious application of these inputs as we are suggesting today for chemically derived nutrients.

References

Abdelfattah A, Wisniewski M, Droby S et al (2016) Spatial and compositional variation in the fungal communities of organic and conventionally grown apple fruit at the consumer point-of-purchase. Hort Res 3:16047

Akhtyamova N (2013) Human pathogens-the plant and useful endophytes. J Med Microbiol Diagnosis 2:e121

Albrectsen BR, Bjorken L, Varad A et al (2010) Endophytic fungi in European aspen (*Populus tremula*) leaves: diversity, detection, and a suggested correlation with herbivory resistance. Fungal Divers 41:17–28

Alzubaidy H, Essack M, Malas TB et al (2016) Rhizosphere microbiome metagenomics of gray mangroves (*Avicennia marina*) in the Red Sea. Gene 576:626–636

Amarante CVT, Steffens CA et al (2008) Yield and fruit quality of apple from conventional and organic production systems. Pesq Agrop Brasileira 43(3):333–340

Aparna K, Pasha MA, Rao DLN et al (2014) Organic amendments as ecosystem engineers: Microbial, biochemical and genomic evidence of soil health improvement in a tropical arid-zone field site. Ecol Eng 71:268–277

Atamna-Ismaeel N, Finkel OM, Glaser F et al (2012) Microbial rhodopsins on leaf surfaces of terrestrial plants. Environ Microbiol 14:140–146

Bacilio-Jimenez M (2003) Chemical characterization of root exudates from rice (*Oryza sativa)* and their effects on the chemotactic response of endophytic bacteria. Plant Soil 249:271–277

Badri DV, Zolla G, Bakker MG et al (2013) Potential impact of soil microbiomes on the leaf metabolome and on herbivore feeding behavior. New Phytol 198:264–273

Bais HP, Park SW, Weir TL et al (2004) How plants communicate using the underground information superhighway. Trends Plant Sci 9:26–32

Bais HP, Weir TL, Perry LG et al (2006) The role of root exudates in rhizosphere interactions with plants and other organisms. Annu Rev Plant Biol 57:233–266

Balsanelli E, Serrato RV, De Baura VA et al (2010) *Herbaspirillum seropedicae* rfbB and rfbC genes are required for maize colonization. Environ Microbiol 12:2233–2244

Bashyal B (1999) *Seimatoantlerium nepalense*, an endophytic taxol producing coelomycete from Himalayan yew (*Taxus wallichiana*). Mycotaxon 72:33–42

Behravesh CB, Williams IT, Tauxe RV (2012) Emerging foodborne pathogens and problems: expanding prevention efforts before slaughter or harvest. In: Institute of Medicine (US). Improving food safety through a one health approach: Workshop summary. Washington (DC): National Academies Press; 2012. A14

Berg G, Grube M, Schloter M et al (2014) Unraveling the plant microbiome: looking back and future perspectives. Front Microbiol 5:148

Bever JD, Broadhurst LM, Thrall PH (2013) Microbial phylotype composition and diversity predicts plant productivity and plant-soil feedbacks. Ecol Lett 16:167–174

Bodenhausen N, Horton MW, Bergelson J (2013) Bacterial communities associated with the leaves and the roots of *Arabidopsis thaliana*. PLoS One 8:e56329

Bonaldi M, Chen X, Kunova A et al (2015) Colonization of lettuce rhizosphere and roots by tagged Streptomyces. Front Microbiol 6. https://doi.org/10.3389/fmicb.2015.00025

Brandl MT, Haxo AF, Bates AH (2004) Comparison of survival of *Campylobacter jejuni* in the phyllosphere with that in the rhizosphere of spinach and radish plants. Appl Environ Microbiol 70:1182–1189

Buckley DH, Schmidt TM (2003) Diversity and dynamics of microbial communities in soil from agro-ecosystem. Environ Microbiol 5:441–452

Bulgarelli D, Rott M, Schlaeppi K et al (2012) Revealing structure and assembly cues for Arabidopsis root-inhabiting bacterial microbiota. Nature 488:91–95

Carbonetto B, Rascovan N, Alvarez R et al (2014) Structure, composition and metagenomic profile of soil microbiomes associated to agricultural land use and tillage systems in argentine pampas. PLoS One 9(6):e99949

Carlos HSG, Luc FM, Jean L et al (2011) Exopolysaccharide production is required for biofilm formation and plant colonization by the nitrogen-fixing endophyte *Gluconacetobacter diazotrophicus*. MPMI 24:1448–1458

Chelius MK, Triplett EW (2001) The diversity of archaea and bacteria in association with the roots of *Zea mays* L. Microb Ecol 41:252–263

Chen K, Pachter L (2005) Bioinformatics for whole-genome shotgun sequencing of microbial communities. PLoS Comp Biol 1(2):e24

Chen L, Zhang QY, Jia M et al (2016) Endophytic fungi with antitumor activities: Their occurrence and anticancer compounds. Crit Rev Microbiol 42:454–473

Chen XH, Koumoutsi A, Scholz R et al (2007) Comparative analysis of the complete genome sequence of the plant growth-promoting bacterium *Bacillus amyloliquefaciens* FZB42. Nat Biotechnol 25:1007–1014

Cheng Y, Jiang Y, Wu Y et al (2016) Soil nitrogen status modifies rice root response to nematode-bacteria interactions in the rhizosphere. PLoS One 11:e0148021

Chi F, Shen SH, Cheng HP et al (2005) Ascending migration of endophytic rhizobia, from roots to leaves, inside rice plants and assessment of benefits to rice growth physiology. Appl Environ Microbiol 71:7271–7278

Compant S, Clément C, Sessitsch A (2010) Plant growth-promoting bacteria in the rhizo- and endosphere of plants: their role, colonization, mechanisms involved and prospects for utilization. Soil Biol Biochem 42:669–678

Compant S, Mitter B, Colli-Mull JG et al (2011) Endophytes of grapevine flowers, berries and seeds: identification of cultivable bacteria, comparison with other plant parts, and visualization of niches of colonization. Microb Ecol 62:188–197

Compant S, Reiter B, Sessitsch A et al (2005) Endophytic colonization of *Vitis vinifera* L. by plant growth promoting bacterium *Burkholderia sp.* strain PsJN. Appl Environ Microbiol 71:1685–1693

Conn VM, Franco CM (2004) Analysis of the endophytic actinobacterial population in the roots of wheat (*Triticum aestivum* L.) by terminal restriction fragment length polymorphism and sequencing of 16S rRNA clones. Appl Environ Microbiol 70:1787–1794

Costa PB, Granada CE, Ambrosini A et al (2014) A model to explain plant growth promotion traits: a multivariate analysis of 2,211 bacterial isolates. PLoS One 9(12):e116020

Curtis TP, Sloan WT, Scannell JW (2002) Estimating prokaryotic diversity and its limits. Proc Natl Acad Sci U S A 99:10494–10499

D'Amico M, Frisullo S, Cirulli M (2008) Endophytic fungi occurring in fennel, lettuce, chicory, and celery-commercial crops in Southern Italy. Mycol Res 112:100–107

Danielsen L, Thurmer A, Meinicke P et al (2012) Fungal soil communities in a young transgenic poplar plantation form a rich reservoir for fungal root communities. Ecol Evol 2:1935–1948

de Souza JT, de Boer M, de Waard P et al (2003) Biochemical, genetic, and zoosporicidal properties of cyclic lipopeptide surfactants produced by *Pseudomonas fluorescens*. Appl Environ Microbiol 69:7161–7172

De Weert S, Vermeiren H, Mulders IH et al (2002) Flagella-driven chemotaxis towards exudate components is an important trait for tomato root colonization by *Pseudomonas fluorescens*. Mol Plant-Microbe Interact 15:1173–1180

DeAngelis KM, Brodie EL, DeSantis TZ et al (2009) Selective progressive response of soil microbial community to wild oat roots. ISME J 3:168–178

Deering AJ, Mauer LJ, Pruitt RE (2012) Internalization of *E. coli* O157:H7 and *Salmonella spp.* in plants: a review. Food Res Int 45:567–575

Ding T, Palmer MW, Melcher U (2013) Community terminal restriction fragment length polymorphisms reveal insights into the diversity and dynamics of leaf endophytic bacteria. BMC Microbiol 13:1

Downing KJ, Leslie G, Thomson JA (2000) Biocontrol of the sugarcane borer *Eldana saccharina* by expression of the *Bacillus thuringiensis cry*1Ac7 and *Serratia marcescens chi*A genes in sugarcane-associated bacteria. Appl Environ Microbiol 66:2804–2810

Dudeja SS, Giri R (2014) Beneficial properties, colonization, establishment and molecular diversity of endophytic bacteria in legume and non-legume. Afr J Microbiol Res 8:1562–1572

Elasri M, Delorme S, Lemanceau P (2001) Acyl-homoserine lactone production is more common among plant-associated *Pseudomonas spp.* than among soilborne *Pseudomonas spp.* Appl Environ Microbiol 67:1198–1209

Fakruddin M, Mannan KSB (2012) Next generation sequencing technologies-principles and prospects. Res Rev Biosci 6(9):240–247

Ferrando L, Manay JF, Scavino AF (2012) Molecular and culture-dependent analyses revealed similarities in the endophytic bacterial community composition of leaves from three rice (*Oryza sativa*) varieties. FEMS Microbiol Ecol 80:696–708

Firakova S, Sturdikova M, Muckova M (2007) Bioactive secondary metabolites produced by microorganisms associated with plants. Biologia 62:251–257

Fisher PJ, Petrini O, Petrini LE et al (1994) Fungal endophytes from the leaves and twigs of *Quercus ilex* L. from England, Majorca and Switzerland. New Phytol 127:133–137

Foster RC (1988) Microenvironments of soil microorganisms. Biol Fertil Soils 6:189–203
Fuentes-Ramirez LE, Caballero-Mellado J, Sepulveda J et al (1999) Colonization of sugarcane by *Acetobacter diazotrophicus* is inhibited by high N-fertilization. FEMS Microbiol Lett 29:117–128
Gaiero JR, McCall CA, Thompson KA (2013) Inside the root microbiome: bacterial root endophytes and plant growth promotion. Am J Bot 100:1738–1750
Gamalero E, Lingua G, Caprì FG (2004) Colonization pattern of primary tomato roots by *Pseudomonas fluorescens* A6RI characterized by dilution plating, flow cytometry, fluorescence, confocal and scanning electron microscopy. FEMS Microbiol Ecol 48:79–87
Gao M, Teplitski M (2003) Production of substances by *Medicago truncatula* that affect bacterial quorum sensing. Mol Plant-Microbe Interact 16:827–834
Garcia de Salamone IE, Hynes RK, Nelson LM (2001) Cytokinin production by plant growth promoting rhizobacteria and selected mutants. Can J Microbiol 47(5):404–411
Gazis R, Chaverri P (2010) Diversity of fungal endophytes in leaves and stems of wild rubber trees (*Hevea brasiliensis*) in Peru. Fungal Ecol 3:240–254
Gehlot P, Singh SK, Pathak R (2012) Morphometric and molecular characterization of fungus Pestalotiopsis using nuclear ribosomal DNA analysis. J Environ Biol 33:897–901
Germida JJ, Siciliano SD (2001) Taxonomic diversity of bacteria associated with the roots of modern, recent and ancient wheat cultivars. Biol Fertil Soils 33:410–415
Ghimire SR, Charlton ND, Bell JD et al (2011) Biodiversity of fungal endophyte communities inhabiting switchgrass (*Panicum virgatum* L.) growing in the native tallgrass prairie of northern Oklahoma. Fungal Divers 47:19–27
Giri R, Dudeja SS (2013) Host specificity of plant endophytic bacterial interactions: root colonization in liquid medium. J. Microbiol Res 1(6):75–82
Gonod LV, Chenu C, Soulas G (2003) Spatial variability of 2,4- dichlorophenoxyacetic acid (2,4-D) mineralisation potential at a millimetre scale in soil. Soil Biol Biochem 35(3):373–382
Govindasamy V, Franco CMM, Gupta VVSR (2014) Endophytic actinobacteria: diversity and ecology. Adv Endophytic Res:27–59
Grayston SJ, Wang S, Campbell CD et al (1998) Selective influence of plant species on microbial diversity in the rhizosphere. Soil Biol Biochem 30:369–378
Gregory PJ (2006) Plant roots: growth, activity and interaction with soils. Blackwell, London
Haichar FZ, Marol C, Berge O et al (2008) Plant host habitat and root exudates shape soil bacterial community structure. ISME J 2:1221–1230
Hallman J, Rodriguez Kabana R, Kloepper JW (1999) Chitin-mediated changes in bacterial communities of the soil, rhizosphere and within roots of cotton in relation to nematode control. Soil Biol Biochem 31:551–560
Hallmann J, Quadt-Hallmann A, Mahaffee WF et al (1997) Bacterial endophytes in agricultural crops. Can J Microbiol 43:895–914
Handelsman J, Rondon MR, Brady SF (1998) Molecular biological access to the chemistry of unknown soil microbes: a new frontier for natural products. Chem Biol 5:245–249
Hardoim PR, van Overbeek LS, Berg G et al (2015) The hidden world within plants: ecological and evolutionary considerations for defining functioning of microbial endophytes. Microbiol Mol Biol Rev 79(3):293–320
Hardoim PR, van Overbeek LS, Elsas JD (2008) Properties of bacterial endophytes and their proposed role in plant growth. Trends Microbiol 16(10):463–471
Hartmann A, Rothballer M, Schmid M (2008) Lorenz H: a pioneer in rhizosphere microbial ecology and soil bacteriology research. Plant Soil 312:7–14
He Z, Deng Y, Van Nostrand JD et al (2010) GeoChip 3.0 as a high-throughput tool for analyzing microbial community structure, composition and functional activity. ISME J 4:1167–1179
Herrmann L, Lesueur D (2013) Challenges of formulation and quality of biofertilizers for successful inoculation. Appl Microbiol Biotechnol 97:8859–8873
Hiltner L (1904) Uber neure erfahrungen und probleme auf dem Gebiet der Boden bakteriologie und unter besondere Beruck-sichtigung der grundungung und Bracke. Arbeiten der Deutschen Landwirtschaft Gesellschaft 98:59–78

Hjort K, Presti I, Elvang A et al (2014) Bacterial chitinase with phytopathogen control capacity from suppressive soil revealed by functional metagenomics. Appl Microbiol Biotechnol 98(6):2819–2828

Holben WE, Jansson JK, Chelm BK et al (1988) DNA probe method for detection of specific microorganisms in the soil bacterial community. Appl Environ Microbiol 54:703–711

Horner-Devine MC, Leibold MA, Smith VH et al (2003) Bacterial diversity patterns along a gradient of primary productivity. Ecol Lett 6:613–622

Humphris SN, Bengough AG, Griffiths BS et al (2005) Root cap influences root colonization by *Pseudomonas fluorescens* SBW25 on maize. FEMS Microbiol Ecol 54:123–130

Hussain Q, Pan GX, Liu YZ et al (2012) Microbial community dynamics and function associated with rhizosphere over periods of rice growth. Plant Soil Environ 58(2):55–61

Ikeda S, Kaneko T, Okubo T et al (2009) Development of a bacterial cell enrichment method and its application to the community analysis in soybean stems. Microb Ecol 58(4):703–714

Ikeda S, Sasaki K, Okubo T et al (2014) Low nitrogen fertilization adapts rice root microbiome to low nutrient environment by changing biogeochemical functions. Microbes Environ 29:50–59

Islam MR, Sultana T, Melvin, et al (2012) Comparisons of direct extraction methods of microbial DNA from different paddy soils. Saudi J Biol Sci 19:337–342

James EK, Gyaneshwar P, Manthan N et al (2002) Infection and colonization of rice seedlings by the plant growth-promoting bacterium *Herbaspirillum seropedicae* Z67. Mol Plant-Microbe Interact 15:894–906

Janssen PH (2006) Identifying the dominant soil bacterial taxa in libraries of 16S rRNA and 16S rRNA genes. Appl Environ Microbiol 72(3):1719–1728

Jiang S, Duan JA, Tao JH et al (2010) Ecological distribution and elicitor activities of endophytic fungi in *Changium smyrnioides*. Chin Tradit Herb Drug 1:121–125

Jiao JY, Wang HX, Zeng Y et al (2006) Enrichment for microbes living in association with plant tissues. J Appl Microbiol 100(4):830–837

Johnston-Monje D, Raizada MN (2011) Conservation and diversity of seed associated endophytes in Zea across boundaries of evolution, ethnography and ecology. PLoS One 6:e20396

Kaur R, Minhas PS, Jain PC et al (2008) Geo-spatial analysis of land water resource degradation in two economically contrasting agricultural regions adjoining national capital territory (Delhi). Environ Monit Assess. doi:10.1007/s10661-008-0378-3

Kieft TL, Soroker E, Firestone MR (1987) Microbial biomass response to a rapid change increase in water potential when dry soil is wetted. Soil Biol Biochem 19:119–126

Kirk JL, Beaudette LA, Hart M et al (2004) Methods of studying soil microbial diversity. J Microbiol Methods 58:169–188

Knee EM, Gong FC, Gao M et al (2001) Root mucilage from pea and its utilization by rhizosphere bacteria as a sole carbon source. Mol Plant Microbe Interact14:775–784

Knietsch A, Bowien S, Whited G et al (2003) Identification and characterization of coenzyme B12-dependent glycerol dehydratase- and diol dehydratase-encoding genes from metagenomic DNA libraries derived from enrichment cultures. Appl Environ Microbiol 69:3048–3060

Koberl M, Müller H, Ramadan EM et al (2011) Desert farming benefits from microbial potential in arid soils and promotes diversity and plant health. PLoS One 6(9):e24452

Kovtunovych G, Lar O, Kamalova S et al (1999) Correlation between pectate lyase activity and ability of diazotrophic *Klebsiella oxytoca* VN13 to penetrate into plant tissues. Plant Soil 215:1–6

Krishnaraj PU, Pasha MA (2017) Metagenomics: concepts, tools and application. In: Kumar P, Bhola R, GJ N Govi (ed) Environmental Science and Engineering. Studium press, New Delhi (In Press)

Kuklinsky-Sobral J, Araujo WL, Mendes R (2004) Isolation and characterization of soybean-associated bacteria and their potential for plant growth promotion. Environ Microbiol 6:1244–1251

Kutter S, Hartmann A, Schmid M (2006) Colonization of barley (*Hordeum vulgare*) with *Salmonella enterica* and *Listeria spp*. FEMS Microbiol Ecol 56:262–271

Lakay FM, Botha A, Prior BA (2007) Comparative analysis of environmental DNA extraction and purification methods from different humic acid-rich soils. J Appl Microbiol 102:265–273

Lazcano C, Gomez-Brandon M, Revilla P et al (2013) Short-term effects of organic and inorganic fertilizers on soil microbial community structure and function: a field study with sweet corn. Biol Fertil Soils 49:723–733

Link HF (1809) Observationes in ordines plantarum naturales, dissertatio prima, complectens anandrarum ordines Epiphytas, Mucedines. Gastromycos et Fungos. Der Gesellschaft Naturforschender Freunde zu Berlin, Berlin

Liu B, Tu C, Hu S et al (2007) Effects of organic, sustainable and conventional management strategies in grower fields on soil physical, chemical and biological factors and the incidence of southern blight. Appl Soil Ecol 37:202–214

Liu M, Hu F, Chen X et al (2009) Organic amendments with reduced chemical fertilizer promote soil microbial development and nutrient availability in a subtropical paddy field: the influence of quantity, type and application time of organic amendments. Appl Soil Ecol 42:166–175

Liu X, Zhang J, Gu T et al (2014) Microbial community diversities and taxa abundances in soils along a seven-year gradient of potato monoculture using high throughput pyrosequencing approach. PLoS One 9(1):e86610

Lucero ME, Barrow JR, Osuna P et al (2008) Enhancing native grass productivity by cocultivating with endophyte-laden calli. Rangel Ecol Manag 61:124–130

Lugtenberg B, Kamilova F (2009) Plant-growth-promoting rhizobacteria. Annu Rev Microbiol 63:541–556

Lugtenberg BJ, Dekkers L, Bloemberg GV (2001) Molecular determinants of rhizosphere colonization by Pseudomonas. Annu Rev Phytopathol 39:461–490

Lundberg DS, Lebeis SL, Paredes SH et al (2012) Defining the core *Arabidopsis thaliana* root microbiome. Nature 488:86–90

Mahmoudi N, Slater RR (2011) Fulthorpe Comparison of commercial DNA extraction kits for isolation and purification of bacterial and eukaryotic DNA from PAH-contaminated soils. Can J Microbiol 57:623–628

Mandimba G, Heulin T, Bally R et al (1986) Chemotaxis of free-living nitrogen-fixing bacteria towards maize mucilage. Plant Soil 90:129–139

Mark GL, Dow JM, Kiely PD et al (2005) Transcriptome profiling of bacterial responses to root exudates identifies genes involved in microbe-plant interactions. Proc Natl Acad Sci U S A 102:17454–17459

Maron PA, Mougel C, Ranjard L (2011) Soil microbial diversity: methodological strategy, spatial overview and functional interest. C R Biol 334:403–411

Mendes R, Kruijt M, de Bruijn I et al (2011) Deciphering the rhizosphere microbiome for disease suppressive bacteria. Science 332:1097–1100

Meng Y, Li Y, Galvani CD et al (2005) Upstream migration of *Xylella fastidiosa* via pilus-driven twitching motility. J Bacteriol 187:5560–5567

Mercado-Blanco J, Prieto P (2012) Bacterial endophytes and root hairs. Plant Soil 361:301–306

Miethling R, Wieland G, Backhaus H et al (2000) Variation of microbial rhizosphere communities in response to crop species, soil origin, and inoculation with *Sinorhizobium meliloti* L33. Microb Ecol 41:43–56

Mirete S, Mora-Ruiz MR, Lamprecht-Grandío M et al (2015) Salt resistance genes revealed by functional metagenomics from brines and moderate-salinity rhizosphere within a hypersaline environment. Front Microbiol 6:1121

Miyamoto T, Kawahara M, Minamisawa K (2004) Novel endophytic nitrogen-fixing *Clostridia* from the grass *Miscanthus sinensis* as revealed by terminal restriction fragment length polymorphism analysis. Appl Environ Microbiol 70(11):6580–6586

Mougel C, Offre P, Ranjard L et al (2006) Dynamic of the genetic structure of bacterial and fungal communities at different developmental stages of *Medicago truncatula*. cv. Jemalong line J5. New Phytol 170:165–175

Nam YJ, Kim H, Lee JH et al (2015) Metagenomic analysis of soil fungal communities on Ulleungdo and Dokdo Islands. J Gen Appl Microbiol 61:67–74

Natvig EE, Ingham SC, Ingham BH (2002) *Salmonella enterica* serovar Typhimurium and *Escherichia coli* contamination of root and leaf vegetables grown in soils with incorporated bovine manure. Appl Environ Microbiol 68:2737–2744

Naveed M, Qureshi MA, Zahir ZA et al (2015) L-Tryptophan-dependent biosynthesis of indole-3-acetic acid (IAA) improves plant growth and colonization of maize by *Burkholderia phytofirmans* PsJN. Ann Microbiol 65:1391–1389

Nielsen TH, Sorensen D, Tobiasen C et al (2002) Antibiotic and biosurfactant properties of cycliclipopeptides produced by fluorescent *Pseudomonas spp.* from the sugar beet rhizosphere. Appl Environ Microbiol 68:3416–3423

Nikolic B, Schwab H, Sessitsch A (2011) Metagenomic analysis of the 1-aminocyclopropane-1-carboxylate deaminase gene (acdS) operon of an uncultured bacterial endophyte colonizing *Solanum tuberosum* L. Arch Microbiol 193(9):665–676

Nunan N, Wu K, Young IM et al (2003) Spatial distribution of bacterial communities and their relationships with the micro-architecture of soil. FEMS Microbiol Ecol 44:203–215

Nunes da Rocha U, van Overbeek L, van Elsas JD (2009) Exploration of hitherto-uncultured bacteria from the rhizosphere. FEMS Microbiol Ecol 69:313–328

Ogram A, Sayler GS, Barkay T (1987) The extraction and purification of microbial DNA from sediments. J Microbiol Methods 7:57–66

Orr J, Hurek T, Reinhold-Hurek B (1998) Type IV pili are involved in plant-microbe and fungus microbe interactions. Mol Microbiol 30:7–17

Partida-Martínez LP, Heil M (2011) The microbe-free plant: fact or artifact? Front Plant Sci 2:100

Pasha MA, Bhat S, Krishnaraj PU et al (2015) Soil bacterial diversity analysis of cotton field under organic and inorganic management using DGGE. J Pure. Appl Microbiol 9(1):631–636

Patrick DS, Handelsman J (2005) Metagenomics for studying unculturable microorganisms: cutting the Gordian knot. Genome Biol 6:229–234

Paulsen IT, Press CM, Ravel J et al (2005) Complete genome sequence of the plant commensal *Pseudomonas fluorescens* Pf-5. Nat Biotechnol 23:873–878

Petrini O (1991) Fungal endophytes of tree leaves. In: Andrews JH, Hirano SS (eds) Microbial ecology of leaves. Springer-Verlag, New York, pp 179–197

Philippot L, Raaijmakers JM, Lemanceau P et al (2013) Going back to the roots: the microbial ecology of the rhizosphere. Nat Rev Microbiol 11(11):789–799

Qin Y, Tan C, Lin J et al (2016) EcoExpress-highly efficient construction and expression of multi-component protein complexes in *Escherichia coli*. ACS Synth Biol 5(11):1239–1246

Raaijmakers JM, Vlami M, de Souza JT (2002) Antibiotic production by bacterial biocontrol agents. Antonie Van Leeuwenhoek 81:537–547

Rasche F, Velvis H, Zachow C et al (2006) Impact of transgenic potatoes expressing anti-bacterial agents on bacterial endophytes is comparable with the effects of plant genotype, soil type and pathogen infection. J Appl Ecol 43:555–566

Rascovan N, Carbonetto B, Perrig D, Díaz M, Canciani W, Abalo M, et al. (2016) Integrated Analysis of Root Microbiomes of Soybean and Wheat from Agricultural Fields. Scientific Reports. 6:28084

Reinhold-Hurek B, Hurek T (1998) Interactions of gramineous plants with *Azoarcus spp.* and other diazotrophs: Identification, localization, and perspectives to study their function. Crit Rev Plant Sci 17:29–54

Reinhold-Hurek B, Hurek T (2011) Living inside plants: bacterial endophytes. Curr Opin Plant Biol 14:435–443

Reinhold-Hurek B, Maes T, Gemmer S et al (2006a) An endoglucanase is involved in infection of rice roots by the not-cellulose-metabolizing endophyte *Azoarcus sp.* strain BH72. Mol Plant-Microbe Interact 19:181–188

Reinhold-Hurek B, Maes T, Gemmer S et al (2006b) An endoglucanase is involved in infection of rice roots by the not-cellulose-metabolizing endophyte *Azoarcus sp.* strain BH72. Mol Plant-Microbe Interact 19:181–188

Rodrigues KF (1994) The foliar fungal endophytes of the Amazonian palm *Euterpe oleracea*. Mycologia 86:376–385
Rodriguez RJ, White JFJ, Arnold AE et al (2009) Fungal endophytes: diversity and functional roles. New Phytol 182:314–330
Rondon MR, August PR, Bettermann AD et al (2000) Cloning the soil metagenome: a strategy for accessing the genetic and functional diversity of uncultured microorganisms. Appl Environ Microbiol 66:2541–2547
Rosenblueth M, Martinez-Romero E (2004) Rhizobium etli maize populations and their competitiveness for root colonization. Arch Microbiol 181(5):337–344
Rosenblueth M, Martínez-Romero E (2006) Bacterial endophytes and their interactions with hosts. Mol Plant-Microbe Interact 19:827–837
Rousk J, Baath E, Brookes PC et al (2010) Soil bacterial and fungal communities across a pH gradient in an arable soil. ISME 4:134–151
Rudrappa T, Czymmek K, Pare PW et al (2008) Root-secreted malic acid recruits beneficial soil bacteria. Plant Physiol 148:1547–1556
Ryan RP, Germaine K, Franks A et al (2008) Bacterial endophytes: recent development and applications. FEMS Microbiol Lett 278:1–9
Saikkonen K, Wäli P, Helander M et al (2004) Evolution of endophyte-plant symbioses. Trends Plant Sci 9:275–280
Schardl CL, Leuchtmann A, Spiering MJ et al (2004) Symbioses of grasses with seedborne fungal endophytes. Annu Rev Plant Biol 55:315–340
Schloss PD, Handelsman J (2003) Biotechnological prospects from metagenomics. Curr Opin Biotechnol 14:303–310
Schneegurt MA, Dore S, Kulpa Jr.CF (2003) Direct extraction of DNA from soils for studies in microbial ecology. Curr Issues Mol Biol 5:1–8
Schulz B, Boyle C (2006) What are endophytes? In Microbial root endophytes Schulz, Barbara JE, Boyle, Christine JC, Sieber Thomas N (Eds.). Springer-Verlag. pp 1–13
Seghers D, Wittebolle L, Top EM et al (2004) Impact of agricultural practices on the *Zea mays* L. endophytic community. Appl Environ Microbiol 70(3):1475–1482
Sessitsch AP, Haroim A, Weilharter A et al (2012) Functional characteristics of an endophyte community colonizing rice roots as revealed by metagenomic analysis. Mol Plant-Microbe Interact 25:28–36
Shamseldin A, El-Sheikh MH, Hassan HAS et al (2010) Microbial bio-fertilization approaches to improve yield and quality of Washington navel orange and reducing the survival of nematode in the soil. J Am Sci 6:264–271
Sieber TN (2007) Endophytic fungi in forest trees: are they mutualists? Fungal Biol Rev 21:75–89
Smalla K, Wieland G, Buchner A et al (2001) Bulk and rhizosphere soil bacterial communities studied by denaturing gradient gel electrophoresis: plant-dependent enrichment and seasonal shifts revealed. Appl Environ Microbiol 67:4742–4751
Smit E, Leeflang P, Gommans S et al (2001) Diversity and seasonal fluctuations of the dominant members of the bacterial soil community in a wheat field as determined by cultivation and molecular methods. Appl Environ Microbiol 67:2284–2291
Smith KP, Goodman RM (1999) Host variation for interaction with beneficial plant associated microbes. Annu Rev Phytopathol 37:473–491
Song S, Otkur M, Zhang Z et al (2007) Isolation and characterization of endophytic microorganisms in *Glycyrrhiza inflat* Bat. from Xinjiang. Microbiology 5:867–870
Soupir ML, Mostaghimi S, Yagow ER et al (2006) Transport of fecal bacteria from poultry litter and cattle manure applied to pasture land. Water Air Soil Pollut 169:125–136
Souza RC, Cantao ME, Vasconcelos ATR et al (2013a) Soil metagenomics reveals differences under conventional and no-tillage with crop rotation or succession. Appl Soil Ecol 72:49–61
Souza SA, Xavier AA, Costa MR et al (2013b) Endophytic bacterial diversity in banana 'Prata Ana' (Musa spp.) roots. Genet Mol Biol 36(2):252–264

Sparling GP (1997) Soil microbial biomass, activity and nutrient cycling as indicators of soil health. In: Pankhurst CE, Doube BM, Gupta VVSR (eds) Biological indicators of soil health. CAB International, Wallingford, UK, pp 97–119

Stark C, Condron L, Stewart A et al (2007) Influence of organic and mineral amendments on microbial soil properties and processes. Appl Soil Ecol 35:79–93

Stierle A, Strobel G, Stierle D et al (1995) The search for a taxol-producing microorganism among the endophytic fungi of the Pacific yew, *Taxus brevifolia*. J Nat Prod 58:1315–1324

Stone JK, Bacon CW, White J (2000) An overview of endophytic microbes: endophytism defined. In: Bacon CW, White J (eds) Microbial endophytes. Taylor & Francis, New York, pp 3–29

Sturz AV, Christie BR, Nowak J (2000) Bacterial endophytes: potential role in developing sustainable systems of crop production. Crit Rev Plant Sci 19:1–30

Sugiyama A, Ueda Y, Zushi T et al (2014) Changes in the bacterial community of soybean rhizospheres during growth in the field. PLoS One 9:e100709

Sun L, Qiu F, Zhang X et al (2008) Endophytic bacterial diversity in rice (*Oryza sativa* L.) roots estimated by 16S rDNA sequence analysis. Microb Ecol 55:415–424

Sun X, Guo LD, Hyde KD (2011) Community composition of endophytic fungi in *Acer truncatum* and their role in decomposition. Fungal Divers 47:85–95

Suryanarayanan TS, Thirunavukkarasu N, Hariharan GN et al (2005) Occurrence of non-obligate microfungi inside lichen thalli. Sydowia 57:119–129

Sylvia A, Carlos M, Luc R et al (2013) The bacterial superoxide dismutase and glutathione reductase are crucial for endophytic colonization of rice roots by *Gluconacetobacter diazotrophicus* PAL5. MPMI 26:937–945

Tan Z, Hurek T, Reinhold-Hurek B (2003) Effect of N-fertilization, plant genotype and environmental conditions on *nifH* gene pools in roots of rice. Environ Microbiol 5:1009–1015

Taylor LL, Leake JR, Quirk J et al (2009) Biological weathering and the long-term carbon cycle: integrating mycorrhizal evolution and function into the current paradigm. Geobiology 7:171–191

Tejesvi MV, Picart P, Kajula M et al (2016) Appl Microbiol Biotechnol 100:9283

Teplitski M (2000) Plants secrete substances that mimic bacterial N-acyl homoserine lactone signal activities and affect population density-dependent behaviors in associated bacteria. Mol Plant-Microbe Interact 13:637–648

Tkacz A, Poole P (2015) Role of root microbiota in plant productivity. J Exp Bot 66(8):2167–2175

Torsvik V, Ovreas L (2002) Microbial diversity and function in soil: from genes to ecosystems. Curr Opin Microbiol 5:240–245

Tsai Y, Olson BH (1991) Rapid method for direct extraction of DNA from soil and sediment. Appl Environ Microbiol 57:1070–1074

Tsurumaru H, Okubo T, Okazaki K et al (2015) Metagenomic analysis of the bacterial community associated with the taproot of sugar beet. Microbes Environ 30(1):63–69

Turnbull GA, Morgan JAW, Whipps JM et al (2001) The role of bacterial motility in the survival and spread of *Pseudomonas fluorescens* in soil and in the attachment and colonization of wheat roots. FEMS Microbiol Ecol 36:21–31

Tyson GW, Lo I, Baker BJ et al (2005) Genome-directed isolation of the key nitrogen fixer *Leptospirillum ferrodiazotrophum* sp. nov. from an acidophilic microbial community. Appl Environ Microbiol 71:6319–6324

Unno Y, Shinano T (2013) Metagenomic analysis of the rhizosphere soil microbiome with respect to phytic acid utilization. Microbes Environ 28:120–127

Uren NC (2000) Types, amount, and possible functions of compounds released into the rhizosphere by soil-grown plants. In: Pinton R, Varanini Z, Nannipieri P (eds) The rhizosphere: biochemistry and organic substances at the soil-plant interface. Marcel Dekker, New York, pp 19–40

Uroz S, Ioannidis P, Lengelle J et al (2013) Functional assays and metagenomic analyses reveals differences between the microbial communities inhabiting the soil horizons of a norway spruce plantation. PLoS One 8(2):e55929

Van Diepeningen AD, de Vos OJ, Korthals GW et al (2006) Effects of organic versus conventional management on chemical and biological parameters in agricultural soils. Appl Soil Ecol 31:120–135

Verma SC, Singh A, Chowdhury SP et al (2004) Endophytic colonization ability of two deep-water rice endophytes, Pantoea sp. and Ochrobactrum sp. using green fluorescent protein reporter. Biotechnol Lett 26:425–429

Whitman WB, Coleman DC, Wiebe WJ (1998) Prokaryotes: the unseen majority. Proc Natl Acad Sci U S A 95:6578–6583

Widmer F, Hartmann M, Frey B et al (2006) A novel strategy to extract specific phylogenetic sequence information from community T-RFLP. J Microbiol Methods 66:512–520

Wu L, Han T, Li W et al (2013) Geographic and tissue influences on endophytic fungal communities of *Taxus chinensis* var. mairei in China. Curr Microbiol 66:40–48

Wu LY, Thompson DK, Li GS et al (2001) Development and evaluation of functional gene arrays for detection of selected genes in the environment. Appl Environ Microbiol 67:5780–5790

Wu T, Chellemi DO, Graham JH et al (2008) Comparison of soil bacterial communities under diverse agricultural land management and crop production practices. Microb Ecol 55:293–310

Xu Z, Hansen MA, Hansen LH et al (2014) Bioinformatic approaches reveal metagenomic characterization of soil microbial community. PLoS One 9(4):e93445

Yuan Z, Druzhinina IS, Labbe J et al (2016) Specialized microbiome of a halophyte and its role in helping non-host plants to withstand salinity. Sci Rep 6:32467

Zakria M, Njoloma J, Saeki Y (2007) Colonization and nitrogen-fixing ability of *Herbaspirillum sp*. strain B501 gfp1 and assessment of its growth-promoting ability in cultivated rice. Microbes Environ 22:197–206

Zolla G, Badria DV, Bakker MG et al (2013) Soil microbiomes vary in their ability to confer drought tolerance to Arabidopsis. Appl Soil Ecol 68:1–9

Unravelling the Interaction of Plant and Their Phyllosphere Microbiome

10

Chetana Roat and Meenu Saraf

Abstract

The phyllosphere is a type of an ecosystem having economical and ecological values and comprises of several microbial population that are present on the aerial parts of the plant. It is a vibrant environment where inhabitant microorganisms have the ability to change in humidity, temperature and heat during the whole day and night. The interaction between the microorganisms in the phyllosphere influences the growth of plants in natural habitat, the productivity of agricultural crops and the protective of horticultural produce for human consumption. Phyllosphere microbial community will help us to understand a deep knowledge of the phyllosphere microorganisms. The focus of this chapter will be (1) diversity study of phyllosphere microbial community; (2) mechanisms of phyllosphere microbe colonization; (3) understanding of the leaf structure, environmental and ecological parameters for growth and survival colonists; (4) understanding of the influences of biotic and abiotic factors on phyllospheric microbiome; (5) adaptations of microorganisms for establishment in the habitat of phyllosphere; and (6) significance of plant genotypic control of phyllosphere communities and its role in plant protection and plant growth. Futhermore, the insights study of phyllosphere microbiota; structure, function and valuable challenges for future research.

Keywords

Phyllosphere • Microbiome • Plant genotype • Biotic and abiotic factors • Host-microbe associations

---

C. Roat (✉) • M. Saraf
Department of Microbiology and Biotechnology, University School of Sciences, Gujarat University, Ahmedabad 380 009, India
e-mail: chetana.roat@gmail.com

R.P. Singh et al. (eds.), *Understanding Host-Microbiome Interactions - An Omics Approach*, DOI 10.1007/978-981-10-5050-3_10

Abbreviations

ARDRA	Amplified ribosomal DNA restriction analysis
EPS	Extracellular polymeric substances
IAA	Indole-3-acetic acid
PCR	Polymerase chain reaction
PMC	Phyllosphere microbial communities
rDNA	Ribosomal DNA
ROS	Reactive oxygen species
RT-PCR	Reverse transcription polymerase chain reaction
VOC	Volatile organic compound

10.1 Introduction

Phyllosphere microbiota comprises of interaction and the relationship between microorganisms, plants and the environment. Plant pathologists and microbiologists have observed the habitat of microorganisms in the aerial part of the plant and called it the phyllosphere, in the 1950s, as few microorganisms improved plant performance while some microorganisms act as pathogens, threatening plant health (Corinn Vacher et al. 2016). The phyllosphere consists of the aerial portions of the plant, the set of photosynthetic leaves that are most common habitats for microorganisms on earth. The phyllosphere consists of different plant parts like the leaves (phylloplane), stem (caulosphere), flowers (anthosphere) and fruits (carposphere), and the presence and composition of the microorganisms in all the parts are significantly different (Junker et al. 2011). The leaf surface of the plant is colonized by microbes like fungi and bacteria from seed, soil, air and water through animal sources (Vorholt 2012).The phyllosphere harbours hundreds of microorganisms which have either beneficial or deleterious effects on the plants (Penuelas et al. 2014).The universal surface area of phyllosphere has been calculated approximately 10^9 km^2, and the presence of bacterial population is approximately 10^{26} cells. The total fungal population estimation of the phyllosphere is estimated to be lower (Lindow et al. 2003). Among all the cellular organisms, bacteria are estimated as the most abundant participant of phyllospheric community with cell density of 10^8 cells/cm^2 of leaf surface (Leveau et al. 2006). The distribution and multiplication of microorganisms are irregular and uneven on phyllosphere due to abiotic and biotic factors and physiological and anatomical behaviour of plant leaf (Remus-Emsermann et al. 2012). Phyllospheric microbiota have an ability to influence and alter both the structure and biochemistry of the plant like leaf functional characters, affect plant growth and affect ecosystem like water and nutrient cycling. Besides these, phyllospheric microorganisms also actively participate in secreting bioactive compounds called secondary metabolites like

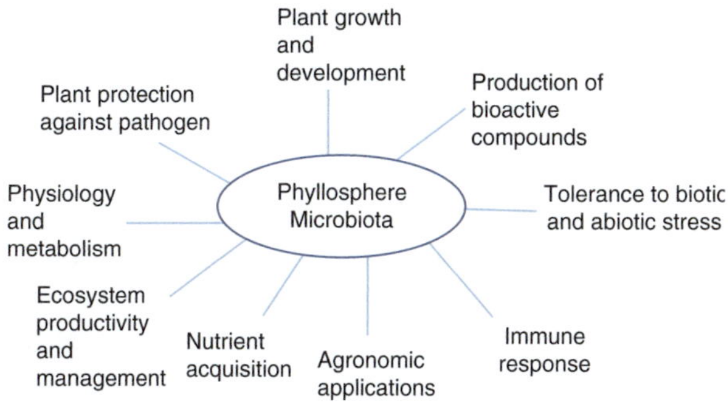

Fig. 10.1 Different roles of phyllosphere microbiodata

polyphenols, terpenoids, alkaloids and organic volatile compounds from the plant due to its locations, inside the leaf tissue or between plant surface and the atmosphere. (Bringel and Couee 2015). The association between the microorganisms and the phyllosphere is temporary due to harsh environmental conditions, and those who resist on phyllosphere are called as "true" occupants which persist, multiply and grow on phyllosphere. Quantitatively the phyllosphere microbiota vary at different developmental and genotypic stages of the plant (Penuelas et al. 2012). Unlike rhizosphere, phyllosphere microorganisms can influence neighbouring environmental ecosystem, and their continued existence is regulated by the plant itself (Barcel et al. 2012). Understanding the mechanism involves microorganisms, plants and the environment for phyllosphere microbiota (Fig. 10.1). This chapter will discuss interaction between the phyllosphere and microorganisms with their structures, functions and mechanisms and applications with important challenges for future research.

10.2 Diversity and Structure of Phyllosphere Microbiota

The microbial phyllosphere is characteristically regarded by 16S and 18S ribosomal RNA (rRNA) gene for bacteria and fungi. The high level of variation in sequences either within the region or in combinations of regions provides taxonomic resolutions that are often comparable with that achieved using whole rRNA gene sequences (Maughan et al. 2012). Earlier the culture-dependent techniques were used (Kuklinsky-Sobral et al. 2004), but currently the advanced molecular tools are being used to improve the technology like polymerase chain reaction (PCR), reverse transcription polymerase chain reaction (RT-PCR), primer designing for amplification of hypervariable regions and amplicon pyrosequencing for amplification of only microbial DNA and analyse the composition of microbial phyllospheric populations (Bulgarelli et al. 2012; Kim et al. 2012).This advancement has opened the door to explore further research for leaf-associate microorganisms like epiphytes and endophytes, by culture-independent methods. Epiphytic microorganisms that display enzyme indicators which were reported in the phyllosphere of tobacco signify that signalling routes may be participating in framing epiphytic microbial communities. The epiphytes also produce exopolysaccharides for adhesion or protection from desiccation (Monier and Lindow 2004) and secrete phytohormones, which enable nutrient secretion from plant tissue and as a result help in relaxation of plant cell wall (Vorholt 2012), while the endophytic microorganism resides within the specific chemical environment of host plant tissue like the leaf, stem, root, etc. and adapt to plant physiology in order to produce plant secondary metabolites (Becon and White 2000). The culture-independent molecular techniques proved the phyllosphere is a composite community of microbes. Bacterial communities at the phylum level, across a large array of farming crops (Fig.10.2) like rice, wheat, lettuce, apple, spinach and naturally occurring trees/plants, are composed of Actinobacteria, Bacteroidetes, Firmicutes and Proteobacteria (Bulgarelli et al. 2013). The study of population composition at the genus level recommends that Methylobacterium, Pantoea, Sphingomonas, Pseudomonas, Bacillus Massilia and Arthrobacter steadily occurred as a part of the phyllosphere microbiota through varied types of plant species. There are examples of phyllosphere studies with molecular methods where spatial and temporal discrepancy in microbial population and the other aspects that come this difference show in Table 10.1.

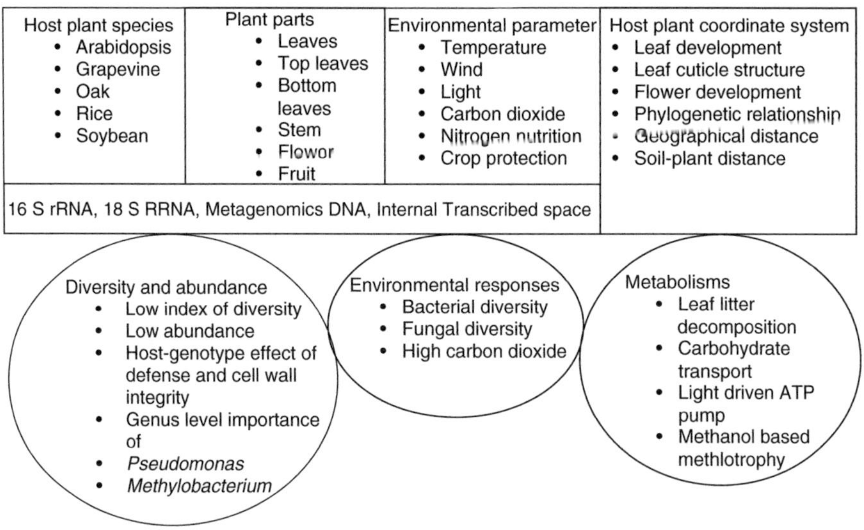

Fig. 10.2 Phyllosphere microbiota biodiversity, metabolism and environmental pliability (Modified and adapted from Bringel and Couee 2015)

Table 10.1 Molecular approaches to study the phyllosphere communities (Rastogi et al. 2013)

Molecular approach	Plant	Major findings	Reference
16S rRNA gene pyrosequencing	Grape	On the surface of leaves and berries from the same grapevine with significantly different bacterial communities	Leveau and Tech (2011)
16S rRNA gene pyrosequencing	Spinach	*Proteobacteria* and *Firmicutes* were the most commonly associated bacteria on field-grown spinach leaves. At genus level, communities were largely composed of *Pseudomonas*	Lopez-Velasco et al. (2011)
16S rRNA gene pyrosequencing	Lettuce	Planting season and irrigation practices (sprinkler/drip) together explained majority of the variation in phyllosphere microbiota composition. *E. coli* O157:H7 inoculation resulted in lower population sizes and induced minor, but lasting changes in microbiota composition	William et al. (2013)
16S rRNA gene pyrosequencing	Pine and other tree	Tree species, not the location, is the major determinant of phyllosphere bacteria community composition	Redford et al. (2010)
16S rRNA gene pyrosequencing, metaproteogenomics	Soyabean, clover, Arabidopsis	Unique metabolic adaptation contributes to the epiphytic fitness of *Sphingomonas* and *Methylobacterium*	Delmotte et al. (2009)
Metaproteogenomics	Rice	Phyllosphere communities were largely composed of *Rhizobium*, *Methylobacterium* and *Microbacterium*. Several methylotrophic enzymes were assigned to *Methylobacterium*, suggesting their role in the carbon cycle	Knief et al. (2012)
ITS pyrosequencing	Oak	Urban and rural management practices affect fungal communities in the oak phyllosphere	

(continued)

Table 10.1 (continued)

Molecular approach	Plant	Major findings	Reference
ITS pyrosequencing	Beech	Fungal communities showed variation even at the smallest spatial scale of individual leaf surfaces. Plant genotype was identified as a major driver of the fungal community composition	Cordier et al. (2012)
ITS pyrosequencing	Balsam poplar	Plant species was found as the major determinant of fungal community composition	

10.3 Process of Colonization, Recognization and Adhesion in the Phyllosphere

The cell and the spores of microorganisms occupying the surface of the leaf first interact with a cuticle, an exogenous and a hydrophobic waxy layer which defends the plant against dryness and other stresses. The formation of long-chain fatty acids with sterols and terpenoids shows 15% of leaf dry weight (Eglinton and Hamilton 1967). The architecture and composition may vary among plant species and environmental conditions. The shiny or crystal appearance of the leaf, resulting in a dull, powdery bloom form, is due to epicuticular waxes (Yeats and Rose 2013). Some microbial communities on the phyllosphere are either affected by the wax phenotype positively or negatively or not affected like the permanent residents of microorganisms like Pseudomonadaceae, Methylobacteriaceae, Flexibacteriaceae, Flavobacteriaceae, Sphingomonadaceae, Rhizobiaceae, Enterobacteriaceae, etc. The phyllosphere microbiota affected the genetic determinants of cuticle formation (Bodenhausen et al. 2014). It revealed that cuticular wax properties played a specific role for adapted microbial communities.

The endophytic and epiphytic microorganisms of the phyllosphere cross the cuticle as well as participate in the abundance of ozone in the environment (Nakamiya et al. 2009, Nadalig et al. 2014). The *Arthrobacter* genus from the phylum Actinobacteria is able to degrade various organic molecules and remain in the phyllosphere. The other species of *Arthrobacter* degrade pesticides like glyphosate, phenylurea herbicides, malathion and aromatic hydrocarbon like phenol, s-triazines, phenanthrene and chlorophenols (Scheublin and Laveau 2013).Phyllospheric microbiome also degrades organic and aromatic molecules, participates in controlling the atmospheric pollution in industrial and urban areas and also shows significant role in sustainable agricultural environments by atmospheric drift of phytosanitary product. Hence, the epiphytic and endophytic microorganisms can also be envisaged for the beneficial effect on plants and could be used as probiotic agents (Bercel 2012). Adhesion plays role to conflict from separation raindrops or by rainfall and is a condition for the EPS formation comprising of mucus that gives defence from dryness.

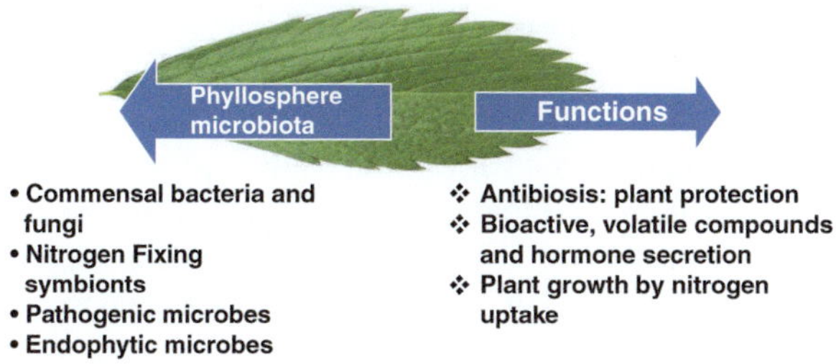

Fig. 10.3 Phyllosphere microbiodata association

Adhesion is effective for leaf expansion and epiphytic colonization. Epiphytic microorganisms can have the capability for aggregate formation and synthesized exopolysaccharide which can improve adhesion and protection from desiccation (Monier and Lindow 2004) and also synthesize and secrete plant hormones, like indole-3-acetic acid (IAA), which enable nutrient secretion from plant tissues and help in relaxation of the plant cell wall. Although the understanding of these adaptive mechanisms remains incomplete (Fig. 10.3).

10.4 Plant Microbe Interaction

10.4.1 Interaction Between Bacteria on Phyllosphere

Most of the bacteria on the leaf surface form large aggregates instead of small groups or single cells. These aggregates are formed at junction of epidermal cells with veins and base of trichomes which are bordered with extracellular polymeric substances. The extracellular polymeric substances maintain moisture around the bacteria and concentrate detoxifying enzymes (Lindow et al. 2003; Baldotto et al. 2008; Monier et al. 2004). The environment of host plant is altered by the bacteria which has been associated with plant pathogenic bacteria, *P. syringae*, which causes diseases in plants related with their epiphytic population on leaves (Stromberg et al. 1999). The epiphytic bacteria present on leaves of plant are swept up into the atmosphere. They have the capability to form precipitation in the clouds by catalysing ice formation (Morris et al. 2014). Methylotrophic bacteria use methanol or methane as a carbon source and are quite rich in the phyllosphere and useful for plant growth like the genus *Methylobacterium* (Abanda-Nkpwatt et al. 2006).The phyllosphere diazotrophic bacteria use atmospheric nitrogen as a source of nitrogen to assess their community structure (Furnkranz et al. 2008). With drought conditions, the nitrogen-fixing bacteria have been raised, suggesting that their involvement may extend the ability of the plant to acclimatize in the environment. The capability of *P. syringae*, a plant pathogenic bacteria, causes disease strongly linked with their epiphytic population size on phyllosphere and may be the subsection of the community that is without symptoms on plant leaves and hence in more close interaction with plant cells (Beattie et al. 1999).

10.4.2 Interaction with Fungi

Filamentous fungi and yeast are massively colonized on the phyllosphere. The density of yeast is 500 cells/cm^2, but it varies from plant to plant species and within the plant (Inacio et al. 2010). Some fungi present in the phyllosphere and act as a biological control due to its ability to outcompete pathogenic microorganisms like *Aureobasidium pullulans* (Cordier et al. 2012). Some fungal pathogens like *Erysiphe*, *Blumeria* and *Podosphaera* elongated their mycelium on the leaf surface, all pervading the cuticle then rupturing the cell wall with particular structure called as appressoria. Once infected, the mycelia grow superficially, covering grey or white coat on the leaves (Glawe 2008). Within the leaf tissue, the endophytic fungal species develop without causing noticeable symptoms, transferred from one adult plant to the next plant, and develop an epiphytic stage before entering into the leaf tissues. (Rodriguez et al. 2009). Some of the endophytic fungi are latent pathogens, for example, palm tree *Iriartea deltoidea*, which produces reactive oxygen species (ROS) by *Diplodia mutila*, an endophytic fungus, and alters it to a pathogen, which confines plantlet persistence to shaded areas (Alverez-Loayza et al. 2011). Some fungi defend the plant from stresses, including herbivores, pathogens and drought. Some fungi are involved in the decomposition of cellulose in senescent leaves and young litter which are then replaced with new colonizers with a higher capacity to decompose lignin that belongs to the higher fungi *Basidiomycota*. The fungi present in riparian ecosystem are termed as aquatic hyphomycetes and play a key role in the functioning of aquatic food webs, as its action enhances the palatability of leaves to invertebrates (Barlocher 2016). In more than 50 plant species, the aquatic hyphomycetes have been found in the phyllosphere (Chauvet et al. 2016). Fungal communities of phyllosphere play a significant role in nutrient cycling and in the functional coupling of aquatic and terrestrial ecosystems. Henceforth, to recognize the procedures of shaping these communities and evaluating their effect to global change is significant.

10.5 Study of the Microbial Community on Leaf

The leaf surface area is a hostile atmosphere for the microbes. The surface of the leaf is affected by relative humidity, sunlight, fluctuating temperature and occurrence and nonoccurrence moisture due to dew and raindrops. On the leaf surface, sometimes microscopic water films occur due to condensation of water vapour which enters into the stomata (Burkhardt and Hunsche 2013). Leaf surfaces are multifaceted microenvironments which show two-dimensional and three-dimensional diverse structures. The dorsal and ventral sides of the leaf surface (Eglinton and Hamilton 1967; Schreiber et al. 2004; Reisberg et al. 2013) are affected by the microbes which live on plant surfaces and the access to nutrients from leaf tissues called as epiphytic microorganism (Bulgarelli et al. 2013), by imparting less or more defence from sunlight (Atamna-Ismaeel et al. 2012) or by showing permission for diffusing in the endosphere of plant (Hirano and Upper

2000; Schreiber et al. 2004). Inside and outside, the live conditions of the leaves are produced by the fluxes of CO_2, O_2 and water vapour resulting from photosynthesis, respiration and evapotranspiration. Many parameters affect the habitat of microbes accomplished by microorganisms on leaves; first, a very thin laminar layer around the leaf, where stomata emitted moisture, can be concealed, thus diminishing the stress of water where epiphytes are revealed. And also, some of the bacteria invade inside the leaf instead of the exposed outside leaf, and to escape the stresses outside of the leaf, they inhabit in sub-stomatal cells or other inner locations (Lindow and Brandl 2003). Many species from the angiosperm group showed the dissimilar structure of leaf both on ventral and the dorsal sides. The ventral side of the leaf consists of a thick cuticle, epidermal cells inside which are tightly occupied sheet of palisade mesophyll cells helpful in optimizing radiance. The transpiration rate and exchange of gases is controlled by the dorsal side part of the leaf which comprise of an epidermal layer with many stomata, beneath which are spongy mesophyll cells loosely arranged (Vacher et al. 2016). The stomata is the site of entry for microorganisms to the inner leaf tissue. Once it enters, it will assemble the interspace between the cells of the spongy mesophyll. As a result, they get nearer to the smaller veins of the leaf where sieve element of the phloem uploads the sugars (Rennie and Turgeon 2009). Plant photoassimilates glucose, fructose and sucrose which are present on leaf surfaces, but fluctuation of day or night alters the changes in the plant metabolite synthesis, which also affects the nutrient availability and growth of epiphytic microorganisms. Among all the metabolites of plant, carbohydrate is the most functional under stress conditions like abiotic and biotic (Trouvelot et al. 2014). Some of the plant metabolites are not directly and freely accessible for epiphytic microorganisms like amino acids, soluble sugars, amines, polyols, VOCs such as isoprenoids, halogenated compounds or alcohols, also salts and plant water. The waxy and lipidic cuticles protect the plant leaf surface which bound metabolite fluxes and water; hence the biochemical exchanges are dependent on different gateways like exudation, leaching, excretion, guttation, infiltration and wounding. All these properties belong to the oligotrophic habitat with limitations in nitrogen and carbon sources. The presence of nutrient containing carbon on the leaves is a main factor of epiphytic colonization. On a well-fertilized plant, the bacterial communities are restricted by nitrogen and carbon accessibility.

10.6 Influence of Biotic and Abiotic Factors on Phyllospheric Microbiome

The microbiome of leaf is affected by stresses like abiotic and biotic factors in determining microbiota composition and dynamics. Phyllospheric microbiome is affected by biological and environmental factors like host genotype, plant development, climate, geographical location and seasonal changes. The phyllosphere is an unlock structure where microbes can penetrate plant leaves by immigration from the other plants, soil, atmosphere, animals and insects. The composition of the phyllosphere microbiome is mainly dependent on the host genotype (de Oliveira Costa et al. 2012; Kim et al. 2012), while the composition of the community is influenced by the geographic location (Rastogi et al. 2012). Some perennial plant communities alter noticeably from 1 year to the next year and have more seasonal changeability, while some perennial plants have season-dependent communities that are more alike from year to year (Jackson and Denney 2010). From the soil-grown plants, phyllosphere communities show similar characteristics to soil communities (Perazzolli et al. 2014) whereas media-grown, sterile plantlets have shown same properties with airborne communities. Birth-, migration- and death-like demographic aspects of the plant can show an important function for the growth of microbes (Vorholt 2012). Overall the plant phyllosphere is very complex and affected by biological and environmental factors on the composition of the population which remains indistinct when all these factors are measured simultaneously. In natural environment, the phyllosphere microbes of seasonal farming plants are mainly affected by temporal population growth, whereas host genotype and spatial division may imitate the exact preference for study of species and the nearby geographic distances between the replicate areas. At the beginning of the summer season, communities' samples strongly resembled the soil population, and, as the season proceeds, the phyllosphere microbiome developed progressively discrete and little wide. Therefore, due to this deviation in population structure, species that is significantly used as biocontrol agents in the beginning of summer season may not be efficient in the end of the summer. To realize that the natural progression in the population is controlled, the development of the microbiome which is useful in biocontrolling farming plantlets will need awareness about population size and shape changes in the continuation of the progressive season (Copeland et al. 2015).

10.7 Establishment of Phyllosphere Habitat by Microbial Adaptations

Adaptation is the ability of the phyllosphere microbiota to withstand stress conditions such as deficient supply of nutrients, production of bioactive compounds like antifungal and antibacterial secondary metabolites and seasonal environmental conditions (Vorholt 2012). Phyllosphere microbiota can be expected to have common strategies to survive these stresses, in addition to species-specific strategies. During photosynthesis, the phyllosphere is exposed to oxygen and the light during daytime. The microbial colonizers prone to reactive oxygen species damage the protein, lipid and nucleic acid. Photolyases, DNA repair caused by ultraviolet light and the production of pigments are being used as protection mechanisms by microbial epiphytic fitness. Catalases and superoxide dismutase enzymes play an active role in reactive oxygen species (ROS) detoxifications (Vorholt 2012). The common environmental factor in phyllosphere is dryness. The secretion of bioactive compounds from microbial masses is helpful to overcome the dryness. Exopolysaccharides are produced by the aggregates, which maintained moisture surrounding the bacteria, and participate in epiphytic fitness. Some phyllosphere bacteria secreted the biosurfactants that increase wettability. To overcome the osmotic stress, the epiphytes like

Pseudomonas spp. and others can defend themselves from the osmotic stress by secreting trehalose or choline or importing plant-derived osmoprotectants or de novo synthesis. Plants have the unique property to produce bioactive compounds which sometimes are antimicrobial in nature and produce antibiotics; hence plant adapted themselves from these pathogenic microorganisms, for example, *P. syringae*, a plant pathogen which is important for evasion of the inhabitant immune reaction and plant reproduction (Stoitsova et al. 2008).

10.8 Impact of Phyllospheric Microorganisms on Plant Growth and Plant Protection

Leaf-associated microorganisms are well known for symbiosis and mutualism relationship with host plant that can influence host plant growth and function, like the production of hormones and growth-promoting nutrients, and also prevent the hosts from infection from disease-producing agents (Innerebner et al. 2011). Under different environmental conditions, phyllosphere microbes have the capability to effect on plant ecosystem and biogeography contribution which effect on plant activities (Fürnkranz et al. 2008), but the discrepancy of bacterial biodiversity in the phyllosphere of the host plants is not well understood. Phyllosphere microbes secrete plant growth regulators, volatile and non-volatile compounds, which may influence the plant growth, morphogenesis and plant immunity. Also they act as phytostimulators, biofertilizers and biopesticides to protect against invading pathogens and for plant growth, development and health. Sometimes plant defence chemicals are degraded by plant foliage-associated bacteria which minimize defence against insect defoliators (Mason et al. 2014). On plant surfaces, the genera *Enterobacter*, *Pseudomonas*, and *Stenotrophomonas* of bacterial symbionts, secreted by the Colorado potato beetle larvae, suppress the anti-herbivore defences and enhance the microbial defence in tomato plant (Chung et al. 2013). The interaction between herbivore masticate insect and its single host plant, *Cardamine cordifolia*, experimental showed that some bacterial strains showed difference on the way of ecologically impacted insect herbivores while some bacterial strain, for example, *Pseudomonas* species, helped host choice by herbivores (Humphrey et al. 2014). Phyllospheric microbiome lives in a sunlight-exposed habitat. The energy formed by photochemical conversion of the sunlight that can organic sources from the host plant could be a beneficial for development in an inadequate amount of nutrient. Study of metagenomic information has showed that in phyllospheric communities, the existence of bacterial rhodopsin genes is there (Bringel and Couee 2015). Retinal-dependent rhodopsin proton pumps which are found in some epiphytic microorganisms stimulated by the wavelength different from the carotenoids and chlorophyll absorption spectra, which participate in photosynthetic processes and production of the plant sugar sources, are finally accessible to epiphytic microorganisms (Atamna-Ismaeel et al. 2012). The light-dependent processes are affected by nitrogen and carbon sources in nutrient regulation and signalling (Moran and Miller 2007). The biosynthetic pathway for the production of plant bioactive molecules in epiphytic bacteria can be affected by the nitrogen and carbon sources and the changes in plant-light interactions and photoassimilate production in the host plant (Sulmon et al. 2011). Manching et al. (2014) have recently discussed universal association between leaf epiphytic bacterial species and plant nitrogen balance in maize crop.

10.9 Phyllospheric Plant–Microorganism–Atmosphere Interactions: Physiological, Ecological and Molecular Studies

The role of physiological, ecological and molecular studies of microbial communities on both sides of the leaves, inside and outside, is likely to affect plant growth and its metabolisms (Lindow et al. 2003; Rastogi et al. 2013), which influence the ecosystem and environmental efficiency eventually. In the phyllosphere, the nitrogen fixation is the key process where nitrogen is added in tropical humid ecosystems (Abril et al. 2005). The tropical rainforest plants, where the formation of phyllospheric populations of nitrogen-fixing microorganisms has been observed but temperate-forest ecosystems have also been observed for the presence of nitrogen-fixing microorganisms, their abundance and diversity vary depending on the availability of water (Pennuelas et al. 2012). Development of plant can also be affected by the foliar microorganisms as they are involved in the production of growth hormones. The external and internal microbiotas can have many other roles, like the indirect defence against pathogens, by the interaction of foliar plant pathogen and commensal bacteria (Vorholt 2012), or secretion and production of different types and quantities of bioactive compounds and emissions of organic volatile compounds (Bulgarelli et al. 2013). Speciation, dispersal, drift and selection are the four processes which shaped the ecological communities. Diversity of the microorganisms that occurs on the leaf surface of plant comes under dispersal and evolutionary diversification. External selection supports those microbes which are mainly adapted to the common situtations like leaf morphology, chemistry and microclimate. The selected microbes then grow and reproduce. The biotic interaction where internal selection is done, like parasitism and competition, controls the shape and structure of the microbial population. Stochastic modification in the population structure is done by drift. The shape and the overall structure of the microbial population are influenced by plant type and structure, the environmental conditions and the population itself. Molecular studies are the best tool, which help to get information about diversity, species richness and analysis of microbial community (Brusseau et al. 1994). Microbial diversity analysis is done by the widely used technique amplified ribosomal DNA restriction analysis (ARDRA), and it is significant to discriminate the species level of the microbes. This technique requires amplification of the 16S and 18S ribosomal DNA (rDNA) region proceeded by restriction enzyme digestion (Heyndrickx et al. 1996). Species richness, occurrence, community structure and diversity in the phyllosphere of the tropical plants are studied by full-length sequencing detail of 16S rDNA, differential carbon-substrate consumption pattern and ARDRA. For example, *Methylobacterium* are classified taxonomically on the basis of 16S rRNA gene sequence data (Tsuji et al. 1990), and the carbon-substrate utilization study of all the species of *Methylobacterium* are well deliberated (Jourand et al. 2004).

10.10 Current Progress and Future Challenge

Microbiological study of phyllosphere is a promising research field at the early stages. The phyllosphere is a best part of the plant to get the basic knowledge of plant structure, growth functioning and environmentalism, especially to perform experiment and visual inspection, and these studies are helpful to have an impact on different aspects of plant like morphological and anatomical condition and physiological process where atmospheric gases participated in phyllosphere functioning. New technologies such as proteogenomics and metaproteogenomics and next-generation sequencing used for community profiling are novel platforms to get knowledge of the shape of microbial population and to explore new objectives for theory of research, escorted by revealing new protein function that is significant for growth, development and survival in the phyllosphere. But still more knowledge is required to know about phyllosphere microorganisms and their interaction in situ and other complementary approaches. Future studies and research work require to explore these dynamic and complex interaction approaches and evaluate their role in the growth of plant and physiology of phyllosphere microorganisms. Ecosystem functioning, the diversity of phyllosphere microbial communities (PMCs) and plant performance have to be deliberated. Hence, the next research needs to understand and identify the correct community parameters of PMCs, ecosystem functions and plant performance. The link between the evolution of PMCs, the dynamics and adaptation mechanism in plants to changing environmental conditions requires further understanding. The effect of environmental change conditions on the plant fitness and diversification of PMCs requires to be explored.

Acknowledgements The authors acknowledge the financial support obtained from the Department of Biotechnology, Government of India and New Delhi, India, for providing DBT BioCARe Women Scientist Fellowship under Grant Ref. BT/Bio-CARe/03/420/2012 & 03-09-2013.

References

Abanda-Nkpwatt D, Mush M, Tschiersch J, Boettner M, Schwab W (2006) Molecular interaction between Methllobacteria extorquens and seedlings: growth promotion, methanol consumption, and localizations of the methanol emission site. J Exp Bot 57(15):4025–4032

Abril AB (2005) The importance of phyllosphere microbial populations in nitrogen cycling in the Chaco semi-arid woodland. J Trop Ecol 21:103–107

Alvarez-Loayza P, White JF, Torres MS, Balslev H, Kristiansen T, Svenning JC, Gil N (2011) Light converts endosymbiotic fungus to pathogen, influencing seedling survival and niche-space filling of a common tropical tree, Iriartea deltoidea. PLoS One 6:e16386
Atamna-Ismaeel N, Finkel OM, Glaser F, Sharon I, Schneider R, Post AF (2012) Microbial rhodopsins on leaf surfaces of terrestrial plants. Environ Microbiol 14:140–146
Bacon CW, White JE (2000) Microbialendophytes. Marcel Dekker New York, USA:4–5
Baldotto LEB, Olivares FL (2008) Phylloepiphytic interaction between bacteria and different plant species in a tropical agricultural system. Can J Microbiol 54:918–931
Barlocher F (2016) Aquatic hyphomycetes in a changing environment. Fungal Ecol 19:14–26
Beattie GA, Lindow SE (1999) Bacterial colonization of leaves: a spectrum of strategies. Phytopathology 89:353–359
Bercel A (2012) Novel techniques and finding in the study of plant microbiota: search for plant probiotics. Plant Sci 193:96–102
Bodenhausen N, Bortfeld-Miller M, Ackermann M, Vorholt JA (2014) A synthetic community approach reveals plant genotypes affecting the phyllosphere microbiota. PLoS Genet 10:100428310–100428137
Bringel F, Couee I (2015) Pivotal roles of phyllosphere microorganisms at the interface between plant functioning and atmosphere trace gas dynamics. Front Microbiol 6:486
Brusseau GA, Bulygina ES, Hanson RS (1994) Phylogenetic analysis and development of probes for differentiating methylotrophic bacteria. Appl Environ Microbiol 60:626–636
Bulgarelli D, Rott M, Schlaeppi K, Ver Loren van Themaat E, Ahmadinejad N, Assenza F, Rauf P, Huettel B, Reinhardt R, Schmelzer E, Peplies J, Gloeckner FO, Amann R, Eickhorst T, Schulze-Lefert P (2012) Revealing structure and assembly cues for *Arabidopsis* root-inhabiting bacterial microbiota. Nature 488:91–95
Bulgarelli D, Schlaeppi K, Spaepen S, Loren V, van Themaat E, Schulze-Lefert P (2013) Structure and functions of the bacterial microbiota of plants. Annu Rev Plant Biol 64:807–838
Burkhardt J, Hunsche M (2013) "Breath figures" on leaf surfaces formation and effects of microscopic leaf wetness. Front Plant Sci 4:422
Chauvet E, Cornut J, Sridhar KR, Selosse MA, Barlocher¨ F (2016) Beyond the water column: aquatic hyphomycetes outside their preferred habitat. Fungal Ecol 19:112–127
Chung SH, Rosa C, Scully ED, Peiffer M, Tooker JF, Hoover K (2013) Herbivore exploits orally secreted bacteria to suppress plant defenses. Proc Natl Acad Sci U S A 110:15728–15733
Copeland JK, Yuan L, Layeghifard M, Wang PW, Guttman DS (2015) Seasonal community succession of the phyllosphere microbiome. Mol Plant-Microbe Interact 28:274–285
Cordier T, Robin C, Capdevielle X, Fabreguettes O, Desprez-Loustau ML, Vacher C (2012) The composition of phyllosphere fungal assemblages of European beech (*Fagus sylvatica*) varies significantly along an elevation gradient. New Phytol 196:510–519
Delmotte N, Knief C, Chaffron S, Innerebner G, Roschitzki B, Schlapbach R, von Mering C, Vorholt JA (2009) Community proteogenomics reveals insights into the physiology of phyllosphere bacteria. P Natl Acad Sci USA 106:16428–16433
De Oliveira Costa L, de Queiroz M, Borges A, de Moraes C, de Araujo E (2012) Isolation and characterization of endophytic bacteria isolated from the leaves of the common bean (*Phaseolus vulgaris*). Braz J Microbiol 43:1562–1575
Eglinton G, Hamilton RJ (1967) Leaf epicuticular waxes. Science 156:1322–1335
Fürnkranz M (2008) Nitrogen fixation by phyllosphere bacteria associated with higher plants and their colonizing epiphytes of a tropical lowland rainforest of Costa Rica. ISME J 2(5):561–570
Glawe DA (2008) The powdery mildews: a review of the world's most familiar (yet poorly known) plant pathogens. Annu Rev Phytopathol 46:27–51
Hirano SS, Upper CD (2000) Bacteria in the leaf ecosystem with emphasis on *Pseudomonas syringae*-a pathogen, ice nucleus, and epiphyte. Microbiol Mol Biol Rev 64:624–653
Heyndrickx M, Vauterin L, Vandamme P, Kersters K, De Vos P (1996) Applicability of combined amplified ribosomal DNA restriction analysis (ARDRA) patterns in bacterial phylogeny and taxonomy. J Microbiol Methods 26:247–259
Humphrey PT, Nguyen TT, Villalobos MM, Whiteman NK (2014) Diversity and abundance of phyllosphere bacteria are linked to insect herbivory. Mol Ecol 23:1497–1515

Inacio´ J, Ludwig W, Spencer-Martins I, Fonseca A (2010) Assessment of phylloplane yeasts on selected Mediterranean plants by FISH with group- and species-specific oligonucleotide probes. FEMS Microbiol Ecol 71(1):61–72

Innerebner G, Knief C, Vorholt JA (2011) Protection of *Arabidopsis thaliana* against leaf-pathogenic*Pseudomonas syringae* by *Sphingomonas* strains in a controlled model system. Appl Environ Microbiol 77(10):3202–3210

Jackson CR, Denney WC (2010) Annual and seasonal variation in the phyllosphere bacterial community associated with leaves of the southern Magnolia (*Magnolia grandiflora*). Microb Ecol 61:113–122

Jourand P, Giraud E, Bena G, Sy A, Willems A, Gillis M (2004) *Methylobacterium nodulans* sp. nov., for a group of aerobic, facultatively methylotrophic, legume root-nodule-forming and nitrogen-fixing bacteria. Int J Syst Evol Microbiol 54:2269–2273

Junker RR (2011) Composition of epiphytic bacterial com- munities differs on petal sand leaves. Plant Biol 13:918–924

Kim M, Singh D, Lai-Hoe A, Go R, Abdul Rahim R, Ainuddin AN, Chun J, Adams JM (2012) Distinctive phyllosphere bacterial communities in tropical trees. Microb Ecol 63:674–681

Knief C, Delmotte N, Chaffron S, Stark M, Innerebner G, Wassmann R, von Mering C, Vorholt JA (2012) Metaproteogenomic analysis of microbial communities in the phyllosphere and rhizosphere of rice. ISME J 6:1378–1390

Kuklinsky-Sobral J, Araújo WL, Mendes R, Geraldi IO, Pizzirani-Kleiner, AA, Azevedo JL, Júlia, K S, and A, Pizzirani-Kleiner, A. (2004) Isolation and characterization of soybean-associated bacteria and their potential for plant growth promotion. Environ Microbiol 6:1244–1251

Leveau JHJ (2006) Microbial communities in the phyllosphere. In: Riederer M, Muller C (eds) Biology of the plant cuticle. Blackwell, Oxford, pp 334–367

Leveau JHJ, Tech JJ (2011) Grapevine microbiomics: bacterial diversity on grape leaves and berries revealed by high-throughput sequence analysis of 16S rRNA amplicons. Acta Hort(ISHS) 905:31–42

Lindow SE, Brandl MT (2003) Microbiology of the phyllosphere. Appl Environ Microbiol 69:1875–1883

Lopez-Velasco G, Welbaum GE, Boyer RR, Mane SP, Ponder MA (2011) Changes in spinach phylloepiphytic bacteria communities following minimal processing and refrigerated storage described using pyrosequencing of 16S rRNA amplicons. J Appl Microbiol 110:1203–1214

Maughan H, Wang PW, Diaz Caballero J, Fung P, Gong Y, Donaldson SL, Yuan L, Keshavjee S, Zhang Y, Yau YCW, Waters VJ, Tullis DE, Hwang DM, Guttman DS (2012) Analysis of the cystic fibrosis lung microbiota via serial Illumina se-quencing of bacterial 16S rRNA hypervariable regions. PLoS One 7:e45791

Manching HC, Balint-Kurti PJ, Stapleton AE (2014) Southern leaf blight disease is correlated with decreased maize leaf epiphytic bacterial species richness and the phyllosphere bacterial diversity decline is enhanced by nitrogen fertilisation. Front Plant Sci 5:403

Mason CJ, Couture JJ, Raffa KF (2014) Plant-associated bacteria degrade defense chemicals and reduce their adverse effects on an insect defoliator. Oecologia 175:901–910

Monier JM, Lindow SE (2004) Frequency, size and localization of bacteria aggregates on bean leaf surface. Appl Environ Microbiol 70:348–355

Moran MA, Miller WL (2007) Resourceful heterotrophs make the most of light in the coastal ocean. Nat Rev Microbiol 5:792–800

Morris CE, Conen F, Huffman JA, Phillips V, Pöschl U, Sands DC (2014) Bioprecipitation: a feedback cycle linking Earth history, ecosystem dynamics and land use through biological ice nucleators in the atmosphere. Glob Chang Biol 20:341–351

Nakamiya K, Nakayama T, Ito H, Shibata Y, Morita M (2009) Isolation and properties of a 2-chlorovinylarsonic acid-degrading microorganism. J Hazard Mater 165:388–393

Penuelas J, Summer season Kim M, Singh D, Lai-Hoe A, Go R, Abdul Rahim R, Ainuddin AN, Chun J, Adams JM (2012) Distinctive phyllosphere bacterial communities in tropical trees. Microb Ecol 63:674–681

Perazzolli M, Antonielli L, Storari M, Storari M, Puopolo G, Pancher M, Giovannini O, Pindo M, Pertot I (2014) Resilience of the natural phyllosphere microbiota of the grapevine to chemical and biological pesticides. Appl Environ Microbiol 80:3585–9356

Rastogi G, Sbodio A, Tech JJ, Suslow TV, Coaker GL, Leveau JHJ (2012) Leaf microbiota in an agroecosystem: Spatiotem-poral variation in bacterial community composition on field-grown lettuce. ISME J 6:1812–1822

Rastogi G (2013) New insights into the structure and function of phyllosphere microbiota through high-throughput molecular approaches. FEMS Microbiol Lett 348:1–10

Redford AJ, Bowers RM, Knight R, Linhart Y, Fierer N (2010) The ecology of the phyllosphere: geographic and phylogenetic variability in the distribution of bacteria on tree leaves. Environ Microbiol 12:2885–2289

Remus-Emesermann MN, Tecon R, Kowalchuk GA, Laveau JH (2012) Variation in local carrying capacity and the individual fate of bacterial colonizers in the phyllosphere. ISME J 6:756–765

Reisberg EE, Hildebrandt U, Riederer M, Hentschel U (2013) Distinct phyllosphere bacterial communities on *Arabidopsis wax* mutant leaves. PLoS One 8:e78613

Rennie EA, Turgeon R (2009) A comprehensive picture of phloem loading strategies. Proc Natl Acad Sci U S A 106:14162–14167

Rodriguez RJ, White JF Jr., Arnold AE, Redman RS (2009) Fungal endophytes: diversity and functional roles. New Phytol 182(2):314–330

Scheublin TR, Leveau JH (2013) Isolation of Arthrobacter spices from the phyllosphere and demonstration of their epiphytic fitness. Microbiology 2:205–2013

Schreiber L, Krimm U, Knoll D (2004) Interactions between epiphyllic microorganisms and leaf cuticles. In: Varma A, Abbott L, Werner D, Hampp R (eds) Plant surface microbiology, Berlin-Heidelberg, Springer-Verlag, pp 145–156

Stoitsova SO, Braun Y, Ullrich MS, Weingart H (2008) Characterization of the RND-type multidrug efflux pump MexAB-OprM of the plant pathogen Pseudomonas syringae. Appl Environ Microbiol 74:3387–3393

Stromberg KD, Kinkel LL, Leonard KJ (1999) Relationship between phyllosphere population sizes of *Xanthomonas translucens* pv. Translucens and bacterial leaf streak severity on wheat seedling. Phytopathology 89:131–135

Sulmon C, Gouesbet G, Ramel F, Cabello-Hurtado F, Penno C, Bechtold N (2011) Carbon dynamics, development and stress responses in *Arabidopsis*: involvement of the APL4 subunit of ADP-glucose pyrophosphorylase (starch synthesis). PLoS One 6:e26855

Trouvelot S, Héloir MC, Poinssot B, Gauthier A, Paris F, Guillier C (2014) Carbohydrates in plant immunity and plant protection: roles and potential application as foliar sprays. Front Plant Sci 5:592

Tsuji K, Tsien HC, Hanson RS, De Palma SR, Scholtz R, LaRoche S (1990) 16S ribosomal RNA sequence analysis for determination of phylogenetic relationship among methylotrophs. J Gen Microbiol 136:1–10

Vacher C, Hampe A, Porte AJ, Sauer U, Compant S, Cindy E, Morris CE (2016) The phyllosphere: microbial jungle at the plant–climate interface. Annu Rev Ecol Evol Syst 47:1–24

Vorholt JA (2012) Microbial life in the phyllosphere. Nat Rev 10:828–840

Williams TR, Moyne A-L, Harris LJ, Marco ML (2013) Season, irrigation, leaf age, and *Escherichia coli* inoculation influence the bacterial diversity in the lettuce phyllosphere. PLoS One 8:e68642

Yeats TH, Rose JKC (2013) The formation and function of plant cuticles. Plant Physiol 163:5–20

A Metagenomic Approach to Identify Distinct Rhizospheric and Endophytic Bacterial Communities from Roots and Root Nodules of *Vigna radiata*

11

Bhagya Iyer and Shalini Rajkumar

Abstract

Soil microbial community encompasses a perplexing range of physiological, metabolic and genomic diversity which can be explored through direct cultivation-based techniques or indirect molecular approaches. As cultivation-based methods are limited to only the culturable microorganisms, information about the majority of the unculturable diversity is missing. To circumvent this problem, molecular approaches based on the analysis of total DNA isolated from the environmental samples, often termed as metagenomic DNA, are employed. These molecular approaches can study the rare bacterial populations present in low abundance in soil. Though molecular approaches can study the soil diversity much better than the traditional culturable approaches, the success of any metagenomic study depends on the quality and quantity of DNA isolated from the metagenomic sample. In this chapter, metagenomic approaches have been employed to study the diversity of rhizospheric, root endophytic and root nodule bacterial communities of *Vigna radiata*. Results indicated that in rhizospheric and root nodule bacterial communities, Proteobacteria were predominant while in root endophytic communities, Actinobacteria were predominant. Deltaproteobacteria predominated rhizospheric community, whereas Gammaproteobacteria and Alphaproteobacteria dominated root endophytic and root nodule communities, respectively. Coupling traditional approaches with advanced next-generation sequencing techniques for accessing bacterial com-

B. Iyer • S. Rajkumar (✉)
Institute of Science, Nirma University, Ahmedabad –382 481, Gujarat, India
e-mail: shalini.rjk@nirmauni.ac.in

R.P. Singh et al. (eds.), *Understanding Host-Microbiome Interactions - An Omics Approach*, DOI 10.1007/978-981-10-5050-3_11

munity ecology and physiology may bring new insights in understanding the microbial life in the rhizosphere and their further progression as root or nodule endophytes.

Keywords
Rhizosphere • OTU • Proteobacteria • Bacterial diversity • Endophyte

11.1 Introduction

The soil zone under the direct influence of plant roots, which acts as a hot spot for a diverse array of bacteria, is termed as rhizosphere (Hiltner 1904). Rhizospheric soil differs from distal soil due to the bio-physicochemical processes which occur as a consequence of root growth, water and nutrient uptake and respiration and rhizodeposition (Hinsinger 2005). As plants secrete up to 40% of their photosynthates into the rhizosphere, microbial population densities in the rhizosphere are much higher but less diverse than in the distal soil. This is termed as "rhizosphere effect" (Costa et al. 2006).

Rhizosphere microbiome harbours multiple bacterial, archaeal, viral and fungal species along with other eukaryotic taxa (Lagos et al. 2015). Some bacteria and fungi can enter the plant and live as endophytes (Azevedo et al. 2000). Depending on the type of influence these microbial communities exert on plant physiology and development, they are classified into three groups. First group comprises of most members of the rhizosphere microbiome which are beneficial to the plant growth followed by second group of pathogenic microorganisms trying to overcome the plant innate defence mechanism, leading to plant diseases. The third group of microbial community consists of the opportunistic human pathogens (Mendes et al. 2013).

Virtually, all tissues of a plant host a microbial community. Bacteria colonizing the soil rhizosphere are termed as rhizobacteria (Schroth and Hancock 1982), and the beneficial plant-associated rhizobacteria are termed as plant growth-promoting rhizobacteria (PGPR) (Kloepper et al. 1989). These PGPR can improve plant growth directly by production of phytohormones or by improving nutrient uptake and indirectly by preventing the effects of phytopathogens (Glick 1995). Current understanding of the complex interaction between plants and microbes is still in its infancy. Complete understanding of how these indigenous bacteria communicate, colonize root environments and compete along the rhizosphere microsites holds great promise for sustainable agriculture. Development of the molecular techniques based on DNA analysis has provided more accurate knowledge of bacterial community composition and has identified abundance of genes involved in rhizospheric processes and the functions of bacteria in the rhizosphere (Lagos et al. 2015).

11.2 Ecology of Bacteria in the Rhizosphere

With the rhizosphere being the transitional zone in between distal soil and plant roots, the rhizospheric microbial communities are crucial for important plant processes like nutrient acquisition, disease protection and plant development (Rocha et al. 2009). One gram of soil can contain more than 1 million distinct bacterial genomes (Gans et al. 2005). Abundance, activity and diversity of microorganisms along rhizosphere depend on multiple biotic and abiotic factors like components of root exudates (Lugtenberg and Kamilova 2009), motility (Capdevila et al. 2004), availability of inorganic compounds like phosphorus (P) and iron (Fe), production of antibiotics (Raaijmaker et al. 1995) and competition between organisms (Lagos et al. 2015). Composition of bacterial communities also differs according to root zone, plant species, stress and disease events (Rovira 1965). NCBI database reports Actinobacteria, Firmicutes, Proteobacteria and Bacteroidetes to be most abundant in the rhizosphere which can potentially promote plant growth and the percentage of 16S rRNA gene sequences of bacterial groups commonly obtained in rhizosphere by culture-dependent and culture-independent techniques, as depicted in Fig. 11.1.

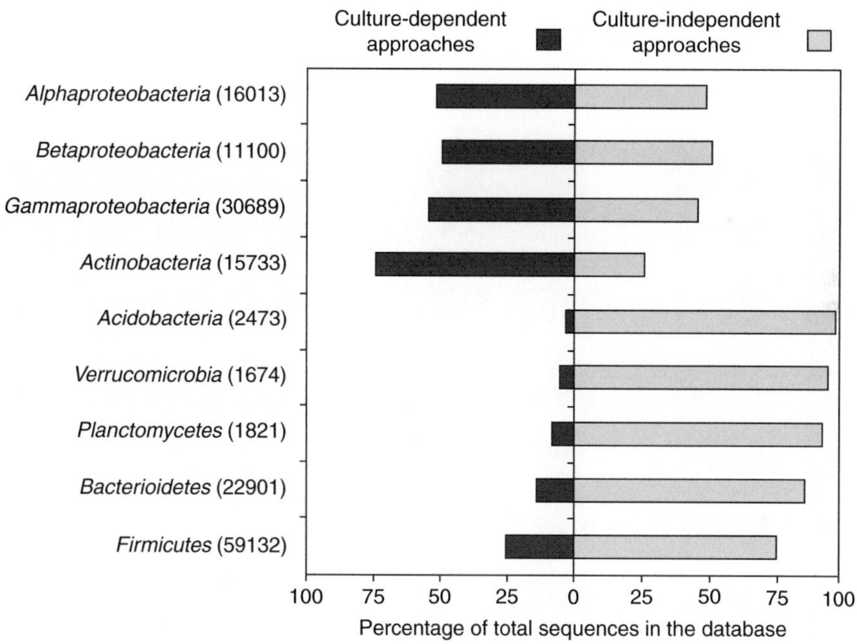

Fig. 11.1 Percentage of 16S rRNA gene sequences from bacterial groups common to the rhizosphere obtained via culture-dependent and culture-independent approaches. All sequences, derived from different ecosystems, were obtained from the Ribosomal Database Project (Cole et al. 2007). In parentheses, total number of sequences used per bacterial group

Traditionally, the components of soil microbiome were characterized by isolating and culturing microbial communities on different media and growth conditions (Kirk et al. 2004). These culture-based techniques missed the vast majority of microbial diversity in an environment or in plant-associated habitats, which are now detectable by modern culture-independent molecular techniques (Lakshmanan et al. 2014). "Great plate count anomaly" states that 95–99% of the environmental microbial communities cannot be accessed by conventional culture-based techniques (Nichols 2007), and hence, the development of efficient molecular approaches is inevitable for studying the soil microbial community. These molecular approaches can contribute unique empirical data to the exploration of rhizosphere function (Cardon and Gage, 2006). During the last few decades, molecular techniques like PCR-based fingerprinting (Berlec 2012), quantitative PCR (qPCR) and gene expression (Deepak et al. 2007) and analysis of transcriptome profile by microarrays have been used to examine the rhizospheric microbial diversity. Figure 11.2 indicates the timeline of different molecular approaches applied to study rhizospheric bacterial traits.

Precise revelation of the microbial population in the rhizospheric zone and on the plant root surface is exclusively dependent on the sampling method and sequencing methods used, and this poses a difficult confront (Lakshmanan et al. 2014). Recent advances in genomic and molecular techniques have enabled characterization of unculturable organisms at an intensity never seen before and have shown that

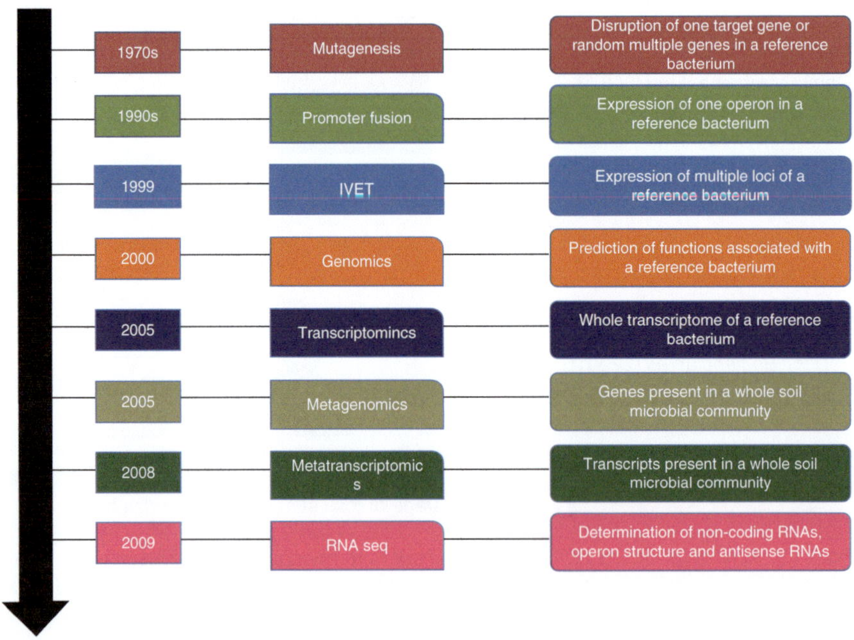

Fig. 11.2 Timeline of molecular approaches applied to decipher rhizospheric bacterial traits (Adapted from Barret et al. 2011)

bacterial diversity of distal and rhizospheric soil is much superior than what was expected. To evaluate activity, distribution and richness of microbial communities, it is imperative to understand the ecological function of each and every bacterial species (Lagos et al. 2015).

11.3 Metagenomics

Metagenomics, the culture-independent genomic analysis of a population of microorganisms, is a rapidly growing area of the genome sciences that aids to define the features of intact microbial communities in their native habitats (Handelsman et al. 1998). It describes the functional and sequence-based analysis of the collective microbial genomes contained in an environmental sample. Sequence-based metagenomics involves sequencing and analysis of metagenomic DNA to assemble genomes, identify genes, find metabolic pathways and compare organisms of different communities. Functional metagenomics involves screening for a particular function or activity like identifying novel antibiotics and proteins involved in antibiotic resistance, vitamin production and xenobiotic degradation. Soil microbial communities have the highest level of prokaryotic diversity, and molecular approaches like metagenomics can improve our access to these communities (Delmont et al. 2010). Metagenomic approaches have been applied to study soil microbial community composition and its diversity, to identify the rare or less abundant bacterial communities, to understand the importance of the rhizosphere microbiome in promoting plant health and to study the impact of root exudates on the rhizosphere soil microbiome (Bakker et al. 2013).

Metagenomic analysis involves isolating DNA from an environmental sample, cloning the DNA into a suitable vector, transforming the clones into a host bacterium and screening the resultant transformants. The clones can be screened for phylogenetic markers like 16S rRNA or *recA* or other conserved genes by hybridization or multiplex PCR (Stein et al. 1996), for expression of specific traits (Courtois et al. 2003) or through random sequencing (Tyson et al. 2004). Soil metagenome study of 16S rRNA gene using next-generation sequencing (NGS) technologies has revealed that 1 g of soil may harbour 33,346 bacterial and archaeal OTUs (operational taxonomic units) (Mendes et al. 2011) and 3320 fungal OTUs (Schmidt et al. 2013). In rhizospheric research area, the drastic changes occur in associated microbial communities during plant growth; thus, time-course data are of utmost importance. Hence, the sequence information will tell only to a brief snap shot of the dominant sequences present in the DNA extract, and information on community development in the rhizospheric zone is often missing (Tyson et al. 2004; Tringe et al. 2005).

Metagenomic approach has been employed to study different types of ecosystems like areas of volcanism (Xie et al. 2011; Kilias et al. 2013), areas of extreme temperature (Bradford et al. 2008; Pearce et al. 2012), alkalinity (Xiong et al. 2012), acidity (Garcia-Moyano et al. 2012; Johnson 2012), low oxygen (Bryant et al. 2012; Stevens and Ulloa 2008) and high heavy metal composition (Chodak et al. 2013; Golebiewski et al. 2014). Though metagenomic approaches have been applied to study range of soil environments, however, the total soil microbial diversity might still be underestimated. This is a consequence of the fact that metagenomic DNA extraction is not always a representation of the microbial diversity, and the dominance of certain groups in a DNA may mask the less abundant species which leads to a biased study of microbial community structure (Tyson et al. 2004). The technique is time consuming and expensive and cannot reconstruct large genomic contigs from organisms that are not highly represented in the environmental DNA (Daniel 2005; Rondon et al. 2000). Hence, the exploration of soil microbial biodiversity requires different protocols of metagenomic approach which extract DNA from all the bacteria and archaea as comprehensively as possible. No single protocol can accurately determine species distribution. Hence, a combination of multiple DNA protocols could be employed, and different DNA pools could then be mixed to maximize the number of different species and to minimize the dominant species ensuring an increase in the final level of metagenomic diversity (Delmont et al. 2010).

11.4 Metagenomic Approach to Study Rhizospheric and Endophytic Bacterial Diversity of *V. radiata*

11.4.1 Sample Preparation

Intact plants of *V. radiata* and the rhizospheric soil samples were collected from the field. The roots of *V. radiata* were shaken vigorously to separate them from loose soil, and the remaining soil closely adhering to the roots (up to 2.5 mm around the root) was pooled and considered as rhizospheric soil. One gram of rhizosphere soil was weighed from each plant and then pooled as a composite rhizosphere sample in sterile laboratory conditions. Similarly, samples were also prepared for the root and root nodules of the host plant. Healthy, symptomless roots and root nodules were collected from ten individual *V. radiata* plants. One gram of root and root nodules from each plant was pooled to make one sample each. Root samples and root nodules were surface-sterilized by washing them with 0.1% $HgCl_2$ followed by repetitive washes with sterile H_2O and processed for further DNA-based analysis. DNA was extracted from 0.5 g soil using the Soil DNA Isolation Kit (Mo Bio, Carlsbad, CA) according to the manufacturer's protocol. Root and root nodule DNA were isolated as described by Ikeda et al. (2009).

11.4.2 Sequence Analysis

Total metagenomic DNA from rhizospheric soil and total root and root nodule were isolated. Bacteria were characterized from the same samples by using primers specific for V3 hypervariable region of 16S rRNA gene. The V3 hypervariable region was amplified with forward primer 341F (5′CCT ACG GGA GGC AGC CAG 3′) and reverse primer 518R (5′ATTACCGCGGCTGCTGG3′). Paired end

Table 11.1 Read summary and pre processing statistics of *V. radiata* rhizosphere, root and root nodule metagenomic DNA

	Rhizosphere	Root	Root nodule
% GC content	57.63%	56.07%	54.94%
Total reads	2,141,549	1,962,389	1,950,073
Pre-processed reads	1,081,041	1,284,999	1,318,029
	Total pre-processed reads		
	Total OTUs		

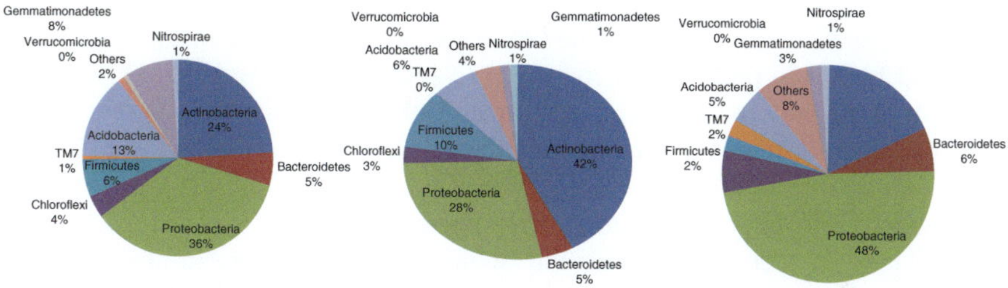

Fig. 11.3 Distribution of OTUs at phylum level in (**a**) rhizosphere, (**b**) root nodules and (**c**) root

sequence from V3 region contains some portion of the conserved region, the spacer and V3 region. Multiple filters were applied to take further only the high-quality V3 region sequences. More than 80% of the paired end reads aligned to each other with zero mismatches with an average contig length of 135–160 bp. More than 90% of the total reads had Phred score greater than 30 and GC content in the range of 50–60%. All reads were processed further and clustered to form a total of 31,025 OTUs from 3,684,069 reads. Read summary and preprocessing statistics are presented in Table 11.1.

11.4.3 Relative Abundance Distribution of OTUs and Reads

Rhizospheric and endophytic bacterial communities exhibited different overall patterns of relative abundance of the major groups at the phylum level. Representative sequences in rhizosphere, roots and root nodules were clustered into ten major phyla: Proteobacteria, Actinobacteria, Acidobacteria, Gemmatimonadetes, Firmicutes, Bacteroidetes, Chloroflexi, Nitrospirae, TM7 and Verrucomicrobia. Overall, the population of Actinobacteria and Proteobacteria was highest in all the three samples. Taxa other than the top ten were categorized as "others".

Figure 11.3 indicates the OTU's distribution at phylum level for rhizosphere, root and root nodules. Rhizosphere bacterial communities were heavily dominated by *Proteobacteria* (36%), Actinobacteria (24%) and Acidobacteria (13%) followed by Gemmatimonadetes (8%) (Fig. 11.3a). Among Proteobacteria, Deltaproteobacteria (11%) heavily colonized the rhizosphere followed by

Alphaproteobacteria (8%), Betaproteobacteria (8%) and Gammaproteobacteria (7%). *V. radiata* showed higher bacteria diversity compared to *A. mangium* and *A. hypogea* (Pongsilp et al. 2012). Root endophytic population was heavily dominated by Actinobacteria (42%) followed by Proteobacteria (28%) and Firmicutes (10%) and Acidobacteria (6%) at the phylum level (Fig. 11.3b). Among Proteobacteria, Gammaproteobacteria (14%) heavily colonized roots and were two fold higher in roots than in the rhizosphere and root nodule as reported (Jin et al. 2014). Root nodule endophytic bacterial communities were heavily dominated by Proteobacteria (48%) at the phylum level, while Actinobacteria comprised only 18% (Fig. 11.3c). Within the Proteobacteria, root nodule endophytic communities were heavily dominated by Alphaproteobacteria (36%) which were four fold higher in root nodule compared to roots and rhizosphere. This was in harmony with the report which stated that Alphaproteobacteria was one of the most abundant bacterial classes on Zea mays L. and potato (Chelius and Triplett 2001; Inceoglu et al. 2011).

Metagenomic analysis of maize rhizosphere microbiome showed that orders Burkholderiales, Oceanospirillales and Sphingobacteriales were in abundance in rhizosphere, whereas phyla Acidobacteria, Chloroflexi, Planctomycetes and Verrucomicrobia were in abundance in distal soil (Peiffer et al. 2013). Rice rhizosphere had predominantly members of Proteobacteria along with a small fraction of Acidobacteria, Firmicutes and Bacteroidetes group (Arjun and Harikrishnan 2011) similar to soyabean rhizosphere (Sugiyama et al. 2014)

At phylum level, Proteobacteria predominated the root nodule followed by the rhizosphere and root, whereas Actinobacteria predominated the root followed by the rhizosphere and root nodule. Chloroflexi were highest in root nodule, and Firmicutes were the maximum in root, while Acidobacteria and Gemmatimonadetes were abundant in rhizosphere (Fig. 11.4a). At class level, Alphaproteobacteria were predominant in nodule followed by rhizosphere and root, Betaproteobacteria were more in rhizosphere followed by root and nodule, and Gammaproteobacteria were maximum in root followed by rhizosphere and nodule, while Deltaproteobacteria were more in rhizosphere followed by nodule and root. Actinobacteria and Acidobacteria predominated the root and rhizosphere, respectively (Fig. 11.4b). None of the bacterial family had high relative OTU abundances in more than one sample. For example, Bacillaceae (13%), had high relative abundance in root followed by rhizosphere and nodule, while Comamonadaceae (3%) and Xanthomonadaceae (2%) were abundant in rhizosphere followed by root and nodule. Cytophagaceae (5.9%) had higher relative abundance in root nodule followed by rhizosphere and root, while Chitinophagaceae (3%) were maximum in rhizosphere followed by nodule and root (Fig. 11.4c). At genus level, *Pseudomonas* were maximum in root followed by rhizosphere. Rhizosphere also harboured *Rubrobacter*, *Nitrospira*, *Bacillus*, *Steroidobacter*, etc. with the maximum bacterial diversity compared to root and root nodules (Fig. 11.5a).

11.4.4 Phylogenetic Tree of Rhizosphere, Root Nodule and Root Microbial Diversity

The phylogenetic trees of the representative sequences of rhizosphere, root and root nodule bacterial diversity are shown in Figs. 11.6, 11.7 and 11.8. As expected, microbial communities were more diverse in rhizosphere followed by root and root nodule. The phylogenetic tree representing rhizosphere bacterial diversity (Fig. 11.6) highlighted the presence of 91 genera with predominance of gram positive *Bacillus* spp. followed by *Peredibacter* spp. and *Pseudoxanthomonas* spp., which represented the major rhizosphere population. The phylogenetic tree representing root endophytic diversity (Fig. 11.7) highlighted the presence of 40 genera with predominant *Glaciecola* spp. populations followed by *Acinetobacter* spp. The phylogenetic tree representing root nodule diversity (Fig. 11.8) highlighted the presence of 17 bacterial genera with none having dominance. Using traditional culture-dependent techniques followed by DNA extraction and 16S rRNA sequencing, we reported predominance of *Bacillus* sp. in root nodules of *V. radiata* (Pandya et al. 2013).

The bacterial taxa observed were similar to the findings from other studies (Bodenhausen et al. 2013; Delmotte et al. 2009; Telias et al. 2011; Uroz et al. 2010). The presence of Bacteroidetes and Chloroflexi in roots and Acidobacteria in rhizosphere sample was in contrast to findings of Jin et al. (2014). The presence of Gemmatimonadetes and Acidobacteria in rhizosphere was two fold higher compared to root nodules and root. These fewer abundant phyla were clearly habitat-specific, and this might have been a sign of a habitat consequence due to the different nutrition prototypes as a result of the diverse organ types (or soil) of *Vigna radiata*. This preference for different habitats has also been demonstrated in other plants as mentioned previously (Bodenhausen et al. 2013; Sessitsch et al. 2002). The top ten predominant phyla in rhizosphere, root and root nodule were the same; however, they differed in abundance pattern. The higher-order classifications of major phyla like Proteobacteria, Actinobacteria and Acidobacteria differed in the abundance dramatically between rhizosphere, roots and root nodules.

Based on the heat map (Fig. 11.9), it was observed that phylum Proteobacteria (Sphingomonadaceae; OTU id: de novo 21380) was predominantly present in rhizosphere followed by phylum Acidobacteria (OTU id: de novo 15222), while in root nodule endophytic population, phylum Cyanobacteria (OTU id: de novo 14454) was predominantly present followed by phylum Proteobacteria (Bradyrhizobiaceae; OTU id: de novo 20939). In root endophytic population, phylum Actinobacteria (OTU id: de novo 21619) was predominantly present followed by phylum Firmicutes (OTU id: de novo 5876).

11.4.5 Richness and Evenness Estimate of Bacterial Diversity

Species diversity is the number of different species that are represented in a given community and consists of two components: species richness and species evenness. Species richness is the number of different species represented in an

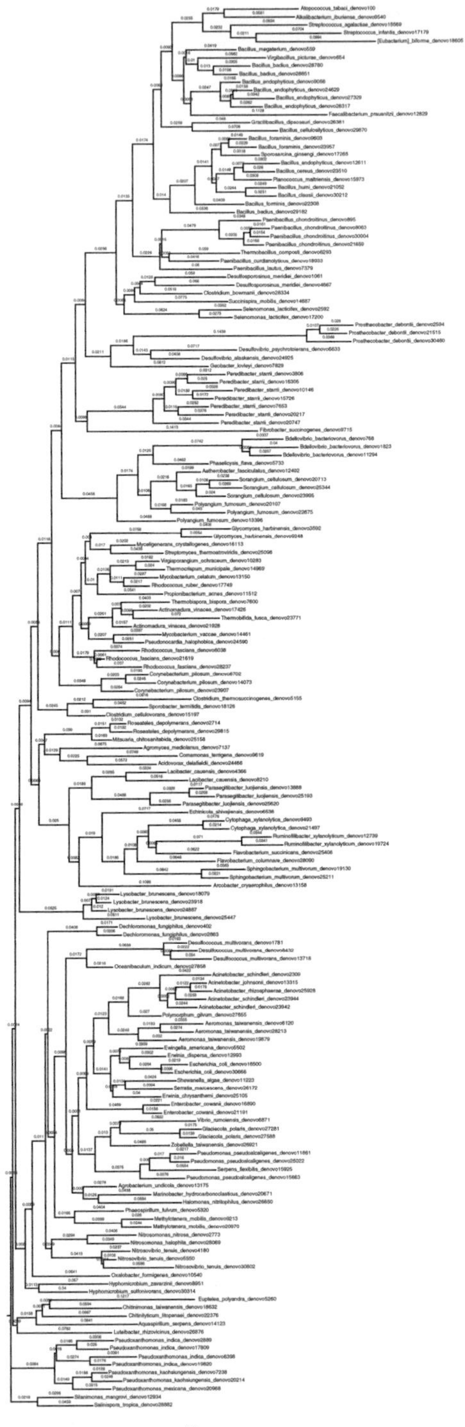

Fig. 11.6 Phylogenetic tree of the rhizospheric bacterial diversity

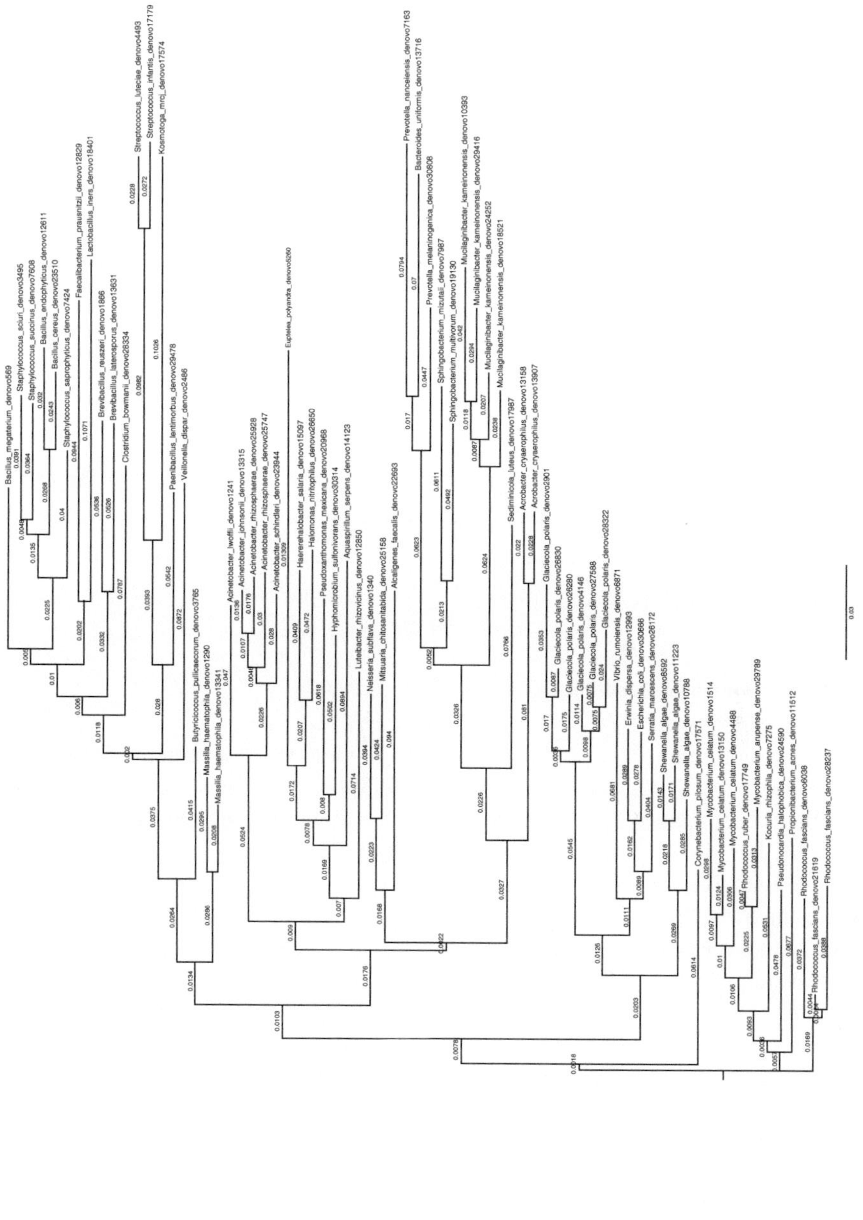

Fig. 11.7 Phylogenetic tree of the root bacterial diversity

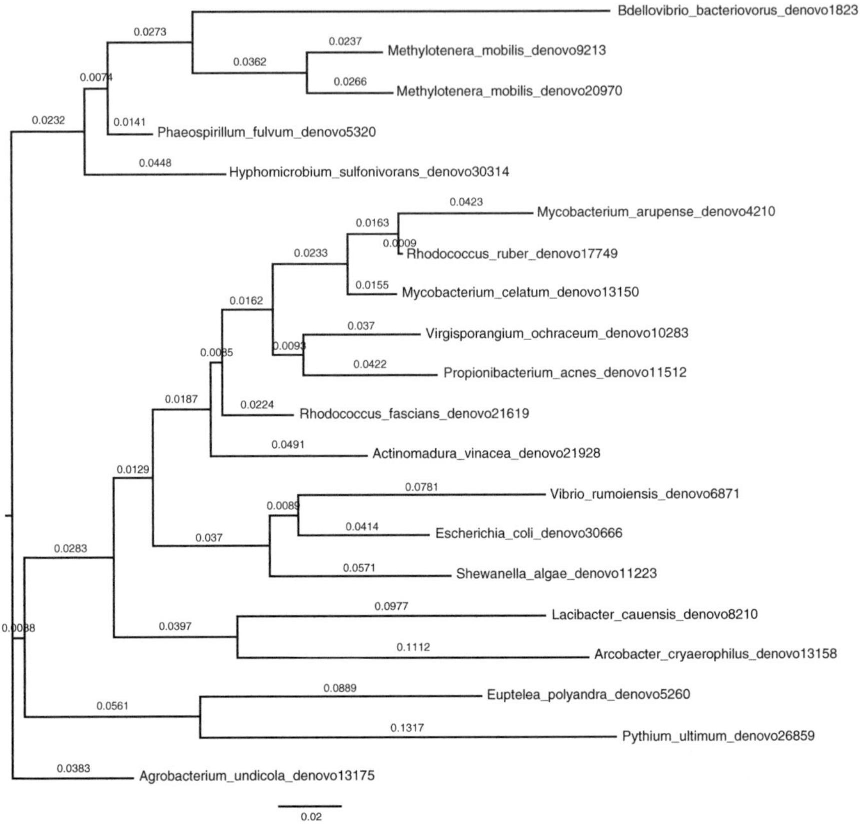

Fig. 11.8 Phylogenetic tree of the root nodule bacterial diversity

ecological community and does not take into account the abundances of the species or their relative abundance distributions. Species evenness is a measure of the relative abundance of the different species making up the richness of an area (Hulbert, 1971).

Microbial diversity can be analysed by calculating Shannon, Chao1 and observed species metrices. The Chao1 metric estimates the species richness while Shannon metric is the measure to estimate observed OTU abundances, accounting for both richness and evenness. The observed species metric is the count of unique OTUs identified in the sample. The Chao1 estimate suggested a trend of highest species richness in rhizosphere. The number of OTUs in rhizosphere was almost 100 times higher than root nodule OTUs and ten times higher than root OTUs (Table 11.2).

Figure 11.10 shows that the rhizosphere sample had highest number of bacterial species compared to roots and nodules. The Shannon metric observed OTU abundances and evaluated both richness and evenness. It revealed that rhizosphere had the highest diversity and evenness followed by root and root nodule although root nodule sample contained the highest number of sequences, followed by root and

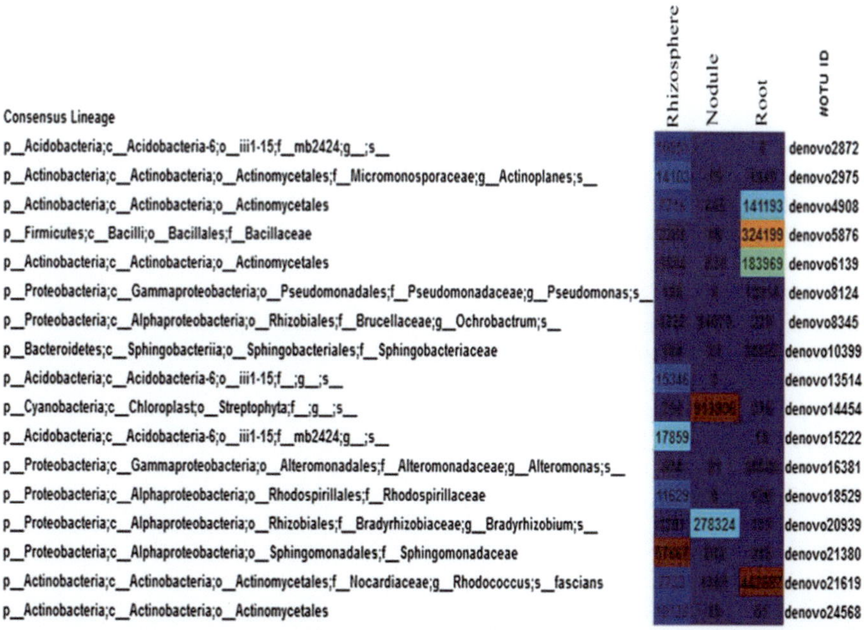

Fig. 11.9 Heat map generated for rhizosphere, root nodule and root samples based on the relative abundances of top OTUs. The OTU heat map displays raw OTU counts per sample, where the counts are coloured based on the contribution of each OTU to the total OTU count present in that sample (*blue*: contributes low percentage of OTUs to sample; *red*: contributes high percentage of OTUs). The heat map is based on filter of 10,000 per OTU

Table 11.2 Comparison of OTU diversity in different samples using Chao1 richness estimate and Shannon diversity index

	Rhizosphere	Root	Root nodule
No. of sequences	1,081,041	1,284,999	1,318,029
No. of OTUs	29,120	2916	394
Chao1 richness	29,107	2906	386
Shannon diversity index	10.37	3.08	1.21

rhizosphere (Fig. 11.10a). This suggested that the number of sequences obtained was not the only indicator for the bacterial diversity levels. Nodules and roots exhibited lower species diversity compared to rhizosphere. The Shannon index value for rhizosphere was almost ten times of root and threefold that of root nodules (Table 11.2). The Chao1 metric estimated the species richness to be highest in rhizosphere (29107) followed by root (2906) and root nodule (386). It can also be observed that the species richness of both roots and nodules was similar (Fig. 11.10b). Observed species richness metric identifies unique OTUs in the sample. It was observed that the rhizosphere sample had more species richness and evenness, compared to root nodule and root (Fig. 11.10c).

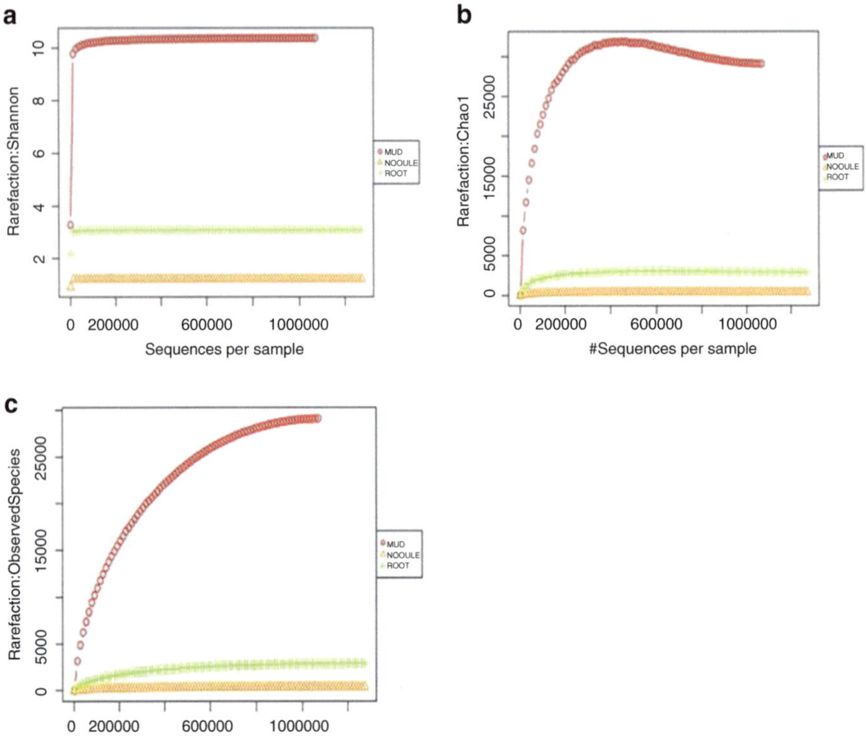

Fig. 11.10 Alpha diversity of the samples: (**a**) Shannon curve, (**b**) Chao1 curve and (**c**) observed species curve

11.4.6 Conclusion, Challenges and Future Aspects

Though metagenomic analysis of bacterial community diversity provides more information than the traditional culture-based techniques, it has its own limitations in terms of quality and quantity of metagenomic DNA sampling. It is indeed a great challenge to develop strategies based on cultivation-dependent and cultivation-independent studies that allow the exploration of the putative roles and function of unculturable microbiota in the rhizosphere and analyse their modes of interactions with plants and rhizospheric microorganisms. Culture-independent molecular strategy such as generation of 16S rRNA library and phylogenetic analysis of prokaryotic communities from widely diverse rhizosphere environment will augment our knowledge on these organisms and help in formulating strategies for improved soil quality, enhanced crop production and protection, to conserve natural resources and ultimately create more sustainable agricultural production and to combat climate change. Advances in soil molecular techniques will continue to improve our understanding of the composition and activities of the soil microbial communities, to target the rare or low-abundance bacterial populations in the rhizosphere and to predict the in situ responses, activities and growth of the bacterial communities.

Even though the "omics" techniques offer great advances in our capabilities to unravel the identity of genes present in the rhizospheric microbiome, the presence of technological barriers impede a meticulous analysis.

References

Arjun JK, Harikrishnan K (2011) Metagenomic analysis of bacterial diversity in the rich rhizosphere soil microbiome. J Biotechnol Bioinfo Bioeng 1(3):361–367

Azevedo JL, Maccheroni W, Pereira JO, de Araujo WL (2000) Endophytic microorganisms: A review on insect control and recent advances on tropical plants. Electron J Biotechnol 3(1):40–65

Bakker P, Berendsen R, Doornbos R, Wintermans P, Pieterse C (2013) The rhizosphere revisited: root microbiomics. Front Plant Sci 165:1–7

Barret M, Morrissey JP, Gara OF (2011) Functional genomics analysis of plant growth promoting rhizobacterial traits involved in rhizosphere competence. Biol Fertil Soils 47:729–743

Berlec A (2012) Novel techniques and findings in the study of plant microbiota: search for plant probiotics. Plant Sci 193:96–102

Bodenhausen N, Horton MW, Bergelson J (2013) Bacterial communities associated with the leaves and the roots of *Arabidopsis thaliana*. PLoS One 8:e56329. doi:10.1371/journal.pone.0056329

Bradford MA, Davies CA, Frey SD et al (2008) Thermal adaptation of soil microbialrespiration to elevated temperature. Ecol Lett 11(12):1316–1327

Bryant JA, Stewart FJ, Eppley JM, DeLong EF (2012) Microbial community phylogenetic and trait diversity declines with depth in a marine oxygen minimum zone. Ecology 93(7):1659–1673

Capdevila S, Martínez-Granero FM, Sanchez-Contreras M, Rivilla R, Martin M (2004) Analysis of *Pseudomonas fluorescens* F113 genes implicated in flagellar filament synthesis and their role in competitive root colonization. Microbiology 150:3889–3897

Cardon ZG, Gaje DJ (2006) Resource exchange in the rhizosphere: molecular tools and the microbial perspective. Annu Rev Ecol Evol Syst 37:459–488

Chelius MK, Triplett EW (2001) The diversity of archaea and bacteria in association with the roots of *Zea mays* L. Microb Ecol 41:252–263

Chodak M, Gołębiewski M, Morawska-Płoskonka J, Kuduk K, Niklińska M (2013) Diversity of microorganisms from forest soils differently polluted with heavy metals. Appl Soil Ecol 64:7–14

Cole JR, Chai B, Farris RJ et al (2007) The ribosomal database project (RDP-II): introducing myRDP space and quality controlled public data. Nucleic Acids Res 35:D169–D172

Costa R et al (2006) Effects of site and plant species on rhizosphere community structure as revealed by molecular analysis of microbial guilds. FEMS Microbiol Ecol 56:236–249

Courtois S, Cappellano CM, Ball M, Francou FX, Normand P et al (2003) Recombinant environmental libraries provide access to microbial diversity for drug discovery from natural products. Appl Environ Microbiol 69:49–55

Da Rocha UN, Van Overbeek L, Van Elsas JD (2009) Exploration of the hitherto-uncultured bacteria from the rhizosphere. FEMS Microbiol Ecol 69:313–328

Daniel R (2005) The metagenomics of soil. Nat Rev Microbiol 3:470–478

Deepak S, Kottapalli K, Rakwal R, Oros G, Rangappa K, Iwahashi H, Masuo Y, Agrawal G (2007) Real–Time PCR: revolutionizing detection and expression analysis of genes. Curr Genomics 8:234–251

Delmont TO, Robe P, Cecillon S, Clark IM, Constancias F, Simonet P, Hirsch PR, Vogel TM (2010) Assessing the soil metagenome for studies of microbial diversity. Appl Environ Microbiol 77(4):1315–1324

Delmotte N et al (2009) Community proteogenomics reveals insights into the physiology of phyllosphere bacteria. Proc Natl Acad Sci U S A 106:16428–16433. doi:10.1073/pnas.090524010

Gans J, Wolinsky M, Dunbar J (2005) Computational improvements reveal great bacterial diversity and high metaltoxicity in soil. Science 309:1387–1390

García-Moyano A, González-Toril E, Aguilera Á, Amils R, Aguilera A (2012) Comparative microbial ecology study of the sediments and the water column of Río Tinto, an extreme acidic environment. FEMS Microbiol Ecol 81(2):303–314

Glick BR (1995) The enhancement of plant growth by free-living bacteria. Can J Microbiol 41:109–117

Gołębiewski M, Deja-Sikora E, Cichosz M, Tretyn A, Wróbel B (2014) 16S rDNA pyrosequencing analysis of bacterial community in heavy metals polluted soils. Microb Ecol 67(3):635–647

Handelsman J, Rondon MR, Brady SF, Clardy J, Goodman RM (1998) Molecular biological access to the chemistry of unknown soil microbes: a new frontier for natural products. Chem Biol 5:R245–R249

Hiltner L (1904) Über neuere Erfahrungen und Probleme auf dem Gebiete der. Bodenbakteriologie unter besonderer Berücksichtigung der Gründüngung und Brache. Arbeiten der Deutschen Landwirtschaftlichen Gesellschaft 98:59–78

Hinsinger P, Gobran GR, Gregory PJ, Wenzel WW (2005) Rhizosphere geometry and heterogeneity arising from root mediated physical and chemical processes. New Phytol 168:293–303

Hulbert SH (1971) The nonconcept of species diversity: a critique and alternative parameters. Ecology 52(4):577–586

Ikeda S, Kaneko T, Okubo T et al (2009) Development of a bacterial cell enrichment method and its application to the community analysis in soybean stems. Microb Ecol 58:703–714

Inceoglu O, Abu Al-Soud W, Salles JF, Semenov AV, van Elsas JD (2011) Comparative analysis of bacterial communities in a potato field as determined by pyrosequencing. PLoS One 6(8):e23321. doi:10.1371/journal.pone. 0023321

Jin H et al (2014) Characterization of rhizosphere and endophytic bacterial communities from leaves, stems and roots of medicinal *Stellera chamaejasme* L. Syst Appl Microbiol 37:376–385

Johnson DB (2012) Geomicrobiology of extremely acidic subsurface environments. FEMS Microbiol Ecol 81(1):2–12

Kilias SP, Nomikou P, Papanikolaou D et al (2013) New insights into hydrothermal vent processes in the unique shallow-submarine arc-volcano, Kolumbo (Santorini), Greece. Sci Rep 3:2421

Kirk JL, Beaudette LA, Hart M, Moutoglis P, Klironomos JN, Lee H, Trevors JT (2004) Methods of studying soil microbial diversity. J Microbiol Methods 58:169–188

Kloepper JW, Lifshitz R, Zablotowicz RM (1989) Free-living bacterial inocula for enhancing crop productivity. Trends Biotechnol 7:39–43

Lagos ML, Maruyama F, Nannipieri P, Mora ML, Ogram A, Jorquera MA (2015) Current overview of the study of bacteria in the rhizosphere by modern molecular techniques: a mini review. J Soil Sci Plant Nutr 15(2):504–523

Lakshmanan V, Gopinath S, Bais HP (2014) Functional soil microbiome: below ground solutions to an Above ground problem. Plant Physiol 166:689–700

Lugtenberg B, Kamilova F (2009) Plant growth promoting rhizobacteria. Annu Rev Microbiol 63:541–556

Mendes R, Kruijt M, de Bruijn I, Dekkers E, van der Voort M et al (2011) Deciphering the rhizosphere microbiome. Science 1097:1097–1100

Mendes R, Garbeva P, Raaijmakers JM (2013) The rhizosphere microbiome: significance of plant beneficial, plant pathogenic, and human pathogenic organisms. FEMS Microbiol Rev 37:634–663

Nichols D (2007) Cultivation gives context to the microbial ecologist. FEMS Microbiol Ecol 60:351–357

Pandya M, Naresh Kumar G, Rajkumar S (2013) Invasion of rhizobial infection thread by non-rhizobia for colonization of *V. radiata* root nodules. FEMS Microbiol Lett 348:58–65. doi:10.1111/1574-6968.12245

Pearce DA, Newsham KK, Thrne MA et al (2012) Metagenomic analysis of a southern maritime Antarctic soil. Front Microbiol 3:403–403

Peiffer JA, Spor A, Koren O, Jin Z, Tringe SG, Dangl JL, Buckler ES, Rey RE (2013) Diversity and heritability of the maize rhizosphere microbiome under field conditions. Proc Natl Acad Sci U S A 110(16):6548–6553

Pongsilp N, Nimnoi P, Lumyong S (2012) Genotypic diversity among rhizospheric bacteria of three legumes assessed by cultivation dependent and cultivation independent techniques. World J Microbiol Biotechnol 28:615–626

Raaijmaker JM, van der Sluis L, Koster M, Bakker PAHM, Weisbeek PJ, Schippers B (1995) Utilization of heterologous siderophores and rhiozsphere competence of florescent *Pseudomonas* spp. Can J Microbiol 41:126–135

Rondon MR, August PR, Bettermann AD, Brady SF, Grossman TH et al (2000) Cloning the soil metagenome: a strategy for accessing the genetic and functional diversity of uncultured microorganisms. Appl Environ Microbiol 66:2541–2547

Rovira A (1965) Interactions between plant roots and soil microorganisms. Annu Rev Microbiol 19:241–266

Schmidt PA, Bálint M, Greshake B, Bandow C, Römbke J, Schmitt I (2013) Illumina metabarcoding of a soil fungal community. Soil Biol Biochem 65:128–132

Schroth MN, Hancock JG (1982) Disease-suppressive soil and root-colonizing bacteria. Science 216:1376–1381

Sessitsch A, Reiter B, Pfeifer U, Wilhelm E (2002) Cultivation-independent population analysis of bacterial endophytes in three potato varieties based on eubacterial and Actinomycetes-specific PCR of 16S rRNA genes. FEMS Microbiol Ecol 39:23–32. doi:10.1111/j.1574-6941.2002.tb00903

Stein JL, Marsh TL, Wu KY, Shizuya H, DeLong EF (1996) Characterization of uncultivated prokaryotes: isolation and analysis of a 40-kilobase-pair genome fragment front a planktonic marine archaeon. J Bacteriol 178:591–599

Stevens H, Ulloa O (2008) Bacterial diversity in the oxygen minimum zone of the eastern tropical South Pacifi. Environ Microbiol 10(5):1244–1259

Sugiyama A, Ueda Y, Zushi T, Takase H, Yazaki K (2014) Changes in the bacterial community of soybean rhizospheres during growth in the field. PLoS One 9(6):e100709

Telias A, White JR, Pahl DM, Ottesen AR, Walsh CS (2011) Bacterial community diversity and variation in spray water sources and the tomato fruit surface. BMC Microbiol 11:81. doi:10.1186/1471-2180-11-81

Tringe SG, Von Mering C, Kobayashi A et al (2005) Comparative metagenomics of microbial communities. Science 308:554–557

Tyson GW, Chapman J, Hugenholtz P et al (2004) Community structure and metabolism through reconstruction of microbial genomes from the environment. Nature 428:37–43

Uroz S, Buée M, Murat C, Frey-Klett P, Martin F (2010) Pyrosequencing reveals a contrasted bacterial diversity between oak rhizosphere and surrounding soil. Environ Microbiol Rep 2:28–288. doi:10.1111/j.1758-2229.2009.00117.x

Wu D, Hugenholtz P, Mavromatis K, Pukall R, Dalin E et al (2009) A phylogeny–driven genomic encyclopaedia of Bacteria and Archaea. Nature 462:1056–1060

Xie W, Wang F, Guo L et al (2011) Comparative metagenomics of microbial communities inhabiting deep-sea hydrothermal vent chimneys with contrasting chemistries. ISME J 5(3):414–426

Xiong J, Liu Y, Lin X et al (2012) Geographic distance and pH drive bacterial distribution in alkaline lake sediments across Tibetan Plateau. Environ Microbiol 14(9):2457–2466

Metagenomics of Plant Rhizosphere Microbiome

12

Ravindra Soni, Vinay Kumar, Deep Chandra Suyal, Lata Jain, and Reeta Goel

Abstract

The rhizosphere is a specific microbial habitat in the soil ecosystem. This is the area where soil swayed by plant roots through plant exudates deposition. Further, rhizosphere and root microbiota provide useful services to their host plant, such as protection from pathogen and enhanced mineral acquirement from nearby soil for plant growth. Microbial communities, usually, interact with each other and their host, so it is important to detain as much of the microbial diversity as possible. It requires the use of modern analytical tools such as metagenomics, which can reveal the functional potential of a rhizosphere microbiome.

Keywords

Rhizosphere • Microbiome • Metagenomics

12.1 Introduction

Apart from many important microbial-mediated processes like plant growth promotion, plant protection, and pathogenesis, there is microbial competition in the rhizosphere, the area of soil adjacent the roots which is most exposed to the influence of

R. Soni
Department of Agricultural Microbiology, College of Agriculture, Indira Gandhi Krishi Vishva Vidyalaya, Raipur, Chhattisgarh, India

V. Kumar • L. Jain
ICAR-National Institute of Biotic Stress Management, Baronda, Chhattisgarh, India

D.C. Suyal • R. Goel (✉)
Department of Microbiology, CBSH, G.B.Pant University of Agriculture and Technology, Pantnagar 263145, Uttaranchal, India
e-mail: rg55@rediffmail.com

R.P. Singh et al. (eds.), *Understanding Host-Microbiome Interactions - An Omics Approach*, DOI 10.1007/978-981-10-5050-3_12

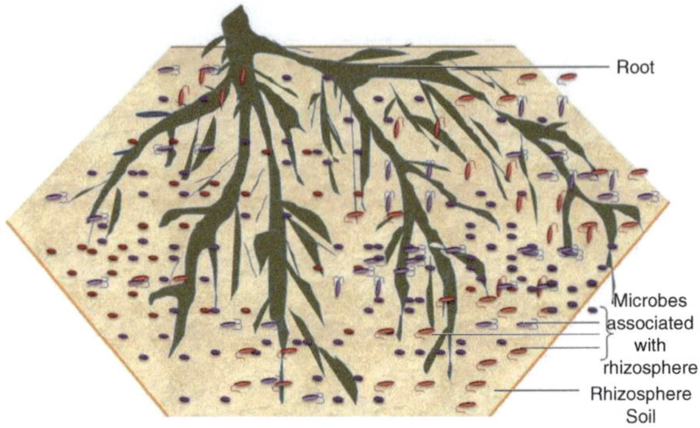

Fig. 12.1 Root, rhizosphere, and its microbiota make a compact association with each other which ultimately affects the plant growth (both positively and negatively)

plant's root exudates (Fig.12.1). In recent years, the rhizosphere microbiology has received significant attention, since it influences plant both directly and indirectly. This microbiome includes the various functional gene pool from prokaryotic to eukaryotic origin associated with various habitats of a plant like rhizosphere and rhizoplane and plays a crucial role in plant protection (Abd-Elsalam et al. 2010; Mendes et al. 2011, Lakshmanan et al. 2014). Rhizosphere microbiota have been well explored for their beneficial effects on plant such as nitrogen fixation, phosphate solubilization, production of plant growth hormones, biocontrol properties, and tolerance to various stresses (Tsurumaru et al. 2015, Elias et al. 2016, Majeed et al. 2015, Massart et al. 2015). The structure of microbial communities in the rhizosphere is largely influenced by soil type, soil texture, plant genotype, cultivars, and developmental stage of the plants (Broeckling et al. 2008). Similarly, plants are affected by rhizospheric microbial population through their involvement in hasty nutrient cycle, dependency of water, and beneficial metabolites (Buscot and Varma 2005). Besides beneficial microbes having effects on plant growth, some microbial pathogen also colonizes the rhizosphere, determined to break through the defensive microbial shield and to overcome the innate plant protection mechanisms in order to cause disease (Sharma et al. 2011; Huang et al. 2013, Mendes et al. 2013).

Further, Curtis et al. (2002) reported that a soil could have up to 4×10^6 different microbial taxa, whereas Gans et al. (2005) predicted that 1 g of soil can contain more than one million diverse bacterial genomes which show the immense microbial diversities present in soil especially in rhizosphere. Despite the immense importance of microbes in rhizosphere, very little information is available about their diversity; for example, what is the actual number of microbial species present in the environment and what each individual species does (its ecological function) (Singh et al. 2010), because more than 99% of microbial species present in the soil are still refractory to cultivation in laboratory growth conditions (Torsvik et al. 2002; Dinsdale et al. 2008; Vieites et al. 2009; Kumar et al. 2015). This was also proved by culture-independent approaches that the soil microbial diversity and rhizosphere microbiomes are highly underestimated. Therefore, substantial opportunities exist for biotechnological applications that alter microbes in order to improve crop productivity and tolerance to environmental conditions (Biteen et al. 2016). Metagenomics takes advantage of the fact that while some microorganisms are culturable under laboratory practices and others are not, all of them are life-forms based on DNA as a carrier of genetic information.

12.2 Metagenomics

Metagenomics is the study of the collective genome of microorganisms (metagenome) from any environmental niche to provide information on the diversity and ecology of microbial forms in a specific environment. The ability to simultaneously analyze the metagenome of all microbes present in a particular environment provides a powerful insight into microbial community structure, the processes mediated by community, and possible complex interactions that may occur. Further, it is a new field combining microbiology, molecular biology, and biotechnology and permits researchers to look into a complex system that can divulge the functional potential of a microbiome such as rhizosphere. The progressive reduction in the cost of nucleotide sequencing and development of high-throughput sequencing techniques made it possible to sequence large quantities of DNA, from mixtures of organisms (Metzker 2010), and offered a deep insight into whole rhizosphere. The high-resolution analyses of the taxonomic composition of rhizosphere soil provide baseline information on the specific microbiome members living in rhizosphere environments (Lagos et al. 2015). To gain a better understanding of the composition and diversity of the rhizospheric soil, metagenomic approach has already been used to examine the phyla and genera that naturally inhabit a niche (Spence et al. 2014). Metagenomics is not only helpful in determining the bacterial diversity but also useful in exploring fungal population in rhizosphere (LeBlanc et al. 2015). Culture-independent fungal community profiling of soil and rhizosphere of field-grown sugarcane showed that the concentration of nitrogen fertilizer strongly modifies the composition but not the taxon richness of fungal communities in soil and rhizosphere (Paungfoo-Lonhienne et al. 2015).

12.2.1 Approach Toward Metagenomics

Metagenomics can be the sequence-based (Soni et al. 2012) or function-based (Rabausch et al. 2013) culture-independent analysis of metagenomes trapped from an environment. The sequence-based approach generally comprises 16S rDNA-based analysis, which is a reliable tool for evaluating the phylogenetic division of a soil metagenome (Soni et al. 2010; Soni and Goel 2010) but does not give insights into the community's metabolic potential, whereas functional capabilities of a microbial community can be analyzed by using functional metagenomics (Aßhauer et al. 2015). Thus, functional metagenomics can be considered as a true discovery tool for identifying and characterizing novel gene families (Nacke et al. 2011; MacGarvey et al. 2012; Craig et al. 2009; Illeghems et al. 2015) from rhizosphere metagenome.

For construction of metagenomic library, the first step is the isolation of high-quality intact DNA from the environment, which is necessary for downstream processes. Several methods have been developed to isolate the high-quality (free from the contaminants and PCR inhibitors) and intact DNA (large size with less fragmentation) (Berry 2003; Bertrand 2005). Major difficulty associated with the isolation of pure DNA is due to the presence of polyphenolic compounds and humic acid which interferes and coprecipitated along with the DNA (Streit and Schmitz 2004).

These coprecipitated compounds create hindrance in the downstream processes including restriction digestion of DNA, ligation, and inhibition of *Taq DNA polymerase* activity during the PCR reaction (Ranjan et al. 2005; Sharma et al. 2007). The DNA extraction procedures have been modified to extract high-quality DNA which is free from the contaminants (Felczykowska et al. 2015, Tanveer et al. 2016). The different methods of DNA extraction and purification yielded noticeably different PCR-DGGE profiles representing different bacterial consortia in rhizosphere (Niemi et al. 2001).

12.2.2 Techniques Coupled with Metagenomics

With the expansion of molecular biology-based techniques, there has been a move toward the characterization of diverse bacterial populations within biomass from the environment. Molecular techniques that have been successfully applied for microbial diversity analysis are the polymerase chain reaction (PCR), cloning and sequencing of ribosomal genes, denaturing gradient gel electrophoresis (DGGE and TGGE), restriction fragment length polymorphism (RFLP), terminal-restriction fragment length polymorphism (T-RFLP), fluorescent in situ hybridization (FISH), and typing of bacterial microbes (Table 12.1). Generally, 16S rRNA gene is used as a phylogenetic marker for microbial diversity analysis because this gene is surprisingly well conserved through several years of evolution (Hugenholtz and Tyson,

Table 12.1 List of advance molecular techniques used for the characterization of rhizosphere microbial communities using metagenomics

Techniques used	Aim of the study	References
Amplicon gene sequencing of conserved marker genes, 16S rRNA	Bacterial and fungal rhizosphere communities in hydrocarbon-contaminated soils	Bell et al. 2014
	Rhizobacterial population of *Arachis hypogaea*	Haldar and Sengupta 2015
Metagenome sequencing	Soybean rhizosphere	Mendes et al. 2013
	Grassland plant community richness and soil edaphics	LeBlanc et al. 2015
	454 pyrosequencing to analyze rhizosphere fungal communities during soybean growth	Sugiyama et al. 2014
Metatranscriptome sequencing	Rhizosphere microbiome assemblage affected by plant development	Chaparro et al. 2014
	Root surface microbiome	Ofek-Lalzar et al. 2014
Metaproteomic profiling	Phyllosphere and rhizosphere of rice	Knief et al. 2012
	Sugarcane rhizospheric	Lin et al. 2013
Metabolomic profiling	Mycorrhizal tomato roots	Tschaplinski et al. 2014
		Rivero et al. 2015

2008). This conservation permits analysis from bacteria and archaea, illuminating the taxonomic distribution and evolutionary associations among microorganisms.

Various next-generation sequencing (NGS) technologies have recently been used for the microbial studies. These include:

1. Amplicon gene sequencing (targeted amplification) of conserved 16S rDNA and internal transcribed spacer (ITS) area to investigate the bacterial and fungal diversity, respectively.
2. Sequencing of metagenome for obtaining information on the genetic diversity and physiological capability of entire microbial communities present in a particular environment. (iii) Metatranscriptomics involves sequencing of cDNA (reverse-transcribed mRNA of functional genes) to measure the level of gene expression in relation to reference metagenomes and gene functions and thus identifies the potential functional activities and metabolically active microbes.
3. Metaproteomic sequencing of protein to assess the expressed proteins and their abundance for giving information on the functional activity between microbial communities and plants.
4. Metabolomic profiling of extracted proteins/metabolites using state-of-the-art mass spectrometry to detect and quantify the abundance of molecules and their probable involvement in metabolic reactions of microbial communities and/or plants.
5. Recently, combination of advance techniques of analytical chemistry and molecular biology especially gas chromatography–mass spectrometry (GC–MS), capillary electrophoresis–mass spectrometry (CE–MS), and liquid chromatography–mass spectrometry (LC–MS) has greatly enhanced the quantitative and qualitative analysis of chemical composition/metabolites of any plant part/tissues, rhizosphere, and environmental niche (Zhang et al. 2012).

12.2.3 Bioinformatic Tools

12.2.3.1 Softwares for Metagenome Analysis

Sequencing of metagenomes generates a huge nucleotide sequence data which need to be further analyzed to get the meaningful results. Bioinformatic software is required to handle the sequences of DNA generated by the Sanger sequencing as well as different next-generation sequencing platform, namely, 454 pyrosequencing and illumina produced long and short reads, respectively. The softwares available for amplicon analysis and used for Sanger and/or 454 ribosomal pyro-tag sequences, namely, mothur (https://www.mothur.org), Quantitative Insights into Microbial Ecology (QIIME), MEGAN, and CARMA, are important and widely used softwares for metagenomic analysis (Gerlach et al. 2009; Caporaso et al. 2014; Gerlach and Stoye, 2011; Huson and Weber 2013). Recently softwares Illumina reads and PacBio reads have been developed for metagenomic analysis of short read sequences and very long sequences, respectively. The PICRUSt software connects the taxonomic classification from meta-profiling results with metabolic information (Langille et al. 2013).

12.2.3.2 Platforms for Metagenomic Analysis

Several metagenomic platforms are available which provide information related to the microbial diversity analysis. A major challenge in the analysis of environmental sequences involves integration of data and a question of how to analyze different types of data in a unified approach which could provide information on both taxonomic and functional analysis. To overcome these imitations, community enabling cloud compatibility platform is available which includes IMG/M (Markowitz et al. 2006), CAMERA (Seshadri et al. 2007), and WebCARMA (Gerlach et al. 2009). The Galaxy framework supports basic metagenomic analyses (Kosakovsky Pond et al. 2009), Cloud Virtual Resource (CloVR) (Angiuoli et al. 2011), and Metagenomics Rapid Annotation using Subsystem Technology (MG-RAST) (Wilke et al. 2011).

12.3 Rhizosphere Metagenomics

The rhizosphere likely provides a richer environment when compared to bulk soil, which is less subjected to chronic energetic stress conditions that are usually associated with archaeal dominance over bacteria (Valentine 2007). Understanding the beneficial interactions between the microorganisms consistently found in the rhizosphere is vital for describing the nature of the soil–plant interface. Appraisal of microbial community structure in the soil is based mainly on the use of cultivation-dependent as well as cultivation-independent methods including soil metagenomics (Daniel. 2005). Among the agriculturally important crops, rice rhizosphere is the most exploited by using metagenomics. In rice rhizosphere, numerous microbial traits such as nitrogen fixation, protein secretion systems, quorum sensing, and their habitat specificity were predicted using metagenomics (Knief et al. 2012). The effect of wild and cultivated rice genotypes on rhizosphere bacterial community composition was recently documented (Shenton et al. 2016) by using culture-independent approach. Bulgarelli et al. (2013) employed a combination of 16S rRNA gene profiling and shotgun metagenome analysis of the microbiota associated with wild and domesticated accessions of barley and reported that the combined action of microbe–microbe and host–microbe interactions drives microbiota differentiation at the root–soil interface. Recently, functional metagenomic approach was applied to identify the trait directly from the microbiome in barley rhizosphere soil which had not received phosphate fertilizer over a 15-year period, and the study revealed that phosphorous solubilization function was mainly linked with the non-culturable microbiome present in rhizospheric soil (Chhabra et al. 2013). Targeted metagenomic approaches were also applied to look for insight into the diversity of soil fungal communities and rhizosphere in gray mangroves from the Red Sea (Abdel-Wahad et al. 2014, Simoes et al. 2015, Alzubaidy et al., 2016). The combination of culture-dependent and culture-independent approaches enhances the better understanding of phenomenon of plant microbe interaction and what maximum benefits can be obtained from the beneficial microflora existing in the rhizosphere (Naz et al. 2014). Metagenomic approach explored microbial wealth of rhizosphere in a better way whenever coupled with other advance molecular techniques (Knief et al. 2011; Unno and Shinano 2013). Recent report by Jin et al. (2016) showed that the rhizosphere soil metagenome can provide a sketch of the functional regions of a protein domain and can be utilized for protein optimization and functional characterization. Recently, metagenomics proves that a small "core" rhizosphere bacterial community that together with an AM fungus and other putative beneficial bioinoculants may interact synergistically to promote plant growth (Valverde et al. 2016).

12.3.1 Major Microbial Groups in Rhizosphere Soil Metagenome

Traditionally, microbial diversity is all about isolation of the microbes on diverse culture media and understanding their metabolic variations. Enrichment media are used to culture the selective microbes under the laboratory conditions, even though only a subpopulation of microbes present in an environmental sample is able to grow, while the rest are not grown even in the different media and growth conditions. Because the majority of bacterial populations remain unculturable, the diversity of complex bacterial communities is inexorably underestimated using typical cultivation methods. Many of these "unculturable" bacteria represent new phylotypes, families, and divisions in domain bacteria and archaea (Sharma et al. 2005). There are, at present, estimated to be 61 distinct bacterial phyla, of which 31 have unculturable representatives (Hugenholtz and Tyson 2008). Furthermore, sequence-based metagenomic studies suggests that rice rhizosphere has the dominance of Acidobacteria, Actinobacteria, Firmicutes, Proteobacteria, Actinomycetes, and Planctomycetes (Arjun and Harikrishnan, 2011; Knief et al. 2011; Mahyarudin and Rusmana 2015; Bhattacharyya et al. 2016), whereas wheat rhizosphere showed affiliation with Chloroflexi and Planctomycetes phyla (Naz et al. 2014) and also associated with Archaea, Actinobacteria, Firmicutes, Proteobacteria, fungi, virus, and unclassified taxa (Hernandez-Leon et al. 2012).

Furthermore, a major portion of soybean rhizosphere is also represented by Actinobacteria, γ-Proteobacteria, and ascomycetous divisions (Bresolin et al. 2010). Among the fungi, Ascomycota is the most important phylum present in the bulk soil and rhizosphere of soybean (Li et al. 2010). Similarly, microbial diversity in sugarcane rhizosphere soil using 16S rRNA genes revealed the abundance of predominating Proteobacteria followed by Acidobacteria, Bacteroidetes, Firmicutes, and Actinobacteria (Pisa et al. 2011). Metagenomic analysis of vegetable rhizosphere identified the majority of members of the Proteobacteria and Bacteroidetes (Jackson et al. 2013). As per the reports available in the public domain, it may be concluded that the diverse group of unculturable bacterial communities present in rhizosphere soil among which the Proteobacteria is major dominant one (Bresolin et al. 2010; Arjun and Harikrishnan 2011; Hernandez-Leon et al. 2012; Unno and Shinano 2013; Shenton et al. 2016; Bhattacharyya et al. 2016). Nonetheless, Schreiter et al. (2014) analyzed the samples from field-grown lettuce rhizosphere and nearby bulk soil and showed that the phyla Proteobacteria, Actinobacteria, Firmicutes, Acidobacteria, and Bacteroidetes were dominant in both bulk soil and rhizosphere. The most common rhizospheric microbial genera revealed through metagenomics (after 2006) are listed in Table 12.2.

Table 12.2 The most common rhizospheric microbial genera revealed through rhizosphere metagenome of some common plant/crop (Studies enlisted were conducted after 2006)

S. no.	Rhizosphere	Common bacterial genera	References
1	Rice	*Geodermatophilus, Actinokineospora, Actinoplanes, Streptomyces, Kocuria*	Mahyarudin and Rusmana 2015
		Acidovorax, Anaeromyxobacter, Azospirillum, Bradyrhizobium, Dechloromonas, Desulfovibrio, Geobacter, Herbaspirillum, Methylobacterium, Magnetospirillum, Methylosinus, Rhodopseudomonas	Knief et al. 2011
2	Wheat plants (*Triticum aestivum*)	*Achromobacter, Bacillus, Cellulomonas, Clostridia, Gallionella, Herbaspirillum, Pseudomonas, Rhizobium, Stenotrophomonas, Sinorhizobium, Burkholderia, Xanthomonas, Pantoea, Enterobacter, Geobacter, Nocardia, Mycobacterium, Microbacterium*, uncultured bacteria	Valverde et al. 2016
		Azoarcus, Balneimonas, Bradyrhizobium, Gemmatimonas, *Lysobacter, Methylobacterium, Mesorhizobium, Microvirga, Rubellimicrobium, Rhodoplanes, Skermanella*	Naz et al. 2014
3	Sugarcane	*Azospirillum, Bacillus, Belnapia, Bradyrhizobium, Cohnella, Chitinophaga, Chryseobacterium, Dactylosporangium, Ferruginibacter, Flavisolibacter, Flavobacterium*, Gemmatimonas, *Hyphomicrobium, Labrys, Lechevalieria, Ktedonobacter, Mesorhizobium, Methylobacterium, Mycobacterium, Mucilaginibacter, Niastella, Nitrospira, Novosphingobium, Nocardioides, Paenibacillus, Pseudolabrys, Rhizobium, Rhodoplanes, Rubrobacter, Sinomonas, Streptomyces, Terrimonas, Tumebacillus,*	Pisa et al. 2011
4	Tobacco (*Nicotiana tabacum*)	*Flavobacterium, Burkholderia, Bordetella*	Brinkmann and Tebbe 2007
5	Lettuce	*Alkanindiges, Sphingomonas, Burkholderia, Novosphingobium, Sphingobium*	Schreiter et al. 2014
6	Maize	*Arthrobacter, Bacillus, Burkholderia, Kitasatospora, Methylobacterium, Pantoea, Streptomyces, Talaromyces*, uncultured bacterium	Oliveira et al. 2009
7	Soybean	*Bacillus, Bradyrhizobium rhizobium, Stenotrophomonas, Streptomyces*	Sugiyama et al. 2014
8	Para grass (*Urochloa mutica*)	*Anaerosinus, Bacillus, Clostridium, Caldilinea*, Chloroflexi, *Microcoleus*	Mukhtar et al. 2016
9	*Arabidopsis thaliana*	*Arthrobacter, Kineosporiaceae, Flavobacterium, Massilia*	Bodenhausen et al. 2013

12.3.2 Our Lead

Since the last decade, we did lots of sincere efforts to explore the rhizosphere bacterial diversity based on 16S rRNA gene amplification of soil metagenome (Soni et al. 2010 Soni et al. 2010). We coupled several other molecular tools with metagenomics to explore the rhizosphere microbiome like PCR-RFLP (Singh et al. 2010), DGGE and TTGE (Soni et al. 2010), and real-time PCR (Premalatha et al. 2009, Soni and Goel, 2010, Suyal et al. 2015). Our group has earlier beautifully painted the prevalence of *csp* (Premalatha et al. 2009) and *nifH* genes (Soni and Goel 2010; Soni et al. 2016) from Western Himalayan region of India. A hypothesis was also proposed by us that *nif* genes like *nifH* may evolve from their nearest genes or adjacent regions and in due course become specific in their functions (Soni and Goel, 2010). To the best of our knowledge, the first major metagenomic effort revealed the presence of diverse diazotrophic microbial assemblages in indigenous red kidney bean (RKB) rhizosphere (Suyal et al. 2014, 2015).

Conclusions

Rhizosphere microbial communities are important for various key processes in plant. Further, the tools of metagenomics offer many openings into a broadened view of the rhizosphere as compared to culture-dependent genomics. Therefore, the novel traits from rhizosphere microbiome from metagenomics are only dependent on the advancement of methodologies.

Acknowledgment The work mentioned in this chapter from author group was supported by the National Bureau of Agriculturally Important Microorganisms; India (NBAIM/ICAR) grant to R.G.

References

Abd-Elsalam KA, Almohimeed I, Moslem MA, Bahkali AH (2010) M13-microsatellite PCR and rDNA sequence markers for identification of Trichoderma (Hypocreaceae) species in Saudi Arabian soil. Genet Mol Res 9(4):2016–2024

Abdel-Wahab MA, Hodhod MS, Bahkali AH, Jones EB (2014) Marine fungi of Saudi Arabia. Bot Mar 57:323–335

Alzubaidy H, Essack M, Malas TB, Bokhari A, Motwalli O, Kamanu FK, Jamhor SA, okhtar NA, Antunes A, Simoes MF, Alam I, Bougouffa S, Lafi FF, Bajic VB, Archer JAC (2016) Rhizosphere microbiome metagenomics of gray mangroves (*Avicennia marina*) in the Red Sea. Gene 576(2 Pt 1):626–636

Angiuoli SV, Matalka M, Gussman A, Galens K, Vangala M et al (2011) CloVR: a virtual machine for automated and portable sequence analysis from the desktop using cloud computing. BMC Bioinforma 12:356

Arjun JK, Harikrishnan K (2011) Metagenomic analysis of bacterial diversity in the rice rhizosphere soil microbiome. Biotechnol Bioinformatics Bioeng 1(3):361–367

Aßhauer KP, Wemheuer B, Daniel R, Meinicke P (2015) Tax4Fun: predicting functional profiles from metagenomic 16S rRNA data. Bioinformatics 31(17):2882–2884

Bell TH, El-Din Hassan S, Lauron-Moreau A, Al-Otaibi F, Hijri M, Yergeau E, St-Arnaud M (2014) Linkage between bacterial and fungal rhizosphere communities in hydrocarbon-contaminated soils is related to plant phylogeny. ISME J 8:331–343. doi:10.1038/ismej.2013.149

Berry AE, ChiocchiniC ST, Sosio M, Wellington EM (2003) Isolation of high molecular weight DNA from soil for cloning into BAC vectors. FEMS Microbiol Lett 223:15–20

Bertrand H, Poly F, Van VT, Lombard N, Nalin R, Vogel TM, Simonet P (2005) High molecular weight DNA recovery from soils prerequisite for biotechnological metagenomic library construction. J Microbiol Methods 62:1–11

Bhattacharyya PKS, Roy M, Dasa S, Raya D, Balachandar S, Karthikeyan AK, Mohapatra NT (2016) Elucidation of rice rhizosphere metagenome in relation to methane and nitrogen metabolism under elevated carbon dioxide and temperature using whole genome metagenomic approach. Sci Total Environ 542:886–898

Biteen JS, Paul C, Blainey CZG, Chun M, George M (2016) Church tools for the microbiome: nano and beyond. ACS Nano 10:6–37

Bodenhausen N, Horton MW, Bergelson J (2013) Bacterial communities associated with the leaves and the roots of *Arabidopsis thaliana*. PLoS One 8(2):e56329. doi:10.1371/journal.pone.0056329

Bresolin JD, Bustamante MMC, Kruger RH, Silva MRSS, Perez KS (2010) Structure and composition of bacterial and fungal community in soil under soybean monoculture in the Brazilian Cerrado. Braz J Microbiol 41:391–403

Brinkmann and Tebbe (2007) Differences in the rhizosphere bacterial community of a transplastomic tobacco plant compared to its non-engineered counterpart. Environ Biosaf Res 6:113–119

Broeckling CD, Broz AK, Bergelson J, Manter DK, Vivanco JM (2008) Root exudates regulate soil fungal community composition and diversity. Appl Environ Microbiol 74:738–744

Bulgarelli D, Schlaeppi K, Spaepen S, Ver E, Themaat LV, Schulze-lefert P (2013) Structure and functions of the bacterial microbiota of plants. Ann rev. Plant Biol 64:807–838

Buscot F, Varma A (2005) Microorganism in soils: roles in genesis and functions. Springer, Germany

Caporaso JG, Kuczynski J, Stombaugh J, Bittinger K, Bushman FD, Costello EK, Fierer N, Peña AG, Goodrich JK, Gordon JI, Huttley GA, Kelley ST, Knights D, Koenig JE, Ley RE, Chaparro JM, Badri DV, Vivanco JM (2014) Rhizosphere microbiome assemblage is affected by plant development. ISME J 8:790–803

Chhabra S, Brazil D, Morrissey J, Burke JI, Gara FO, Dowling DN (2013) Characterization of mineral phosphate solubilization traits from a barley rhizosphere soil functional metagenome. Microbiol Open 2(5):717–724

Craig JW, Chang FY, Brady SF (2009) Natural products from environmental DNA hosted in Ralstonia metallidurans. ACS Chem Biol 4:23–28

Curtis TP, Sloan WT, Scannell JC (2002) Estimating prokaryotic diversity and its limits. Proc Natl Acad Sci U S A 99:10494–10499

Daniel R (2005) The metagenomics of soil. Nat Rev Microbiol 3:470–478

Dinsdale EA, Edwards RA, Hall D, Angly F, Breitbart M, Brulc JM et al (2008) Functional metagenomic profiling of nine biomes. Nature 452:629. –U628.

Elias F, Woyessa D, Muleta D (2016) Phosphate Solubilization potential of rhizosphere fungi isolated from plants in Jimma zone, Southwest Ethiopia. Int J Microbiol 2016:5472601

Felczykowska A, Krajewska A, Zielińska S, Los JM (2015) Sampling, metadata and DNA extraction important steps in metagenomic studies. Acta Biochim Pol 62(1):151–160

Gans J, Wolinsky M, Dunbar J (2005) Computational improvements reveal great bacterial diversity and high metal toxicity in soil. Science 309:1387e1390

Gerlach W, Jünemann S, Tille F, Goesmann A, Stoye J (2009) WebCARMA: a web application for the functional and taxonomic classification of unassembled metagenomic reads. BMC Bioinformatics 10:430. doi:10.1186/1471-2105-10-430

Gerlach W, Stoye J (2011) Taxonomic classification of metagenomic shotgun sequences with CARMA3. Nucleic Acids Res 39:e91

Haldar S, Sengupta S (2015) Impact of plant development on the rhizobacterial population of *Arachis hypogaea*: a multifactorial analysis. J. J Basic Microbiol 55:922–928

Hernandez-Leon R, Martinez-Trujillo M, Valencia-Cantero E et al (2012) Construction and characterization of a metagenomic DNA library from the rhizosphere of wheat (*Triticum aestivum*). Phyton Int J Exp Bot 81

Huang Q, Jiang H, Briggs BR, Wang S, Hou W, Li G, Wu G, Solis R, Arcilla CA, Abrajano T, Dong H (2013) Archaeal and bacterial diversity in acidic to circumneutral hot springs in the Philippines. FEMS Microbiol Ecol 85(3):452–464

Hugenholtz P, Tyson GW (2008) Microbiology: metagenomics. Nature 455:481–483

Huson DH, Weber N (2013) Microbial community analysis using MEGAN. Methods Enzymol 531:465–485. doi:10.1016/B978-0-12-407863-5. 00021-6.

Illeghems K, Weckx S, De Vuyst L (2015) Applying meta-pathway analyses through metagenomics to identify the functional properties of the major bacterial communities of a single spontaneous cocoa bean fermentation process sample. Food Microbiol 50:54–63

Jackson CR, Randolph KC, Osborn SL, Tyler HL (2013) Culture dependent and independent analysis of bacterial communities associated with commercial salad leaf vegetables. BMC Microbiol 13:274

Jin Z, Di Rienzi SC, Janzon A, Werner JJ, Angenent LT, Dangl JL, Fowler DM, Ley RE (2016) Novel rhizosphere soil alleles for the enzyme 1- aminocyclopropane-1-carboxylate deaminase queried for function with an in vivo competition assay. Appl Environ Microbiol 82:1050–1059

Knief C, Delmotte N, Chaffron S, Stark M, Innerebner G, Wassmann R, von Mering C, Vorholt A (2011) Metaproteogenomic analysis of microbial communities in the phyllosphere and rhizosphere of rice. ISME J:1–13

Knief C, Delmotte N, Chaffron S, Stark M, Innerebner G, Wassmann R, von Mering C, Vorholt JA (2012) Metaproteogenomic analysis of microbial communities in the phyllosphere and rhizosphere of rice. ISME J 6(7):1378–1390

Kumar S, Krishnani KK, Bhushan B, Brahmane MP (2015) Metagenomics: retrospect and prospects in high throughput age. Biotechnol Res Int:121735. http://dx.doi.org/10.1155/2015/121735

Lagos ML, Maruyama F, Nannipieri P, Mora ML, Ogram A, Jorquera MA (2015) Current overview on the study of bacteria in the rhizosphere by modern molecular techniques: a mini review. J Soil Sci Plant Nutr 15(2):04–523

Lakshmanan V, Selvaraj G, Bais HP (2014) Functional soil microbiome: belowground solutions to an aboveground problem. Plant Physiol 166:689–700

LeBlanc N, Kinkel LL, Kistler HC (2015a) Soil fungal communities respond to grassland plant community richness and soil edaphics. Microb Ecol 70:188–195

LeBlanc N, Kinkel LL, Kistler HC (2015b) Soil fungal communities respond to grassland plant community richness and soil edaphics. Microb Ecol 70:188–195

Li C, Weidong XL, Ying K, Wang WJ (2010) Effect of monoculture soybean on soil microbial community in the Northeast China. Plant Soil 330:423

Lin W, Wu L, Lin S, Zhang A, Zhou M, Lin R, Wang H, Chen J, Zhang Z, Lin R (2013) Metaproteomic analysis of ratoon sugarcane rhizospheric soil. BMC Microbiol 13:135

Mahyarudin I, Rusmana YL (2015) Metagenomic of actinomycetes based on 16S rRNA and nifH genes in soil and roots of four Indonesian rice cultivars using PCR-DGGE. Hayati J Biosci 22:113e121

Majeed A, Abbasi KM, Hameed S, Imran A, Rahim N (2015) Isolation and characterization of plant growth-promoting rhizobacteria from wheat rhizosphere and their effect on plant growth promotion. Front Microbiol 6:198

Massart S, Martinez-Medina M, Jijakli MH (2015) Biological control in the microbiome era: challenges and opportunities. Biol Control 89(98):108

Mendes R, Garbeva P, Raaijmakers JM (2013) The rhizosphere microbiome: significance of plant beneficial, plant pathogenic, and human pathogenic microorganisms. FEMS Microbiol Rev 37:634–663. doi:10.1111/1574-6976.12028.

Mendes R, Kruijt M, de Bruijn I, Dekkers E, van der Voort M, Schneider JH, Piceno YM, DeSantis TZ, Andersen GL, Bakker PA, Raaijmakers JM (2011) Deciphering the rhizosphere microbiome for disease-suppressive bacteria. Science 332:1097–1100

Metzker ML (2010) Sequencing technologies - the next generation. Nat Rev Genet 11:31–46. 10.1038/nrg2626

Mukhtar S, Mirza MS, Awan HA, Asma M, Mehnaz S, Malik K (2016) Microbial diversity and metagenomic analysis of the rhizosphere of para grass (urochloamutica) growing under saline conditions. Pak J Bot 48(2):779–791

Nacke H, Will C, Herzog S, Nowka B, Engelhaupt M, Daniel R (2011) Identification of novel lipolytic genes and gene families by screening of metagenomic libraries derived from soil samples of the German Biodiversity Exploratories. FEMS Microbiol Ecol 78:188–201

Naz I, Mirza MS, Bano A (2014) Molecular characterization of rhizosphere bacterial communities associated with wheat (*Triticum aestivum* L.) cultivars at flowering stage. J Animal Plant Sci 24(4):1123–1134

Niemi RM, Heiskanen I, Wallenius K, Lindstrom K (2001) Extraction and purification of DNA in rhizosphere soil samples for PCR-DGGE analysis of bacterial consortia. J Microbiol Methods 45:155–165

Ofek-Lalzar M, Sela N, Goldman-Voronov M, Green SJ, Hadar Y, Minz D (2014) Niche and host-associated functional signatures of the root surface microbiome. Nat Commun 5:4950

Oliveira CA, VMC A, Marriel IE, Gomes EA, Scotti MR, Carneiro NP, Guimarães CT, Schaffert RE, NMH S (2009) Phosphate solubilizing microorganisms isolated from rhizosphere of maize cultivated in an oxisol of the Brazilian Cerrado biome. Soil Biol Biochem 41(2009):1782–1787

Paungfoo-Lonhienne C, Yeoh YK, Kasinadhuni NRP, Lonhienne TGA, Robinson N, Hugenholtz P, Ragan MA, Schmidt S (2015) Nitrogen fertilizer dose alters fungal communities in sugarcane soil and rhizosphere. Sci Rep 5:8678. doi:10.1038/srep08678

Pisa G, Magnani GS, Weber H, Souza EM, Faoro H, Monteiro RA, Daros E, Baura V, Bespalhok JP, Pedrosa FO, Cruz LM (2011) Diversity of 16S rRNA genes from bacteria of sugarcane rhizosphere soil. Braz J Med Biol Res 44:1215–1221

Langille MGI, Zaneveld J, Caporaso JG, McDonald D, Knights D, Reyes JA et al (2013) Predictive functional profiling of microbial communities using 16S rRNA marker gene sequences. Nat Biotechnol 31:814–821

Premalatha K, Soni R, Khan M, Marla SS, Goel R (2009) Exploration of Csp gene(s) from temperate and glacier soils of Indian Himalaya and in silico analysis of encoding proteins. Curr Microbiol 58:343–348

Rabausch U, Juergensen J, Ilmberger N, Böhnke S, Fischer S, Schubach B et al (2013) Functional screening of Metagenome and genome libraries for detection of novel flavonoid-modifying enzymes. Appl Environ Microbiol 79(15):4551–4563

Ranjan R, Grover A, Kapardar RK, Sharma R (2005) Isolation of novel lipolytic genes from uncultured bacteria of pond water. Biochem Biophys Res Commun 335:57–65

Rivero J, Gamir J, Arocal R, Pozo MJ, Flors V (2015) Metabolic transition in mycorrhizal tomato roots. Front Microbiol 6:598

Schreiter S, Ding G, Heuer H, Neumann G, Sandmann M, Grosch R et al (2014) Effect of the soil type on the microbiome in the rhizosphere of field-grown lettuce. Front Microbiol 5:144

Sharma PK, Capalash N, Kaur J (2007) An improved method for single step purification of metagenomic DNA. J Mol Biotechnol 36:61

Sharma R, Ranjan R, Kapardar RK, Grover A (2005) Unculturable bacterial diversity: an untapped resource. Curr Sci 89(1):72–77

Sharma SK, Johri BN, Ramesh A, Joshi OP, Prasad SV (2011) Selection of plant growth-promoting pseudomonas spp. that enhanced productivity of soybean wheat cropping system in central India. J Microbiol Biotech 21:1127–1142

Shenton M, Iwamoto C, Kurata N, Ikeo K (2016) Effect of wild and cultivated rice genotypes on rhizosphere bacterial community composition. Rice 9:42

Simoes MF, Antunes A, Ottoni CA, Amini MS, Alam I, Alzubaidy H, NA Mokhtar JA, Archer C, Bajic VB (2015) Soil and rhizosphere associated fungi in gray mangroves (*Avicennia marina*) from the Red Sea- a metagenomic approach. Genomics Proteomics Bioinformatics 13:310–320

Singh C, Soni R, Jain S, Roy S, Goel R (2010) Diversification of nitrogen fixing bacterial community using nifH gene as a biomarker in different geographical soils of Western Indian Himalayas. J Environ Biol 31:553–556

Soni R, Acahrya C, Primalatha K et al (2012) Metagenomics technology. In: Kumar A, Pareek A, Gupta SM (eds) Biotechnology in medicine and agriculture principles and practice. IK International, New Delhi, p 835

Soni R, Shaluja B, Goel R (2010) Bacterial community analysis using temporal gradient gel electrophoresis of 16 S rDNA PCR products of soil metagenomes. Ekologija 56(3&4):94–98
Soni R, Goel R (2010) Triphasic approach for assessment of bacterial population in different soil systems. Ekologija 56(3&4):99–104
Spence C, Alff E, Johnson C, Ramos C, Donofrio N, Sundaresan V, Bais H (2014) Natural rice rhizospheric microbes suppress rice blast infections. BMC Plant Biol 14:130
Streit WR, Schmitz RA (2004) Metagenomics- the key to the uncultured microbes. Curr Opin Microbiol 7:492–498
Sugiyama A, Ueda Y, Zushi T, Takase H, Yazaki K (2014) Changes in the bacterial community of soybean rhizospheres during growth in the field. PLoS One 9(6):e100709
Suyal DC, Yadav A, Shouche Y, Goel R (2015) Bacterial diversity and community structure of Western Indian Himalayan red kidney bean (*Phaseolus vulgaris*) rhizosphere as revealed by 16S rRNA gene sequences. Biologia 70(3):305–313
Suyal DC, Yadav A, Shouche Y, Goel R (2014) Diversified diazotrophs associated with the rhizosphere of Western Indian Himalayan native red kidney beans (*Phaseolus vulgaris* L.) 3 Biotech 5(4):433–441
Tanveer A, Yadav S, Yadav D (2016) Comparative assessment of methods for metagenomic DNA isolation from soils of different crop growing fields. 3 Biotech 2:220
Tsurumaru H, Okubo T, Okazaki K, Hashimoto M, Kakizaki K, Hanzawa E, Hiroyuki T, Noriyuki A, Fukuyo T, Yasuyo S, Seishi I, Minamisawa K (2015) Metagenomic analysis of the bacterial community associated with the taproot of sugar beet. Microbes Environ 30(1):63–69
Torsvik V, Daae FL, Sandaa RA, Øvreås L (2002) Microbial diversity and function in soil: from genes to ecosystems. Curr Opin Microbiol 5:240–245
Tschaplinski TJ, Plett JM, Engle NL, Deveau A, Cushman KC, Martin MZ, Doktycz MJ, Tuskan GA, Brun A, Kohler A, Martin F (2014) *Populus trichocarpa* And *Populus deltoides* exhibit different metabolomic responses to colonization by the symbiotic fungus Laccaria bicolor. Mol Plant-Microbe Interact 27:546–556
Unno Y, Shinano T (2013) Metagenomic analysis of the rhizosphere soil microbiome with respect to phytic acid utilization. Microbes Environ 28(1):120–127
Valentine DL (2007) Adaptations to energy stress dictate the ecology and evolution of the Archaea. Nat Rev Microbiol 5(4):316–323
Valverde A, De-Maayer P, Oberholster T, Henschel J, Louw MK, Cowan D (2016) Specific microbial communities associate with the rhizosphere of *Welwitschia mirabilis*, a living fossil. PLoS One 11(4):e0153353
Vieites JM, Guazzaroni ME, Beloqui A, Golyshin PN, Ferrer M (2009) Metagenomics approaches in systems microbiology. FEMS Microbiol Rev 33:236–255
Wilke A, Wilkening J, Glass EM, Desai NL, Meyer F (2011) An experience report: porting the MG-RAST rapid metagenomics analysis pipeline to the cloud. Concurrency Computat Pract Exper 23:2250–2257
Zhang AH, Sun H, Wang P, Han Y, Wang XJ (2012) Modern analytical techniques in metabolomics analysis. Analyst 137:293–300

Plant–Pathogen Interactions: A Proteomic Approach

13

Amanpreet Kaur, Anil Kumar, and M. Sudhakara Reddy

Abstract

Changing environmental conditions have resulted in the outbreak of many new plant diseases, which are becoming a threat to food security. To overcome this challenge, there is an urgent need to develop resistant varieties through traditional breeding and/or biotechnological approaches. For an effective crop improvement programmes, a thorough understanding of plant–pathogen interaction is a prerequisite. Pathogenesis is generally effected by host innate and/or systemic acquired resistance, which includes cell wall lignification, release of oxidative enzymes, biosynthesis of phytoalexins and pathogenesis-related (PR) proteins. Studies of the proteins induced during disease development have a potential to throw light on complex defence mechanisms adopted by plants. Recently, proteomics has gained popularity in understanding host–pathogen interactions. Proteomics helps in studying the dynamics of important proteins during host–pathogen interaction and disease development. Apart from protein identification, it also allows to determine the role and molecular structure of a specific protein or a group of proteins during pathogenesis. Present article will review the role of proteomics with the special emphasis on PR proteins.

Keywords

Pathogenesis-related proteins • Mass spectrometry • 2D–PAGE • 2D–DIGE

A. Kaur (✉) • A. Kumar • M. Sudhakara Reddy
Department of Biotechnology, TIFAC-Centre of Relevance and Excellence in Agro and Industrial Biotechnology (CORE), Thapar University, Patiala 147001, India
e-mail: amanbath07@yahoo.com

R.P. Singh et al. (eds.), *Understanding Host-Microbiome Interactions - An Omics Approach*, DOI 10.1007/978-981-10-5050-3_13

Abbreviations

MS	Mass spectrometry.
MS/MS	Tandem mass spectrometry.
2D–PAGE	Two-dimensional polyacrylamide gel electrophoresis.
2D–DIGE	Two-dimensional differential gel electrophoresis.
PR	Pathogenesis related.

13.1 Introduction

Host and pathogen interaction is a complicated process, which involves alterations in the composition of various biomolecules such as proteins, glycoproteins, carbohydrates, lipids, etc. resulting from altered metabolism of plant as well as pathogen (Gupta et al. 2015). Unlike mammals, plants do not possess mobile defence cells but have innate immunity to defend against pathogenic invasions (Jones and Dangl 2006). The beginning of host–pathogen interaction is marked by the induction of various biomolecules and elicitors such as flagellin, chitin, etc. resulting in successful establishment of pathogenicity (Mehta et al. 2008, Lodha et al. 2013). These elicitors also known as pathogen-associated molecular patterns (PAMPs) are recognized by membrane-localized pattern recognition receptors (PRRs) of plant which act as first line of defence (Zipfel 2014). This kind of plant immunity is termed as PAMP-triggered immunity (PTI).

PTI induces certain cellular changes in plant, resulting in cell wall lignifications, apoplastic acidification, kinase activation, production of oxidative species, release of phytoalexins and pathogenesis-related (PR) proteins (Borad and Sriram 2008). Among immunity-associated factors, the release of PR proteins is considered as one of the most important innate responses of plant to pathogenic invasions. The PR proteins are known to get accumulated in intercellular spaces of infected plants, as a defensive response to disease initiation (Cohen 2001). Levels of PR protein accumulation are also correlated with the degree of resistance incurred by a plant against pathogen (Bol et al. 1990). Apart from their role in biotic stress management, PR proteins are also known for their important role during different abiotic stresses (Antoniw et al. 1983). Studies conducted by Antoniw et al. (1983) on tobacco revealed the isolation and characterization of ten important PR proteins involved in fighting biotic and abiotic stresses (Antoniw et al. 1983). The number of characterized PR proteins has now increased to 17 (Sinha et al. 2014). Few of the important classes of PR proteins are discussed in this review article (Table 13.1).

PR-1 proteins are reported to possess antifungal activity against a number of fungi and also oomycetes (Niderman et al. 1995). Though their exact mode of action is still unclear, PR-1-like protein, "helothermine", extracted from lizard is found to inhibit release of calcium ions to confer fungal resistance (Monzingo et al. 1996). β-1,3-Glucanases belonging to PR-2 proteins are also found in wide number of plants such as peanut, tobacco, etc. (Roy-Barman et al. 2006). They possess antifungal activities through hydrolytic cleavage of glycosidic linkages in β-1,3-glucans present in cell

Table 13.1 Characteristic of pathogenesis-related (PR) proteins and their role and mode of action in plants

Class of PR proteins	Name of PR protein	Size (kDa)	Role	Mode of action	References
PR-1 PR-1 like	Not specified Helothermine	14–17 (basic nature)	Antifungal	Interact with membrane channel proteins of target cells, inhibit release of Ca^{+2}	Monzingo et al. (1996)
PR-2	β-1,3-Glucanase	33–44 (basic and acidic nature)	Antifungal	Hydrolytic cleavage of 1,3-glucosidic linkages in 1,3-glucans	Dassi et al. (1998)
PR-3	Chitinase (I, II, IV, V, VI, VII)	15–43 (basic and acidic nature)	Antifungal	Hydrolyse chitin	Neuhaus (1999)
PR-4	Chitin-binding proteins	9–30 (basic nature)	Antifungal Antibacterial	Increase substrate accessibility	Van-Loon and Van Strein (1999)
PR-5	Osmotin Thaumatin-like	25 18–25 (acidic nature)	Antifungal, antiviral and antibacterial	Increase membrane permeability, enhance cytotoxicity	Pierpoint et al. (1987)
PR-6	Proteinase inhibitor	8.13	Antioomycete	Inhibits trypsin; cysteine proteinase; cathepsin D, B, H and L; papain; and metalloproteinase	Koiwa et al. (1997); Graham et al. (2003)
PR-7	Endoproteinase	69–70	Antifungal and antioomycete	Accumulate to provide resistance against fungal pathogen	Vera and Conejero (1988)
PR-8	Chitinase (III)	29 (acidic and basic nature)	Antifungal	Hydrolyse chitin	Van-Loon et al. (1994)
PR-9	Peroxidase	39.3	Antimicrobial	Induce hypersensitivity in plants	Lagrimini et al. (1987)
PR-10	Ribonuclease	15–20 (acidic nature)	Antimicrobial	Unknown	Casanal et al. (2013)
PR-11	Chitinase (V)	41.43	Antifungal	Hydrolyse chitin	Van-Loon et al. 1994

(continued)

Table 13.1 (continued)

Class of PR proteins	Name of PR protein	Size (kDa)	Role	Mode of action	References
PR-12	Defensin	5 (basic nature)	Antifungal	Reduce hyphal elongation by increasing Ca^{+2} uptake by plant	Terras et al. (1995)
PR-13	Thionin	45–50 amino acids	Toxic to animal and insect cells	Cell wall permeabilization, inhibit sugar uptake, induce ion leakage from cells, inhibit insect amylase	Florack and Stiekema (1994)
PR-14	Lipid transfer protein	9 kDa	Role in cutin formation, somatic embryogenesis, adaptation events of higher plants	Transfer of lipids between membranes	Cheng et al. (2004)
PR-15–PR-16	Oxalate oxidase	95–100	Antifungal	Induction of hypersensitive response, signal transduction	Zhou et al. (1998)

wall of oomycetes as well as fungi. Chitinases, the PR-3 proteins, have been expressed in plants to achieve resistance against fungal invasions (Jiang et al. 2013; Ebrahim et al. 2011). In fungi, chitinases are reported to disturb cell wall morphology (Jiang et al. 2013). Chitinases from plants and insects have been cloned and expressed in different plants such as potato, banana, apple, wheat, mustard, etc. to enhance resistance against fungal diseases (Kovacs et al. 2013, Shin et al. 2008, Hirai et al. 2004, Bolar et al. 2000, Grison et al. 1996). Recent studies also showed enhanced abiotic stress tolerance of transgenic plants by overexpressing chitinases, but underlying mechanism is still not clear (Dana et al. 2006). The vacuolar isoforms of osmotin (PR-5 protein) are known to possess membrane-disturbing properties and are reported to inhibit hyphal growth in vitro and cause sporangial lysis (Abdin et al. 2011). With the help of various molecular biology tools, genes encoding osmotin have been cloned and extensively studied for their expression (Singh et al. 1989; Nelson et al. 1992; LaRosa et al. 1992). In light of various studies conducted, several factors such as tissue injury, fungal or viral pathogen attack, osmotic stress, UV radiations, etc. resulted in transcriptional activation of osmotin gene (Nelson et al. 1992; Kononowicz et al. 1992). Osmotin is also reported to have antifungal activities against a wide range of fungi including oomycetes such as *P. infestans*, *Candida albicans*, *Neurospora crassa and Trichoderma reesei* (Woloshuk et al. 1991; Stintzi et al. 1991; Vigers et al. 1992). Overexpression of osmotin gene in exotic cultivars of potato has been reported to impart some degree of resistance against fungal pathogens (Abdin et al. 2011) through weakening of cell wall and cytotoxicity (Abdin et al. 2011). Further, overexpression of osmotin to attain enhanced tolerance against abiotic stress such as cold, salt and drought has also been reported (Das et al. 2011; Rajam et al. 2007). Another PR protein belonging to PR-10, a ribosome-inactivating protein (RIP), has been isolated from various plants (Borad and Sriram 2008). It is reported to confer resistance to various bacteria and fungi such as *Fusarium oxysporum* and *Pseudomonas solanacearum* (Kim et al. 2001). Transgenic potatoes with defensin, a PR-12 protein, have also been reported to show resistance against *Alternaria solani* and *Fusarium culmorum* (Rao et al. 1999). Overexpression of defensin isolated from *Neurospora megalosiphon* is reported to provide high levels of resistance against *P. infestans* (Portieles et al. 2010).

In addition to PR proteins, various macromolecules are associated with antipathogenic activities triggered by plants (Zipfel 2008). Thus, in order to effectively understand the host–pathogen interactions, "omic" techniques are seen as an alternative to conventional experimental methodologies (Lodha and Basak 2012).

13.2 Proteomic Approaches Used in Plants

Recent advancements in modern biology have helped in the exploitation of genomics, proteomics and more recently metabolomics to understand various changes occurring in plant or pathogen during the course of interaction (Wilkins et al. 1996, Gonzalez-Fernandez et al. 2010). After only 20 years of its introduction as "protein complement of genome", proteomics is seen as the most important research module among all other "omic" techniques due to its advantages over genome analysis (Gonzalez-Fernandez et al. 2010). Proteomics has successfully provided insights about interactions between different proteins during pathogenic invasions as well as cellular processes occurring as a result of growth of an organism (Picotti et al. 2009).

In plants, a typical proteome study starts with the extraction of total protein content; however, this is itself a challenging task. The presence of other molecules such as phenols, lipids, sugars, lipopolysaccharides and abundant proteins such as RUBISCO (ribulose bisphosphate carboxylase-oxygenase) makes extraction of less abundant proteins a quite cumbersome process (Chen and Harmon 2006, Rakwal and Agrawal 2003). Therefore, different extraction processes involving the use of TCA (trichloroacetic acid), phenols, etc. have been standardized for different plant tissues (Carpentier et al. 2005).

Proteomic techniques are generally classified as gel-based (two-dimensional polyacrylamide gel electrophoresis (2D–PAGE), two-dimensional differencial gel electrophoresis (2D–DIGE)) and non-gel-based methods (stable isotope labelling with amino acids in cell culture (SILAC), multidimensional protein identification technology (MudPIT), isotope-coded affinity tag (ICAT), etc.) (Fig.13.1). However, gel-based 2D–PAGE is the most commonly used technique in analysis of plant proteome, but development of 2D–DIGE has allowed the identification of two different proteins on the same gel by differential labelling of protein samples with structurally similar fluorescent dyes followed by gel scanning based on excitation wavelength of each dye (Marouga et al. 2005). Resulting images are analysed using softwares such as BioNumerics and Gel2DE (Li and Seiller-Moiseicuitsch 2011).

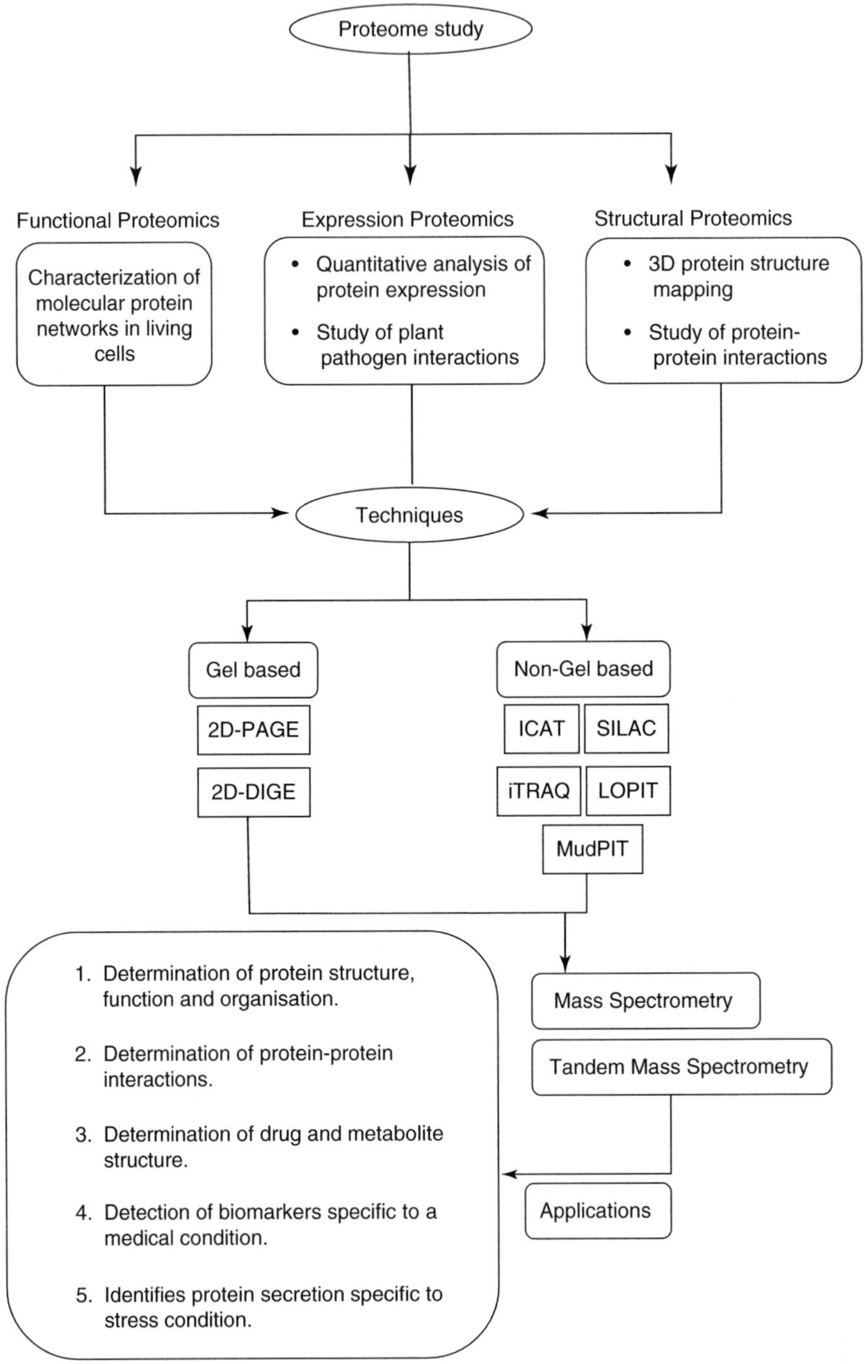

Fig. 13.1 Proteomics: applications and various techniques involved in protein identification and characterization

Despite innumerable applications and importance of gel-based proteomic techniques, there are several associated disadvantages such as separation limitation (Zhu et al. 2003), poor reproducibility, etc. (Lodha et al. 2013). These disadvantages paved the way for introduction of new proteomic techniques with improved features. Alternative robust techniques such as ICAT, SILAC, MudPIT, localization of organelle proteins by isotope tagging (LOPIT), etc. allowing analysis of low-copy-number proteins and eliminating variation in results caused by the use of different gels are seen as a revolution in proteomic studies. These techniques followed by mass spectrometry or tandem mass spectrometry have various applications ranging from the study of proteins present in specific organelle (Dunkley et al. 2004) to determination of qualitative changes in protein levels during different stress conditions (Gygi et al. 1999).

Proteomics has also been successfully used to study interactions of plants with a wide range of pathogens (Casado-vela et al. 2006; Li et al. 2012; Shah et al. 2012). During plant–pathogen interactions, changes were observed in the plant at cellular, biochemical and apoplast levels (Delaunois et al. 2014). One of the most common changes involves lowering of plant photosynthetic capabilities (Delaunois et al. 2014). This response of plant reflects the increased energy demand for defence-related activities including cell wall lignifications and release of reactive oxidative species (ROS) and defence-related proteins (Mathesius 2009). During the process of infection, both host and pathogen are present in close proximity to each other; thus, it becomes very difficult to differentiate between if the proteins secreted by pathogen or from that of the host (Mathesius 2009). Therefore, numerous scientific approaches using one actor at one time have been deployed to overcome such challenges (Brunet et al. 2003). Few of the approaches are explained below:

13.2.1 The Use of Plant Extracts

In order to understand the secretome induced by pathogen during establishment of pathogenicity, crude leaf extracts from the plant are used instead of host itself. For instance, in a study conducted on *Xanthomonas axonopodis* pv. Passiflora to study proteins induced in pathogen in the presence of the host (Tahara et al. 2003), leaf extracts of passionflower plant were inoculated with the pathogen. Proteomic study of the pathogen revealed the induction of few hypothetical proteins (function not clear) and membrane-related proteins and enzymes such as pyrophosphatase. Another similar study was conducted by Mehta and Rasato (2001). In this study, *X. axonopodis* pv. citri was cultured on the medium containing the leaf extracts from resistant and sensitive host (sweet orange) plants as well as a nonhost plant. Sulphate-binding proteins in pathogen were found to get upregulated in the presence of leaf extract from susceptible host plant, whereas no such response was observed in the presence of extracts from resistant and nonhost plants.

Secretory analysis of pathogens in the presence of host plant extract has revealed the role of cellulose, pectin acetyltransferase, lyases and other cell wall-degrading enzymes in successful establishment of pathogenicity (Kazemi-Pour et al. 2004; Yajima and Kav 2006).

13.2.2 Cell Labelling and Pathogen Elicitors

To avoid interference of proteins of pathogenic origin with host proteome during plant–pathogen interaction studies, pathogen-based elicitors were used instead of pathogen itself (Rajjou et al. 2006). Green fluorescent protein (GFP)-tagged *Oryza sativa* cell suspension culture was challenged with compatible and noncompatible strains of *Xanthomonas oryzae*, a causative agent of rice bacterial blight. After incubation, plant cells were filtered out for proteome analysis by identification of GFP expression in plasma membranes. About eleven proteins were identified which includes proteins such as prohibitin, antioxidants, etc. (Chen et al. 2007). Similar studies were also conducted using radiolabelled cell suspension of *Arabidopsis* inoculated with pathogen elicitors such as flagellin (Peck et al. 2001). Phosphorylation of proteins was found to be directly involved in pathogenesis-related defensive response of plant (Peck et al. 2001).

13.2.3 Mutated Pathogenic Strains

In this kind of approach, plant is infected with the mutant pathogen lacking a specific gene for pathogenecity. Proteome of plant, post-inoculation with mutant strain and control strain, is compared to identify the proteins induced in plant specifically against control. For instance, a study using a model plant, *Arabidopsis*, concluded that a plant challenged with virulent strain of bacterial pathogen *Pseudomonas syringae* secretes proteins involved in hypersensitive response, whereas no such proteins were secreted upon infection with non-virulent or mutant strains of the pathogen (Kaffarnik et al. 2009).

13.3 Importance of Proteomics in the Study of Plant–Pathogen Interactions

- In eukaryotic cells, however DNA remains uniform, but a set of protein components varies in each and every cell based on the function of tissue or organ. Even under the conditions of stress, protein expression levels vary from one cell type to another. Therefore, the study of proteome rather than genome holds complete information regarding cellular behaviour in response to biotic or abiotic stress conditions.
- Proteomics also reveals the information about posttranslational modifications occurring within the cells (Duley and Grover 2001).
- Expressional proteomics has the potential to reveal information regarding host–pathogen interactions which is not possible with genomics.

13.4 Proteome Basis of Plant–Pathogen Interactions

The use of proteomics in the study of plant–pathogen interactions is relatively new (Quirino et al. 2010). It has been only few decades when proteomics was used for the first time for protein profiling of *Pinus* spp. (Ekramoddoullah and Hunt 1993).

Proteins involved in susceptibility or resistance of the plant to *Cronartium ribicola* (a causative agent of an important fungal disease of pine, white pine blister) were identified using 2D–PAGE. It was reported that the expression of proteins was different in resistant and susceptible *Pinus* seedlings (Ekramoddoullah and Hunt 1993). Since this report in 1993, numerous studies have been carried out on proteome analysis of host and pathogen during the course of interaction (Coaker et al. 2004, Katam et al. 2015; Yang et al. 2015; Brizard et al. 2006). Apart from the identification of changes in levels of proteins in plant cell during pathogenic interaction, the power of recognition possessed by a plant to differentiate between friendly and pathogenic signals has also been reported in a model plant, *Medicago truncatula* (genome size 454–526 mega base pairs) (Mathesius et al. 2001). In this study, the plant was challenged with a pathogen, *Pseudomonas aeruginosa*, and a non-pathogenic bacteria, *Sinorhizobium meliloti* (a nitrogen-fixing bacteria). Comparative analysis was carried out between induction of proteins in host in response to N-acyl homoserine lactone (AHL) signals from both types of bacteria, and about one-third of total proteins induced during interaction were found to be specific to pathogenic bacteria (Mathesius et al. 2001).

Apart from the proteins expressed within the host or pathogen during interaction, analysis of extracellular proteins has also revealed some of the interesting facts. In a study conducted by Bouchart et al. (2007), a comparison was made between the secretome of soft rot pathogen, *Erwinia chrysanthemi*, treated with *Chrysanthemum* (host) and control (nonhost) leaf extracts. It was observed that in the presence of control, bacteria secrete flagellin and proteases as major proteins, whereas treatment with host extract induces secretion of 14 proteins such as pectin lyases and acetyl esterases. These proteins were found to be specific to interaction of *E. chrysanthemi* and *Chrysanthemum* spp. (Bouchart et al. 2007). Similar proteome and secretome studies have been carried out in numerous fungal, bacterial, viral as well as nematodal pathogen-infecting plants (Gao et al. 2003; Huang et al. 2003; Houterman et al. 2007; Campo et al. 2004). Few of the case studies are discussed in this review.

13.4.1 Plant and Viral Interaction

Virus exists inactive under natural conditions, until they enter into a living host cell. In the process of infection, virus enters into plant cell through a vector or wound, multiplies in host cell and finally makes its way to plant vascular system (Mehta et al. 2008). During infection, virus mainly depends upon the plant/host protein machinery for its multiplication. During this process, virus faces numerous challenges from host defence system (Mehta et al. 2008). Proteomic approaches have been used for analysis and identification of plant proteins induced specifically to suppress PAMP-triggered immunity (PTI) (Whitham et al. 2006; Lee et al. 2006; Elvira et al. 2008).

The dynamic proteomic analysis of plant following viral infection by 2D–PAGE, followed by MALDI-TOF-based mass spectrometry, was carried out in tobacco mosaic virus (TMV)-resistant cultivar of *Capsicum annuum* following infection with virus strain TMV-$P_{0.}$ Defence response triggered in plant includes induction of proteins related with apoptosis, endocytosis, membrane trafficking and biotic stress (Whitham et al. 2006). A similar study was carried out to analyse protein induction in apoplast of *Prunus persica* following infection with plum pox potyvirus (PPV) (Diaz-Vivancos et al. 2006). A total of 22 proteins were analysed which mainly include antioxidant enzymes, pathogenesis-related proteins and flavoproteins such as mandelonitrile lyases (MDL). Another proteome study of rice species susceptible to rice yellow mottle sobemovirus reveals induction of 223 proteins which mainly include metabolic proteins (proteins essential for energy production), transcription factors (essential for translational activities) and defence-related proteins (maintenance of oxido-reduced environment) (Brizard et al. 2006). Proteomic studies conducted on plants with certain degree of resistance to the viral pathogen revealed the induction of pathogenesis-related (PR) proteins as a major defence mechanism (Casado-Vela et al. 2006; Table 13.2). During interaction between *Lycopersicon esculentum* and tobacco mosaic virus, about 16 proteins were found to be induced, out of which the maximum belongs to PR protein family (Casado-Vela et al. 2006). In *Capsicum chinense* plants infected with two different strains of pepper mild mottle virus (PMMoV-I (Italian strain) and PMMoV-S (Spanish strain)), different levels of PR protein accumulation were reported (Elvira et al. 2008). Infection with Italian strain results in compatible interaction, whereas hypersensitivity response (HR) was induced in plants upon infection with Spanish strain which resulted in incompatible interaction. Proteome analysis found the induction of PR proteins of family PR-1, PR-2, PR-3, PR-5, PR-9, PR-16 and PR-17 which led to necrosis to limit infection (Elvira et al. 2008).

13.4.2 Plant and Bacterial Interaction

Bacterial secretion systems (I–V) are key to establish successful pathogenic interaction with plants (Mehta et al. 2008). Out of five secretion systems, type III is involved in successful disease establishment in various plants (Puhler et al. 2004). Secretion systems allow bacteria to influence host cellular responses using virulence factors and effector proteins (Noel et al. 2001; Arlat et al. 1994). Type III system effector proteins generally include avirulence (Avr) factors released by bacterial pathogen for suppression of plant immune response (Schechter et al. 2004). Previously, most of the bacteria-plant interaction studies revolved around the genomics for analysis of various genes responsible for pathogenicity. Recently, the analysis of protein profiles from bacteria during host interaction is seen as a simplified model for understanding protein induction in host as well as pathogen upon close interaction with each other.

Due to bacterial colonizing within the plant, it becomes extremely difficult to differentiate between proteins expressed by bacteria and the plant. Thus, few approaches used to separately study the proteome induced during interaction in pathogen include the use of plant extract to activate bacterial genes. The proteome study from *Xanthomonas axonopodis* pv. Passiflora induced using leaf extracts of

Table 13.2 The pathogenesis-related (PR) proteins identified during different plant-pathogen interactions

Class of PR protein	Name of PR protein	Pathogenic organism	Plant counterpart	References
PR-3	Chitinase	*Clavibacter michiganensis ssp.*	*Lycopersicon hirsutum*	Coaker et al. (2004)
PR-5	Thaumatin-like	*Xanthomonas oryzae*	*Oryza sativa*	Mahmood et al. (2006)
PR-5	Thaumatin-like	*PPV*	*Prunus persica*	Diaz-Vivancos et al. (2006)
PR-3	Chitinase	*TMV*	*Lycopersicon esculentum*	Casado-Vela et al. (2006)
PR-2 PR-3 PR-5	β-1,3-Glucanases Chitinase Thaumatin-like	*Xylella fastidiosa*	*Vitis vinifera*	Katam et al. (2015)
PR-1 PR-2 PR-3 PR-5 PR-9 PR-16 PR-17	β-1,3-Glucanases Chitinase Osmotin-like Peroxidases Germin-like protein pRp27	*Pepper mild mottle virus (PMMoV)*	*Capsicum chinense*	Elvira et al. (2008)
PR-2 PR-5 PR-9 PR-10	β-1,3-Glucanases Thaumatin-like Peroxidases	*Magnaporthe grisea*	*Oryza sativa*	Kim et al. (2004)
PR-2 PR-5 PR-9 PR-3	β-1,3-Glucanases Thaumatin-like Peroxidase Chitinase	*Fusarium graminearum*	*Triticum aestivum*	Zhou et al. (1998)
PR-10		*Aphanomyces euteiches*	*Medicago truncatula*	Colditz et al. (2004)
PR-2	β-1,3-Glucanases	*Fusarium verticillioides*	*Zea mays*	Campo et al. (2004)
PR-2 PR-5 PR-9 PR-3	β-1,3-Glucanases Thaumatin-like Peroxidase Chitinase	*Fusarium oxysporum*	*Tomato*	Houterman et al. (2007)
PR-3	Chitinase	*Fusarium graminearum*	*Humulus lupulus*	Phalip et al. (2005)
PR-9	Peroxidases	*Rice yellow mottle virus*	*Oryza sativa*	Brizard et al. (2006)
PR-2 PR-5 PR-9 PR-1	β-1,3-Glucanases Thaumatin-like Peroxidase	*Zymoseptoria tritici*	*Triticum aestivum*	Yang et al. (2015)

host plant, *Passiflora edulis*, showed the upregulation of outer membrane protein in pathogen (Tahara et al. 2003), indicating the role of membrane proteins in establishing pathogenicity. In another study, the role of outer membrane protein in pathogenicity of *Dickeya dadantii* (a pathogen of soft rot disease in tuber vegetables and ornamental plants) was studied (Babujee et al. 2007). Functional proteins induced during biotic stress include porins and membrane translocation proteins (Babujee et al. 2007).

Apart from bacterial secretome studies, the effect of bacteria-plant interactions on plant proteome has also been studied using pathogenic elicitors (Rajjou et al. 2006) and mutant strains of bacterium (Jones et al. 2004). In a study conducted on the model plant, *Arabidopsis thaliana*, for proteome analysis against infection caused by *Pseudomonas syringae*, two strains of pathogen i.e. a compatible strain (DC3000) and a mutant strain (DC3000Hrp) lacking protein machinery for pilus synthesis were used (Jones et al. 2004). Comparison between proteome and transcriptome of plants challenged with both strains of *P. syringae* showed that proteins, glutathione S-transferases and peroxiredoxins are of high importance in inducing HR in plants (Jones et al. 2004). After 2 years of the above study, the same experiment was repeated, but proteome analysis was carried out on organellar fractions isolated from plants post-inoculation with the pathogen (Jones et al. 2006). Results showed the alterations in metabolic and antioxidant enzymes in addition to changes in components of Calvin-Benson cycle.

In several studies conducted on plants challenged with bacterial pathogens, the involvement of defence proteins in general and pathogenesis-related proteins in particular was identified during pathogenic interactions. In an experimentation studying the interaction between rice and an important bacteria, *Xanthomonas oryzae*, the accumulation of pathogenesis-related proteins was observed (Mahmood et al. 2006). Post-inoculation proteome studies of protein fractions from rice using 2D–DIGE followed by tandem mass spectrometry showed the accumulation of peroxidases (PR-9) and thaumatin-like protein (PR-5) in addition to antioxidative enzymes such as superoxide dismutase (SOD).

A similar study was carried out in susceptible and resistant lines of tomato (*Lycopersicon hirsutum*) infected with *Clavibacter michiganensis*, a causal agent of bacterial canker. Among 26 differentially regulated proteins identified using proteomic approach, 12 were identified as pathogenesis-related proteins and others were antioxidants (Coaker et al. 2004).

13.4.3 Plant and Fungus Interaction

Plant and fungus associations are believed to be about 400 million years old (Remy et al. 1994). Association between plants and fungus partners may be beneficial or pathogenic, and both are reported to result in numerous changes in host cellular metabolism (Jones and Dangl 2006). Fungus makes contact with plant through spore germination (Tucker and Talbot 2001), haustoria or appressorium formation (Grenville-Briggs et al. 2005) which are used for nutrient uptake by pathogen as well as for introducing various lytic enzymes and effector proteins to host cell. Plants respond to fungal invasion by altering its physiochemical process such as nutrient translocation, respiration, defence and photosynthesis (Zeilinger et al. 2016). The availability of genome sequences of about 25 important pathogenic fungi and few important model plants has opened new horizons for expressional analysis involved in establishment of plant and fungus interaction (Tuskan et al. 2006, Velasco et al. 2007, Mehta et al. 2008).

Secretome studies conducted on important fungus, *Fusarium graminearum*, to identify the proteins involved in establishing successful contact with host showed that fungi regulate their secretory system according to the availability of substrate (Phalip et al. 2005, Meijer et al. 2006). Proteomic studies of *F. graminearum*, growing on the cell wall of *Humulus lupulus*, showed that fungi mostly secrete proteins (phospholipases, proteinases, etc.) actively participating in the cell wall degradation. When the same fungi were provided with glucose as a substrate, an alteration in protein induction was observed, and secretome mostly involves proteins corresponding enzymes of primary metabolism (Phalip et al. 2005).

Study of plant proteomics in association with fungal pathogens usually revealed the induction of proteins involved in defence, stress management, metabolism, energy production and signal transduction (Mehta et al. 2008). Among all, defence-related proteins (pathogenesis-related proteins) hold immense importance in pathogenic interactions occurring between plant and fungi (Cui et al. 2014). Few of the case studies are discussed below.

First proteomic-based study of fungus-plant interaction was carried out in 2001 in rice leaf blades inoculated with pathogen *Magnaporthe grisea* (Konishi et al. 2001). Protein profiles of plant were also correlated with availability of nitrogen as disease severity increases at higher concentrations of nitrogen (Long et al. 2000). Proteome analysis indicates that different levels of infection were marked with differential accumulation of proteins belonging to PR-5 family of pathogenesis-related proteins. In an attempt to further elaborate, Kim et al. (2003) carried out the same experiment on rice cell suspension instead of leaf blades and observed involvement of 12 new proteins including PR-10 proteins. This was followed by identification of increased level of peroxidases (PR-9), β-1,3-glucanases (PR-2), etc. in rice during fungal attacks (Kim et al. 2004).

Limitations of gel-based proteomic techniques such as insufficiency in identification of proteins with low copy number have supported new techniques involving isotopic tagging followed by mass spectrometry (Lodha et al. 2013). In an isobaric tag for relative and absolute quantitation (iTRAC) labelling-based proteomic study of wheat inoculated with *Zymoseptoria tritici*, protein upregulation was observed during necrotrophic stage in comparison to biotrophic stage in the plant. Expression cluster involved the accumulation of pathogenesis-related proteins belonging to family PR-1, PR-2, PR-3, PR-5, PR-9 and PR-17 in addition to carbohydrate metabolic proteins. This study revealed the importance of PR proteins in resistance of wheat against *Z. tritici* (Yang et al. 2015). Similar study was also conducted in maize (Mohammadi et al. 2011). In this study, attempts were made to investigate proteins imparting resistance to *Gibberella* ear rot disease caused by *F. graminearum* in maize; protein profiling was carried out after 2 days of inoculation using iTRAC followed by tandem mass spectrometry. Cultivars with resistant and susceptibility towards disease showed different protein profiles. About 96 proteins were identified which include PR-10, PR-5, PR-9 and PR-3 proteins (Mohammadi et al. 2011).

Apart from the leaves, proteome studies have also been conducted on various other organs such as roots, embryos and vascular systems for investigation of role of pathogenesis-related proteins in fungus-plant interactions (Campo et al. 2004, Houterman et al. 2007, Colditz et al. 2004). A proteomic study carried out during interaction of plant with *Xylella fastidiosa*, a pathogen living in the xylem (Chakraborty et al. 2016), revealed the accumulation of various pathogenesis-related protein families such as PR-2, PR-3, PR-5, etc. along with various heat-shock proteins (Zhang et al. 2015). The upregulation and induction of these proteins have also been reported as a major defensive response against Pierce's disease (PD) of grapevine (Katam et al. 2015).

Conclusion

For in-depth understanding of cellular-level changes occurring in plant as well as pathogen, "omic" approaches have shown a tremendous potential. Advancements in the proteomics through development of novel and improved techniques allow easy protein profiling of organisms of great importance. Proteomics also helps in assigning functions to the newly identified protein components (Muturi et al. 2010). In the present review, attempts have been made to present proteomic studies in relation with plant–pathogen interactions. In case of bacteria and fungus, pathogenicity was established by secretion of proteins such as cellulases, proteases, etc. In addition to these secretion proteins, antioxidants were also found to be upregulated in pathogen as a defensive mechanism to protect it against reactive oxygen species (ROS) produced by plant as a result of oxidative stress. In plants, mostly pathogenesis-related (PR) proteins such as chitinases, osmotin, peroxidases, etc. were found to be associated with the defence system. Studies conducted on various plants revealed that PR proteins are upregulated in response to pathogenic invasions. Although the role of PR proteins in abiotic as well as biotic stress is well established, still there is a need to understand the mechanisms of PR protein regulation during pathogenic interaction.

References

Abdin MZ, Kira U, Alam A (2011) Analysis of osmotin, a PR protein as metabolic modulator in plants. Bioinformation 5:338–340

Antoniw JF, Oooms G, White RF et al (1983) Pathogenesis related proteins in plants and tissue of *Nicotiana tabaccum* transformed by *Agrobacterium tumefaciens*. Plant Mol Biol 2:317–320

Arlat M, Van Gijsegem F, Huet JC et al (1994) PopA1, a protein which induces a hypersensitivity-like response on specific *Petunia* genotypes, is secreted via the Hrp pathway of *Pseudomonas solanacearum*. EMBO J 13:543–553

Babujee L, Venkatesh B, Yamazaki A et al (2007) Proteomic analysis of the carbonate insoluble outer membrane fraction of the soft-rot pathogen *Dickeya dadantii* (syn. Erwinia chrysanthemi) strain 3937. J Proteome Res 6:62–69

Bol JF, Linthorst HJ, Cornelissen BJ (1990) Plant pathogenesis-related proteins induced by virus infection. Annu Rev Phytopathol 28:113–138

Bolar JK, Norelli JL, Wong KW et al (2000) Expression of endochitinase from *Trichoderma harzianum* in transgenic apple increases resistance to apple scab and reduces vigor. Pytophathol 90:72–76

Borad V, Sriram S (2008) Pathogenesis related proteins for the plant protection. Asian J Exp Sci 22:189–196

Bouchart F, Delangle A, Lemoine J et al (2007) Proteomic analysis of a non-virulent mutant of the phytopathogenic bacterium *Erwinia chrysanthemi* deficient in osmoregulated periplasmic glucans: change in protein expression is not restricted to the envelope, but affects general metabolism. Microbiology 153:760–767

Brizard JP, Carapito C, Delalande F et al (2006) Proteome analysis of plant–virus interactome: comprehensive data for virus multiplication inside their hosts. Mol Cell Proteomics 5:2279–2297

Brunet S, Thibault P, Gagnon E et al (2003) Organelle proteomics: looking at less to see more. Trends Cell Biol 13:629–638

Campo S, Carrascal M, Coca M et al (2004) The defense response of germinating maize embryos against fungal infection: a proteomics approach. Proteomics 4:383–396

Carpentier SC, Witters E, Laukens K et al (2005) Preparation of protein extracts from recalcitrant plant tissues: an evaluation of different methods for two-dimensional gel electrophoresis analysis. Proteomics 5:2497–2507

Casado-Vela J, Selles S, Martinez RB (2006) Proteomic analysis of tobacco mosaic virus-infected tomato (*Lycopersicon esculentum M.*) fruits and detection of viral coat protein. Proteomics 6(Suppl. 1):S196–S206

Casanal A, Zander U, Munoz C et al (2013) The strawberry (PR-10) Fra a proteins control flavonoid biosynthesis by binding to metabolic intermediates. J Biol Chem 288:35322–35332

Chakraborty S, Nascimento R, Zaini PA et al (2016) Sequence/structural analysis of xylem proteome emphasizes pathogenesis related proteins, chitinases and β-1,3-glucanases as key players in grapevine defense against *Xylella fastidiosa*. Peer J. doi:10.7717/peerj-2007

Chen F, Yuan Y, Li Q et al (2007) Proteomic analysis of rice plasma membrane reveals proteins involved in early defense response to bacterial blight. Proteomics 7:1529–1539

Chen S, Harmon AC (2006) Advances in plant proteomics. Proteomics 6:5504–5516

Cheng HC, Cheng PT, Peng P et al (2004) Lipid binding in rice non-specific lipid transfer protein-1 complexes from *Oryza sativa*. Protein Sci 13:2304–2315

Coaker GL, Willard B, Kinter M et al (2004) Proteomic analysis of resistance mediated by Rcm 2.0 and Rcm 5.1, two loci controlling resistance to bacterial canker of tomato. Mol Plant-Microbe Interact 17:1019–1028

Cohen E (2001) Chitin synthesis and inhibition. Pest Manag Sci 57:946–950

Colditz F, Nyamsuren O, Niehaus K et al (2004) Proteomic approach: identification of *Medicago truncatula* proteins induced in roots after infection with the pathogenic oomycete *Aphanomyces euteiches*. Plant Mol Biol 55:109–120

Cui P, Zhang S, Ding F et al (2014) Dynamic regulation of genome wide pre-mRNA splicing and stress tolerance by the Sm-like protein LSm5 in *Arabidopsis*. Genome Biol 15:R1

Dana M, Pintor TJA, Cubero B (2006) Transgenic tobacco plants overexpressing chitinases of fungal origin show enhanced resistance to biotic and abiotic stress agents. Plant Physiol 142:722–730

Das M, Chauhan H, Chhibbar A et al (2011) High-efficiency transformation and selective tolerance against biotic and abiotic stress in mulberry, *Morus indica* cv. K2, by constitutive and inducible expression of tobacco Osmotin. Transgenic Res 20:231–246

Dassi B, Dumas-Gaudot E, Gianinazzi S (1998) Do pathogenesis-related (PR) proteins play a role in bioprotection of mycorrhizal tomato roots towards *Phytophthora parasitica*? Physiol Mol Plant Pathol 52:167–183

Delaunois B, Jeandet P, Clement C et al (2014) Uncovering plant-pathogen crosstalk through apoplastic proteomic studies. Front Plant Sci 5:1–18

Diaz-Vivancos P, Rubio M, Mesonero V et al (2006) The apoplastic antioxidant system in *Prunus*: response to long-term plum pox virus infection. J Exp Bot 57:3813–3824

Duley H, Grover A (2001) Current initiatives in proteomics research: the plant perspective. Curr Sci 80:262–269

Dunkley TP, Watson R, Griffin JL et al (2004) Localization of organelle proteins by isotope tagging (LOPIT). Mol Cell Proteomics 3:1128–1134

Ebrahim S, Usha K, Singh B (2011) Pathogenesis related (PR) proteins in plant defense mechanism. Science against microbial pathogens: communicating current research and technological advances. Sci Against Microb Pathog 2:1043–1054

Ekramoddoullah AKM, Hunt RS (1993) Changes in protein profile of susceptible and resistant sugar-pine foliage infected with the whitepine blister rust fungus *Cronartium ribicola*. Can J Plant Pathol 15:259–264

Elvira MI, Galdeano MM, Gilardi P et al (2008) Proteomic analysis of pathogenesis related protein (PRs) induced by compatible and incompatible interactions of pepper mild mottle virus (PMMoV) in *Capsicum chinense* L3 plants. J Exp Bot 59:1253–1265

Florack DE, Stiekema VJ (1994) Thionins: properties, possible biological roles and mechanism of action. Plant Mol Biol 26:25–37

Gao B, Allen R, Maier T et al (2003) The parasitome of the phytonematode *Heterodera glycines*. Mol Plant-Microbe Interact 16:720–726

Gonzalez-Fernandez R, Prats E, Jorrin-Novo JV (2010) Proteomics of plant pathogenic fungi. J Biomed Biotechnol. doi:10.1155/2010/932527

Graham MY, Weidner J, Wheeler K et al (2003) Induced expression of pathogenesis-related protein genes in soybean by wounding and the *Phytophthora sojae* cell wall glucan elicitor. Physiol Mol Plant Pathol 63:141–149

Grenville-Briggs LJ, Avrova AO, Bruce CR et al (2005) Elevated amino acid biosynthesis in *Phytophthora infestans* during appressorium formation and potato infection. Fungal Genet Biol 42:244–256

Grison R, Bruno GB, Michel S et al (1996) Field tolerance to fungal pathogens of *Brassica napus* constitutively expressing a chimeric chitinase gene. Nat Biotechnol 14:643–648

Gupta R, Lee SE, Agrawal GK et al (2015) Understanding the plant-pathogen interactions in the context of proteomics-generated apoplastic proteins inventory. Front Plant Sci 6:352. doi:10.3389/fpls.2015.00352

Gygi SP, Rist B, Gerber SA et al (1999) Quantitative analysis of complex protein mixtures using isotope-coded affinity tags. Nat Biotechnol 17:994–999

Hirai D, Suzuki T, Yanagida D et al (2004) An evaluation of disease resistance of *Agrobacterium*-mediated transgenic potato (*Solanum tuberosum* L.) containing the chicken lysozyme gene or the wild spinach chitinase gene. Bulletin of the Hokkaido Forest Experiment Station 86:19–26

Houterman PM, Speijer D, Dekker HL et al (2007) The mixed xylem sap proteome of *Fusarium oxysporum*-infected tomato plants. Mol Plant Pathol 8:215–221

Huang G, Gao B, Maier T et al (2003) A profile of putative parasitism genes expressed in the esophageal gland cells of the root knot nematode *Meloidogyne incognita*. Mol Plant-Microbe Interact 16:376–381

Jiang C, Huang RF, Song JL, Huang M, Xu L (2013) Genomewide analysis of the chitinase gene family in *Populus trichocarpa*. J Genet 92:121–125

Jones AM, Thomas V, Bennett MH, et al. (2006) Modifications to the *Arabidopsis* defence proteome occur prior to significant transcriptional change in response to inoculation with *Pseudomonas syringae*. Plant Physiology142:1603–1620

Jones AM, Thomas V, Truman B et al (2004) Specific changes in the *Arabidopsis* proteome in response to bacterial challenge: differentiating basal and R-gene mediated resistance. Phytochemistry 65:1805–1816

Jones JDG, Dangl JL (2006) The plant immune system. Nature 444:323–329

Kaffarnik FA, Jones AM, Rathjen JP et al (2009) Effector proteins of the bacterial pathogen Pseudomonas syringae alter the extracellular proteome of the host plant, *Arabidopsis thaliana*. Mol Cell Proteomics 8:145–156

Katam R, Chibanguza K, Latinwo LM et al (2015) Proteome biomarkers in xylem reveal pierce's disease tolerance in grape. J Proteomics Bioinform 8:217–224

Kazemi-Pour N, Condemine G, Hugouvieux-Cotte-Pattat N (2004) The secretome of the plant pathogenic bacterium *Erwinia chrysanthemi*. Proteomics 4:3177–3186

Kim H, Mun JH, Byun H et al (2001) Molecular cloning and characterization of the gene encoding osmotin protein in *Petunia hybrida*. Plant Sci 162:745–752

Kim ST, Cho KS, Yu S et al (2003) Proteomic analysis of differentially expressed proteins induced by rice blast fungus and elicitor in suspension-cultured rice cells. Proteomics 3:2368–2378

Kim ST, Kim SG, Hwang DH et al (2004) Proteomic analysis of pathogen-responsive proteins from rice leaves induced by rice blast fungus, *Magnaporthe grisea*. Proteomics 4:3569–3578

Koiwa H, Bressan RA, Hasegawa PM (1997) Regulation of protease inhibitors and plant defense. Trends Plant Sci 2:379–384

Konishi H, Ishiguro K, Komatsu S (2001) A proteomics approach towards understanding blast fungus infection of rice grown under different levels of nitrogen fertilization. Proteomics 1:1162–1171

Kononowicz AK, Nelson DE, Singh NK et al (1992) Regulation of the osmotin gene promoter. Plant Cell 4:513–524

Kovacs G, Sagi L, Jacon G et al (2013) Expression of a rice chitinase gene in transgenic banana confers resistance to black leaf streak disease. Transgenic Res 22:117–130

Lagrimini LM, Burkhart W, Moyer M et al (1987) Molecular cloning of complementary DNA encoding the lignin-fonning 5 7 7-5 83 peroxidase from tobacco: molecular analysis and tissue specific expression. Proc Natl Acad Sci U S A 84:7542–7546

LaRosa PC, Chen Z, Nelson DE et al (1992) Osmotin gene expression is posttranscriptionally regulated. Plant Physiol 100:409–415

Lee BJ, Kwon SJ, Kim SK et al (2006) Functional study of hot pepper 26S proteasome subunit RPN7 induced by tobacco mosaic virus from nuclear proteome analysis. Biochem Biophys Res Commun 351:405–411

Li F, Seiller-Moiseicuitsch F (2011) RegStatGel: proteomic software for identifying differentially expressed proteins based on 2D gel images. Bioinformation 6:389–390

Li H, Goodwin PH, Han Q et al (2012) Microscopy and proteomic analysis of the non-host resistance of *Oryza sativa* to the wheat leaf rust fungus *Puccinia triticina f sp Tritici*. Plant Cell Rep 31. doi:10.1007/s00299-011-1181-0

Lodha TD, Basak J (2012) Plant-Pathogen interaction: what microarray tells about it? Mol Biotechnol 50:87–97

Lodha TD, Hembram P, Tep N et al (2013) Proteomics: a successful approach to understand the molecular mechanism of plant-pathogen interaction. Am J Plant Sci 4:1212–1226

Long DH, Lee FN, TeBeest DO (2000) Effect of nitrogen fertilization on disease progress of rice blast on susceptible and resistant cultivars. Plant Dis 84:403–409

Mahmood T, Jan A, Kakishima M et al (2006) Proteomic analysis of bacterial-blight defence-responsive proteins in rice leaf blades. Proteomics 6:6053–6065

Marouga R, David S, Hawkins E (2005) The development of the DIGE system: 2D fluorescence difference gel analysis technology. Anal Bioanal Chem 382:669–678

Mathesius U (2009) Comparative proteomic studies of root-microbe interactions. J Proteome 72:353–366

Mathesius U, Keijzers G, Natera SH et al (2001) Establishment of a root proteome reference map for the model legume *Medicago truncatula* using the expressed sequence tag database for peptide mass fingerprinting. Proteomics 1:1424–1440

Mehta A, Rosato YB (2001) Differentially expressed proteins in the interaction of *Xanthomonas axonopodis* pv. *citri* with leaf extract of the host plant. Proteomics 1:1111–1118

Mehta A, Brasileiro ACM, Souza DSL et al (2008) Plant-pathogen interactions: what is proteomics telling us? FEBS J 275:3731–3746

Meijer HJ, van de Vondervoort PJ, Yin QY et al (2006) Identification of cell wall-associated proteins from *Phytophthora ramorum*. Mol Plant-Microbe Interact 19:1348–1358
Mohammadi M, Anoop V, Gleddie S et al (2011) Proteomic profiling of two maize inbreds during early gibberella ear rot infection. Proteomics 11:3675–3684. doi:10.1002/pmic.201100177
Monzingo AF, Marcotte EM, Hart PJ et al (1996) Chitinases, chitosanases, and lysozymes can be divided into procaryotic and eucaryotic families sharing a conserved core. Nat Struct Biol 3:133–140
Muturi PW, Mwololo JK, Munyiri SW et al (2010) A perspective on proteomics: current applications, challenges and potential uses. Agric Biol J N Am 1:916–918
Nelson DE, Raghothama KG, Singh NK et al (1992) Analysis of structure and transcriptional activation of an osmotin gene. Plant Mol Biol 19:577–588
Neuhaus JM (1999) Plant chitinases (PR-3, PR-4, PR-8, PR-11). In: Datta SK, Muthukrishanan S (eds) Pathogenesis-related proteins in plants. CRC Press, Boca Raton, FL, pp 75–105
Niderman T, Genetet I, Buryere T et al (1995) Pathogenesis-related PR-1 proteins are antifungal. Isolation and characterization of three 14-Kilodalton proteins of tomato and of a basic PR-1 of tobacco with inhibitory activity against *Phytophthora infestans*. Plant Physiol 108:17–27
Noel L, Thieme F, Nennstiel D et al (2001) cDNA-AFLP analysis unravels a genome-wide hrpG-regulon in the plant pathogen *Xanthomonas campestris* pv. *vesicatoria*. Mol Microbiol 41:1271–1281
Peck SC, Nuhse TS, Hess D et al (2001) Directed proteomics identifies a plant specific protein rapidly phosphorylated in response to bacterial and fungal elicitors. Plant Cell 13:1467–1475
Phalip V, Delalande F, Carapito C et al (2005) Diversity of the exoproteome of *Fusarium graminearum* grown on plant cell wall. Curr Genet 48:366–379
Picotti P, Bodenmiller B, Mueller LN et al (2009) Full dynamic range proteome analysis of *S. cerevisiae* by targeted proteomics. Cell 138:795–806
Pierpoint WS, Tatham AS, Pappin DJC (1987) Identification of the virus-induced protein of tobacco leaves that resembles the sweet-protein thaumatin. Physiol Mol Plant Pathol 31:291–298
Portieles R, Ayra C, Gonzales E et al (2010) NmDef02, novel antimicrobial gene isolated from *Nicotiania megalosiphon* confers high-level pathogen resistance under greenhouse and field conditions. Plant Biotechnol J 8:678–690
Puhler A, Arlat M, Becker A et al (2004) What can bacterial genome research teach us about bacteria–plant interactions? Curr Opin Plant Biol 7:137–147
Quirino BF, Candido ES, Campos PF et al (2010) Proteomic approaches to study plant-pathogen interactions. Phytochemistry 71:351–362
Rajam MV, Chandola N, Saiprasad P et al (2007) Thaumatin gene confers resistance to fungal pathogen as well as tolerance to abiotic stresses in transgenic tobacco plants. Biol Plant 51:135–141
Rajjou L, Belghazi M, Huguet R et al (2006) Proteomic investigation of the effect of salicylic acid on *Arabidopsis* seed germination and establishment of early defense mechanisms. Plant Physiol 141:910–923
Rakwal R, Agrawal GK (2003) Rice proteomics: current status and future perspectives. Electrophoresis 24:3378–3389
Rao GU, Kaur M, Verma A et al (1999) Genetic engineering of crop plants for resistance to fungal pathogens. J Plant Biol 26:31–42
Remy W, Taylor TN, Hass H et al (1994) Four hundred million year old vesicular arbuscular mycorrhizae. Proc Natl Acad Sci U S A 91:11841–11843
Roy-Barman S, Sautter C, Chattoo BB (2006) Expression of the lipid transfer protein Ace-AMP1 in transgenic wheat enhances antifungal activity and defense responses. Transgenic Res 15:435–446
Schechter LM, Roberts KA, Jamir Y et al (2004) *Pseudomonas syringae* type III secretion system targeting signals and novel effectors studied with a Cya translocation reporter. J Bacteriol 186:543–555
Shah P, Powell ALT, Orlando R et al (2012) Proteomic analysis of ripening tomato fruit infected by *Botrytis cinerea*. J Proteome Res 11:2178–2192

Shin S, Mackintosh CA, Lewis J et al (2008) Transgenic wheat expressing a barley class II chitinase gene has enhanced resistance against *Fusarium graminearum*. J Exp Bot 59:2371–2378

Singh NK, Nelson DE, Kuhn D et al (1989) Molecular cloning of osmotin and regulation of its expression by ABA and adaptation to low water potential. Plant Physiol 90:1096–1101

Sinha M, Singh RP, Kushwaha GS et al (2014) Current overview of allergens of plant pathogenesis related protein families. Sci World J. doi:10.1155/2014/543195

Stintzi A, Heitz T, Kauffman S et al (1991) Identification of a basic pathogenesis related, thaumatin like protein of virus infected tobacco as osmotin. Physiol Mol Plant Pathol 38:137–146

Tahara ST, Mehta A, Rosato YB (2003) Proteins induced by *Xanthomonas axonopodis* pv. *passiflorae* with leaf extract of the host plant (*Passiflorae edulis*). Proteomics 3:95–102

Terras FRG, Eggermont K, Kovaleva V et al (1995) Small cysteine-rich antifungal proteins from radish: their role in host defense. Plant Cell 7:573–588

Tucker SL, Talbot NJ (2001) Surface attachment and pre-penetration stage development by plant pathogenic fungi. Annu Rev Phytopathol 39:385–417

Tuskan GA, Difazio S, Jansson S et al (2006) The genome of black cottonwood, *Populus trichocarpa* (Torr. And Gray). Science 313:1596–1604

Van Loon LC, Pierpoint WS, Boller T et al (1994) Recommendations for naming plant pathogenesis related proteins. Plant Mol Biol Report 12:245–264

Van Loon LC, Van-Strein EA (1999) The families of pathogenesis-related proteins, their activities, and comparative analysis of PR-1 type proteins. Physiol Mol Plant Pathol 55:85–97

Velasco R, Zharkikh A, Troggio M et al (2007) A high quality draft consensus sequence of the genome of a heterozygous grapevine variety. PLoSONE 2:e1326

Vera O, Conejero V (1988) Pathogenesis-related (PR) proteins of tomato. P69 as an alkaline endoproteinase. Plant Physiol 87:58–63

Vigers AJ, Wiedemann S, Roberts WK et al (1992) Thaumatin-like pathogenesis related proteins are antifungal. Plant Sci 83:155–161

Whitham SA, Yang C, Goodin MM (2006) Global impact: elucidating plant responses to viral infection. Mol Plant-Microbe Interact 19:1207–1215

Wilkins MR, Sanchez JC, Gooley AA et al (1996) Progress with proteome projects: why all proteins expressed by a genome should be identified and how to do it. Biotechnol Genet Eng Rev 13:19–50

Woloshuk CP, Muelenhoff JS, Sela-Buurlage M et al (1991) Pathogen induced proteins with inhibitory activity towards *Phytophthora infestans*. Plant Cell 3:619–628

Yang F, Li W, Derbyshire M (2015) Unraveling incompatibility between wheat and the fungal pathogen Zymoseptoria tritici through apoplastic proteomics. BMC Genomics 16:362–374

Yajima W, Kav NNV (2006) The proteome of the phytopathogenic fungus *Sclerotinia sclerotiorum*. Proteomics 6:5995–6007

Zeilinger S, Gupta VK, Dahms TES et al (2016) Friends or Foes? Emerging insights from fungal interactions with plants. FEMS Microb Rev 40:182–207

Zhang C-X, Yi T, Cong P (2015) Proteome analysis of pathogen responsive proteins from apple leaves induced by the Alternaria blotch *Alternaria alternata*. PLOSOne. doi:10.1371/journal.pone.0122233

Zhou F, Zhang Z, Gregersen PP et al (1998) Molecular characterization of oxalate oxidase involved in the response of barley to the powdery mildew fungus. Plant Physiol 117:33–41

Zhu H, Bilgin M, Snyder M (2003) Proteomics. Annu Rev Biochem 72, No. 1, 2003, pp. 783-812. doi:10.1146/annurev.biochem.72.121801.161511

Zipfel C (2008) Pattern-recognition receptors in plant innate immunity. Curr Opin Immunol 20:10–16

Zipfel C (2014) Plant pattern-recognition receptors. Trends Immunol 35:345–351

Biochemical and Proteomics Analysis of the Plant Growth-Promoting Rhizobacteria in Stress Conditions

14

Kalpna D. Rakholiya, Mital J. Kaneria, Satya P. Singh, V.D. Vora, and G.S. Sutaria

Abstract

Among the emerging environmental threats of the twentieth century, the effect of biotic and abiotic stress on agricultural soils has been considered as one of the most alarming threats in both developed and developing countries. Among them, salt stress is a major problem, and cost associated with the salt salinity is potentially enormous affecting agriculture, food quality, safety, biodiversity, and environments. Several bacteria present in rhizosphere have great potential in improving crop production. Among these bacteria, plant growth-promoting rhizobacteria (PGPR) are the most important. PGPR are able to provide the plant with essential elements, ammonia, growth hormone, and hydrolytic enzymes helping against plant pathogens in salinity and improving soil fertility. The present review aims to establish the conception of the rhizospheric bacteria and to elucidate the mechanisms of rhizobacteria-mediated plant growth promotion. Recent tools available to analyze gene expression and metabolites under the larger umbrella of the genomics and proteomics will also be discussed.

Keywords

Plant growth-promoting rhizobacteria (PGPR) • 1-Aminocyclopropane-1-carboxylic acid (ACC) • Stress • Rhizosphere • Plant-microbe interaction

K.D. Rakholiya
Institute of Biotechnology, Department of Biosciences, Saurashtra University, Rajkot 360005, Gujarat, India

M.J. Kaneria • S.P. Singh (✉)
Department of Biosciences (UGC-CAS), Saurashtra University, Rajkot 360005, Gujarat, India
e-mail: satyapsingh@yahoo.com

V.D. Vora • G.S. Sutaria
Main Dry Farming Research Station, Junagadh Agricultural University, Targhadia, Rajkot 360003, Gujarat, India

R.P. Singh et al. (eds.), *Understanding Host-Microbiome Interactions - An Omics Approach*, DOI 10.1007/978-981-10-5050-3_14

14.1 Introduction

Plants are exposed to a number of environmental stresses which directly affect agriculture and crop production (Szymańska et al. 2013; Gontia-Mishra et al. 2014; Soussi et al. 2015). Salinity is one of the major environmental constraints to crop productivity in the arid and semiarid regions of the world. Due to the high salinity, various deleterious effects on plant growth such as nutritional imbalance, drought stress, and deficiency of specific ions are visible. A number of corrective measures are used to overcome the negative impacts of the salinity that include organic matter amendments or farmyard manure, gypsum applications, calcium chloride, and leaching of salts from root zone (Nadeem et al. 2016). With reference to the biological means to alleviate salinity, development of the salt-tolerant crop varieties and the use of soil management practices that promote microbial activity and high population densities of plant growth-promoting rhizobacteria (PGPR) are important.

With respect to the above scenario, an alternative strategy to improve the salt tolerance in plants could be the introduction of salt-tolerant bacteria enhancing the plant growth and protection against drought, salinity, and phytopathogens. However, only few soil bacteria can survive and resist the stress conditions due to osmotic strength and toxic effects. A new biological approach "plant-microbe interaction" to address salinity problem has recently gained momentum. The development of salt-tolerant crops is not an economical approach for sustainable agriculture, whereas microbial inoculation to alleviate salt stress is a better option as it minimizes production costs and environmental hazards. An alternative to alleviate salt stress is to inoculate crop seeds and seedlings with salt-tolerant plant growth-promoting rhizobacteria. This chapter focuses on the aspects of the rhizosphere bacteria with special reference to salt stress.

14.2 Plant Growth-Promoting Rhizobacteria

Application of plant growth-promoting bacteria is an alternate and potential approach reducing the dependence on the chemical pesticides, fertilizers, herbicides, and fungicides (Grover et al. 2011; Noor and Feroz 2015; Gontia-Mishra et al. 2016). The plant-beneficial bacteria can interact with different parts of the plants. Some reside on leaves and are referred to as phyllosphere bacteria, while others live in thin plant tissues (bacterial endophytes) or bind to root (rhizosphere bacteria). Among these interactions, the highest concentrations of the microorganisms typically exist around the roots in the rhizosphere due to the presence of diverse organic nutrients released from the root (Glick et al. 2007; Lambers et al. 2009; Dodd and Perez-Alfocea 2012; Kumar et al. 2016). A long time ago, Lorenz Hiltner (1904) defined the term rhizosphere, recognizing that plant roots affect the microorganisms in the soil surrounding them, and the term PGPR was coined by Kloepper and Schroth in the late 1970s. The rhizosphere is the narrow zone surrounding and influenced by plant roots, and it houses a rich microbial community, comprising up to 10^{10} bacteria per grams of soil comprising huge diversity of taxa (Vacheron et al.

2013; Baig et al. 2014). PGPR are divided into two groups on the basis of their relationship with the host plants: (1) nonsymbiotic rhizobacteria (free living) that exist outside plant cells called extracellular PGPR and (2) symbiotic rhizobacteria, which survive within the cells and invade the interior of the cells (intracellular PGPR), e.g., nodule bacteria *Rhizobium* (Reinhold-Hurek et al. 2015).

14.3 Species of PGPR

Microorganisms can impart a certain degree of tolerance to plants toward abiotic stresses like drought, chilling injury, salinity, metal toxicity, and high temperature. For the isolation of such organisms, glucose asparagine agar, Ashby's mannitol agar, King's B agar, yeast extract mannitol agar, rose Bengal agar, nutrient agar, tryptone soya agar (TSA), Jenson's medium, DF (Dworkin and Foster) minimal salt medium, Pikovskaya agar, and NBRIP media are generally used (Table 14.1).

During the last decade, a wide range of Gram-positive and Gram-negative bacteria belonging to different genera including *Bacillus*, *Pseudomonas*, *Arthrobacter*, *Halomonas*, *Nitrinicola*, *Achromobacter*, *Polymyxa*, *Serratia*, *Enterobacter*, *Microbacterium*, *Lysinibacillus*, *Citrobacter*, *Shigella*, *Klebsiella*, *Pantoea*, *Paenibacillus*, *Burkholderia*, and *Citrobacter* have been reported to provide tolerance to host plants against salinity in different countries (Table 14.1). *Bacillus* and *Pseudomonas* species identified from rhizosphere of various host plants were studied under both *in vitro* and *in vivo* conditions (Table 14.1). These bacteria play a significant role in stress management caused by high salinity. The understanding of the tolerance could be used for successful development of agriculture production.

14.4 Alleviation of Abiotic and Biotic Stress in Plants by the Rhizosphere Bacteria

Abiotic and biotic stresses are highly diverse and affect the plants in different ways. Among the abiotic factors, salinity is one of the most brutal environmental factors limiting the productivity of agricultural crops (Saleem and Moe 2014; Shrivastava and Kumar 2015). Nowadays, management of biotic and abiotic stresses through microorganisms is gaining importance. Adaptation of microbes to various stresses is complex process where a number of genes may be involved (English et al. 2010; Grover et al. 2011). Certain species living under harsh environmental conditions, such as halophiles and thermophiles, have been recently described (Pandey et al. 2012). In order to cope with the stress conditions, plants produce different enzymes, such as deaminase, peroxidase, catalase, superoxide dismutase, and glutathione reductase (Ahemad and Kibret 2014; Hassan et al. 2016). Most of the rhizospheric bacteria produce some osmoprotectants, phytohormones, and antibiotics as a defense strategy (Gouffi and Blanco 2000; Paul 2013). Paul and Nair (2008) found in their study that rhizobacteria, *Pseudomonas fluorescens* MSP-393, accumulate aspartate, serine, and glutamate in increasingly high concentrations in response to

Table 14.1 Role of plant growth-promoting bacteria in salinity stress alleviation in plants

No.	Host plant/ habitat	Media used for isolation	Method employed for identification of PGPR	Identified PGPR strain	Applicated plant	Country	Reference(s)
1	*Gynura pseudochina* (L.) DC.	Nutrient agar, *Pseudomonas* agar F	ST (2, 4, 6, 8, 10% w/v NaCl), HS, HCE, HCA, PS, NF, IAA, BI, GT, PE	*Pseudomonas* sp.	*Oryza sativa* L. cv.	Thailand	Nakbanpote et al. (2014)
2	Halophytic plants	Tryptone soya agar medium	ST (5%) NF, SOP, IAA, NH_3, PS, ZS, HS, HCE, HCA, HCH, HP, HG, PU, BI, ACC	*Brevibacterium epidermidis* RS15, *Micrococcus yunnanensis* RS222, *Bacillus aryabhattai* RS341	*Brassica napus* L.	Korea, India	Siddikee et al. (2010)
3	*Triticum aestivum*	–	IAA, PS, gibberellin, siderophores, proline content, reducing sugar and total soluble sugar production at 8% NaCl, GT, PE	*B. subtilis*	*Triticum aestivum*	India	Upadhyay and Singh (2015)
4	Soil sample	Nutrient agar	ST (5, 10, 15, 20, 25% w/v NaCl), BI, siderophore, HCN, IAA, PS, GT at 15% NaCl	*Hallobacillus* sp. SL3 and *Bacillus halodenitrificans* PU62	*Triticum aestivum*	India	Ramadoss et al. (2013)
5	*Triticum aestivum*	King's medium B	GT, antagonistic activity, BI, ST, TR, HCN, AP	*Pseudomonas aeruginosa*, *Staphylococcus saprophyticus*, *Bacillus cereus*, *Enterobacter hormaechei*, *Pantoea agglomerans*, *Alcaligenes faecalis*	*Triticum aestivum*	Netherlands	Egamberdieva et al. (2008)

6	*Lycopersicon esculentum*	Nutrient agar	ST (1. 2.5, 5, 8, 10% w/v), IAA, HCN, PS, NH_3, siderophore	*Bacillus* sp.	–	India	Singh and Lal (2015)
7	Wheat, Sambhar Salt Lake, Rajasthan (India)	King's medium B, nutrient agar	BI, PS, IAA, NH_3, siderophore , 2–25% w/v NaCl, proline content, DRS, polysaccharides as soluble sugar, choline oxidase, ACC, PT, HPLC	*Bacillus pumilus*, *Pseudomonas mendocina*, *Arthrobacter* sp., *Halomonas* sp., *Nitrinicola lacisaponensis*	Wheat cultivar PBW-343	India	Tiwari et al. (2011)
8	Wheat	Nutrient agar	GT, BI, HCN, lipase, protease, glucanase, HCE, ST (3, 4, 5, 6), antagonistic activity, IAA, ACC, GT, PT	*P. putida*, *P. extremorientalis*, *P. chlororaphis*, *P. extremorientalis*, *P. aurantiaca*	Wheat	Finland	Egamberdieva and Egamberdieva (2009)
9	*Lycium shawii*	–	ACC, *ethylene production*	*Achromobacter piechaudii*	*Lycopersicon esculentum* Mill cv. F144)	Israel, Canada	Mayak et al. (2004)
10	*S. fruticosa*	Glucose asparagine agar, Ashby's mannitol agar, King's B agar, yeast extract mannitol agar, rose Bengal agar, nutrient agar 5% NaCl	IAA, PS, NH_3, siderophore, HCN, chitinase, antifungal, PT	*Bacillus licheniformis*	*Arachis hypogea* L.	India	Goswami et al. (2014)
11	*Salicornia europaea*	–	Biolog phenotype microarray technique	–	–	Poland, Germany	Szymańska et al. (2014)

(continued)

Table 14.1 (continued)

No.	Host plant/ habitat	Media used for isolation	Method employed for identification of PGPR	Identified PGPR strain	Applicated plant	Country	Reference(s)
12	Alkaline soil	King's medium	BI, PT, IAA, abscisic acid, field experiment	*Pseudomonas putida* Rs-198	*Gossypium hirsutum* L. var. Xinluzao13	China	Yao et al. (2010)
13	Melon and maize	–	IAA, acetylene reduction, protease, amylase, salt tolerance (4%), temperature resistance (40 °C), biocontrol assay, PT	*Pseudomonas alcaligenes* PsA15, *Bacillus polymyxa* BcP26, and *Mycobacterium phlei* MbP18 h	Maize	Uzbekistan	Egamberdiyeva (2007)
14	*Aster tripolium* L.	–	Biolog phenotype microarray technique	–	–	Poland	Szymańska et al. (2013)
15	*Cynodon dactylon*	NA + 10% NaCl	BI, PS, siderophore production, chitinase activity, IAA protease production, FT	*Bacillus cereus*	*Vigna radiata*, *Cicer arietinum*, and *Oryza sativa*	India	Chakraborty et al. (2011)
16	Millets and sunflower	King's B	Bacterial growth under water stress, BI, NH_3, siderophore production, HCN, IAA, PS	*Pseudomonas putida* GAP-P45	Sunflower	India	Sandhya et al. (2009)
17	Maize	King A and King B agar, N Agar	BI, HCN, NH_3	*B. polymyxa*, *B. pantothenticus*, *B. anthracis*, *B. thuringiensis*, and *B. circulans*; *P. cichorii*, *P. putida*, and *P. syringae*; and *Serratia marcescens*	–	Benin	Agbodjato et al., 2015

18	Halophytes	Nutrient agar, King's base medium and Jensen's media	ST (5, 10%), PS, IAA, siderophore production, amylase, cellulose, protease, lipase, chitinase, BI	*Bacillus cereus*, *B. thuringiensis*, *Alcaligenes faecalis*, *Microbacterium resistens*, *Enterobacter* sp., *Lysinibacillus* sp., *Bacillus pumilus*, *Bacillus safensis*	–	India	Natarajan et al., 2016
19	*H. ovalis* (R. Br.) Hook. and *H. pinifolia* (Miki) Hartog (sea grass)	National Botanical Research Institute's phosphate growth medium (NBRIP), Pikovskaya agar, hydroxyapatite medium	PS, BI, 16S rDNA sequencing, and phylogenetic analysis	*Citrobacter* sp., *Shigella* sp., and *Klebsiella* sp., *Bacillus circulans*	–	India	Ghosh et al., 2012
20	Crop fields	NBRIP agar plates	16S rRNA gene amplification, sequencing, and phylogenetic analysis, PS	*Bacillus* spp. *Pseudomonas* sp., *Acinetobacter* sp., *Rhizobium* sp.	–	USA, China	Liu et al., 2015

(continued)

Table 14.1 (continued)

No.	Host plant/ habitat	Media used for isolation	Method employed for identification of PGPR	Identified PGPR strain	Applicated plant	Country	Reference(s)
21	*Zingiber officinale* Rosc.	Tryptone soya agar (TSA), nutrient agar (NA), basal medium amended with glucose, mannitol, sorbitol, inositol, and sucrose	BI; temperatures (28 °C, 37 °C, 41 °C, 50 °C, and 60 °C) and salt concentrations (1, 2, 5, 7, and 10% NaCl); IAA; NH_3; HCN; solubilization of P, K, and Zn and Si-amylase, cellulase, pectinase, and protease; antagonism; 16S rRNA gene amplification and phylogenetic analysis; and field trial	*B. amyloliquefaciens* and *S. marcescens*	Zingiber	India, USA	Dinesh et al. (2015)
22	*Cucumis sativus*	–	DNA extraction, PCR-DGGE, and phylogenetic analysis, two salinity levels, 3 and 8 dS/m; three B concentrations of 0.7, 5, and 8 mg/L; and two pH levels where solutions were frequently adjusted to 6.5 and 8	–	*Cucumis sativus*	USA	Ibekwe et al. (2010)
23	*C. jamacaru*, *P. gounellei*, and *Melocactus* sp., members of the Cactaceae	Tryptone soya agar (TSA) medium (10%) at 28 °C	Bacterial growth under reduced water availability, exopolysaccharide production, IAA, PS, NH_3, HCN, cellulose, PT	*Bacillus* spp. and *Pantoea* sp.	*Zea mays*	Brazil	Kavamura et al. (2013)

24	Bean	Nutrient agar	BI, 16S rRNA gene sequencing, IAA, PS, phytase production, HCN, siderophore production, organic acid production, K, Zn, ACC, cell wall-degrading enzymes, antagonistic activity	*Bacillus* sp.	–	India	Kumar et al. (2012)
25	*Sorghum bicolor*	DF (Dworkin and Foster) minimal salt medium supplemented with 3.0 mM ACC	ACC, PS, IAA, gibberellic acid, NH_3, siderophore production, antagonistic activity, ST (0.5–10%), temperature tolerance (25–50 °C), BI, test of motility	*Klebsiella* sp.	*Triticum aestivum* C-309	India	Singh et al. (2015a, b)
26	*Brassica oleracea* var. alboglabra	L-G (N-free) medium	Seed surface sterilization, acetylene reduction assay (ARA), IAA, ACC, PS, PT, DNA isolation, PCR amplification, and DGGE analyses	*Bacillus* sp., *Pseudomonas* sp.	*Brassica oleracea* var. alboglabra	Thailand, Japan	Piromyou et al. (2013)
27	Potato	–	ST (50–400 mM), heavy metal toxicity $ZnCl_2$ (10–50 mM), polyethylene glycol (PEG6000), a tryptic soy broth (TSB) of varying PEG concentrations (0–50%), ACC, IAA, PS, proline content, determination of hydrogen peroxide levels and enzyme activity	*Bacillus pumilus*, *Bacillus firmus*	Potato	Korea, India	Gururani et al. (2012)

(continued)

Table 14.1 (continued)

No.	Host plant/ habitat	Media used for isolation	Method employed for identification of PGPR	Identified PGPR strain	Applicated plant	Country	Reference(s)
28	Avocado orchard	DF minimal salts containing ACC (3 mM)	BI, IAA, ACC, PS, siderophore production, HCN, ST (5, 10%)	*Pseudomonas* spp.	–	Pakistan, USA	Nadeem et al. (2012)
29	Halophytes	–	Luria-Bertani (LB) media supplemented with different NaCl concentrations (0, 100, 200, 400, 600, 800 mM), bacterial proline content, IAA, PS, antioxidant enzyme	*Paenibacillus* strains	*Pisum sativum* L.	Pakistan, Germany	Ali et al. (2015)
30	Rice	NA	Nitrogen-fixing activity, PS, IAA, ST (0–10%), pH (4–10), drought tolerance, BI, germination assay	*Bacillus* sp. and *Citrobacter* sp.	Rice	Malaysia, Bangladesh	Habib et al. (2015)

ST salt tolerance assays, *PS* phosphate solubilization, *NF* N_2 fixation, *IAA* indole-3-acetic acid production, *GT* germination testing, *PE* pot experiment, *BI* bacterial identification, *HS* hydrolysis of starch, *HCE* hydrolysis of cellulose, *HCA* hydrolysis casein, *SOP* sulfur-oxidizing potentials, NH_3 Ammonia production, *ZS* zinc-solubilizing ability, *HCH* hydrolysis of chitin, *HP* hydrolysis of pectin, *HG* hydrolysis of gelatine, *PU* production of urea, *ACC* 1-Aminocyclopropane-1-carboxylic acid (ACC) deaminase activity, *HCN* hydrogen cyanide production, *TR* temperature resistance, *AP* auxin production, *DRS* direct reducing sugar increasing concentrations of salt. The increased activities of antioxidant enzymes and phytohormones positively correlate with the resistance in many plants against biotic and abiotic stresses (Ahemad and Kibret 2014).

14.5 Mechanism of Plant Growth Promotion by Rhizospheric Bacteria

The modes of action of plant growth-promoting rhizobacteria (PGPR) involve complex mechanisms to promote plant growth, development, and protection. PGPR can enhance the growth and development of plant through either direct or indirect mechanisms (Kloepper and Beauchamp 1992; Azarmi et al. 2016) (Fig. 14.1). PGPR include a large number of different taxa that express at least one trait but typically two or multiple traits responsible for plant growth and development (Kloepper et al. 1989; Nadeem et al. 2016). PGPR are involved in nitrogen fixation, phosphate solubilization (Seshadri et al. 2000), potassium solubilization (Liu et al. 2015; Saha et al. 2016), and zinc solubilization (Kumar et al. 2012). It is also involved in the

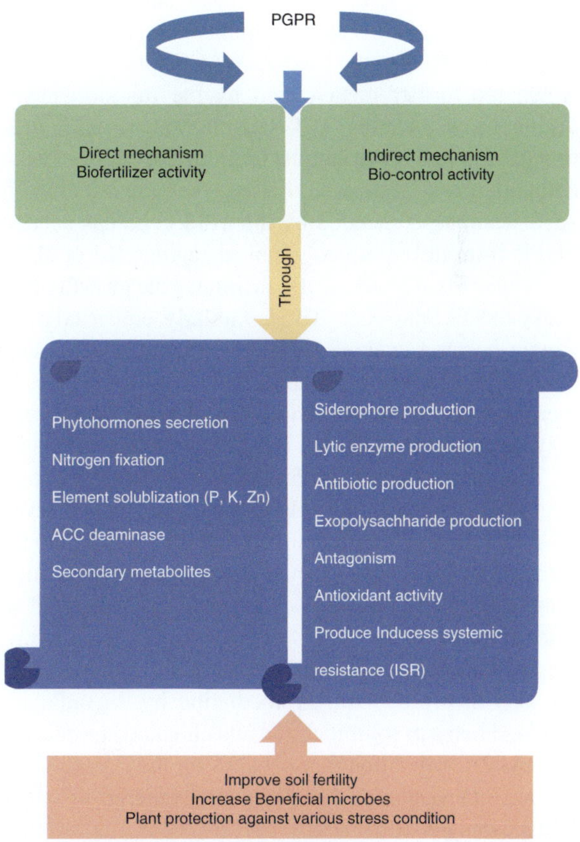

Fig. 14.1 Mechanism of plant growth promotion by PGPR

production of ACC deaminase (Akhgar et al. 2014; Hassan et al. 2016), siderophore (Scagliola et al. 2016; Vejan et al. 2016), and HCN (Walia et al. 2014). Further, heavy metal absorption (Gontia-Mishra et al. 2016) and production of secondary metabolites are among the PGPR traits (Godino et al. 2016; Tiwari et al. 2017).

In phyto-stimulation, PGPR directly facilitate plant growth through the secretion of phytohormones like indole-3-acetic acid (IAA) production (Nadeem et al. 2016) and some auxins synthesized using tryptophan present in root exudates, ethylene and cytokinins, and gibberellins (Bose et al. 2016; Adams et al. 2017). Inoculation of PGPR with the ability to produce IAA can be an indicator for potential efficacy of the strain to reduce the effects of osmotic stress on plants (Boiero et al. 2006). This effect is due to enhanced root growth and root hairs leading to enlarged root surface for better water and nutrient uptake. Positive effect of the phytohormones producing PGPR is well documented in various crops, such as pearl millet (Anatala et al. 2015), rice (Rashedul et al. 2009), wheat (Singh et al. 2015a, b), maize (Zerrouk et al. 2016), and tomato (Singh and Lal 2015).

The PGPR also facilitate plant growth and development by reducing the stress response by decreasing the ethylene level. Ethylene, a gaseous plant hormone endogenously produced by majority of the plants, is an essential growth regulator as stress hormone. The production of ethylene can be induced by abiotic and biotic stress factors. Certain PGPR carry 1-aminocyclopropane-1-carboxylate (ACC) deaminase, which convert ethylene precursor ACC into 2-oxobutanoate and ammonia, both of which are further metabolized by the microorganism. Plant ethylene levels are decreased by lowering ACC deaminase and ACC, thus, relieving the inhibition of root growth and rendering the plants more resistant against various stress conditions and pathogenic bacteria and fungi (Nascimento et al. 2014; Gamalero et al. 2016). ACC deaminase encoded by the acdS gene has been studied in various species of PGPR from different geographical regions (Li et al. 2011). ACC deaminase from *Pseudomonas stutzeri* A1501 facilitates the growth of rice in the presence of salt and heavy metals (Han et al. 2015). Similarly, enhanced exopolysaccharide production in certain bacteria enables plants to better withstand stressful conditions (Upadhyay et al. 2011; Qurashi and Sabri 2012; Vurukonda et al. 2016). It is hypothesized that measurement of the bacterial ACC-deaminase expression under stress conditions may be a better indicator of the stress tolerance potential.

The rhizospheric bacteria produce metal-chelating agents called siderophores, which play an important role in the acquisition of several metal ions enhancing the bioavailability of soil-bound iron (Jing et al. 2007). This implies that siderophore-producing PGPR can enhance the plants growth in high level of heavy metals. It is established that plants grown in metal-contaminated soils are often iron deficient. Therefore, production of siderophores by plant growth-promoting bacteria may help plants obtain sufficient iron by chelating action. Siderophores producing PGPR are also involved in rhizoremediation by degrading soil pollutants (Gupta et al. 2015). It helps in solubilization and enhanced mobility of metals such as Cd, Cu, Ni, Pb, and Zn, a function that depends on the ligand binding (Ahmed and Holmstrom 2014). Siderophores, therefore, may emerge as a useful tool in bioremediation (Kuppusamy et al. 2017).

PGPR act as biocontrol agents against plant pathogens and indirectly stimulate the plant growth through the production of hydrolytic enzymes, i.e., protease (Sharma et al. 2016), lipase, and chitinase (Passari et al. 2016). Glucanase lyse fungal cell wall and thus prevent and/or reduce pathogenic diseases. The rhizosphere provides frontline defense for plant roots against soilborne pathogens (Mendes et al. 2013). Further, PGPR antibiotics (oligopeptides) inhibit cell wall synthesis in pathogens at initial stages and interfere with the protein synthesis by inhibiting the binding of the small subunit of ribosomes. PGPR provide a broad-spectrum activity against pathogenic Gram-positive and Gram-negative bacteria and fungi.

It is also noteworthy that PGPR-associated plants are benefited through activation of defense mechanism—induced systemic resistance (ISR)—against abiotic and biotic stress (Beneduzi et al. 2012). Exudates produced by PGPR stimulate ISR by activating components such as lipid peroxidases, lipoxygenases, and reactive oxygen species conferring protection against diseases caused by different organisms. The ISR activation is dependent on jasmonic acid and ethylene. PGPR that stimulate ISR in one plant species may not do so in another due to interaction specificity between the plant and rhizobacteria (Egamberdieva et al. 2008). Systemic acquired resistance (SAR) also enhances resistance against diseases mediated by the salicylic acid signaling. Several bacteria produce AHL-degrading enzymes that interfere with the quorum sensing system and may be potentially useful against plant pathogens (Perez-Montano et al. 2014).

14.6 Use of PGPR in Plants

One PGPR strain can harbor several plant-beneficial properties, which may or may not be co-regulated. Within the rhizosphere, the expression of PGPR's plant-beneficial properties is affected by both abiotic (pH, oxygen, clay mineralogy, heavy metals, etc.) and biotic factors (compounds produced by plants or the rhizo-microbiome), leading to a distinct expression pattern (Piccoli and Bottini 1994; Pothier et al. 2008; Dinesh et al. 2015). Therefore, inoculation of plants with PGPR can provide a method to increase the population density in the rhizosphere and to develop tolerance against stress. As an alternate option, different PGPR can function together, as consortia, with a possible synergistic effect.

14.7 A Proteomics Perspective of PGPR

The proteomic analysis revealed that expression levels of diverse proteins involved in plant growth promotion, plant pathogen inhibition, photosynthesis and antioxidative processes, and transportation across membranes were influenced in the presence of PGPR (Qin et al. 2016). *Pseudomonas fluorescens* (Gammaproteobacteria) helped canola (*Brassica* spp.) to endure salinity by the enrichment of proteins related to energy metabolism and cell division (Banaei-Asl et al. 2015).

With the advancement in techniques, several methodological approaches have recently been applied in rhizosphere research, which primarily includes 16S rRNA gene analysis for the microbial community, metagenome sequencing of DNA to judge the genomic diversity and physiological potential of the whole microbial communities, and metatranscriptome sequencing of reverse-transcribed mRNA to assess gene expression in relation to meta-genomes. Further, the techniques of the "meta-omics" consortium do not entirely depend on the DNA/RNA sequencing but also on the state-of-the-art methods of the extraction of the proteins/metabolites. Metaproteomics measures abundance of the expressed proteins, providing the status on the functional activities of the microbial communities and plants, while metabolomic profiling analyzes and quantifies the abundance of the molecules involved in metabolic reactions of the microbial communities and/or plants (Oburger and Schmidt 2016).

The interaction among the biocontrol agents, a plant pathogen, and a plant brings significant changes in the plant proteome and metabolism. Among the various molecular techniques, high-throughput whole-genome gene expression tools, viz., microarrays and proteomics, will add to our knowledge on the gene(s) and pathways induced during host-PGPR interaction. 2D-PAGE has been widely used in understanding stress responses. It provides the broad spectrum of proteins produced during the interaction. It further allows the detection of the signal transduction pathways and posttranslational modifications of proteins, understanding the function of the protein (Kandasamy et al. 2009). Toward this end, Kwon et al. (2016) characterized *Paenibacillus polymyxa* and *Arabidopsis thaliana* interactive proteins. Further, with the help of affinity enrichment and high-resolution LC-MS/MS, Fan et al. (2017) analyzed lysine malonylation during the interaction of plant growth-promoting rhizobacteria (PGPR) and *Bacillus amyloliquefaciens* FZB42. Recently, the role of sRNAs in the regulation of biocontrol traits in fluorescent *Pseudomonas* strain Psd has been emphasized (Upadhyay et al. 2017).

14.8 Concluding Remarks and Future Prospects

In this chapter, plant growth-promoting rhizobacteria (PGPR) have been described with respect to their ability to mitigate negative impacts of the salinity stress on plant growth. The PGPR found in the rhizosphere region of roots enhance plant growth and inhibit plant pathogens. PGPR play a positive role in crop productivity by improving plant nutrition and being useful in biocontrol and bioremediation. In order to achieve sustainable crop production, the management of various environmental stresses would be extremely important. With the help of current leads, concerted efforts are required to understand the cross talks between rhizospheric bacteria and plants under various stress conditions.

Acknowledgments We thank the Science and Engineering Research Board (SERB) for providing financial support under National Postdoctoral Fellowship (NPDF) scheme (File No. PDF/2015/000430/LS, 06 June, 2016), Department of Science and Technology, New Delhi, India.

Conflicts of Interest There is no conflict of interest.

References

Adams J, Wright M, Wagner H, Valiente J, Britta D, Anderson A (2017) Cu from dissolution of CuO nanoparticles signals changes in root morphology. Plant Physiol Biochem 110:108–117

Agbodjato NA, Noumavo PA, Baba-Moussa F, Salami HA, Sina H, Sezan A, Bankolé H, Adjanohoun A, Baba-Moussa L (2015) Characterization of potential plant growth promoting rhizobacteria isolated from Maize (*Zea mays* L.) in Central and Northern Benin (West Africa). Appl Environ Soil Sci. doi:10.1155/2015/901656

Ahemad M, Kibret M (2014) Mechanisms and applications of plant growth promoting rhizobacteria: current perspective. J King Saud Univ Sci 26:1–20

Ahmed E, Holmstrom SJM (2014) Siderophores in environmental research: roles and applications. J Microbial Biotechnol 7:196–208

Akhgar AR, Arzanlou M, Bakker PAHM, Hamidpour M (2014) Characterization of 1-Aminocyclopropane-1-Carboxylate (ACC) deaminase-containing *Pseudomonas* spp. in the rhizosphere of salt-stressed canola. Pedosphere 24:461–468

Anatala TJ, Gajera HP, Mandavia MK, Dave RA, Kothari VV, Golakiya BA (2015) Leaf proteome alterations in tolerant pearl millet (*Pennisetum glaucum* L.) genotype under water stress. Int J Agric Environ Biotechnol 8:539–549

Azarmi F, Mozaffari V, Hamidpour M, Abbaszadeh-Dahaji P (2016) Interactive effect of fluorescent *Pseudomonads* rhizobacteria and Zn on the growth, chemical composition, and water relations of Pistachio (*Pistacia vera* L.) seedlings under NaCl stress. Commun Soil Sci Plant Anal 47:955–972

Baig KS, Arshad M, Khalid A, Hussain S, Abbas MN, Imran M (2014) Improving growth and yield of maize through bioinoculants carrying auxin production and phosphate solubilizing activity. Soil Environ 33:159–168

Banaei-Asl F, Bandehagh A, Uliaei ED, Farajzadeh D, Sakata K, Mustafa G, Komatsu S (2015) Proteomic analysis of canola root inoculated with bacteria under salt stress. J Proteomics 124:88–111

Beneduzi A, Ambrosini A, Passaglia LMP (2012) Plant growth-promoting rhizobacteria (PGPR): their potential as antagonists and biocontrol agents. Genet Mol Biol 35:1044–1051

Boiero L, Perrig D, Masciarelli O, Penna C, Cassán F, Luna V (2006) Phytohormone production by three strains of *Bradyrhizobium japonicum*, and possible physiological and technological implications. Appl Microbiol Biotechnol 74:874–880

Bose A, Kher MM, Nataraj M, Keharia H (2016) Phytostimulatory effect of indole-3-acetic acid by *Enterobacter cloacae* SN19 isolated from *Teramnus labialis* (L. f.) Spreng rhizosphere. Biocatal Agric Biotechnol 6:128–137

Chakraborty U, Roy S, Chakraborty AP, Dey P, Chakraborty B (2011) Plant growth promotion and amelioration of salinity stress in crop plants by a salt-tolerant bacterium. Recent Res Sci Technol 3:61–70

Dinesh R, Anandaraj M, Kumar A, Bini YK, Subila KP, Aravind R (2015) Isolation, characterization, and evaluation of multi-trait plant growth promoting rhizobacteria for their growth promoting and disease suppressing effects on ginger. Microbiol Res 173:34–43

Dodd IC, Perez-Alfocea F (2012) Microbial amelioration of crop salinity stress. J Exp Bot 63:3415–3428

Egamberdieva D, Egamberdieva Z (2009) Selection for root colonising bacteria stimulating wheat growth in saline soils. Biol Fertil Soils 45:563–571

Egamberdieva D, Kamilova F, Validov S, Gafurova L, Kucharova Z, Lugtenberg B (2008) High incidence of plant growth-stimulating bacteria associated with the rhizosphere of wheat grown on salinated soil in Uzbekistan. Environ Microbiol 10:1–9

Egamberdiyeva D (2007) The effect of plant growth promoting bacteria on growth and nutrient uptake of maize in two different soils. Appl Soil Ecol 36:184–189

English MM, Coulson TJD, Horsman SR, Patten CL (2010) Overexpression of hns in the plant growth-promoting bacterium *Enterobacter cloacae* UW5 increases root colonization. J Appl Microbiol 108:2180–2190

Fan B, Li Y, Li L, Peng X, Bu C, Wu X, Borriss R (2017) Malonylome analysis of rhizobacterium *Bacillus amyloliquefaciens* FZB42 reveals involvement of lysine malonylation in polyketide synthesis and plant-bacteria interactions. J Proteomics 154:1–12

Gamalero E, Marzachi C, Galetto L, Veratti F, Massa N, Bona E, Novello G, Glick BR, Ali S, Cantamessa S, D'Agostino G, Berta G (2016) An 1-Aminocyclopropane-1-carboxylate (ACC) deaminase-expressing endophyte increases plant resistance to flavescencedoree phytoplasma infection. Plant Biosyst. doi:10.1080/11263504.2016.1174172

Ghosh U, Subhashini P, Dilipan E, Raja S, Thangaradjou T, Kannan L (2012) Bacteria from seagrass rhizosphere soil. J Ocean Univ China 11:86–92

Glick R, Todorovic B, Czarny J, Cheng Z, Duan J, McConkey B (2007) Promotion of plant growth by bacterial ACC deaminase. Crit Rev Plant Sci 26:227–242

Godino A, Principe A, Fischer S (2016) A ptsP deficiency in PGPR *Pseudomonas fluorescens* SF39a affects bacteriocin production and bacterial fitness in the wheat rhizosphere. Res Microbiol 167:178–189

Gontia-Mishra I, Sasidharan S, Tiwari S (2014) Recent developments in use of 1-aminocyclopropane-1-carboxylate (ACC) deaminase for conferring tolerance to biotic and abiotic stress. Biotechnol Lett doi: 10.1007/s10529-014-1458-9

Gontia-Mishra I, Sapre S, Sharma A, Tiwari S (2016) Alleviation of mercury toxicity in wheat by the interaction of mercury-tolerant plant growthpromoting rhizobacteria. J Plant Growth Regul. doi: 10.1007/s00344-016-9598-x

Goswami D, Dhandhukia P, Patel P, Thakker JN (2014) Screening of PGPR from saline desert of Kutch: growth promotion in *Arachis hypogea* by *Bacillus licheniformis* A2. Microbiol Res 169:66–75

Gouffi K, Blanco C (2000) Is the accumulation of osmoprotectant the unique mechanism involved in bacterial osmoprotection? Int J Food Microbiol 55:171–174

Grover M, Ali SZ, Sandhya V, Rasul A, Venkateswarlu B (2011) Role of microorganisms in adaptation of agriculture crops to abiotic stresses. World J Microbiol Biotechnol 27:1231–1240

Gupta G, Parihar SS, Ahirwar NK, Snehi SK, Singh V (2015) Plant growth promoting rhizobacteria (PGPR): current and future prospects for development of sustainable agriculture. J Microb Biochem Technol 7:096–102

Gururani MA, Upadhyaya CP, Baskar V, Venkatesh J, Nookaraju A, Park AW (2012) Plant growth-promoting rhizobacteria enhance abiotic stress tolerance in *Solanum tuberosum* through inducing changes in the expression of ros-scavenging enzymes and improved photosynthetic performance. J Plant Growth Regul. doi:10.1007/s00344-012-9292-6

Habib SH, Kausar H, Saud HM, Ismail MR, Othman R (2015) Molecular characterization of stress tolerant plant growth promoting rhizobacteria (PGPR) for growth enhancement of rice. Int J Agric Biol. doi:10.17957/IJAB/15.0094

Han Y, Wang R, Yang Z, Zhan Y, Ma Y, Ping S, Zhang L, Lin M, Yan Y (2015) 1-Aminocyclopropane-1-carboxylate deaminase from *Pseudomonas stutzeri* A1501 facilitates the growth of rice in the presence of salt or heavy metals. J Microbiol Biotechnol 25:1119–1128

Hassan W, Bashir S, Ali F, Ijaz M, Hussain M, David J (2016) Role of ACC-deaminase and/or nitrogen fixing rhizobacteria in growth promotion of wheat (*Triticum aestivum* L.) under cadmium pollution. Environ Earth Sci 75:267

Hiltner L (1904) Uber neuere Erfahrungen und Probleme auf dem Gebiete der Bodenbakteriologie unter besonderer Berucksichtigung der Gründüngung und Brache. Arb Landwirtsch Ges 98:59–78

Ibekwe AM, Poss JA, Grattan SR, Grieve CM, Suarez D (2010) Bacterial diversity in cucumber (*Cucumis sativus*) rhizosphere in response to salinity, soil pH, and boron. Soil Biol Biochem 42:567–575

Jing Y, He Z, Yang X (2007) Role of soil rhizobacteria in phytoremediation of heavy metal contaminated soils. J Zhejiang Univ Sci B 8:192–207

Kandasamy S, Loganathan K, Muthuraj R, Duraisamy S, Seetharaman S, Thiruvengadam R, Ponnusamy B, Ramasamy S (2009) Understanding the molecular basis of plant growth promotional effect of *Pseudomonas fluorescens* on rice through protein profiling. Proteome Sci 7:47

Kavamura VN, Santos SN, da Silva JL, Parma MM, Avila LA, Visconti A, Zucchi TD, Taketani RG, Andreote FD, de Melo IS (2013) Screening of Brazilian cacti rhizobacteria for plant growth promotion under drought. Microbiol Res 168:183–191

Kloepper JW, Beauchamp CJ (1992) A review of issues related to measuring colonization of plant roots by bacteria. Can J Microbiol 38:1219–1232

Kloepper JW, Lifshitz R, Zablotowicz RM (1989) Free living bacterial inocula for enhancing crop productivity. Trends Biotechnol 7:39–44

Kumar P, Dubey RC, Maheshwari DK (2012) *Bacillus* strains isolated from rhizosphere showed plant growth promoting and antagonistic activity against phytopathogens. Microbiol Res 167:493–499

Kumar P, Dubey RC, Maheshwari DK, Park Y, Bajpai VK (2016) Isolation of plant growth-promoting *Pseudomonas* sp. PPR 8 from the rhizosphere of *Phaseolus vulgaris* L. Arch Biol Sci 68:363–374

Kuppusamy S, Thavamani P, Venkateswarlu K, Lee YB, Naidu R, Megharaj M (2017) Remediation approaches for polycyclic aromatic hydrocarbons (PAHs) contaminated soils: technological constraints, emerging trends and future directions. Chemosphere 168:944–968

Kwon YS, Lee DY, Rakwal R, Baek SB, Lee JH, Kwak YS, Seo JS, Chung WS, Bae DW, Kim SG (2016) Proteomic analyses of the interaction between the plant-growth promoting rhizobacterium *Paenibacillus polymyxa* E681 and *Arabidopsis thaliana*. Proteomics 16:122–135

Lambers H, Mougel C, Jaillard B, Hinsinger P (2009) Plant-microbe-soil interactions in the rhizosphere: an evolutionary perspective. Plant and Soil 321:83–115

Li Z, Chang S, Lin L, Li Y, An Q (2011) A colorimetric assay of 1-aminocyclopropane-1-carboxylate(ACC) based on ninhydrin reaction for rapid screening of bacteria containing ACC deaminase. Lett Appl Microbiol 53:178–185

Liu Z, Li YC, Zhang S, Fu Y, Fan X, Patel JS, Zhang M (2015) Characterization of phosphate-solubilizing bacteria isolated from calcareous soils. Appl Soil Ecol 96:217–224

Mayak S, Tirosh T, Glick BR (2004) Plant growth-promoting bacteria confer resistance in tomato plants to salt stress. Plant Physiol Biochem 42:565–572

Mendes R, Garbeva P, Raaijmakers JM (2013) The rhizosphere microbiome: significance of plant beneficial, plant pathogenic, and human pathogenic microorganisms. FEMS Microbiol Rev 37:634–663

Nadeem SM, Ahmad M, Naveed M, Imran M, Zahir ZA, Crowley DE (2016) Relationship between in vitro characterization and comparative efficacy of plant growth promoting rhizobacteria for improving cucumber salt tolerance. Arch Microbiol. doi:10.1007/s00203-016-1197-5

Nadeem SM, Shaharoona B, Arshad M, Crowley DE (2012) Population density and functional diversity of plant growth promoting rhizobacteria associated with avocado trees in saline soils. Appl Soil Ecol 62:147–154

Nakbanpote W, Panitlurtumpai N, Sangdee A, Sakulpone N, Sirisom P, Pimthong A (2014) Salt-tolerant and plant growth-promoting bacteria isolated from Zn/Cd contaminated soil: identification and effect on rice under saline conditions. J Plant Interact 9:379–387

Nascimento FX, Rossi MJ, Soares CRFS, McConkey BJ, Glick BR (2014) New insights into 1-aminocyclopropane-1-carboxylate (ACC) deaminase phylogeny, evolution and ecological significance. PLoS One 9:99168

Natarajan A, Kumar K, Madhuri K, Usharani GK (2016) Isolation and characterization of salt tolerant plant growth promoting rhizobacteria from plants grown in Tsunami affected regions of Andaman and Nicobar Islands. Geomicrobiol J. doi:10.1080/01490451.2015.1128994

Noor R, Feroz F (2015) Requirements for microbiological quality management of the agricultural products. Nutr Food Sci 45:808–816

Oburger E, Schmidt H (2016) New methods to unravel rhizosphere processes. Trends Plant Sci 21:243–255

Pandey S, Rakholiya KD, Raval H, Singh SP (2012) Catalysis and stability of an alkaline protease from a haloalkaliphilic bacterium under non-aqueous conditions as a function of pH, salt and temperature. J Biosci Bioeng 114:251–256

Passari AK, Mishra VK, Gupta VK, Yadav MK, Saikia R, Singh BP (2016) *In vitro* and *in vivo* plant growth promoting activities and DNA fingerprinting of antagonistic endophytic actinomycetes associates with medicinal plants. PLOS One. doi: 10.1371/journal.pone.0139468

Paul D (2013) Osmotic stress adaptations in rhizobacteria. J Basic Microbiol 53:101–110

Paul D, Nair S (2008) Stress adaptations in a plant growth promoting rhizobacterium [PGPR] with increasing salinity in the coastal agricultural soils. J Basic Microbiol 48:378–384

Perez-Montano F, Alías-Villegas C, Bellogín RA, del Cerro P, Espuny MR, Jiménez-Guerrero I, López-Baena FJ, Ollero FJ, Cubo T (2014) Plant growth promotion in cereal and leguminous agricultural important plants: from microorganism capacities to crop production. Microbiol Res 169:325–336

Piccoli P, Bottini R (1994) Effects of C/N ratio, N-content, pH, and incubation time on growth and gibberellin production by *Azospirillum lipoferum*. Symbiosis 17:229–236

Piromyou P, Noisangiam R, Uchiyama H, Tittabutr P, Boonkerd N, Teaumroong N (2013) Indigenous microbial community structure in rhizosphere of Chinese kale as affected by plant Growth-promoting rhizobacteria inoculation. Pedosphere 23:577–592

Pothier JF, Prigent-Combaret C, Haurat J, Menne-Loccoz Y, Wisniewski-Dye F (2008) Duplication of plasmidborne nitrite reductase gene nirK in the wheat-associated plant growth-promoting rhizobacterium *Azospirillum brasilense* Sp245. Mol Plant Microbe Interact 21:831–842

Qin Y, Druzhinina IS, Pan X, Yuan Z (2016) Microbially mediated plant salt tolerance and microbiome-based solutions for saline agriculture. Biotechnol Adv 15:1245–1259

Qurashi AW, Sabri AN (2012) Bacterial exopolysaccharide and biofilm formation stimulate chickpea growth and soil aggregation under salt stress. Braz J Microbiol 43:1183–1191

Ramadoss D, Lakkineni VK, Bose P, Ali S, Annapurna K (2013) Mitigation of salt stress in wheat seedlings by halotolerant bacteria isolated from saline habitats. SpringerPlus 2(1):6

Rashedul IM, Madhaiyan M, DekaBoruah HP, Yim W, Lee G, Saravanan VS et al (2009) Characterization of plant growth-promoting traits of free-living diazotrophic bacteria and their inoculation effects on growth and nitrogen uptake of crop plants. Microbiol Biotechnol 19:1213–1222

Reinhold-Hurek B, Bunger W, Burbano CS, Sabale M, Hurek T (2015) Roots shaping their microbiome: global hotspots for microbial activity. Annu Rev Phytopathol 53:403–424

Saha M, Maurya BR, Meena VS, Bahadur I, Kumar A (2016) Identification and characterization of potassium solubilizing bacteria (KSB) from Indo-Gangetic Plains of India. Biocatal Agric Biotechnol. doi:10.1016/j.bcab.2016.06.007

Saleem M, Moe LA (2014) Multitrophic microbial interactions foreco- and agro-biotechnological processes: theory and practice. Trends Biotechnol 32:529–537

Sandhya V, Ali SKZ, Grover M, Reddy G, Venkateswarlu B (2009) Alleviation of drought stress effects in sunflower seedlings by the exopolysaccharides producing *Pseudomonas putida* strain GAP-P45. Biol Fertil Soils 46:17–26

Scagliola M, Pii Y, Mimmo T, Cesco S, Ricciuti P, Crecchio C (2016) Characterization of plant growth promoting traits of bacterial isolates from the rhizosphere of barley (*Hordeum vulgare* L.) and tomato (*Solanum lycopersicon* L.) grown under Fe sufficiency and deficiency. Plant Physiol Biochem 107:187–196

Sharma R, Sharma P, Chauhan A, Walia A, Shirkot CK (2016) Plant growth promoting activities of rhizobacteria isolated from *Podophyllum hexandrum* growing in North-West regions of the Himalaya. Proc Natl Acad Sci India B Biol Sci. doi:10.1007/s40011-016-0722-2

Shrivastava P, Kumar R (2015) Soil salinity: a serious environmental issue and plant growth promoting bacteria as one of the tools for its alleviation. Saudi J Biol Sci 22:123–131

Siddikee MA, Chauhan PS, Anandham R, Han G, Sa T (2010) Isolation, characterization, and use for plant growth promotion under salt stress, of ACC Deaminase-producing halotolerant bacteria derived from coastal soil. J Microbiol Biotechnol 20:1577–1584

Singh P, Singh P, Singh MP (2015a) Assessment of antifungal activity of PGPR (plant growth-promoting rhizobacterial) isolates against *Rhizoctonia solani* in wheat (*Triticum aestivum* L.) Int J Adv Res 3:803–812

Singh RP, Jha P, Jha PN (2015b) The plant-growth-promoting bacterium *Klebsiella* sp. SBP-8 confers induced systemic tolerance in wheat (*Triticum aestivum*) under salt stress. J Plant Physiol 184:57–67

Singh Y, Lal N (2015) *In vitro* screening of salt tolerant *bacillus* from rhizosphere of tomato (*Lycopersicon esculentum*) showing plant growth promoting traits. Ind J Biol 2:55–60

Soussi A, Ferjani R, Marasco R, Guesmi A, Cherif H, Rolli E, Mapelli F, Ouzari HI, Daffonchio D, Cherif A (2015) Plant-associated microbiomes in arid lands: diversity, ecology and biotechnological potential. Plant and Soil doi: 10.1007/s11104-015-2650-y

Szymańska S, Piernik A, Baum C, Złoch M, Hrynkiewicz K (2014) Metabolic profiles of microorganisms associated with the halophyte *Salicornia europaea* in soils with different levels of salinity. Ecosci 21:114–122

Szymańska S, Piernik A, Hrynkiewicz K (2013) Metabolic potential of microorganisms associated with the halophyte *Aster tripolium* L. in saline soils. Ecol Quest 18:9–19

Tiwari S, Pandey S, Chauhan PS, Pandey R (2017) Biocontrol agents in co-inoculation manages root knot nematode [*Meloidogyne incognita* (Kofoid & White) Chitwood] and enhances essential oil content in *Ocimum basilicum* L. Ind Crop Prod 97:292–301

Tiwari S, Singh P, Tiwari R, Meena KK, Yandigeri M, Singh DP, Arora DK (2011) Salt-tolerant rhizobacteria-mediated induced tolerance in wheat (*Triticum aestivum*) and chemical diversity in rhizosphere enhance plant growth. Biol Fertil Soils 47:907–916

Upadhyay A, Kochar M, Upadhyay A, Tripathy S, Rajam MV, Srivastava S (2017) Small RNAs regulate the biocontrol property of fluorescent *Pseudomonas* strain Psd. Microbiol Res 196:80–88

Upadhyay SK, Singh DP (2015) Effect of salt-tolerant plant growth-promoting rhizobacteria on wheat plants and soil health in a saline environment. Plant Biol 17:288–293

Upadhyay SK, Singh JS, Singh DP (2011) Exopolysaccharide-producing plant growth-promoting rhizobacteria under salinity condition. Pedosphere 21:214–222

Vacheron J, Desbrosses G, Bouffaud M, Touraine B, Moenne-Loccoz Y, Muller D, Legendre L, Wisniewski-Dye F, Prigent-Combaret C (2013) Plant growth-promoting Rhizobacteria and root system functioning. Front Plant Sci 4:1–19

Vejan P, Abdullah R, Khadiran T, Ismail S, Boyce AN (2016) Role of plant growth promoting rhizobacteria in agricultural sustainability—a review. Molecules 21:573

Vurukonda SKP, Vardharajula S, Shrivastava M, Skz A (2016) Enhancement of drought stress tolerance in crops by plant growth promoting rhizobacteria. Microbiol Res 184:13–24

Walia A, Mehta P, Chauhan A, Shirkot CK (2014) Effect of *Bacillus subtilis* strain CKT1 as inoculum on growth of tomato seedlings under net house conditions. Proc Natl Acad Sci India B Biol Sci 84:145–155

Yao L, Wu Z, Zheng Y, Kaleem I, Li C (2010) Growth promotion and protection against salt stress by *Pseudomonas putida* Rs-198 on cotton. Eur J Soil Biol 46:49–54

Ali Z, Ullah N, Naseem S, Inam-Ul-Haq M, Jacobsen H (2015) Soil bacteria conferred a positive relationship and improved salt stress tolerance in transgenic pea (*Pisum sativum* L.) harboring Na^+/H^+ antiporter. Turk J Bot 39:962. doi:10.3906/bot-1505-50

Zerrouk IZ, Benchabane M, Khelifi L, Yokawa K, Ludwig-Müller J, Baluska F (2016) A *Pseudomonas* strain isolated from date-palm rhizospheres improves root growth and promotes root formation in maize exposed to salt and aluminum stress. J Plant Physiol 191:111–119

Endophytic Actinobacteria and Their Interactions with Plant Host Systems

15

Sangeeta D. Gohel, Amit K. Sharma, Foram J. Thakrar, and Satya P. Singh

Abstract

Actinobacteria are widely distributed in terrestrial and aquatic ecosystems. Soil actinobacteria play a crucial role in nutrient recycling by decomposing complex mixtures of polymers of dead plants, animals and fungal materials. Rhizospheric actinobacteria are extensively found in the agro-environment, while endophytic actinobacteria are relatively less explored. Endophytic microorganisms exist within the plant tissues without adversely affecting the host. Further, the endophytic actinobacteria from the saline habitats are rarely explored. These microbes have huge potential to synthesize numerous novel compounds of pharmaceutical and agricultural significance. They produce a broad range of biologically active metabolites and affect plant growth. Endophytic actinobacteria associated with plants can provide a greater insight into the plant-endophyte interactions. In this chapter, we describe endophytic actinobacteria, its diversity and distribution, mechanism of colonization and current molecular techniques to study them.

Keywords

Saline habitats • Endophytic actinobacteria • Microbial diversity • Microbial colonization

---

S.D. Gohel • A.K. Sharma • F.J. Thakrar • S.P. Singh (✉)
Department of Biosciences (UGC-CAS), Saurashtra University,
Rajkot 360005, Gujarat, India
e-mail: satyapsingh@yahoo.com

R.P. Singh et al. (eds.), *Understanding Host-Microbiome Interactions - An Omics Approach*, DOI 10.1007/978-981-10-5050-3_15

15.1 Introduction

Endophytic actinobacteria are ubiquitous in most plant species. Several actinobacteria are well known to interact with plants that include endophytic and plant pathogenic species (Tian et al. 2004; Bouizgarne and Aouamar 2014; Farrar et al. 2014; Francis et al. 2010; Trujillo et al. 2015; Akshatha et al. 2016). The first actinobacterial endophyte *Frankia*, a nitrogen-fixing microorganism, stimulates nodulation in many angiospermic plant families and has gained attention due to its role in nitrogen economy of its hosts (Callaham et al. 1978; Verma et al. 2009). These microbes synthesize numerous novel compounds useful in pharmaceutical, agricultural and other industries. Diverse endophytic actinobacteria and their association with medicinal plants can provide insight into the plant-endophyte interactions (Golinska et al. 2015).

Among the actinomycetes, *Streptomyces*, *Rhodococcus*, *Corynebacterium*, *Nocardiopsis* and *Mycobacterium* have been greatly studied and explored for various applications. Our research group at Saurashtra University has been working on the haloalkaliphilic actinomycetes with respect to their diversity, phylogeny and biocatalytic potential and reported several alkaline proteases and different antimicrobial compounds (Gohel et al. 2015; Gohel and Singh 2012a, b, 2013, 2015, 2016; Singh et al. 2013; Thumar and Singh 2009, 2011; Singh et al. 2010). A database was created on salt-tolerant actinomycetes, highlighting their diversity, biocatalytic potential and phylogeny (Sharma et al. 2012).

Many novel endophytic actinobacteria have been isolated from the saline and alkaline habitats. Qin et al. (2013) isolated *Modestobacter roseus* sp. nov., an endophytic actinomycete from the coastal halophyte, *Salicornia europaea* Linn. and *Streptomyces halophytocola* sp. nov., an endophytic actinomycete from the surface-sterilized stems of a coastal halophyte *Tamarix chinensis* Lour., collected from east of China and south-west China, respectively. Similarly, Zhao et al. (2011) isolated *Pseudonocardia kunmingensis* sp. nov., an actinobacterium, from surface-sterilized roots of *Artemisia annua* L., while novel actinomycetes, *Pseudonocardia rhizophila* sp. nov. and *Pseudonocardia sichuanensis* sp. nov., are reported from a rhizosphere soil and root of *Jatropha curcas* L., respectively (Qin et al. 2011; Li et al. 2010). Actinobacteria as host for the production of recombinant proteins have been reported (Nakashima et al. 2005). *Streptomyces* are well known to produce various types of antibiotics (Weber et al. 2003), while some *Rhodococcus* spp. are used for the industrial production of acrylamide (Komeda et al. 1996). *Nocardiopsis* spp. from the saline habitats are reported to produce various alkaline proteases (Gohel and Singh 2012a, b).

The endophytic actinobacteria in plants are variable and often capable of eliciting physiological changes in plants affecting growth and development. The host-actinobacteria interactions are quite exciting and a newly emerging field of research.

15.2 Diversity of Actinobacteria in Different Plant Host Systems

It is well known that actinobacteria are present in diverse habitats (Gohel et al. 2016; Kim et al. 2015; Antony et al. 2014; Gohel and Singh 2012; Kikani et al. 2010; Thumar and Singh 2007). Actinobacteria are also associated with wide variety of hosts. Some actinobacteria benefit their host by producing secondary metabolites, the process of symbioses being highly complex. Five new anti-trypanosomal macrolides, actinoallolides A–E (1–5), produced by an endophytic actinobacteria, *Actinoallomurus fulvus* MK10-036, have recently been reported (Inahashi et al. 2015).

15.2.1 Actinobacteria Associated with Mangrove Plants

Mangroves grow at the interface of land and sea in tropical and sub-tropical latitudes under extreme conditions of salinity, tides, strong winds, high temperatures and muddy, anaerobic soils (Kathiresan and Bingham 2001). Ravikumar et al. (2010) investigated five Indian mangrove plants (*Rhizophora apiculata*, *Rhizophora mucronata*, *Bruguiera cylindrica*, *Ceriops decandra*, *Avicennia marina*) to evaluate the antibacterial activity against UTI bacterial pathogens. Preliminary phytochemical analysis of the plant parts indicated the presence of active compounds such as flavonoids, anthraquinone, phenolic group, alkaloids and triterpenoids. The studied plants might be potential sources of anti-UTI bacterial pathogens. As endophytic bacteria, few endophytic actinobacteria are also associated with different mangrove plants. Ding et al. (2012) isolated endophytic *Streptomyces* sp. HKI0595 from the mangrove stem. Five novel eudesmene-type sesquiterpenes, kandenols A–E (1–5), were isolated from *Streptomyces* sp. HKI0595. Among them, kandenol E is the first bacterial agarofuran, which belongs to an important group of antibiotics (agarofurans: cytotoxic, antibacterial and anti-inflammatory).

15.2.2 Actinobacteria Associated with Legume and Cereal Plants

Legumes being nitrogen-fixing crop play a significant role in soil health. Besides the rhizospheric bacteria, many endophytic bacteria are also reported from legumes. A number of actinobacterial strains from lentil (*Lens esculentus*), chickpea (*Cicer arietinum* L.), pea (*Pisum sativum*), faba bean (*Vicia faba*) and wheat (*Triticum vulgare*) have been isolated from Paskeville, South Australia (Misk and Franco 2011). Most of these actinomycetes belong to *Streptomyces*, while some are identified with *Microbispora*. Endophytic *Streptomycetes* from a Thai jasmine rice plant (*Oryza sativa* L. cv. KDML105) have been isolated (Rungin et al. 2012). Similarly, 38 strains of endophytic actinobacteria identified with the genera *Streptomyces*, *Microbispora*, *Micromonospora* and *Nocardioides* are described from healthy cereal plants (Coombs et al. 2004). In general, endophytic actinobacteria are advantageous as biological control agents (Coombs et al. 2004).

15.2.3 Actinobacteria Associated with Mandarin and Blooms

Mandarin, a small citrus tree, harbours endophytic actinomycete (Shutsrirung et al. 2013). Based on spore morphology, cell wall and 16S rRNA gene sequences, the isolates are classified into six genera: *Streptomyces*, *Nocardia*, *Nocardiopsis*,

Spirillospora, *Microbispora* and *Micromonospora*, the most frequent isolates being *Streptomyces*. Similarly, endophytic actinomycetes belonging to nine different genera are isolated from *Combretum latifolium* Blume (Combretaceae) of the Western Ghats of Southern India (Rao et al. 2015). This study has well indicated the dominance of *Streptomyces* followed by *Nocardiopsis* and *Micromonospora*.

15.2.4 Endophytic Actinobacteria Associated with Medicinal Plants

Medicinal plants reflect enormous diversity of endophytic actinobacteria. The plant endosphere consists of a large variety of microbial endophytes, which constitute a complex micro-ecosystem (El-Shatoury et al. 2013). The genes responsible for the production of secondary metabolites are located as a cluster in the genome and referred as biosynthetic gene clusters. The genes within *Mycobacterium*, *Streptomyces* and *Frankia* suggest its crucial roles in natural product synthesis (Doroghazi and Metcalf 2013). Gangwar et al. (2011) isolated 40 endophytic actinomycetes from roots, stems and leaves of three medicinal plants, viz. *Mentha*, *Aloe vera* and *Ocimum sanctum*. The majority of the isolates were *Streptomyces* spp. with a wide spectrum activity against different fungal phytopathogens. Similarly, 53 isolates, 31 from leaves and 22 from roots of maize, are obtained, with the genus *Microbispora* followed by *Streptomyces* and *Streptosporangium* being the most prominent. Analysis of 16S rRNA gene sequences revealed their affiliation with *Streptomyces*, *Amycolatopsis*, *Actinoallomurus*, *Kribbella* and *Microbispora* (Bunyoo et al. 2009). Forty endophytic strains are reported from leaves, stems and roots of maize (Costa et al. 2013). The morphological properties and fatty acid methyl ester (FAME) profile suggested the prominence of the *Streptomyces* genus.

Chankhamhaengdecha et al. (2013) reported quorum quenching activity from 68 endophytic actinomycetes. These endophytes show acyl homoserine lactone-degrading enzymes. The organisms and their biotopes, subjected to steady environmental interactions, produce more secondary metabolites (Singh and Dubey 2015). The endophytic actinomycetes have been described from three medicinal plants, *Ocimum sanctum*, *Azadirachta indica* and *Phyllanthus amarus* (Shenpagam et al. 2012). An antagonistic activity of endophytic actinobacteria was tested against different bacterial pathogens (*Staphylococcus aureus*, *Streptococcus pyogenes*, *Klebsiella pneumoniae* and *Pseudomonas aeruginosa*) and the fungi *Rhizopus*. They established that UV-mutated endophytic actinobacteria increased antibiotic production as compared to non-mutated endophytic actinobacteria. Table 15.1 provides an account of different endophytic actinobacteria associated with pla.

Table 15.1 Endophytic actinobacteria recently reported from various plant sp. and their potential biological applications

Isolates	Host plant	Tissue	Role of actinobacteria in hosts	References
Arthrobacter				
Arthrobacter sp. (JQ926171)	*Leucas ciliata*	Stem	Antioxidant activity	Akshatha et al. (2016)
Actinoallomurus				
Actinoallomurus caesius (GMKU 931)	*Acacia auriculiformis*	Root	Antibacterial activity	Bunyoo et al. (2009)
Amycolatopsis				
Amycolatopsis tolypomycina (GMKU 932)	Wattle tree (*Acacia auriculiformis* A. Cunn. ex Benth.)	Root	Antibacterial activity	Bunyoo et al. (2009)
Actinoallomurus				
Actinoallomurus coprocola (GKMU 943)	*Acacia auriculiformis* A. Cunn. ex Benth.)	Root	Antibacterial activity	Bunyoo et al. (2009)
Actinopolyspora				
Actinopolyspora spp.	*Ocimum sanctum* and *Mentha arvensis*	Root	Antagonistic activity against one or more phytopathogenic fungi	Gangwar et al. (2014)
Kribbella				
Kribbella jejuensis (GMKU 938)	*Acacia auriculiformis* A. Cunn. ex Benth.	Root	Antagonistic activity against one or more phytopathogenic fungi	Bunyoo et al. (2009)
Microbispora				
Microbispora (TGsR-01-08, TGsR-02-11)	Mandarin	Root	Plant growth-promoting activity	Shutsrirung et al. (2013)
Microbispora corallina (GMKU 936)	*Acacia auriculiformis* A. Cunn. ex Benth.	Root	Antibacterial activity	Bunyoo et al. (2009)
Micromonospora				
Micromonospora tulbaghiae (TGsR-02-01)	Mandarin	Root	Plant growth-promoting activity	Shutsrirung et al. (2013)
Micromonospora echinospora (TGsR-02-17)				
Micromonospora tulbaghiae (TGsR-02-18)				

(continued)

Table 15.1 (continued)

Isolates	Host plant	Tissue	Role of actinobacteria in hosts	References
Micromonospora O-14	Three medicinal plants, viz. *Aloe vera*, *Mentha* and *Ocimum sanctum*	Root, stem and leaves	Antagonistic activity against plant pathogenic fungi	Gangwar et al. (2011)
Micromonospora A9				
Microbispora mesophila (GMKU 941)	*Acacia auriculiformis* A. Cunn. ex Benth (wattle tree)	Root	Antibacterial activity	Bunyoo et al. (2009)
Micromonospora spp.	Roots of *Aloe vera*, *Mentha* and *Ocimum sanctum*, *leaves* and *stem of Mentha* and *Ocimum sanctum*	Root	Antagonistic activity against one or more phytopathogenic fungi	Gangwar et al. (2014)
Nocardia				
Nocardia aobensis (TGsR-01-12)	Mandarin	Root	Plant growth-promoting activity	Shutsrirung et al. (2013)
Nocardiopsis				
Nocardiopsis (mhce0814)	*Triticum aestivum* (wheat) plant	Root	Improve plant growth	Jog et al. (2014)
Nocardiopsis alba (TGcL-04-28, TGcL-04-56, TGcL-04-60, TGsL-02-05	Mandarin	Leaves	Plant growth-promoting activity	Shutsrirung et al. (2013)
Nocardiopsis umidischolae (INA01099)	*Aloe arborescens*	Leaves	Antibiotic activity against Gram-positive bacteria	Machavariani et al. (2014)
Nocardiopsis quinghaiensis (INA01100)				
Nocardiopsis sp. (INA01101)				
Nocardiopsis quinghaiensis (INA01102)	*Mentha arvensis*			
Nocardiopsis exhalans (INA01103)				
Nocardiopsis tropica (INA01104)	*Lysimachia nummularia*			
Nocardiopsis dassonvillei (INA01097)	*Fragaria vesca*			
Nocardiopsis viridoflava (INA01105)	*Arctium lappa*			

Table 15.1 (continued)

Rhodococcus				
Rhodococcus (mhcr0825)	*Triticum aestivum* (wheat) plant	Root	Improve plant growth	Jog et al. (2014)
Saccharopolyspora				
Saccharopolyspora O-9	Three medicinal plants, viz. *Aloe vera*, *Mentha* and *Ocimum sanctum*	Root, stem and leaves	Antioxidant activity	Akshatha et al. (2016)
Saccharopolyspora spp.	*Aloe vera*, *Mentha* and *Ocimum sanctum*	Root	Antagonistic activity against one or more phytopathogenic fungi	Gangwar et al. (2014)
Streptomyces				
Streptomyces globosus (JQ926176)	*Rauwolfia densiflora*	Stem	Antioxidant activity	Akshatha et al. (2016)
Streptomyces coelicoflavus (GMKU 937)	Wattle tree (*Acacia auriculiformis* A. Cunn. ex Benth.)	Root	Antibacterial activity	Bunyoo et al. (2009)
Streptomyces regensis (GMKU 939)				
Streptomyces sioyaensis (GMKU 944)				
Streptomyces cheonanensis (TGsS-01-09)	Mandarin	Branch	Plant growth-promoting activity	Shutsrirung et al. (2013)
Streptomyces antibioticus (TGsR-01-04)		Root		
Streptomyces violaceorectus (TGsR-03-04)				
Streptomyces cellulosae (TGcB-01-01)		Branch		
Streptomyces violascens (TGcB-01-07)				
Streptomyces puniceus (TGcB-02-27)				
Streptomyces luteogriseus (TGcL-01-09)		Leaves		
Streptomyces bellus (TGcR-02-01)		Root		
Streptomyces viridis A3	*Aloe vera*, *Mentha* and *Ocimum sanctum*	Stem, root and leaves	Antagonistic activity against plant pathogenic fungi	Gangwar et al. (2011)
Streptomyces albosporus A4				
Streptomyces cinereus A6, O-1				
Streptomyces albosporus O-11				

(continued)

Table 15.1 (continued)

Isolates	Host plant	Tissue	Role of actinobacteria in hosts	References
Streptomyces species (*19 isolates*)	Maize plant	Leaf	Biocontrol of phytopathogenic fungi	Costa et al. (2013)
Streptomyces species (*6 isolates*)		Stem		
Streptomyces species (*9 isolates*)		Root		
Streptomyces albosporus	*Aloe vera*, *Mentha arvensis* and *Ocimum sanctum*	Root	Antagonistic activity against one or more phytopathogenic fungi	Gangwar et al. (2014)
Streptomyces aureus	*Ocimum sanctum*	Root		
	Mentha arvensis	Root and stem		
Streptomyces globisporus	*Aloe vera* and *Mentha arvensis*	Root		
	Mentha arvensis	Stem		
Streptomyces griseofuscus	*Aloe vera*, *Mentha arvensis* and *Ocimum sanctum*	Root		
Streptomyces roseosporus	*Ocimum sanctum*	Root		
	Mentha arvensis	Leaf		
Streptomyces viridis	*Aloe vera* and *Mentha arvensis*	Root		
	Ocimum sanctum	Stem		
Streptomyces griseorubruviolaceus	*Mentha arvensis* and *Ocimum sanctum*	Root		
	Mentha arvensis	Root and stem		
Streptomyces (mhcr0810, mhcr0816, mhcr0817, mhcr0824, mhce0811)	*Triticum aestivum* (wheat) plant	Root	Improve plant growth	Jog et al. (2014)

15.3 Colonization of Endophytic Actinobacteria

Microbes associated with plants are termed rhizospheric or endophytic based on their localization outside or inside the plant, respectively. Endophytes may originate from the rhizosphere or phyllosphere (Dudeja et al. 2012). Studies suggest the richness of endophytic actinobacterial species with various plant sp. (Araújo et al. 2000; Coombs and Franco 2003a, b; Ryan et al. 2008; Rogers et al. 2012; Qin et al. 2013). The actinobacteria are relatively less studied for their potential to secrete novel natural products significant in medicine, agriculture and industries (Wan et al. 2008; Verma et al. 2011; Palaniyandi et al. 2013; Inahashi et al. 2015; Nimaichand et al. 2016).

15.3.1 Novel Endophytic Actinobacteria from Saline and Alkaline Habitats

Recently, many novel species from endophytic actinobacteria have been retrieved from different saline habitats. Krishnan et al. (2016) isolated *Arthrobacter pokkalii* sp. nov., a novel plant-associated *Actinobacterium*, from saline-tolerant Pokkali rice, Kerala, India. Xing et al. (2012) isolated a novel endophytic actinomycete *Kibdelosporangium phytohabitans* sp. nov., from oilseed plant *Jatropha curcas* L. containing 1-aminocyclopropane-1-carboxylic acid deaminase. A novel endophytic actinomycete, *Streptomyces phytohabitans* sp. nov., from medicinal plant, *Curcuma phaeocaulis*, is also described (Bian et al. (2012). A novel endophytic actinomycete, *Nocardioides panzhihuaensis* sp. nov., has been studied in a medicinal plant, *Jatropha curcas* L. (Qin et al. 2012a, b). Bian et al. (2012) isolated *Kineococcus endophytica* sp. nov., a novel endophytic actinomycete from a coastal halophyte in Jiangsu, China. Moreover, Xing et al. (2012) isolated *Pseudonocardia nantongensis* sp. nov., a novel endophytic actinomycete from the coastal halophyte *Tamarix chinensis* Lour. Also, Xing et al. (2013) isolated *Amycolatopsis jiangsuensis* sp. nov., a novel endophytic actinomycete from a coastal plant in Jiangsu, China. In addition, Zhang et al. (2013) reported *Saccharopolyspora dendranthemae* sp. nov., a halotolerant endophytic actinomycete from a coastal salt marsh plant in Jiangsu, China. Overall, endophytic actinobacteria are originated from diverse plant sp. along the saline and alkaline environments.

15.3.2 The Entry of Endophytic Actinobacteria into the Plant Host

The significant aspect of the plant-microbe interactions is how endophytic actinomycetes enter and colonize inside plant tissues. The bacterial cells are usually not able to penetrate intact epidermal cells (Huang 1986). Bacteria usually enter into the plant tissues through stomata, wounds, lenticels, projecting areas of lateral roots and broken trichomes. These organisms stimulate seed germination and promote plant establishment under unfavourable conditions (Hurek et al. 2002; Ryan et al. 2008; Taj and Rajkumar 2016; Reid and Greene 2012; Nimaichand et al. 2016).

Endophytes are categorized into three main categories: obligate, facultative and passive endophytes (Hardoim et al. 2008). Although Rosenblueth and Martinez-Romero (2006) have emphasized the importance of endophytic bacteria in plants, detailed accounts of the localization of the endophytic actinobacteria are still limited (Hasegawa et al. 2006). The endophytic bacteria, as compared to the rhizospheric/epiphytic bacteria, develop stable interactions with plants (Compant et al. 2010; Malfanova et al. 2011).

The passive endophytes are equipped with genes essential for the maintenance and protection of plant-endophyte associations. Earlier, it was considered that plant-associated microbes mainly comprise Gram-negative bacteria. However, later it was known that the representatives of high and low G+C Gram-positives belonging to the phyla Actinobacteria and Firmicutes are also able to promote plant growth. A genetically modified *Streptomyces* strain acts as potential biological control agents (BCA) in lettuce root sand rhizosphere (Bonaldi et al. 2015). The spatiotemporal dynamics of colonization by *Streptomyces* spp. was studied to understand the rhizospheric establishment.

The endophytic actinomycetes were microscopically examined from the roots of various plant species of northwestern Italy (Sardi et al. 1992). The study revealed the dominance of *Streptomyces* followed by the presence of *Streptoverticillium*, *Nocardia*, *Micromonospora* and *Streptosporangium*. Similarly, Tokala et al. (2002) studied novel plant-microbe rhizosphere interaction involving *Streptomyces lydicus* WYEC108 and the pea plant (*Pisum sativum*). Coombs and Franco (2003a, b) isolated and identified actinobacteria from the surface-sterilized wheat roots. Cao et al. (2004) isolated and characterized endophytic *Streptomyces* strains from surface-sterilized tomato (*Lycopersicon esculentum*) roots. Further, auxin production has been described for endophytic and free-living *Streptomyces* in rhizosphere (Coombs et al. 2004). The induction of systemic resistance in the plant due to internal colonization of plant tissue by the endophyte has been described. The rhizosphere soil and root exudates are the natural source of tryptophan essential for rhizospheric microorganisms (Khamna et al. 2009). Endophytic *Streptomyces* spp. stimulates nutrient assimilation in their plant host (Tokala et al. 2002; Seipke et al. 2012).

Actinorhizal associations contribute to the global biological nitrogen fixation. Trujillo et al. (2010) hypothesized that endophytic *Micromonspora* populations were innate inhabitants of nitrogen-fixing root nodules of *Lupinus angustifolius* and fix nitrogen in symbiosis with their host. GFP-tagged actinobacterium coated on seed stimulated colonization (Coombs and Franco 2003a).

15.3.3 Colonization by Pathogenic Endophytic Actinobacteria

Actinobacteria play a significant role related to the health of plants (Shimizu 2011). Actinobacterial colonization of plant roots and rhizosphere depends on the ecological strategies used by the respective genera and the characteristics of the host plant. The survival of the environmental isolates of actinomycetes within the seeds of plants in soil depends on many factors (Merzaeva and Shirokikh 2006). Different strategies of colonization of the rhizosphere were followed using *Streptomyces*, *Micromonospora* and *Streptosporangium*, under the extreme climate rhizosphere. The plants of winter rye (*Secale cereale* L.) inoculated with actinomycetes displayed growth advantages, while the cow clover plants (*Trifolium pratense* L.) had no effect on growth.

Streptomycetes produce two-thirds of the valuable antibiotics and cause infections in humans and plants. The phytopathogenic *Streptomyces* spp. has been studied well (Loria et al. 1997). *Streptomyces scabies* produces extensive hyphae on the surface of potato tubers. After penetration, the pathogen grows in the intercellular spaces and feeds on these cells as a saprophyte (Loria et al. 1997). *Streptomyces griseofuscus* being the commonest member of endophytic actinomycetes represents largest antagonistic communities. The endophytic actinomycetes including *S. griseofuscus* and *S. hygroscopicus* are antagonistic to the rice pathogens, *Magnaporthe grisea*, *Xanthomonas oryzae* pv. *oryzae*, *R. solani* and *F. moniliforme*. The endophytic actinomycetes produce active antagonistic metabolites, such as antibiotics, enzymes and siderophores in plant tissues. In wheat, Coombs and Franco (2003b) demonstrated that endophytic *Streptomyces* sp. with GFP expression system could colonize seed embryos and emerging radicals.

15.3.4 Genes Involved in the Colonization and Maintenance of Plant Endophyte Association

Actinobacteria support plant growth by nutrient mobilization, growth hormone secretion, siderophore production and stimulation of beneficial rhizospheric microbes. The interaction of endophytic actinobacteria *Micromonospora* with nitrogen-fixing plant has been recently reported (Trujillo et al. 2015). *Frankia*, a unique genus amongst the actinobacteria, fix atmospheric nitrogen, both under free-living and associative state. The vesicle of *Frankia* sets it apart from *Rhizobium* and protects against ambient oxygen levels. Certain nodulating bacteria have evolved specific adaptation strategies for active penetration of the root system, a specific phenomenon mediated by chemotaxis towards flavonoid exudates and microbial signals and nod factors (Hardoim et al. 2008). Genome analysis of the facultative symbiotic nodule-forming *Frankia* confirms the role of nod genes in nodulation (Normand et al. 2006).

The endophytic actinobacteria increase plant growth by producing high level of auxin and indole acetic acid, significant in disease control. Molecular characterization of a plasmid from the *S. caviscabies* strain traces genes, involved in regulation of host plant genes (Coombs and Franco 2003a, b). The genes involved in the detoxification of reactive oxygen species, protein secretion and motility are important determinants for successful plant colonization (Gaiero et al. 2013). Similarly, it is established that the genes associated with the production of siderophores, abscisic acid and indole acetic acid and QS autoinduction are associated with biocontrol, phytostimulation and colonization (Gaiero et al. 2013). The proteomic studies of plant-bacterial interactions have added to our understanding of this association (Cheng et al. 2010). Understanding plant protein expression related to hormone production and defence response can elucidate how plants control the phenotype of their endophytic partners (Gaiero et al. 2013). Further, the metagenomic sequencing is useful in recognizing the protein coding sequences associated with colonization, competence and plant growth promotion (Barret et al. 2011; Sessitsch et al. 2012). The genes involved in the colonization of beneficial endophytic actinobacterial communities are incredibly significant in plant-actinobacterial association.

15.4 Techniques to Analyse Plant-Actinobacteria Interactions

Isolation and cultivation methods for endophytic actinobacteria are crucial to understand the host-microbe interactions. Therefore, developing specific protocols for the isolation of the endophytic actinobacteria of a plant is extremely important (Coombs and Franco 2003a, b; Hallmann et al. 2006). Various methods to study such interactions include immunological detection of bacteria, fluorescence tagging and confocal laser scanning microscopy (Chelius and Triplett 2000; Hartmann et al. 2000; Verma et al. 2004). In addition, specific oligonucleotide probes can be used to analyse bacteria residing inside plants (Hartmann et al. 2000). During the last 20 years, combination of both molecular and microscopic techniques has greatly contributed to probe the microbe-host interactions (Cardinale 2015).

Among the molecular methods, omics-based methodologies have been developed in recent years to investigate the microbiome of an environment. The complex interaction between host and a microbial pathogen can be understood by following the molecular details of the interaction using virulence-associated microbial genes and their regulation (Cummings and Relman 2000). While extensive information is available on the molecular mechanisms of the bacteria-plant interactions, only limited work on the endophyte-host molecular interactions is reported (Cummings and Relman 2000; Lugtenberg et al. 2002; Oldroyd and Downie 2004).

The complex interactions between a microbial pathogen and its host define the underlying basis of the infectious. Virulence associated with the microbial genes and defence adaptation of the plant can be elucidated by understanding molecular aspects. High-density DNA microarrays are a basic tool in many fields that include cellular physiology, cancer biology and pharmacology (Cummings and Relman 2000). The importance of this application relates to microarray, which can be used in gene expression and its response profiling and detection of the polymorphism in sequences (Cummings and Relman 2000).

At the cellular level, CLSM (confocal laser microscopy) techniques reveal the plant-microbe interactions (Cardinale 2015). CLSM is based on the detection of fluorescent light. However, it differs from the conventional fluorescence microscopy in that it acquires the fluorescent signal(s) exclusively from the focal plane as a pinhole excluding out-of-focus light (Cardinale 2015). Several tools are available for the qualitative and quantitative analysis of CLSM stacks (Schneider et al. 2012). The advantageous features of the CLSM include three objects: (a) molecule, cells and tissues stained with one or more fluorochromes, (b) genetically modified organisms (GMO) that express fluorescent proteins and (c) autofluorescent cells, tissues and substrates. The CLSM was used by Schloter et al. (1993) to demonstrate the interactions between wheat roots and *Azospirillum brasilense* SP7, a plant growth-promoting rhizobacterium (PGPR). The CLSM in combination with other techniques or microscopic methods could provide added advantages. For instance, CLSM techniques coupled with scanning probe systems such as AFM (atomic force microscopy) are an important approach that can be used to elucidate plant-microbe interaction (Haupt et al. 2006).

Pyrosequencing was used to study the endophytic bacterial community in the roots of potato varieties (Manter et al. 2010). The next-generation sequencing and culture-independent approaches are among the upcoming approaches to study microbial diversity of soil and rhizosphere microbiomes (Zhang et al. 2010; Mendes et al. 2011; Jansson et al. 2012). Majority of the studies have focused on the number and diversity of the bacterial taxa rather than on other rhizosphere inhabitants. Based on the analysis of OTUs in the oat rhizosphere, it was concluded that root growth was dominated by Alphaproteobacteria, Firmicutes and Actinobacteria (DeAngelis et al. 2009). The bacterial communities in the rhizosphere of some plant species were analysed by PhyloChip (Mendes et al. 2011). The presence of actinobacteria in rhizosphere microbiome was reported (DeAngelis et al. 2009; Mendes et al. 2011). Similar to rhizospheric bacteria, the endophytic actinobacterial communities can also be investigated using PhyloChip analysis in the future.

Metagenomics and metatranscriptomics, high-throughput molecular methods, have been used in plant-microbe interaction studies in many cases (Kint et al. 2010; Röling et al. 2010; Zhang et al. 2010; Jansson et al. 2012). Omics methodologies based on the extraction of biomolecules directly from the environment increase the detection limit broadening the search of organisms in the rare microbiome (Cardinale 2015). A recent study with *Arthrobacter pokkalii* sp. nov., a plant-associated Actinobacterium from saline-tolerant Pokkali rice, Kerala, India, highlighted plant-beneficial properties (Krishnan et al. 2016). Genome fingerprinting, GC content and DNA-DNA hybridization followed by chemotaxonomy analysis are useful to identify the organism associated with soil rhizosphere. The diversity and antimicrobial activity of the endophytic actinomycetes from *Azadirachta indica* A. Juss has been reported (Verma et al. 2009). The study revealed the dominance of the genus *Streptomyces*.

Many molecular typing techniques such as BOX-PCR, ARDRA, RFLP and RAPDS have been employed to investigate genetic variation of actinobacteria (Cerda 2008; Carro 2009; Alonsodela Vega 2010; Trujillo et al. 2010; Martinez-Hidalgo et al. 2014). The significance of small RNAs in elucidation of the host-microbe interactions has been emphasized (Katiyar-Agarwal and Jin 2010). Small RNAs of 20–40-nucleotide-long non-coding RNA molecules, present in the eukaryotic organism, play an important role in the regulation of gene expression. This approach is efficiently used to study the symbiotic nitrogen fixation.

15.5 Effects of Plant-Actinobacteria Interactions

Besides soil and rhizosphere microbial community, a diverse population of microbes, broadly termed as endophyte, live within plants without any adverse effect (Zinniel et al. 2002). Endophytic bacteria usually have lower population densities in the host plant tissues as compared to the pathogens; however, they can protect the host against pathogens. The endophytic bacteria belonging to Proteobacteria, Firmicutes and Actinobacteria enter into the plants and established mutualistic associations (Azevedo et al. 2000). Actinomycetes recognized as a potential group of rhizobacteria influence plant growth and nutrient uptake (Taj and Rajkumar 2016). Actinobacteria with plant growth-promoting traits are reported in cereals and to a lesser extent in legumes (Mishra et al. 1987; Nimaichand et al. 2016).

Endophytic actinobacteria usually produce novel bioactive compounds (Qin et al. 2011). Cell-free lysate of endophytic bacteria degrades the QS molecules and suppresses biofilm formation in *Pseudomonas* spp. PA01 (Rajesh and Ravishankar 2013).

Some actinobacterial endophytes release metal-mobilizing metabolites into the contaminated soil and mobilize Zn and/or Cd, enhancing the metal accumulation in the leaves of *Salix caprea* (Kuffner et al. 2010). *Nocardiopsis* sp. from the *Hibiscus rosa-sinensis* leaves degrade biosurfactant, polythene, plastic and diesel (Singh and Sedhuraman 2015). Actinobacteria suppress disease and promote growth (Palaniyandi et al. 2013). The root-associated beneficial endophytic microbes interact with a variety of plants. Based on the nature of the interactions, they are referred as antagonistic (pathogen) and plant growth-promoting microbes. *Streptomyces* is reported to reduce the incidence and the severity of leaf/seedling blight of rice (Wan et al. 2008). Several fungi and bacteria are known to exhibit hyperparasitism on several other pathogenic fungi, in which they feed. *Nocardiopsis dassonvillei* possess antibiotic, mycolytic and parasitic activities against the vegetative hyphae of *Fusarium* species (Sabaou et al. 1983).

PGPR strains promote the growth of plant by regulating the plant growth regulators. The most commonly reported plant growth regulators are indole acetic acid (IAA) and natural auxin (Palaniyandi et al. 2013). IAA and plant growth promotion has been reported for several actinobacteria (El-Tarabily 2008; Khamna et al. 2010; Palaniyandi et al. 2013). In wheat plants, *Streptomyces olivaceoviridis*, *S. rimosus* and *S. rochei* produce auxins, gibberellins and cytokinin-like substances and enhance the growth (Aldesuquy et al. 1998). Endophytic *Streptomyces* strains from *Azadirachta indica*, with the ability to produce a high amount of IAA, are reported to promote growth of tomato plants (Verma et al. 2011). The actinobacteria from yam rhizosphere produce IAA and promote the growth of *Arabidopsis* (Palaniyandi et al. 2013). The negative impact of IAA production is also known (Vereecke et al. 1997). IAA production has been demonstrated as a virulence factor in few plant pathogenic actinobacteria. *Streptomyces scabies* causes scab disease in most tap root and tuber crops (Hsu 2010). Siderophore produced by *Streptomyces acidiscabies* E13 promoted growth of *Vigna unguiculata* under nickel stress (Dimkpa et al. 2009). During siderophore production in E13 strain, the level of nickel uptake is reduced by *V. unguiculata*. *Streptomyces tendae* F4 promoted the growth and enhanced the uptake of cadmium (Cd) by sunflower plants, a significant function in phytoremediation (Dimkpa et al. 2009).

Several actinobacteria including *Streptomyces* and *Rhodococcus* are reported to play an important role in nitrogen fixation, via symbiosis, and to promote symbiosis between plants and mycorrhiza (Palaniyandi et al. 2013). Hasegawa et al. (2006) reported the role of some endophytic actinobacteria in drought stress tolerance. In present scenario, development of actinobacterial formulations to promote sustainable agriculture for crop protection and production is necessary (Palaniyandi et al. 2013). The molecular tools to probe the actinobacterial interactions should be developed further.

Conclusion

We have described biological functions of the plant-associated actinobacteria. The endophytic association of actinobacteria with the host plants triggers a cascade of signal transduction reactions, which induce a variety of defence responses at biochemical and molecular levels. Further, analysis of the sequenced genomes and identification of genes expressed during the colonization will add to our understanding of this association. In this chapter, we further described how cultivation-dependent and cultivation-independent methods can be employed to recognize the dynamics of the actinobacterial endophytic communities of different plants. The modern molecular techniques facilitate the screening of the endophytic communities. Overall, the exploration of the endophytic actinobacteria will contribute to the development of methodologies and novel technologies to elucidate the plant-actinobacterial association.

Acknowledgement Our work on actinomycetes has been supported under various programmes of UGC, including the current CAS Programme, DST-FIST, DST-Women Scientist Programme, UGC Start-Up Research Project, DST-SERB Research Project, MoES Networking project and Saurashtra University. The UGC-BSR Meritorious Fellowship and CSIR Research Associateship to S.D.G. and UGC-BSR Meritorious Fellowship to A.S. and F.T. are duly acknowledged.

References

Akshatha JV, Prakash HS, Nalini MS (2016) Actinomycete endophytes from the ethno medicinal plants of southern India: antioxidant activity and characterization studies. J Biol Active Prod Nat 6(2):166–172

Aldesuquy HS, Mansour FA, Abo-Hamed SA (1998) Effect of the culture filtrates of *Streptomyces* on growth and productivity of wheat plants. Folia Microbiol 43(5):465–470

Alonsodela Vega P (2010) Distribucion, Caracterizacione Importancia Ecologica de *Micromonosporaen* Nódulos Fijadores de Nitrógeno de Lupinus. Ph.D. thesis, Universidadde Salamanca, Salamanca

Antony CP, Shimpi GG, Cockell CS, Patole MS, Shouche YS (2014) Molecular characterization of prokaryotic communities associated with lonar crater basalts. Geomicrobiol J 31(6):519–528

Araújo JMD, Silva ACD, Azevedo JL (2000) Isolation of endophytic actinomycetes from roots and leaves of maize (*Zea mays* L.) Braz Arch Biol Technol 43(4). doi:10.1590/S1516-89132000000400016

Azevedo JL, Maccheroni W Jr, Pereira JO, de Araujo WL (2000) Endophytic microorganisms: a review on insect control and recent advances on tropical plants. Electron J Biotechnol 3(1):15–16

Barret M, Morrissey JP, and O'Gara F (2011) Functional genomics analysis of plant growth-promoting rhizobacterial traits involved in rhizosphere competence. Biology and Fertility of Soils 47:729–743

Bian GK, Feng ZZ, Qin S, Xing K, Wang Z, Cao CL, Liu CH, Dai CC, Jiang JH (2012) *Kineococcus endophytica* sp. nov., a novel endophytic actinomycete isolated from a coastal halophyte in Jiangsu, China. Antonie Van Leeuwenhoek 102:621–628

Bonaldi M, Chen XYL, Kunova A, Pizzatti C, Saracchi M, and Cortesi P (2015) Colonization of lettuce rhizosphere and roots by tagged Streptomyces. Front. Microbiol 6:25

Bouizgarne B, Aouamar AA (2014) Diversity of plant associated actinobacteria. In: Bacterial diversity in sustainable agriculture. Springer International Publishing, Switzerland, pp 41–99

Bunyoo C, Duangmal K, Nuntagij A, Thamchaipenet A (2009) Characterisation of endophytic actinomycetes isolated from wattle trees (*Acacia auriculiformis* A. Cunn. ex Benth.) in Thailand. Thai J Genet 2:155–163

Callaham D, Deltredici P, Torrey JG (1978) Isolation and cultivation in vitro of the actinomycete causing root nodulation in Comptonia. Science 199(4331):899–902

Cao L, Qiu Z, You J, Tan H, Zhou S (2004) Isolation and characterization of endophytic *Streptomyces* strains from surface-sterilized tomato (*Lycopersicon esculentum*) roots. Lett Appl Microbiol 39:425–430

Cardinale M (2015) Scanning a microhabitat: plant-microbe interactions revealed by confocal laser microscopy. The plant microbiome and its importance for plant and human health. Front Microbiol 22:14

Carro L (2009) Avancesenla Sistemáticadel Género *Micromonospora*: Estudio de Cepasaisladasde la Rizosferay Nódulosde *Pisum Sativum*. Ph.D. thesis, Universidadde Salamanca, Salamanca

Cerda E (2008) Aislamientode *Micromonosporade* Nódulosde Leguminos as Tropicalesy Análisisde SuInterés Como Promotordel Crecimiento Vegetal. Ph.D. thesis, Universidadde Salamanca, Salamanca
Chankhamhaengdecha S, Hongvijit S, Srichaisupakit A, Charnchai P, Panbangred W (2013) Endophytic actinomycetes: a novel source of potential acyl homoserine lactone degrading enzymes. Biomed Res Int 2013:782847
Chelius MK, Triplett EW (2000) Immunolocalization of dinitrogenase reductase produced by *Klebsiella pneumoniae* in association with *Zea mays* L. Appl Environ Microbiol 66(2):783–787
Cheng Z, Mcconkey BJ and Glick BR (2010) Proteomic studies of plant–bacterial interactions. Soil Biology & Biochemistry 42:1673–1684
Compant S, Clément C, Sessitsch A (2010) Plant growth-promoting bacteria in the rhizo-and endosphere of plants: their role, colonization, mechanisms involved and prospects for utilization. Soil Biol Biochem 42(5):669–678
Coombs JC, Franco CMM (2003a) Isolation and identification of actinobacteria from surface-sterilized wheat roots. Appl Environ Microbiol 69:5603–5608
Coombs JC, Franco CMM (2003b) Visualization of an endophytic *Streptomyces* species in wheat seed. Appl Environ Microbiol 69:4260–4262
Coombs JT, Michelsen PP, Franco CM (2004) Evaluation of endophytic actinobacteria as antagonists of *Gaeumannomyces graminis* var. tritici in wheat. Biol Control 29(3):359–366
Costa FG, Zucchi TD, Melo ISD (2013) Biological control of phytopathogenic fungi by endophytic actinomycetes isolated from maize (*Zea mays* L.) Braz Arch Biol Technol 56(6):948–955
Cummings CA, Relman DA (2000) Using DNA microarrays to study host-microbe interactions. Emerg Infect Dis 6(5):513
DeAngelis KM, Brodie EL, DeSantis TZ, Andersen GL, Lindow SE and Firestone MK (2009) Selective progressive response of soil microbial community to wild oat roots. The ISME Journal 3:168–178
Dimkpa CO, Merten D, Svatoš A, Büchel G, Kothe E (2009) Siderophores mediate reduced and increased uptake of cadmium by *Streptomyces tendae* F4 and sunflower (*Helianthus annuus*), respectively. J Appl Microbiol 107(5):1687–1696
Ding L, Maier A, Fiebig HH, Lin WH, Peschel G, Hertweck C (2012) Kandenols A–E, eudesmenes from an endophytic *Streptomyces* sp. of the mangrove tree Kandelia candel. J Nat Prod 75(12):2223–2227
Doroghazi JR, Metcalf WW (2013) Comparative genomics of actinomycetes with a focus on natural product biosynthetic genes. BMC Genomics 14(1):1
Dudeja SS, Giri R, Saini R, Suneja MP, Kothe E (2012) Interaction of endophytic microbes with legumes. J Basic Microbiol 52:248–260
El-Shatoury SA, El-Kraly OA, Trujillo ME, El-Kazzaz WM, El-Din ESG, Dewedar A (2013) Generic and functional diversity in endophytic actinomycetes from wild Compositae plant species at South Sinai-Egypt. Res Microbiol 164(7):761–769
El-Tarabily KA (2008) Promotion of tomato (*Lycopersicon esculentum Mill.*) plant growth by rhizosphere competent 1-aminocyclopropane-1-carboxylic acid deaminase-producing *streptomycete* actinomycetes. Plant Soil 308(1–2):161–174
Farrar K, Bryant D, Cope-Selby N (2014) Understanding and engineering beneficial plant–microbe interactions: plant growth promotion in energy crops. Plant Biotechnol J 12(9):1193–1206
Francis S, Holsters M, Vereecke D (2010) The Gram positive side of plant-microbe interactions. Environ Microbiol 12:1–12
Gaiero JR, McCall CA, Thompson KA, Day NJ, Best AS and Dunfield KE (2013) Inside the root microbiome: Bacterial root endophytes and plant growth promotion. Am J Bot 100:1738–1750
Gangwar M, Dogra S, Sharma N (2011) Antagonistic bioactivity of endophytic actinomycetes isolated from medicinal plants. J Adv Lab Res Biol 2(4):154–157
Gangwar M, Dogra S, Gupta UP, Kharwar RN (2014) Diversity and biopotential of endophytic actinomycetes from three medicinal plants in India. Afr J Microbiol Res 8(2):184–191
Gohel SD, Singh SP (2012) Purification strategies, characteristics and thermodynamic analysis of a highly thermostable alkaline protease from a salt-tolerant alkaliphilic actinomycete, *Nocardiopsis alba* OK-5. J Chromatogr B 889:61–68

Gohel SD, Singh SP (2013) Characteristics and thermodynamics of a thermostable protease from a salt-tolerant alkaliphilic actinomycete. International Journal of Biological Macromolecules. 56:20–27

Gohel SD, Singh SP (2015) Thermodynamics of a Ca^{2+} dependent highly thermostable alkaline protease from haloalkliphilic actinomycetes. Int J Biol Macromol 72:421–429

Gohel SD, Singh SP (2016) Morphological, cultural and molecular diversity of the salt-tolerant alkaliphilic actinomycetes from the saline habitats. In: Microbial biotechnology technological challenges and developmental trends. Apple Academic Press; distributed by CRC Press Taylor and Francis group. ISBN online: 978-1-77188-333-7

Gohel SD, Sharma AK, Dangar KG, Thakrar FJ, Singh SP (2015) Antimicrobial and biocatalytic potential of haloalkaliphilic actinobacteria. In: Halophiles. Springer International Publishing, Switzerland, pp 29–55

Gohel SD, Sharma AK, Dangar K, Thakrar F, Singh SP (2016) Biology and applications of halophilic/halo-alakliphilic actinobacteria. In: Extremophiles: from biology to biotechnology. CRC Publishers, USA

Golinska P, Wypij M, Agarkar G, Rathod D, Dahm H, Rai M (2015) Endophytic actinobacteria of medicinal plants: diversity and bioactivity. Antonie Van Leeuwenhoek 108(2):267–289

Hallmann J, Berg G, Schulz B (2006) Isolation procedures for endophytic microorganisms. In: Microbial root endophytes. Springer, Berlin, Heidelberg, pp 299–319

Hardoim PR, Van Overbeek LS, Van Elsas JD (2008) Properties of bacterial endophytes and their proposed role in plant growth. Trends Microbiol 16:463–471

Hartmann A, Stoffels M, Eckert B, Kirchhof G, Schloter M (2000) Analysis of the presence and diversity of diazotrophic endophytes. In: Prokaryotic nitrogen fixation: a model system for analysis of a biological process. Horizon Scientific Press, Wymondham, UK, pp 727–736

Hasegawa S, Meguro A, Shimizu M, Nishimura T, Kunoh H (2006) Endophytic actinomycetes and their interactions with host plants. Actinomycetologica 20(2):72–81

Haupt BJ, Pelling AE, Horton MA (2006) Integrated confocal and scanning probe microscopy for biomedical research. Sci World J 15:1609–1618. doi:10.1100/tsw.2006.269

Hsu SY (2010) IAA production by *Streptomyces scabies* and its role in plant microbe interaction. Doctoral dissertation, Cornell University

Huang JS (1986) Ultra structure of bacterial penetration in plants. Annu Rev Phytopathol 24:141–157

Hurek T, Handley LL, Reinhold HB, Piche Y (2002) *Azoarcus* grass endophytes contribute fixed nitrogen to the plant in an unculturable state. Mol Plant Microbe Interact 15:233–242

Inahashi Y, Iwatsuki M, Ishiyama A, Matsumoto A, Hirose T, Oshita J, Sunazuka T, Panbangred W, Takahashi Y, Kaiser M, Otoguro K (2015) Actinoallolides A–E, new anti-trypanosomal macrolides, produced by an endophytic actinomycete, actinoallomurusfulvus MK10-036. Org Lett 17(4):864–867

Jansson JK, Neufeld JD, Moran MA, Gilbert JA (2012) Omics for understanding microbial functional dynamics. Environ Microbiol 14:1–3. doi:10.1111/j.1462-2920.2011.02518.x

Jog R, Pandya M, Nareshkumar G, Rajkumar S (2014) Mechanism of phosphate solubilization and antifungal activity of *Streptomyces* spp. isolated from wheat roots and rhizosphere and their application in improving plant growth. Microbiology 160(4):778–788

Kathiresan K, Bingham BL (2001) Biology of mangroves and mangrove ecosystems. Adv Mar Biol 40:81–251

Katiyar-Agarwal S, Jin H (2010) Role of small RNAs in host-microbe interactions. Annu Rev Phytopathol 48:225

Khamna S, Yokota A, Lumyong S (2009) Actinomycetes isolated from medicinal plant rhizosphere soils: diversity and screening of antifungal compounds, indole-3 acetic acid and siderophore production. World J Microbiol Biotechnol 25:649–655

Khamna S, Yokota A, Peberdy JF, Lumyong S (2010) Indole-3-acetic acid production by *Streptomyces* sp. isolated from some Thai medicinal plant rhizosphere soils. Eur Asia J Biosci 4:23–32

Kikani BA, Shukla RJ, Singh SP (2010) Biocatalytic potential of thermophilic bacteria and actinomycetes. In: Current research, technology and education topics in applied microbiology and microbial biotechnology, A. Mendez-Vilas (Ed), Formatex vol 2, pp 1000–1007

Kim SH, Shin Y, Lee SH, KB O, Lee SK, Shin J, DC O (2015) Salternamides A–D from a halophilic *Streptomyces* sp. actinobacterium. J Nat Prod 78(4):836–843

Kint G, Fierro C, Marchal K, Vanderleyden V (2010) Integration of omics data, does it lead to new insights in to host–microbe interactions? Future Microbiol 5:313–328. doi:10.2217/fmb.10.1

Komeda H, Kobayashi M, Shimizu S (1996) Characterization of the gene cluster of high-molecular-mass nitrile hydratase (H-NHase) induced by its reaction product in *Rhodococcus* rhodochrous J1. Proc Natl Acad Sci U S A 93(9):4267–4272

Krishnan R, Menon RR, Tanaka N, Busse HJ, Krishnamurthi S, Rameshkumar N (2016) Arthrobacter pokkalii sp. nov., a novel plant associated actinobacterium with plant beneficial properties, isolated from saline tolerant Pokkali rice, Kerala, India. PLoS One 11(3):e0150322

Kuffner M, De Maria S, Puschenreiter M, Fallmann K, Wieshammer G, Gorfer M, Strauss J, Rivelli AR, Sessitsch A (2010) Culturable bacteria from Zn-and Cd-accumulating Salix caprea with differential effects on plant growth and heavy metal availability. J Appl Microbiol 108(4):1471–1484

Li J, Zhao GZ, Huang HY, Zhu WY, Lee JC, Kim CJ, Xu LH, Zhang LX, Li WJ (2010) Pseudonocardia rhizophila sp. nov., a novel actinomycete isolated from a rhizosphere soil. Antonie van Leeuwenhoek 98:77–83

Loria R, Bukhalid RA, Fry BA, King RR (1997) Plant pathogenecity in the genus *Streptomyces*. Plant Dis 81:836–846

Lugtenberg BJ, Chin-A-Woeng TF, Bloemberg GV (2002) Microbe–plant interactions: principles and mechanisms. Antonie Van Leeuwenhoek 81(1–4):373–383

Machavariani NG, Ivankova TD, Sineva ON, Terekhova LP (2014) Isolation of endophytic actinomycetes from medicinal plants of the Moscow region, Russia. World Appl Sci J 30(11):1599–1604

Malfanova N, Kamilova F, Validov S, Shcherbakov A, Chebotar V, Tikhonovich I et al (2011) Characterization of *Bacillus subtilis* HC8, a novel plant-beneficial endophytic strain from giant hogweed. Microb Biotechnol 4:523–532

Manter DK, Delgado JA, Holm DG, Stong RA (2010) Pyrosequencing reveals a highly diverse and cultivar-specific bacterial endophyte community in potato roots. Microb Ecol 60(1):157–166

Martinez-Hidalgo P, Galindo-Villardón P, Trujillo ME, Igual JM, Martínez-Molina E (2014) *Micromonospora* from nitrogen fixing nodules of alfalfa (*Medicago sativa L.*) a new promising plant probiotic bacteria. Sci Rep 4:6389

Mendes R, Kruijt M, Bruijni DE, Dekkers M, Voort VD, Schneider HM et al (2011) Deciphering the rhizosphere microbiome for disease-suppressive bacteria. Science 332:1097–1100

Merzaeva OV, Shirokikh IG (2006) Colonization of plant rhizosphere by actinomycetes of different genera. Mikrobiologiia 75:271–276

Mishra SK, Keller JE, Miller JR, Heisey RM, Nair MG, Putnam AR (1987) Insecticidal and nematicidal properties of microbial metabolites. J Ind Microbiol 5:267–276

Misk A, Franco C (2011) Biocontrol of chickpea root rot using endophytic actinobacteria. BioControl 56(5):811–822

Nakashima N, Mitani Y, Tamura T (2005) Actinomycetes as host cells for production of recombinant proteins. Microb Cell Factories 4(1):7

Nimaichand S, Devi AM, Li WJ (2016) Direct plant growth-promoting ability of actinobacteria in grain legumes. In: Plant growth promoting actinobacteria. Springer, Singapore, pp 1–16

Normand P, Lapierre P, Tisa LS, Gogarten JP, Alloisio N, Bagnarol E et al (2006) Genome characteristics of facultatively symbiotic Frankia sp.strains reflect host range and host plant biogeography. Genome Research 17(1):7–15

Oldroyd GE, Downie JA (2004) Calcium, kinases and nodulation signalling in legumes. Nat Rev Mol Cell Biol 5(7):566–576

Palaniyandi SA, Yang SH, Zhang L, Suh JW (2013) Effects of actinobacteria on plant disease suppression and growth promotion. Appl Microbiol Biotechnol 97(22):9621–9636

Qin S, Xing K, Jiang J, Xu L, Li W (2011) Biodiversity, bioactive natural products and biotechnological potential of plant-associated endophytic actinobacteria. Appl Microbiol Biotechnol 89:457–473

Qin S, Chen HH, Zhao GZ, Li J, Zhu WY, Xu LH et al (2012a) Abundant and diverse endophytic actinobacteria associated with medicinal plant *Maytenus austroyunnanensis* in Xishuangbanna tropical rain forest revealed by culture-dependent and culture-independent methods. Environ Microbiol Rep 4:522–531

Qin S, Yuan B, Zhang YJ, Bian GK, Tamura T, Sun BZ, Li WJ, Jiang JH (2012b) *Nocardioides panzhihuaensis* sp. nov., a novel endophytic actinomycete isolated from medicinal plant *Jatropha curcas* L. Antonie Van Leeuwenhoek 102:353–360

Qin S, Bian GK, Zhang YJ, Xing K, Cao CL, Liu CH, Dai CC, Li WJ, Jiang JH (2013) *Modestobacter roseus* sp. nov., an endophytic actinomycete isolated from the coastal halophyte

Salicornia europaea Linn., and emended description of the genus *Modestobacter*. Int J Syst Evol Microbiol 63:2197–2202

Rajesh PS, Ravishankar RV (2013) Quorum quenching activity in cell-free lysate of endophytic bacteria isolated from *Pterocarpus santalinus* Linn., and its effect on quorum sensing regulated biofilm in *Pseudomonas aeruginosa* PAO1. Microbiol Res 169(7–8):561–568

Rao HY, Rakshith D, Satish S (2015) Antimicrobial properties of endophytic actinomycetes isolated from Combretum latifolium Blume, a medicinal shrub from Western Ghats of India. Front Biol 10(6):P528–P536

Ravikumar S, Gnanadesigan M, Suganthi P, Ramalakshmi A (2010) Antibacterial potential of chosen mangrove plants against isolated urinary tract infectious bacterial pathogens. Int J Med Med Sci 2(3):94–99

Reid A, Greene SE (2012) How microbes help feed the world. American Academy of Microbiology, Washington, DC

Rogers A, McDonald K, Muehlbauer MF, Hoffman A, Koenig K, Newman L, Taghavi S, Lelie D (2012) Inoculation of hybrid poplar with the endophytic bacterium *Enterobacter* sp. 638 increases biomass but does not impact leaf level physiology. GCB Bioenergy 4(3):364–370

Röling WFM, Ferrer M, Golyshin PN (2010) Systems approaches to microbial communities and their functioning. Curr Opin Biotechnol 21:532–538. doi:10.1016/j.copbio.2010.06.007

Rosenblueth M, Martinez-Romero E (2006) Bacterial endophytes and their interactions with hosts. Mol Plant Microbe Interact 19:827–837

Rungin S, Indananda C, Suttiviriya P, Kruasuwan W, Jaemsaeng R, Thamchaipenet A (2012) Plant growth enhancing effects by a siderophore-producing endophytic streptomycete isolated from a Thai jasmine rice plant (*Oryza sativa* L. cv. KDML105). Antonie Van Leeuwenhoek 102(3):463–472

Ryan RP, Germaine K, Franks A, Ryan DJ (2008) Bacterial endophytes: recent developments and applications. FEMS Microbiol 278:1–9

Sabaou N, Bounaga N, Bounaga D (1983) Mycolytic and parasitic and antibiotic action of 2 actinomycetes in *Fusarium oxysporum* f. sp. *albedinis* and other formae speciales. Can J Microbiol 29(2):194–199

Sardi P, Saracchi M, Quaroni S, Petrolini B, Borgonovi GE, Merli S (1992) Isolation of endophytic *Streptomyces* strains from surface-sterilized roots. Appl Environ Microbiol 58:2691–2693

Schloter M, Borlinghaus R, Bode W, Hartmann A (1993) Direct identification, and localization of *Azospirillum* in the rhizosphere of wheat using fluorescence labelled monoclonal antibodies and confocal scanning laser microscopy. J Microsc 171(2):173–177

Schneider CA, Rasband WS, Eliceiri KW (2012) NIH image to image J: 25 years of image analysis. Nat Methods 9:671–675. doi:10.1038/nmeth.2089

Seipke RF, Kaltenpoth M, Hutchings MI (2012) *Streptomyces* as symbionts: an emerging and widespread theme? FEMS Microbiol Rev 36:862–876

Sessitsch A, Ardoim PH, Weilharter A, Krause A, Woyke TB, Mitter L et al (2012) Functional characteristics of an endophyte community colonizing rice roots as revealed by metagenomic analysis. Molecular Plant-Microbe Interactions 25:28–36

Sharma AK, Gohel S, Singh SP (2012) Actinobase: database on molecular diversity, phylogeny and biocatalytic potential of salt tolerant alkaliphilic actinomycetes. Bioinformation 8(11):535–538

Shenpagam HN, Devi DK, Sinduja G, Sandhya R (2012) Isolation of endophytic actinomycetes from medicinal plants and its mutational effect in biocontrol activity. Int J Pharm Sci Res 3(11):4338

Shimizu M (2011) In: Maheshwari DK (ed) Bacteria in agrobiology: plant growth responses. Springer-Verlag, Berlin, Heidelberg, pp 201–220. doi:10.1007/978-3-642-20332-9_10

Shutsrirung A, Chromkaew Y, Pathom-Aree W, Choonluchanon S, Boonkerd N (2013) Diversity of endophytic actinomycetes in mandarin grown in northern Thailand, their phytohormone production potential and plant growth promoting activity. Soil Sci Plant Nutr 59(3):322–330

Singh R, Dubey AK (2015) Endophytic actinomycetes as emerging source for therapeutic compounds. Indo Global J Pharm Sci 5(2):106–116

Singh MJ, Sedhuraman P (2015) Biosurfactant, polythene, plastic, and diesel biodegradation activity of endophytic *Nocardiopsis* sp. *mrinalini9* isolated from *Hibiscus rosasinensis* leaves. Bioresour Bioprocessing 2(1):1

Singh SP, Thumar JT, Gohel SD, Purohit MK (2010) Molecular diversity and enzymatic potential of salt-tolerant alkaliphilic actinomycetes. In: Mendez-Vilas A (ed) Current research, technology and education topics in applied microbiology and microbial biotechnology. Formatex Research Center, Badajoz, Spain, pp 280–286

Singh SP, Thumar JT, Gohel S, Kikani BA, Shukla R, Sharma A, Dangar K (2013) Actinomycetes from marine habitats and their enzymatic potential. In: Marine enzymes for bicatalysis. Woodhead Publishing Series in Biomedicine (Oxford) Ltd, Oxford (Library Congress Number: 2013948122), pp 191–214

Taj ZZ, Rajkumar M (2016) Perspectives of plant growth-promoting actinomycetes in heavy metal phytoremediation. In: Plant growth promoting actinobacteria. Springer, Singapore, pp 213–231

Thumar JT, Singh SP (2007) Secretion of an alkaline protease from a salt-tolerant and alkaliphilic, Streptomyces clavuligerus strain MIT-1. Braz J Microbiol 38(4):766–772

Thumar JT, Singh SP (2009) Organic solvent tolerance of an alkaline protease from salt-tolerant alkaliphilic *Streptomyces clavuligerus* strain Mit-1. J Ind Microbiol Biotechnol 36(2):211–218

Thumar JT, Singh SP (2011) Repression of alkaline protease in salt-tolerant alkaliphilic *Streptomyces clavuligerus* strain Mit-1 under the influence of amino acids in minimal medium. Biotechnol Bioprocess Eng 16(6):1180–1186

Tian XL, Cao LX, Tan HM, Zeng QG, Jia YY, Han WQ, Zhou SN (2004) Study on the communities of endophytic fungi and endophytic actinomycetes from rice and their antipathogenic activities in vitro. World J Microbiol Biotechnol 20:303–309

Tokala RK, Strap JL, Jung CM, Crawford DL, Salove MH, Deobald LA et al (2002) Novel plant-microbe rhizosphere interaction involving *Streptomyces lydicus* WYEC108 and the pea plant (*Pisumsativum*). Appl Environ Microbiol 68:2161–2171

Trujillo ME, Alonso-Vega P, Rodríguez R, Carro L, Cerda E, Alonso P, Martínez-Molina E (2010) The genus *Micromonospora* is widespread in legume root nodules: the example of *Lupinus angustifolius*. ISME J 4(10):1265–1281

Trujillo ME, Raul R, Patricia B, Lorena C (2015) Endophytic actinobacteria and the interaction of *Micromonospora* and nitrogen fixing plants. Front Microbiol 6:1341

Vereecke D, Messens E, Klarskov K, De Bruyn A, Van Montagu M, Goethals K (1997) Patterns of phenolic compounds in leafy galls of tobacco. Planta 201(3):342–348

Verma SC, Singh A, Chowdhury SP, Tripathi AK (2004) Endophytic colonization ability of two deep-water rice endophytes, *Pantoea* sp. and *Ochrobactrum* sp. using green fluorescent protein reporter. Biotechnol Lett 26(5):425–429

Verma VC, Gond SK, Kumar A, Mishra A, Kharwar RN, Gange AC (2009) Endophytic actinomycetes from *Azadirachta indica A. Juss.*: isolation, diversity, and anti-microbial activity. Microb Ecol 57(4):749–756

Verma VC, Singh SK, Prakash S (2011) Bio-control and plant growth promotion potential of siderophore producing endophytic *Streptomyces* from *Azadirachta indica A. Juss.* J Basic Microbiol 51(5):550–556

Wan M, Li G, Zhang J, Jiang D, Huang HC (2008) Effect of volatile substances of *Streptomyces platensis* F-1 on control of plant fungal diseases. Biol Control 46(3):552–559

Weber T, Welzel K, Pelzer S, Vente A, Wohlleben W (2003) Exploiting the genetic potential of polyketide producing *streptomycetes*. J Biotechnol 106(2):221–232

Xing K, Qin S, Bian GK, Zhang YJ, Zhang WD, Dai CC, Liu CH, Li WJ, Jiang JH (2012) *Pseudonocardia nantongensis* sp. nov., a novel endophytic actinomycete isolated from the coastal halophyte *Tamarix chinensis* Lour. Antonie Van Leeuwenhoek 102:659–667

Xing K, Liu W, Zhang YJ, Bian GK, Zhang WD, Tamura T, Lee JS, Qin S, Jiang JH (2013) Amycolatopsis jiangsuensis sp. nov., a novel endophytic actinomycete isolated from a coastal plant in Jiangsu, China. Antonie van Leeuwenhoek 103:433–439

Zhang F, She Y, Zheng Y, Zhou Z, Kong S, Hou D (2010) Molecular biologic techniques applied to the microbial prospecting of oil and gas in the Ban 876 gas and oil field in China. Appl Microbiol Biotechnol 86(4):1183–1194

Zhang YJ, Zhang WD, Qin S, Bian GK, Xing K, Li YF, Cao CL, Jiang JH (2013) *Saccharopolyspora dendranthemae* sp. nov. a halotolerant endophytic actinomycete isolated from a coastal salt marsh plant in Jiangsu, China. Antonie Van Leeuwenhoek 103:1369–1376

Zinniel DK, Lambrecht P, Harris NB, Feng Z, Kuczmarski D, Higley P, Ishimaru CA, Arunakumari A, Barletta RG, Vidaver AK (2002) Isolation and characterization of endophytic colonizing bacteria from agronomic crops and prairie plants. Appl Environ Microbiol 68(5):2198–2208

Metatranscriptomic Studies of the Plant Rhizosphere for Finding Biological Agents

16

Vishal Kothari, Charmy Kothari, Jalpa Rank, Anjali Joshi, Ravindra Pal Singh, and Ramesh Kothari

Abstract

Soilborne diseases are accountable for most important crop losses globally. There are two ways to control this ailment. The judicious use of synthetic agrochemicals harms ecosystem and beneficial soil microbes and alternatively promotes the use of biological agents to control plant pathogens. Rhizospheric beneficial microbes play a crucial role in disease suppression process in the soil ecosystems by guarding plants from infections that they likely get from soilborne pathogens. The microorganisms and mechanisms involved in the disease suppression in soils are poorly known till date. Therefore, development of meta-omics techniques would provide deeper understanding of this association. Metatranscriptomics, a study of the total content of gene transcripts (RNA copies of the genes) in a microbial community, is a right approach to unveil the role of genes responsible for the disease suppression mechanism at molecular level as this technique is considered as a unique entity at a specific moment of sampling. This technique is applied to obtain the whole expression profile in a community and to follow the dynamics of gene expression patterns over time and/or various environmental parameters. The microbial communities in several environments have been extensively studied to reveal their roles in plant-microbes interaction and disease suppression in rhizospheric soils.

V. Kothari (✉)
Department of Biotechnology, Junagadh Agricultural University, Junagadh 362001, Gujarat, India
e-mail: kotharivishal87@gmail.com

C. Kothari
Department of Biotechnology, Christ College, Rajkot 360005, Gujarat, India

J. Rank • A. Joshi • R.P. Singh • R. Kothari
Department of Biosciences (UGC-CAS), Saurashtra University, Rajkot 360005, Gujarat, India

R.P. Singh et al. (eds.), *Understanding Host-Microbiome Interactions - An Omics Approach*, DOI 10.1007/978-981-10-5050-3_16

Keywords
Meta-transcriptomics • Metagenomics • Bioinformatics tools • Rhizosphere

16.1 Introduction

Plants depend on their rhizosphere microbiome for proper functions and other traits related to its own growth, development and then health (Berendsen et al. 2012; Mendes et al. 2013); Cook et al. (1995) hypothesized that natural selection occurs due to their own tendency to develop genetic resistance variety against below-ground pathogens. Microbial communities of the rhizosphere microbiome dock a range of beneficial properties to host nutrient acquisition, enhance stress tolerance and host immune regulation thereby protecting against soilborne pathogens (Berendsen et al. 2012; Bakker et al. 2013). However, it is known that less than 1% of microorganisms are cultivable under laboratory conditions from most environments (Rondon et al. 1999). Therefore, there is a challenge to identifying whole microbial community members present in a particular environment by a particular molecular method. Diversity and structure of microbial communities are examined using various techniques like i.e. direct counting, molecular fingerprinting such as amplified rDNA restriction analysis (ARDRA), phospholipid fatty acid (PLFA) analysis, fluorescent in situ hybridization (FISH), terminal restriction fragment length polymorphism (T-RFLP) analysis and denaturing gradient gel-electrophoresis (DGGE). These can provide information on the structure, diversity and compositional change of microbial communities and some idea about functional potential (Schink 2000). However, these methods cannot provide complete information about their phylogenetic and functional activity which can occur in response to different environmental cues, for instance, defence against plant pathogen on root surface, uptake of nutrient, flow of energy, degradation of substrates and distributions of metabolic pathways. Therefore, there is a strong need for the alternative approaches like community-wide analysis of gene content (metagenomics) and functional gene (metatranscriptomics) in a specific microbial environment at a given point of time. Metagenomics tells us which microbes are present and what genomic potential they have. Metatranscriptomics was used to determine quantitative abundance and their metabolism of microbes in soil and the rhizosphere (Mendes et al. 2013).

16.2 Meta-omics of Plant Rhizosphere and Controlling Fungal Pathogen

Fungal pathogens are one of the detrimental factors of the plant host that leads to loss of crops and postharvest of fruits all over the world (Chen et al. 2008). There are so many synthetic chemical fungicides are being used to prevent and kill pathogenic fungi of plant in various environments. However, fungi have high frequency of mutation in their genome that lead to high diversity and increased resistance to frequently used fungicides. Because of this tendency, several important chemical fungicides have lost their efficacy against pathogenic fungi in the field, such as demethylation inhibitors, anilinopyrimidine, benzimidazoles, dicarboximide, phenylpyrrole and strobilurin (Yang et al. 2008). There are always demands to discover and develop novel fungicides to minimize the risk of crop disease and augment the safety of food in the different environment (Coloretti et al. 2007). In addition to using microorganisms to prevent fungal diseases, it would be a best alternative to deal with harmful fungal pathogens; thus, there are currently gaining interest globally for reducing potential negative effects of chemical fungicides (Prema et al. 2008). Applying meta-omics technologies would be beneficial to find out those bacterial communities that can suppress growth of the fungal pathogens.

Plant-microbes interaction is adequately analysed by metatranscriptomic tool particularly in the rhizosphere. Applying metatranscriptomic analysis on the rhizosphere would provide evidence about how they do influence host metabolism and will provide deeper insight on developing association of parasitic relationship. Notably, rhizosphere is the zone where soilborne pathogens establish a parasitic relationship with their host (Chapelle et al. 2015). To study competition, antagonism and process of disease suppression in the rhizosphere as well as to consider a unique entity at a specific moment of sampling responsible for disease suppression requires functional analysis of the rhizosphere which can be possible with metatranscriptome. Kim and Liesack (2015) studied paddy soil microbiomes and metatranscriptome on oxic and anoxic zone and identified some of the members related to Cyanobacteria, Fungi, Xanthomonadales, Myxococcales and Methylococcales in the oxic zone, whilst Clostridia, Actinobacteria, Geobacter, Anaeromyxobacter, Anaerolineae and methanogenic archaea conquered the anoxic zone. They were stably maintained throughout the incubation period. Moreover, methane oxidation and photosynthesis were carried out by Methylococcales and Cyanobacteria, respectively, which are unique to the oxic zone (Kim and Liesack 2015). On the contrary, methane production by methanogenic archaea and degradation of aromatic compound by Anaeromyxobacter were characteristics of the anoxic zone. Chapelle et al. (2015) studied on the plant pathogenic fungi *Rhizoctonia solani.* Particularly, the study found that during hyphal growth toward the plant root, it produces oxalic and phenylacetic acid which promote shift in the specific rhizobacterial families present in the suppressive rhizosphere microbiome. Therefore, they proposed that pathogenic fungus may directly or indirectly increase transcript of oxidative stress-related genes in those rhizobacterial families by ppGpp signalling pathway. This trend could be responsible for the shift in microbial communities and inhibiting growth of the fungal pathogens. Luo et al. (2015) studied de novo metatranscriptomics of plant and fungi in the symbiotic roots of sewage-cleaning *Eichhornia crassipes* and interpret genes down- and upregulation during formation of their symbiotic process. In the presence of fungi, they found changes in genetic pathways of plant and fungal strain and postulated that these pathways could be responsible for dealing with an environmental pollutant.

Unno and Shinano (2013) used a metagenomic approach on the rhizosphere soil microbiome, particularly for utilization of phytic acid in flowering and non-flowering plants, and found changes in transcripts of some genes that may be responsible for the utilization of phytic acid such as alkaline phosphatase and citrate synthase. The secondary metabolite production, including genes related to the production of antibiotic compounds and plant hormone-like compounds, accounts for a higher proportion in the flowering plant than that of the non-flowering plants (fourfold). Knief et al. (2012) analysed microbial communities in the phyllosphere and rhizosphere of rice (*Oryza sativa*) and then used protein-coding marker genes for deciphering the phylogenetic information on both zones at functional level. This permits evaluating the relative functional profusion of the members in the microbial community. Similarly, Stark et al. (2010) and Arjun et al. (2011) studied on phyllosphere and rhizosphere of rice cultivar IR-72 and discovered the dominance of Alphaproteobacteria and Actinobacteria in the phyllosphere, whilst Alphaproteobacteria, Betaproteobacteria and Deltaproteobacteria were most abundant in the rhizospheric zone. The study also found abundant taxa including the Firmicutes, Actinobacteria, Gammaproteobacteria and the Deinococcus-Thermus. With regard to Archaea, they were abundant in the rhizosphere than in the phyllosphere. Conrad (2007) reviewed on the activity of methanogenic archaea and summarized that archaeal rhizosphere is comprised in particular of diverse methanogens and even unknown taxa which are higher in the rhizosphere.

16.3 Metagenomics and Metatranscriptomics

Over the past few decades, next-generation nucleotide sequencing technologies have significantly advanced and become widespread with the adoption of several platforms depending on the question to be addressed (Parmar et al. 2017). Since these approaches are generated, large amount of data raised a comparable set of new challenges for experimental strategy, data analysis and their explanation. With reduction in the DNA and RNA sequencing costs from 60$ to 1$, these technologies offer huge datasets with great complexity. Therefore, there is always a demand to develop a reliable bioinformatic tools to competently convert the raw data into a biologically meaningful manner (Shendure and Ji 2008; Goodwin et al. 2016). Therefore, in this chapter, we are also summarizing different bioinformatic tools used to analyse metagenomics and metatranscriptomics with special influence on rhizospheric microbial communities.

Bailly et al. (2007) did the first metatranscriptomic study of soil, particularly focusing on the functional diversity of fungal communities by selective sequencing of polyadenylated transcripts. A metatranscriptome is a set of the total RNA from a microbial community which gives real-time gene expression of a community at functional level. Profiling of rRNA of a metatranscriptome permits vigorous phylogenetic profiling of organisms present in particular microbiome across all domains of life. This approach has been adopted to study phylogenetic relevance and determination of functional activity in the oceans (Ottesen et al. 2011; Shi et al. 2010), soil (Urich et al. 2008), phyllosphere (Knief et al. 2012) and the rhizospheres microbiome of different crops (Knief et al. 2012; Chaparro et al. 2013). Metatranscriptomic

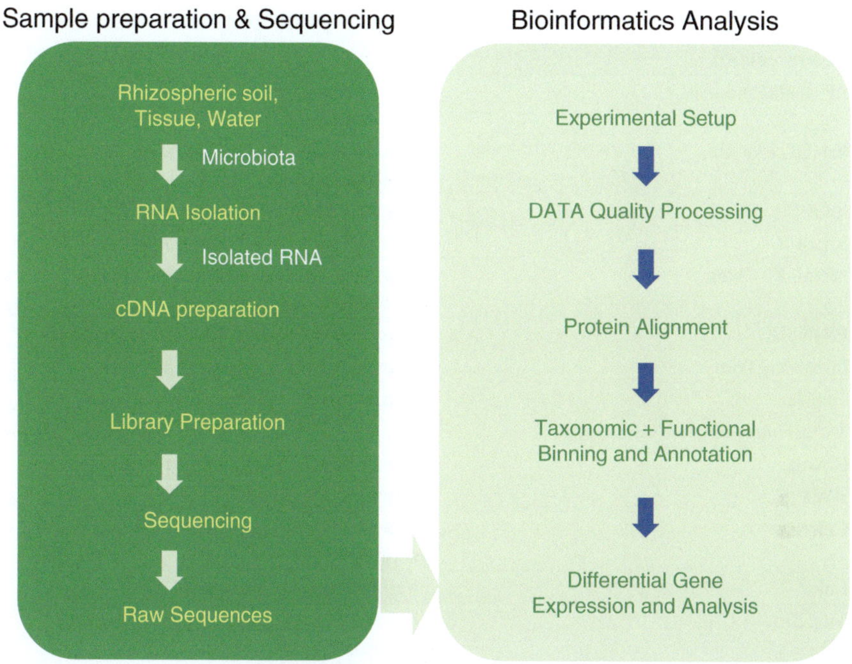

Fig. 16.1 Major step in metatranscriptomic analysis

studies also allow us to determine the expression level of non-coding and small RNA species (ncRNA and sRNA; Akiyoshi et al. 2009) in a microbiome which have been reported to have important regulatory roles in bacteria (Sanguin et al. 2006; Narberhaus and Vogel 2009). Particularly, metatranscriptome unveils information of the active metabolic pathways of the microbiome in a given condition in an environment.

For metagenomic and metatranscriptomic analysis of microbial communities, majority of steps are the same excluding DNA required in metagenome whilst RNA in metatranscriptome as shown in Fig. 16.1. Metatranscriptome gives differential gene expression and analysis for the expressed genes present in the community at the time of sampling. However, this approach has not been widely used in the rhizosphere in previous decade due to the instability of mRNAs and difficulties in their extraction from complex ecosystems. Other difficulties are short half-lives of mRNA, separation of mRNA from other RNA types (i.e. tRNA, rRNA, miRNA) and interference of humid compounds that co-extract with nucleic acids from soil (Simon and Daniel 2011). In current decade better availability of commercial kits that successfully provide high quality of RNA and other RNA stability agents makes it easier to study them in great detail (Buschmann et al. 2016). Additionally, current bioinformatic analysis of metagenome and metatranscriptomics is going to be easier as compared to past years due to the availability of the tools (Table 16.1).

Table 16.1 Tools used for the metagenomics and metatranscriptomics

Quality control	Link
EP_metagenomics	http://www.computationalbioenergy.org/qc-chain.html
FastQC tool kit	http://www.bioinformatics.babraham.ac.uk/projects/fastqc
fitGCP	http://sourceforge.net/projects/fitgcp
MESER	http://biotech.jejunu.ac.kr/~abl/16s
Meta-QC-Chain	http://www.computationalbioenergy.org/qc-chain.html
PRINSEQ	http://sourceforge.net/projects/prinseq/files
StreamingTrim	https://github.com/GiBacci/StreamingTrim
TaxMan	http://www.ibi.vu.nl/programs/taxman
Metagenome assembly/mapping	
Bowtie	http://bowtie-bio.sourceforge.net/index.shtml
BWA	http://bio-bwa.sourceforge.net
Celera	http://www.cbcb.umd.edu/research/assembly.shtml#software
Euler	http://nbcr.sdsc.edu/euler/JAZZ
MetaSim	http://ab.inf.uni-tuebingen.de/software/metasim
TAG	http://omics.informatics.indiana.edu/TAG
Velvet	http://www.ebi.ac.uk/~zerbino/velvet
Gene calling	
FragGeneScan	http://omics.informatics.indiana.edu/FragGeneScan
GeneMark.hmm	http://exon.gatech.edu/GeneMark
MetaGeneAnnotator	http://metagene.nig.ac.jp
MetaGeneMark	http://exon.gatech.edu/Genemark/meta_gmhmmp.cgi
Orphelia	http://orphelia.gobics.de
Microbial diversity analysis	
MLST	http://www.mlst.net
EstimateS	http://viceroy.eeb.uconn.edu/EstimateS
Mothur	http://www.mothur.org
PHACCS	http://phaccs.sourceforge.net
QIIME	http://qiime.org/install/virtual_box.html
Binning/functional annotation/comparative analysis	
TETRA	http://www.megx.net/tetra/index.html
AmphoraNet	http://pitgroup.org/amphoranet
AmrPlusPlus	https://megares.meglab.org/amrplusplus/latest/html
Anvi'o	http://merenlab.org/software/anvio
FunGene Pipeline	http://fungene.cme.msu.edu/FunGenePipeline

Table 16.1 (continued)

Quality control	Link
Galaxy	http://main.g2.bx.psu.edu/u/aun1/w/metagenomic-analysis
IMG/M	http://img.jgi.doe.gov/cgi-bin/m/main.cgi
IM-Tornado	https://readthedocs.org/projects/imtornado
iVirus	http://ivirus.us
MEGAN	http://ab.inf.uni-tuebingen.de/software/megan
MePIC	https://mepic.nih.go.jp/cgi-bin/mepic/index.cgi
MetaGene	http://www.metagene.de
MetaLook	http://www.megx.net/metalook/index.php
MetaMine	http://www.megx.net/metamine
MetaStats	http://metastats.cbcb.umd.edu/detection.html
MEX (motif extraction)	http://adios.tau.ac.il/SPMatch
MG-RAST	http://metagenomics.anl.gov
MG-RAST	http://metagenomics.anl.gov
MOCAT	http://mocat.embl.de
Mothur	https://www.mothur.org
PanGEA	http://www.kofler.or.at/bioinformatics/PanGEA
Parallel-META	http://www.computationalbioenergy.org/parallel-meta.html
Phylopathia	http://cbcsrv.watson.ibm.com/phylopythia.html
Phyloseq	https://joey711.github.io/phyloseq/index.html
Phymm	http://www.cbcb.umd.edu/software/phymm
QIIME	http://qiime.org/1.6.0/index.html
RAMMCAP	http://weizhong-lab.ucsd.edu/rammcap/cgi-bin/rammcap.cgi
RDP pipeline	https://rdp.cme.msu.edu/
RTG Metagenomics	http://realtimegenomics.com/products/metagenomics-1.0
ShotgunFunctionalizeR	http://shotgun.math.chalmers.se/
SOrt-ITEMS	http://metagenomics.atc.tcs.com/binning/SOrt-ITEMS
SURPI	https://github.com/chiulab/surpi
UniFrac	http://bmf.colorado.edu/unifrac
Vegan	https://cran.r-project.org/web/packages/vegan/index.html
WebMGA	http://weizhong-lab.ucsd.edu/metagenomic-analysis
Metatranscriptome	
COMAN	http://sbb.hku.hk/COMAN/

(continued)

Table 16.1 (continued)

Quality control	Link
IDBA-MT	http://i.cs.hku.hk/~alse/hkubrg/projects/idba_mt/index.html
MetaModules	https://github.com/njsmith/metamodule
MetaTrans	http://www.metatrans.org
rRNAFilter	http://hulab.ucf.edu/research/projects/rRNAFilter/rRNAFilter.html
SAMSA	http://creativecommons.org/licenses/by/4.0

16.4 Concluding Remarks

Metatranscriptomics of rhizosphere microbiomes helps to re-constructing knowledge of dynamic nature of rhizobacterial communities with the eventual goal to explicate if and how these rhizobacteria confine pathogenic fungal infection of the plant host and development of metabolic pathways capable of dealing with an environmental pollutant. This strategy can be also expanded to study symbiont bacteria, archaea and new insights into the complex nature of the rhizosphere and facilitate further studies on plants-microbes interactions.

References

Akiyoshi DE, Morrison HG, Lei S et al (2009) Genomic survey of the non-cultivatable opportunistic human pathogen, *Enterocytozoon bieneusi*. PLoS Pathog 5:1–10

Arjun JK, Harikrishnan K (2011) Metagenomic analysis of bacterial diversity in the rice rhizosphere soil microbiome. Biotechnol Bioeng 1:361–367

Bakker PA, Doornbos RF, Zamioudis C, Berendsen RL, Pieterse CM (2013) Induced systemic resistance and the rhizosphere microbiome. Plant Pathol J 29(2):136

Bailly J, Fraissinet-Tachet L, Verner MC, Debaud JC, Lemaire M, Wésolowski-Louvel M, Marmeisse R (2007) Soil eukaryotic functional diversity, a metatranscriptomic approach. The ISME journal 1(7):632–642

Berendsen RL, Corn? Pieterse MJ, Bakker PAHM (2012) The rhizosphere microbiome and plant health. Trends in Plant Science 17(8):478–486

Buschmann D, Haberberger A, Kirchner B, Spornraft M, Riedmaier I, Schelling G, Pfaffl MW (2016) Toward reliable biomarker signatures in the age of liquid biopsies – how to standardize the small RNA-Seq workflow. Nucleic Acids Res 44(13):5995–6018

Chaparro JM, Badri DV, Vivanco JM (2013) Rhizosphere microbiome assemblage is affected by plant development. ISME J 8:790–803

Chapelle E, Mendes R, Bakker PH, Raaijmakers JM (2015) Fungal invasion of the rhizosphere microbiome. ISME J 10:1–4

Chen H, Wang L, Su CX, Gong GH, Wang P et al (2008) Isolation and characterization of lipopeptide antibiotics produced by *Bacillus subtilis*. Lett Appl Microbiol 47:180–186

Coloretti F, Carri S, Armaforte E, Chiavari C, Grazia L et al (2007) Antifungal activity of lactobacilli isolated from salami. FEMS Microbiol Lett 271:245–250

Cook RJ, Thomashow LS, Weller DM, Fujimoto D, Mazzola M, Bangera G, Kim DS (1995) Molecular mechanisms of defense by rhizobacteria against root disease. Proceedings of the National Academy of Sciences 92(10):4197–4201

Conrad R (2007) Microbial ecology of methanogens and methanotrophs. Adv Agron 96:1–63
Goodwin S, McPherson JD, McCombie WR (2016) Coming of age: ten years of next-generation sequencing technologies. Nat Rev Genet 17:333–351
Kim Y, Liesack W (2015) Differential assemblage of functional units in paddy soil microbiomes. PLoS One 10:1–20
Knief C, Delmotte N, Chaffron S et al (2012) Metaproteogenomic analysis of microbial communities in the phyllosphere and rhizosphere of rice. ISME J 6:1378–1390
Luo B, Gu W, Zhong J et al (2015) Revealing crosstalk of plant and fungi in the symbiotic roots of sewage-cleaning *Eichhornia crassipes* using direct de novo metatranscriptomic analysis. Sci Rep 5:15407
Mendes R, Garbeva P, Raaijmakers JM (2013) The rhizosphere microbiome: Significance of plant beneficial, plant pathogenic, and human pathogenic microorganisms. FEMS Microbiol Rev 37:634–663
Narberhaus F, Vogel J (2009) Regulatory RNAs in prokaryotes: here, there and everywhere. Mol Microbiol 74:261–269
Ottesen EA, Marin R, Preston CM, Young CR, Ryan JP, Scholin CA et al (2011) Metatranscriptomic analysis of autonomously collected and preserved marine bacterioplankton. ISME J 5:1881–1895
Parmar KM, Gaikwad SL, Dhakephalkar PK, Kothari R, Singh RP (2017) Intriguing interaction of bacteriophage-host association: an understanding in the era of omics. Front Microbiol 8:559
Prema P, Smila D, Palavesam A, Immanuel G (2008) Production and characterization of an antifungal compound (3-Phenyllactic Acid) produced by *Lactobacillus plantarum* Strain. Food Bioprocess Technol 3:379–386
Rondon MR, Goodman RM, Handelsman J (1999) The Earth's bounty: assessing and accessing soil microbial diversity. Trends in biotechnology 17:403–409
Sanguin H, Remenant B, Dechesne A et al (2006) Potential of a 16S rRNA-based taxonomic microarray for analyzing the rhizosphere effects of maize on Agrobacterium spp. and bacterial communities. Appl Environ Microbiol 72:4302–4312
Schink B (2000) Syntrophism among prokaryotes. The Prokaryotes 1:276–299
Shendure J, Ji H (2008) Next-generation DNA sequencing. Nat Biotechnol 26:1135–1145
Shi Y, Tyson GW, Eppley JM, DeLong EF (2010) Integrated metatranscriptomic and metagenomic analyses of stratified microbial assemblages in the open ocean. ISME J 5:999–1013
Simon C, Daniel R (2011) Metagenomic analyses: past and future trends. Applied and environmental microbiology 77(4):1153–1161
Stark M, Berger SA, Stamatakis A, von Mering C (2010) MLTreeMap – accurate Maximum Likelihood placement of environmental DNA sequences into taxonomic and functional reference phylogenies. BMC Genomics 11:461
Unno Y, Shinano T (2013) Metagenomic analysis of the rhizosphere soil microbiome with respect to phytic acid utilization. Microbes Environ 28:120–127
Urich T, Lanzen A, Qi J, Huson DH, Schleper C, Schuster SC (2008) Simultaneous assessment of soil microbial community structure and function through analysis of the meta- transcriptome. PLoS One 3:e2527
Yang J-H, Liu H-X, Zhu G-M, Pan Y-L, Guo J-H (2008) Diversity analysis of antagonists from rice-associated bacteria and their application in biocontrol of rice diseases. J Appl Microbiol 104:91–104

Part IV

Animal Microbiome

Canine and Feline Microbiomes

17

Corrin V. Wallis, Zoe V. Marshall-Jones, Oliver Deusch, and Kevin R. Hughes

Abstract

There is an increasing appreciation for the importance of the symbiotic relationship between microbes and their mammalian hosts in modulating companion animal health and nutrition. Indeed, the colonisation dynamics and influence of microorganisms inhabiting such body systems as the gastrointestinal tract (from mouth to anus) show many similarities between dogs and cats and their human counterparts. However, given the evolutionary divergence of these host species, as well as inherent differences in their diet and lifestyle, disparities in both the microbial communities and their impact on host health do exist. These differences are perhaps best exemplified in the oral cavity where microbial communities, host physiology and dietary influences result in differences in disease prevalence and phenotypic characteristics. Companion animals such as dogs and cats rarely experience dental caries which are commonly found in humans. However, periodontal disease, a destructive inflammation of the gums and periodontal soft tissues which is modulated by the microbiota, has an alarming prevalence within both the dog and cat population. Similarly, within the gastrointestinal tract, asymptomatic colonisation of dogs and cats by species known to be pathogenic within the human host is widely reported. This highlights the need to better understand these host microbial interactions in species-specific models.

Insights into the microbial species that inhabit these ecological niches have been achieved through targeted 16S rDNA gene sequencing of the microbiota. Meanwhile, shotgun sequencing approaches have generated further novel insights linking taxonomy with functionality and have started to help delineate links between such functionality and possible outcomes for host health and disease.

C.V. Wallis (✉) • Z.V. Marshall-Jones • O. Deusch • K.R. Hughes
The WALTHAM Centre for Pet Nutrition, Melton Mowbray, Leicestershire LE14 4RT, UK
e-mail: Corrin.wallis@effem.com; Zoe.marshall-jones@effem.com; Oliver.deusch@effem.com; Kevin.r.hughes@effem.com

R.P. Singh et al. (eds.), *Understanding Host-Microbiome Interactions - An Omics Approach*, DOI 10.1007/978-981-10-5050-3_17

In this chapter we review recent findings describing the microbiota and microbiomes of the dog and cat with particular focus on the oral and gastrointestinal microbial communities and their interplay with nutritional influences and the health of the host.

Keywords
Canine • Feline • Microbiota • Microbiome • Gastrointestinal • Oral • Periodontal • Skin • Health

17.1 Introduction

Domesticated dogs (*Canis familiaris*) and cats (*Felis catus*) are monogastric mammals that possess some physiological similarities with their human owners. As in humans, these mammals have co-evolved with a symbiotic microbiota which can rapidly respond to selective pressures due to their short generation time. This symbiotic relationship results in the holobiont, a term used to describe a host and its symbionts, and, at the genetic level, the hologenome, which describes the total gene pool of the holobiont. Characteristic differences do however exist between companion animals and humans, and these differences are likely to influence the colonisation dynamics and ecology of symbiotic microorganisms colonising these hosts. Humans, in the main, are omnivorous with diverse and complex dietary intakes and a healthy body temperature of 36.5–37.5 °C. Dogs are also considered omnivorous with a healthy body temperature of 38.3–39.2 °C, whereas cats are obligate carnivores with a healthy body temperature of 37.7–39.1 °C. Such differences may lead to unique evolutionary pressures that influence characteristics of the colonising microbiota, where disparate bacterial species can take advantage of the different niches existing in these hosts.

The role of the microbiota in the development and maintenance of health in mammals is increasingly appreciated. In humans, their involvement in outcompeting pathogenic organisms, in the digestion of nutrients and harvesting of energy for the intestinal epithelium, in the development of the immune system and maintenance of immune function has been described (Garrett et al. 2010; Clemente et al. 2012). The microbiota of the dog and cat are presumed to be involved in similar functions with similarities identified in the microbiome and increasing evidence of the interactions between diet and the gastrointestinal (GI) microbiota (Swanson et al. 2011; Turnbaugh et al. 2006, 2009). Although the mechanisms of action are largely unreported and are frequently linked to measures of diversity rather than specific taxa, the microbiota associated with health and with multiple disease conditions implicate the microbiome as being vital to physiological homeostasis (Kostic et al. 2015; Turnbaugh et al. 2006; Jackson et al. 2016; O'Mahony et al. 2015). The dog in particular represents an interesting model for the application of omics technologies towards the description of the microbiome and its innate variability as well as in uncovering its interaction with the host and factors influencing the functional and phylogenetic composition of the microbiome. Unlike the inherent variation in human genetics, behaviours and nutritional intakes, individual canine breeds often possess a relatively narrow genetic base, have a comparatively unvaried nutritional intake and are often subject to a less variable daily regime than their human counterparts. The relatively narrow genetic base of distinct canine breeds can, in some instances, be used to assess genetic polymorphisms underlying breed-associated predispositions and in some cases provide insights into host-microbiome interactions. The lower prevalence of purebred cats yields greater genetic variation in the feline population, and variation in nutritional intake may be greater, or less easily controlled, in the cat due to varied hunting behaviours. Despite these potential confounding influences, the feline microbiome is of fundamental interest due to cat's obligatory carnivorous lifestyle.

In the following review, recent findings detailing the microbiome of the dog and cat are described with particular focus on the oral and GI microbial communities and their interaction with the nutrition and health of the host.

17.2 The Canine and Feline Gastrointestinal Microbiome

The GI tract has long been regarded a rich and highly complex ecosystem, with changes in nutrients, pH, bile salts and other factors that control the growth of microorganisms, linked to differences in the microbiome along its length. Furthermore the proximity to the gut-associated lymphoid tissue (GALT) lends itself to a dynamic and symbiotic relationship in which crosstalk between the microbes and the host immune system occurs (Eckburg et al. 2005; Ley et al. 2008; Spor et al. 2011). The intestinal microbiota is considered vital in the development of the immune system and in providing competition for incoming pathogens (Smith et al. 2007). More recently, it has also been found to influence the development of the microstructure in the intestinal epithelium (Al-Asmakh and Zadjali 2015) and to have a role in homeostasis towards maintaining the health of the host (Sommer and Backhed 2013).

Traditional microbial culture-based approaches are variously described to be capable of identifying only 10–50% of the total population (Zoetendal et al. 2004; Sommer and Backhed 2013). This is particularly significant for the microbial populations of the canine and feline GI tract, which, although often closely related to their human counterparts at the genus level, the species represented frequently differ and possess inherently distinct growth characteristics. Thus, while traditional culture techniques are limited in their ability to support identification of bacterial species in humans, the absence of specific culture conditions for feline and canine microbial species would be expected to further impact proportions of bacterial species able to be detected using these techniques. Indeed, early studies attempting to review the microbial ecology of the canine gut surmised that traditional culture methods failed to reflect the bacterial diversity present and that agar selectivity was poor with the 16S rDNA gene sequences of many of the isolates not correlating with known species (Greetham et al. 2002).

Prior to the advancements in high-throughput sequencing (HTS) methodologies, the microbiology of the GI tract in dogs and cats was to a large extent restricted to the study of the microbial content in conditions such as antibiotic-responsive diarrhoea or small intestinal dysbiosis (previously described as small intestinal bacterial overgrowth) (German et al. 2003) or of specific pathogens causing diarrhoeal diseases. Pathogens such as *Campylobacter*, *Salmonella*, *Escherichia coli*, *Clostridium difficile and Clostridium perfringens* were of particular interest due to the perceived potential for zoonotic transfer. Such studies revealed higher bacterial numbers reside in the small intestine of healthy cats than in canines and humans. Bacterial counts of 10^5–10^9 cfu/ml were observed in the small intestinal contents of healthy cats, and the bacterial load was not indicative of small intestinal bacterial overgrowth or antibiotic-responsive diarrhoea (Johnston et al. 2001; Johnston and Batt 1993). Cats also possessed higher proportions of obligate anaerobes compared with humans and dogs. Meanwhile, similar to humans, the colon was identified as the major site of microbial activity with estimations of the total bacterial numbers in the large intestine of dogs and cats ranging between 10^9 and 10^{11} cfu/g with similar levels of total counts found in faeces (Greetham et al. 2002). The major cultivable groups observed differed between research studies, being described as Bacteroides, Clostridium, Lactobacillus, Bifidobacterium and Enterobacteriaceae by some (Benno et al. 1992; Mentula et al. 2005), with others finding no evidence of groups such as Bifidobacterium (Greetham et al. 2002). These differences are likely due to the conditions employed for microbial culture.

With the development of molecular techniques, substantial headway has been made in uncovering the true complexity of the GI microbiome and in identifying changes during health and GI disease. Deep-sequencing techniques, facilitating the analysis of the total population of 16S rDNA molecules, or the total load of genes present in a sample have become mainstay research tools.

17.2.1 Gastrointestinal Microbiota in Health

At the phylum level, taxonomic classifications of bacteria by HTS have described the canine and feline GI microbiota to resemble that of other monogastric mammals (Swanson et al. 2011). In the human host, seven phyla make up the majority of the GI microbiota: the Firmicutes, Bacteroidetes, Proteobacteria, Fusobacteria, Verrucomicrobia, Cyanobacteria and Actinobacteria, with the Firmicutes and Bacteroidetes comprising over 90% of the gut microbiota (Backhed et al. 2005; Eckburg et al. 2005; Tap et al. 2009). Studies in healthy animals describe around 10–12 phyla present in the GI tract and faeces of dogs and cats (summarised in Tables 17.1 and 17.2). The major portion of the microbiota includes organisms from the Firmicutes, Bacteroidetes and Fusobacteria, which together constitute around 95% of the total bacterial population with Proteobacteria and Actinobacteria comprising approximately 1–5% of the microbiota (Handl et al. 2011; Middelbos et al. 2010; Minamoto et al. 2012). Although numerically lower in their representation, the Proteobacteria are frequently notable due to their association with pathogenicity

Table 17.1 Summary of publications that have used high-throughput sequencing technologies to characterise the canine gastrointestinal and faecal microbiota

Reference	Beloshapka et al. (2013)	Chaban et al. (2012)	Foster et al. (2013)	Garcia-Mazcorro et al. (2012)	Guard et al. (2015)	Hand et al. (2013)	Handl et al. (2011)	Middelbos et al. (2010)	Minamoto et al. (2015)	Panasevich et al. (2015)	Suchodolski et al. (2012a)	Suchodolski et al. (2012b)	Swanson et al. (2011)
Number and breed of animals	Six healthy female adult beagles	Various and mixed breed dogs Seven healthy Nine diarrhoeic	12 healthy dogs and 7 dogs with acute diarrhoea Various and mixed breeds	Eight healthy 9-month-old dogs	13 healthy dogs and 13 dogs with acute diarrhoea	11 healthy adult Miniature Schnauzer dogs	12 healthy pet dogs and 12 pet cats	Six healthy adult dogs	Ten healthy dogs and 12 dogs with IBD before and after 3 weeks of therapy	Ten healthy adult female dogs	Six healthy dogs, seven dogs with moderate IBD and seven dogs with severe IBD	32 healthy dogs 12 dogs with acute non-haemorrhagic diarrhoea 13 dogs with acute haemorrhagic diarrhoea 9 dogs with active IBD 10 dogs with therapeutically controlled idiopathic IBD	Six healthy adult dogs
Type of samples	Faecal	Faecal	Faecal	Endoscopic biopsies from the stomach and duodenum	Faecal, serum and urine	Faecal	Faecal	Faecal	Faecal	Faecal	Small intestinal mucosal biopsies	Faecal	Faecal
Target gene	16S rDNA	cpn60	18S rDNA	16S rDNA gene	16S rDNA gene	16S rDNA	16SrDNA and 18S rDNA genes	16S rDNA	16S rDNA genes	16S rDNA gene	16S rDNA gene	16S rDNA gene	Shotgun
Sequencing platform	454 pyrose-quencing	454 pyrose-quencing	FLX Titanium pyrose-quencing	454 pyrosequencing	454 pyrose-quencing	454 FLX pyrose-quencing	FLX Titanium pyrosequencing	454 pyrose-quencing	454 pyrose-quencing	454 pyrose-quencing	454 pyrose-quencing	454 pyrose-quencing	454 pyrosequencing
Number of sequence reads	5016 sequencing reads/sample	6216 sequencing reads/ sample	A total of 57,179 sequences 2800 sequencing reads/sample	A total of 142,026 (stomach); 133,449 (duodenum) sequences ~4000 sequencing reads/sample	A total of 297,315 sequences, 9013 sequencing reads/sample	17,899 sequencing reads/sample	A total of 120,406 sequences, bacterial 5017 sequencing reads/sample Fungal 5359 sequencing reads	A total of 77,771 sequencing reads	A total of 297,619 sequences An average of 8753 sequencing reads/sample	A total of 1,040,107 sequences An average of 20,802 sequencing reads/sample	A total of 73,998 sequences An average of 3700 sequencing reads/sample	A total of 189,138 sequencing reads/sample	A total of 1,008,341 sequences: 503,280 control and 505,061 test diet

(continued)

Table 17.1 (continued)

Reference	Beloshapka et al. (2013)	Chaban et al. (2012)	Foster et al. (2013)	Garcia-Mazcorro et al. (2012)	Guard et al. (2015)	Hand et al. (2013)	Handl et al. (2011)	Middelbos et al. (2010)	Minamoto et al. (2015)	Panasevich et al. (2015)	Suchodolski et al. (2012a)	Suchodolski et al. (2012b)	Swanson et al. (2011)
Number of OTUs	1176 (at 97%)	834	Not reported	Stomach—median 36 OTUs Duodenum—median 173 OTUs	268	500–1500	Dogs 85 and cats 113	129 OTU	138	Not reported	Control group 112 (range 70–156); IBD Group 84 (range 54–165)	242	Not reported
Number of phyla	5 (predominant)	4	5	Stomach 3 (+) Duodenum 7	4	11	Not reported	3	11	6	9	5	17 (5 predominant)
Number of genera/species	37 predominant species	Not reported	219	Not reported	42 genera	500–1000 species (OTU 97%)	Cats bacterial genera: 85 Cats fungal genera: 17 Dogs bacterial: 113 Dogs fungal genera: 33	~90–150 genera	138 genera	Min 24 genera	Min 22 genera	Not reported	Not reported
Predominant phyla (% of sequence reads):	Fusobacteria 45.26 Firmicutes 32.61 Bacteroidetes 14.78 Proteobacteria 5.56 Actinobacteria 1.72	Actinobacteria 7.45 Bacteriodetes 29.0 Firmicutes 26.2 Proteobacteria 29.55	Ascomycota 97.9 Basidiomycota 1 Chytridiomycota not reported and Neocallimastigomycota not reported Microsporidia not reported	Stomach Proteobacteria > 90% Firmicutes data not reported Fusobacteria data not reported Duodenum (data not reported) Firmicutes < Proteobacteria> Bacteroidetes	Bacteroidetes 32.6(12.9–48.4) Firmicutes 60.9(41.3–86.6) Fusobacteria 4.5(0.1–12.7) Proteobacteria 0.1(0.0–0.3) unclass. Bacteria 0.2 (0.0–0.8)	Fusobacteria 39.17 Bacteroidetes 33.36 Firmicutes 15.81 Proteobacteria 11.31 Actinobacteria 0.33 Tenericutes 0.01 Acidobacteria 0 (detected) Deferribacteres 0 (detected) Spirochaetes 0 (detected) Chloroflexi 0 (detected) Cyanobacteria 0 (detected)	Dogs bacterial: Firmicutes 95.36 Bacteroidetes 2.25 Actinobacteria 1.81 Fusobacteria 0.3 Dogs fungal: Ascomycota 99.62 Basidiomycota 10 sequences Glomeromycota 3 sequences Zygomycota 1 sequence	Fusobacteria (23–40) Firmicutes (14–28) and Bacteroidetes (31–34)	Firmicutes (Not reported) Bacteroidetes (Not reported) Actinobacteria (Not reported) Fusobacteria (Not reported) Proteobacteria (Not reported)	Firmicutes (22.63–94.21) Fusobacteria (0.53–58.18) Tenericutes (1.78–40.96) Actinobacteria (0–3.47) Bacteroidetes (0–9.06) Proteobacteria (0–12.53)	Actinobacteria 0.79 Bacteroidetes 28.80 Firmicutes 15.29 Fusobacteria 14.56 Proteobacteria 31.58	Firmicutes 96.6 Proteobacteria 0.30 Bacteroidetes detected Actinobacteria 1.8 Fusobacteria 0.8	Bacteroidetes/Chlorobi group ~35 Firmicutes ~35 Proteobacteria 13–15 and Fusobacteria 7–8%

Table 17.2 Summary of publications that have used high-throughput sequencing technologies to characterise the feline gastrointestinal and faecal microbiota

Reference	Bermingham et al. (2013)	Barry et al. (2012)	Deusch et al. (2014)	Deusch et al. (2015)	Handl et al. (2011)	Hooda et al. (2013)	Ramadan et al. (2014)	Suchodolski et al. (2015)	Tun et al. (2012)
Number and breed of animals	16 mixed sex, neutered short-haired domestic cats	Four male healthy adult cats	12 healthy kittens	30 kittens	12 healthy pet cats (and 12 healthy pet dogs)	14 healthy kittens	15 diarrheic adult domestic short-haired cats	21 healthy cats, 19 cats with acute diarrhoea and 29 cats with chronic diarrhoea	Four domestic short-haired cats and one domestic long-haired cat
Type of samples	Faecal	Faecal	Faecal	Faecal	Faecal	Faecal	Faecal	Faecal	Faecal
Target gene	16S rDNA	16S rDNA	Shotgun	Shotgun	16S rDNA and 18SrDNA genes	16S rDNA gene	16S rDNA gene	16S rDNA gene	Shotgun
Sequencing platform	454 titanium pyrosequencing	454 titanium pyrosequencing	Illumina	Illumina	FLX Titanium pyrosequencing	454 pyrosequencing	454 pyrosequencing	Ion torrent PGM	Junior Titanium pyrosequencing
Number of sequence reads	4616 reads/ sample	4,192,192 (total)	95.9 million/ sample	4.8 billion, 55 million/ sample	120,406 pyrosequencing reads for bacteria (mean 5017) 5359 sequences for fungi	A total of 384,588 sequences (average 9374/ sample)	A total of 556,366 reads An average of 11,591 reads/ sample	A total of 2,748,939 and an average of 39,752 reads/ sample	A total of 152,494 pyrosequencing reads
Number of OTUs	3927	Not reported	1114 species	1113 species	113 OTUs in cats	Not reported	Not reported	3900	Not reported
Number of phyla	5	4	32	33	3	5	8	6	17 (bacterial)

(continued)

Table 17.2 (continued)

Reference	Bermingham et al. (2013)	Barry et al. (2012)	Deusch et al. (2014)	Deusch et al. (2015)	Handl et al. (2011)	Hooda et al. (2013)	Ramadan et al. (2014)	Suchodolski et al. (2015)	Tun et al. (2012)
Number of genera/ species	46	Not reported	578	605	Bacterial genera: 85 Fungal genera: 17	50	96 genera 146 species	Not reported	Not reported
Predominant phyla (% of sequence reads)	Fusobacteria 11.7 Firmicutes 65.6 Bacteriodetes 12.3 Proteobacteria 0.75 Actinobacteria 8.3	Firmicutes 36.3 Bacteroidetes/ Chlorobi 36.1 Proteobacteria 12.4 Actinobacteria 7.7%	Firmicutes 53.91 Bacteroidetes 28.27 Actino-bacteria 10.32 Proteo-bacteria 3.76 Fusobacteria 1.19 Spirochaetes 1.06	Data (at 18 weeks of age) Actino-bacteria 23.5 Bacteroi-detes 10.2 Firmicutes 52.8 Proteo-bacteria 9.3 Spiro-chaetes 0.8	Cats bacterial: Firmicutes 92.10 Actinobacteria 7.31 Fusobacteria 0.04 Cats fungal: Ascomycota 100.00	Firmicutes 80.10 Actinobacteria 18.21 Fusobacteria 0.10 Bacteroidetes 0.55 Proteobacteria 1.06	Firmicutes 34.34 Bacteroidetes 30.05 Fusobacteria 18.81 Proteobacteria 7.66 Tenericutes 6.56 Actinobacteria 2.57 Cyanobacteria 0.0034 TM7 0.0003	Actinobacteria 0.11 Bacteroidetes 33.34 Firmicutes 50.58 Fusobacteria 1.15 Proteobacteria 4.49 Actinobacteria 0.09	Bacteroidetes/Chlorobi group 67.54% Firmicutes 12.98 Bacteroidetes 8.68 Proteobacteria 5.85 Actinobacteria 1.16 Fusobacteria 0.68 Synergistetes 0.58 Cyanobacteria 0.51 Thermotogae 0.48 Spirochaetes 0.41 Chlamydiae/ Verrucomicrobia group 0.34 Fibrobacteres/ Acidobacteria group 0.27 Deinococcus-Thermus 0.27 Chloroflexi 0.14 Planctomycetes 0.05 Aquificae 0.02 Chlorobi 0.02

and infection. Represented within the microbiota but still less abundant are organisms from the phyla Spirochaetes, Tenericutes, Verrucomicrobia, TM7, Cyanobacteria, Planctomycetes and Chloroflexi, which make up less than 1% of the 16S rDNA sequences identified. Although researchers generally agree on the dominant groups in the canine GI tract, the observed relative abundance of the bacterial groups differ, sometimes substantially, between studies. For example, despite an agreement that the Firmicutes, Bacteroides and Fusobacteria represent the dominant bacterial phyla in the canine and feline colon and faeces (Desai et al. 2009; Ritchie et al. 2008, 2010; Suchodolski 2011; Swanson et al. 2011), the Firmicutes phylum was found to represent between 25% and 95% of the 16S rDNA sequences identified in two 454-pyrosequencing studies on the canine faecal microbiota (Handl et al. 2011; Middelbos et al. 2010). Thus, despite overcoming the differences inherent in bacterial culture techniques, biases may also be introduced in molecular analyses of the microbiota. DNA extraction techniques are known to produce differences in DNA yield dependent on the bacterial species in question, and this is particularly true when comparing the Gram-positive and Gram-negative species in complex microbial ecologies (Zoetendal et al. 2001b). Furthermore, despite its prevalence in methods to define taxonomic groups, the 16S ribosomal RNA (rRNA) gene region analysed, based on primer selection around the genes' variable and hypervariable regions, are key influencers of the taxonomic units uncovered (Jumpstart Consortium Human Microbiome Project Data Generation Working 2012). Following the publication of protocols and data by the National Institute for Health-funded Human Microbiome Project Consortium (HMP) in 2012, most researchers appear to have aligned in targeting the most descriptive variable regions three to five (V3–5) of the 16S rDNA gene (Human Microbiome Project 2012). In particular, the analysis of lactic acid bacteria including the *Bifidobacterium* and *Lactobacillus* spp. belonging to the phylum Actinobacteria is often confounded, with these genera frequently under-represented within complex populations (Farris and Olson 2007). However, the use of *Lactobacillus* and *Bifidobacterium* specific primers or use of the chaperonin 60 gene for taxonomic analysis has been found to enhance the representation of both genera within feline faeces and of the Epsilonproteobacteria in canine faeces (Desai et al. 2009; Ritchie et al. 2010; Chaban et al. 2012). These differentials introduced by the inherent disparities within the experimental design mean that comparison across research studies and particularly across research groups can be confounded.

17.2.2 The Microbiota Throughout the Gastrointestinal Compartments

As an easily available and pet-friendly sample to collect, faeces is frequently used as the sample of choice for analysis of the GI microbiota; hence much of the available data has been generated from faeces as a substitute for the in vivo microbial community. However, it should be borne in mind that the varied conditions and niches along the GI tract are likely, for some regions of the GI tract, to make this a poor surrogate marker. There is also an increasing appreciation for the variation in ecology within the luminal and mucosal-associated microbiota (Suchodolski et al. 2008b). Several studies have considered the microbiota present across the various niches within the GI tract of companion animals. The stomach is considered to be sparsely colonised with microorganisms due to the gastric microbicidal barrier, which includes innate defences such as low pH, migrating motor complex and the entero-salivary circulation of nitrate. Though few studies have been published, multiple species of *Helicobacter* including *Helicobacter baculiformis*, *H. salomonis*, *H. bizzozeronii*, *H. felis*, *H. cynogastricus* and 'Candidatus *H. heilmannii*' have been detected within feline or canine gastric mucosal samples (Baele et al. 2008) and are apparently commensal within their host (Norris et al. 1999; Washabau and Day 2012). A 454-pyrosequencing study on endoscopic stomach biopsies demonstrated that the microbiota of the canine stomach comprises over 98% *Helicobacter* sp. although *Lactobacillus* species were also detected (Garcia-Mazcorro et al. 2012). In the study a median of 36 operational taxonomic units (OTUs) representative of species-level classifications (97% sequence identity) were detected in the stomach per dog, while approximately 190 OTUs were detected in duodenal biopsies. Unsurprisingly the authors summarised that a distinctive microbiota was present in each of the evaluated segments of the GI tract (Garcia-Mazcorro et al. 2012). Several studies describing the diversity of the small and large intestinal microbiota of the canine and feline host have been conducted (Ritchie et al. 2008; Suchodolski et al. 2008a). The analysis of intestinal contents from the duodenum, jejunum, ileum, and colon of six healthy dogs was conducted with spatial differences detected in the microbiota between regions. Overall, although the depth of HTS techniques was not reached in this cloning and Sanger sequencing study, bacterial sequence types detected were from four bacterial phyla, the Firmicutes, Fusobacteria, Bacteroidetes and Proteobacteria. Indices of bacterial diversity increased along the intestinal tract from the duodenum to the colon. In the duodenum Clostridiales predominated, representing 40% of clones, while similarly this group represented the major bacterial group in the jejunum (39%), although not predominating in the ileum (25%) and colon (26%), the Clostridiales remained abundant. Organisms from *Clostridium cluster XI* dominated the proximal small intestinal microbiota, while the colon was dominated by organisms from the *Clostridium cluster XIVa*. Bacterial orders Fusobacteriales and Bacteroidales were the highest in abundance in ileal (33%) and colonic (30%) intestinal contents, while Enterobacteriales were higher in abundance in the small intestine compared to the colon. The Lactobacillales were ubiquitous throughout the intestine (Suchodolski et al. 2008a). A similar molecular cloning and sequencing approach was used to detect the bacterial content of the feline stomach, duodenum, jejunum, ileum, and colon of five healthy cats (Ritchie et al. 2008). Across the feline GI tract five phyla were identified, being, in order of abundance, Firmicutes (68%), Proteobacteria (14%), Bacteroidetes (10%), Fusobacteria (5%) and Actinobacteria (4%). The majority of sequences were assigned to the Clostridiales order (54%), with the Lactobacillales, Bacteroidales, Campylobacterales and Fusobacteriales being detected in lower abundances (14%, 11%, 10% and 6%, respectively). Clostridiales were the most abundant organisms within the Firmicutes phylum and crossed six clusters of clostridia. *Clostridium* clusters I (58%) and XIVa (27%) predominated, with *Clostridium* cluster I phylotypes identified across all intestinal sites sampled and peaking in relative abundance in the colon. This differs from the human intestinal microbiota where *Clostridium* cluster I are observed in low levels (Delgado et al. 2006; Wang et al. 2003). *Clostridium* cluster XIVa had the greatest species richness and was isolated predominantly in the colon, while sequences from *Clostridium* cluster IV were identified exclusively in samples from the colon in cats. The Lactobacillales represented the second major group of Firmicutes, with highest relative abundance in the jejunum and colon. Bacteroidales and Fusobacteriales were also detected in the jejunum but were observed in highest abundance in the ileum and colon. The Proteobacteria phylum was most commonly detected in the small intestine rather than the large intestine, while sequences assigned to the phylum Actinobacteria were isolated in greater frequency in the ileum and colon. Similar to the findings of a study of the faecal microbiome of captive cheetahs (Becker et al. 2014), *Bifidobacterium* spp. were completely absent; however in this study of domestic felids, the absence is likely to have been due to bias in the 16S rDNA gene libraries since studies have shown *Bifidobacterium* spp. to be present in feline faeces using group-specific primers and fluorescent in situ hybridisation (FISH) analysis with species-specific probes (Inness et al. 2007; Ritchie et al. 2010; Handl et al. 2011). Unlike in dogs, comparison of the sequences identified by UniFrac analysis (a method developed to determine the level of difference/similarity in the phylogeny of complex microbial communities) revealed that in cats, the samples were more similar by individual as opposed to by the intestinal site sampled. This may reflect differences in the microbiota observed within cats by culture-based methods, whereby the microbial content of the small intestine is greater in abundance than that observed in the canine small intestine and that of humans (Johnston et al. 1993; Papasouliotis et al. 1998).

17.2.3 The Faecal Microbiota

Studies of the faecal microflora in dogs reveal this rich community may harbour as many as 500–1500 OTUs (at 97% sequence identity), which, under the sequence length, obtained approximates to species-level identifications (Hand et al. 2013), although other studies suggest a more conservative estimate of 85 OTUs at the genus level in canine and 113 OTUs in feline faeces (Handl et al. 2011). Again, technical differences in the methods used for these analyses are likely to drive some inherent variation in the results obtained. An apparent underestimation of the OTUs present was suggested by Handl et al., while Hand et al. utilised Roche-454-pyrosequencing methodologies, previously described to introduce errors through homopolymeric runs, which are considered to lead to an overestimate of predicted diversity (Quince et al. 2009). Despite this the authors of both studies found substantial inter-animal variation was evident at both the phylum and genus levels and particularly in the abundance of the major genera represented (Fig. 17.1). In the

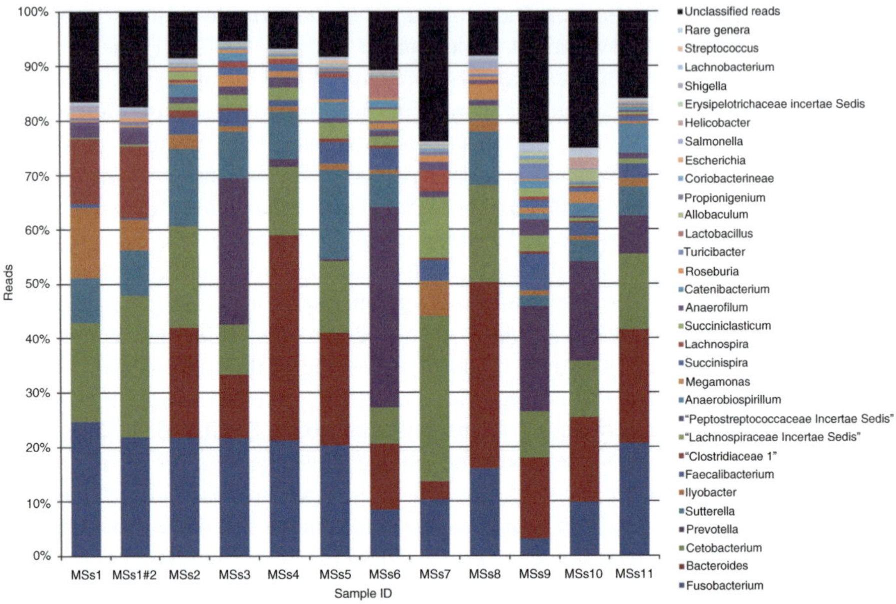

Fig. 17.1 Ribosomal database project (RDP) classification of reads to the genus level. Sample number is shown on the *x* axis and percentage reads classified on the *y* axis. Genera with fewer than 100 reads in all samples were pooled and are shown as 'rare genera'. MSs1#2 denotes replicated read data on the same DNA sample for dog MSs1. Taken from Hand et al. (2013)

majority of the cohort, the five major genera (in terms of abundance) accounted for around 60–80% of the total population (Hand et al. 2013). However, the microbial taxa represented were highly variable between animals even within these genera. Several Fusobacteria (*Fusobacterium*, *Cetobacterium* and *Ilyobacter* represented abundant groups in all animals; however no other genera were highly abundant across the cohort of dogs. This individual variation in the taxa detected between animals suggests that a 'core microbiota' of universally present taxa does not exist in the canine host and is in agreement with several studies in cats and in humans (Desai et al. 2009; Turnbaugh et al. 2009; Biagi et al. 2010; Claesson et al. 2011). Indeed, in humans, the distinct characteristics of the intestinal microbiota have been described as a unique characteristic as the fingerprint (Qin et al. 2010; Nielsen et al. 2014) contradicting earlier suggestions that a core microbiota may exist (Tap et al. 2009). Despite this inter-animal variation in the microbial phylogenies detected, one study describes the microbiota of genetically related dogs to be more similar than that of unrelated individuals. Hand et al. demonstrated clustering of faecal samples from littermates in principal component analyses of 454-pyrosequencing data, which was apparently driven by one set of littermates having higher abundances of a group of genera including *Lactobacillus*, *Streptococcus*, *Faecalibacterium*, *Bacteroides* and *Lachnospira* and representatives from the families Lachnospiraceae, Coriobacterineae and Erysipelotrichaceae, while another set of littermates possessed the genera Roseburia, Lachnobacterium, Propionigenium,

Anaerofilum and Cetobacterium and organisms from the family Peptostreptococcaceae at higher abundances (Hand et al. 2013). These findings are replicated in several human gut microbiota studies, where genetically related humans have been found to exhibit more similar gut microbial communities (Zoetendal et al. 2001a; Ley et al. 2005).

17.2.4 Linking Phylogeny to Function in the Gastrointestinal Microbiota

A small number of studies have attempted to link phylogeny to function in the canine and feline GI microbiota. However, because such metagenome studies are expensive, in order to enhance the study of the microbiota and function, predictive computational tools such as PICRUSt (Phylogenetic Investigation of Communities by Reconstruction of Unobserved States) have been developed. This in silico predictive tool aims to generate potential metagenomes and understand the functional capacity of a microbial community by reconstruction from the 16S rDNA gene sequences detected (Langille et al. 2013). However, it is strongly dependent on the availability of genomes of closely related bacteria. A metagenomic study of canine faeces described the Bacteroidetes/Chlorobi group and Firmicutes to represent around 35% of sequences, with the Proteobacteria (15%), Fusobacteria (8%) and Actinobacteria (1%) more minor elements of the microbiome (Swanson et al. 2011). In the study the metabolic capacity of the canine gut microbiota was assessed in dogs fed with a base diet and a diet supplemented with sugar beet pulp, which is commonly used as a fibre and prebiotic source in animal feeds. Despite being the first study of the microbiome in dogs, approximately half of all sequences were classified into metabolic functional categories, supporting the theory of a core metagenome with conservation of functional processes, despite divergent microbial ecologies. These metabolic pathways were linked to protein, carbohydrate and DNA metabolism; vitamins, cofactors, prosthetic groups and pigments; amino acids and derivatives; and cell wall and capsules as well as bacterial virulence. The distribution of the metabolic pathways was not affected by diet and was phylogenetically and functionally similar to murine and human metagenomes (Swanson et al. 2011; Turnbaugh et al. 2009). Since the metabolite profiles and functional metagenomes were found to be similar despite disparate microbial phylogeny, it was hypothesised that a core functionality with redundancy in the actual species exists where functional roles may be fulfilled by a number of alternative species within the same niche (Dethlefsen et al. 2008; Turnbaugh et al. 2009). Alternative or additional hypotheses suggest the existence of a more generalised symbiotic relationship between the microbiota and its host with microbe-elicited reactive oxygen species (ROS) considered responsible for the influence of the gut microbiota on intestinal epithelial physiology and function (Neish 2013). In these interactions commensal bacteria stimulate the production of ROS within enterocytes, thereafter signalling is mediated by the rapid and transient oxidation of thiol groups on sensor regulatory proteins. Indeed ROS-dependent mechanisms are involved in stimulating cellular proliferation, motility and modulating innate immunity through their role as signalling molecules in diverse transduction pathways (Neish and Jones 2014).

Analysis of the feline faecal metagenome using a 454-pyrosequencing approach demonstrated Firmicutes to be the most highly abundant phylum (approximately 40%) with the Bacteroidetes and Proteobacteria representing approximately 30% and 12%, respectively (Barry et al. 2012). Functional metabolome sequences determined using the KEGG pathway database (http://www.genome.jp/kegg/) were linked to carbohydrates, clustering-based subsystems, protein metabolism and amino acids and derivatives. Unlike in the canine study, an effect of diet was detected within the feline faecal metagenome, with in-depth analysis of carbohydrate-active enzymes demonstrating changes in glycoside hydrolases, glycosyl transferases and carbohydrate-binding molecules during supplementation with pectin and fructooligosaccharides, ingredients frequently used as prebiotics in pet food products (Barry et al. 2012). A second study of the faecal metagenome in cats described the Bacteroides/Chlorobi group to be the most abundant bacterial phylum representing ~68% of the total diversity classified, with the Firmicutes (~13%) and Proteobacteria (~6%) representing prevalent but numerically less dominant taxa. Archaea, fungi and viruses were detected as only minor communities (Tun et al. 2012). Similar to the metagenome from dogs, carbohydrate and protein metabolism; cell wall and capsule; cofactors, vitamins, prosthetic groups and pigments; DNA and RNA metabolism and amino acids and derivatives; as well as bacterial virulence were the predominant metagenomic elements detected (Swanson et al. 2011; Tun et al. 2012). The analysis identified the presence of 41 species from 12 genera of putative zoonotic pathogens within the feline faecal microbiome. These mostly represented primarily food-borne opportunistic pathogens detected at low levels within the total microbial community. Multiple antimicrobial resistance genes were detected with multidrug resistance efflux pumps, fluoroquinolone resistance genes and beta-lactamase sequences all detected. A clustering analysis was employed for the comparison of data from nine GI metagenomes detected in five monogastric hosts including that from dogs, humans, mice, cats and chickens. Amongst these the feline metagenome was most closely related to that from chickens both in the phylogenetic and functional metabolic pathways detected (>80%) (Tun et al. 2012).

17.2.5 Eukaryotic, Viral and Archaeal Elements Within the Gastrointestinal Tract

Much of the research into the microbiota of dogs and cats has focussed on the bacterial and archaeal content as indicated by the diversity and abundance of 16S rDNA gene variants within a sample. However, to fully appreciate the influence of the microbiome on the health of the host, the microscopic eukaryotes and viruses should also be considered. A HTS study of the 18S rDNA in faecal samples from dogs with acute diarrhoea compared to healthy dogs described fungal taxa within all 19 individuals (Handl et al. 2011; Foster et al. 2013). Individual differences were observed in the taxa present. However, multiple taxa were detected in all animals (median 28 genera). Overall five phyla of fungi were detected with around 98% of sequences present representing species from the large and diverse phylum Ascomycota and the Basidiomycota representing up to 1% of the total sequences. The phyla, Chytridiomycota, Neocallimastigomycota and Microsporidia, were detected in less than half of the study cohort. No significant differences were detected in the fungal taxa present in healthy dogs and those with acute diarrhoea (Foster et al. 2013). In another study of fungal 18S rDNA sequences in 24 healthy dogs and cats, the Ascomycota were again found to represent the dominant phylum with 100% of sequences in cats and over 99% of 18S rDNA sequences in dogs assigned to this group (Handl et al. 2011). The most predominant fungal class in cats and dogs was Saccharomycetes, with *Candida castelli*, the most predominant species in dogs, and a novel species from the genus *Saccharomyces* most abundant in cats. Although the total fungal load was not assessed, a metagenome study by Swanson et al. (2011) estimated that sequences of fungal origin represented approximately 0.01% of all sequences obtained. This is likely an under-representation of the actual fungal content, since only three phylotypes were detected in the cohort of six dogs, all belonging to the Dikarya subkingdom of which the Ascomycota represents a major division. However it is possible, perhaps even likely, that the fungal portion of the microbiota is a minor element within the total microbial load. Previous studies have detected differences in the mycobiota of the lumen and the mucosal border with the latter enriched for fungal DNA (Suchodolski et al. 2008b). Therefore it is likely that the sampling methods impose differences on the composition detected. Deeper sequencing efforts and the assessment of different micro-ecosystems within the gut may therefore enhance the diversity and possibly the relative abundance of the mycobiota uncovered in the GI microbiome of dogs and cats. To fully understand the fungal content of the microbiota, studies should also consider the DNA extraction techniques used, since even within the mycobiota, differences exist in the efficiency of DNA extraction techniques with species (Fredricks et al. 2005). Viruses apparently represent a minor element in the canine microbiome. Virus detection and quantification is challenging to investigate by means other than metagenome studies due to the range of genetic material, diversity of the genes carried and a lack of conserved genetic elements. However, a metagenomic study of the canine intestinal microbiome revealed only 0.4% of sequences representing the faecal viral load with over 99% of these representing genetic material from bacteriophage (Swanson et al. 2011). The Archaea are obligatory anaerobic microorganisms phylogenetically distinct from both the bacterial and eukaryotic subdivisions. Although their role in health is unclear, archaea have been detected in the GI tract of monogastric mammals and of ruminants (Eckburg et al. 2005) and are metabolically important in their effect on GI function. Methanobacteria are the most abundant phylotype within the human GI tract and are involved in the metabolism of fermentation end products and production of methane (Rajilic-Stojanovic et al. 2007). Similar to humans, methanogens represent the most predominant and evolutionarily diverse group of archaea detected in the faeces from both dogs and cats (Swanson et al. 2011; Tun et al. 2012). Their role in hydrogen consumption is vital to GI homeostasis, removing this metabolic end product and hence supporting the maintenance of the fermentative environment and controlling the generation of hydrogen sulphide. Fermentation within the colonic microbiota is considered key to host health involving the production of short-chain fatty acids; these bacterial end products comprise a source of energy for the colonic epithelium. In their metagenome study, Swanson et al. identified archaeal DNA sequences to represent around 1% of the total metagenome (Swanson et al. 2011).

17.2.6 The Effect of Nutrition on the Gastrointestinal Microbiome

Studying the impact of nutrition on the microbiome in companion animals benefits from a reduced impact of individual dietary preference as a confounding factor as complete and balanced diets may be fed as the sole source of food for extended periods of time. Dietary variables studied so far include changes in macronutrient composition (e.g. varying protein content) (Deusch et al. 2014; Hooda et al. 2013), changes in diet format (e.g. dry vs. wet) (Bermingham et al. 2013) and the addition of dietary fibre (Beloshapka et al. 2013; Middelbos et al. 2010; Panasevich et al. 2015). Since faeces are most commonly used as a surrogate sample for the colonic microbiome of companion animals, the effect of diet on the various regions of the GI tract and on the mucus layer is not well understood. Dietary effects are commonly studied in healthy animals, and the long-term interactions between diet, the microbiome and health are yet to be discovered.

Complete and balanced commercial cat and dog diets have minimum requirements for the content of the macronutrients protein, carbohydrate and fat, and they may be influenced by factors such as life stage. Despite this, different commercially available diets meeting these requirements vary greatly in their macronutrient profile, and it is of interest to investigate the effect on the microbiome. When cats fed with a high-protein diet were compared to those on a moderate-protein diet, 56% of the identified bacterial genera showed statistically significant differences in their relative abundances (Deusch et al. 2014). The genera with the strongest enrichment for the high-protein diet were *Eubacterium*, *Streptobacillus*, *Desulfovibrio*, *Ilyobacter* and *Fusobacterium*. *Megasphaera*, *Bifidobacterium*, *Selenomonas* and *Acidaminococcus* showed the strongest enrichment for the moderate-protein diet. The high-protein microbiome also had increased Shannon diversity and was enriched in genes for biochemical pathways of amino acid metabolism (Deusch et al. 2014).

Although cats have evolved on a strictly carnivorous diet, the feline hindgut is capable of fermenting a range of dietary plant fibres (De Godoy et al. 2013; Sunvold et al. 1995). Many commercially available diets for domestic cats contain carbohydrates and plant fibre as a consequence of the manufacturing process. Dog diets typically contain more plant material and carbohydrates reflecting a more omnivorous lifestyle. For both domestic cats and dogs, research into the effects of fibre on the microbiome has been of interest towards the improvement of dietary performance and particularly on the effect of diet on faeces consistency. Non-digestible carbohydrates may act as prebiotics in the colon, which are selectively fermented by organisms considered beneficial to the health of the host and as such can improve the balance of the intestinal microflora and the production of short-chain fatty acids. In dogs no strong prebiotic effect was observed when 1.4% inulin or yeast cell wall extract (YCW) was added to experimental raw chicken and beef diets (Beloshapka et al. 2013). Inulin however increased *Megamonas* and decreased Enterobacteriaceae vs. control and decreased *Escherichia* vs. YCW. Animals fed with the YCW-supplemented diet had increased *Bifidobacterium* vs. inulin and control, while those on the inulin-supplemented diet had increased *Lactobacillus* vs. YCW. Animals on beef-based diets had greater *Escherichia* and decreased *Anaerobiospirillum* compared to chicken-based diets indicating a potential effect of the animal protein source.

When beet pulp fibre (7.5% beet pulp) was added to an experimental dog diet, the faecal Firmicutes/Fusobacteria ratio changed significantly in favour of Firmicutes, possibly reflecting a more complex fermentative activity (Middelbos et al. 2010). No significant differences were observed for *Bacteroidetes*, the third most dominant phylum. Another study found similar results of increased Firmicutes and decreased Fusobacteria when different levels of potato fibre (0%, 1.5%, 3%, 4.5% or 6% added fibre) were added to an experimental diet (Panasevich et al. 2015). At the genus level, *Faecalibacterium*, *Bifidobacterium* and *Lactobacillus* spp. increased with higher potato starch concentrations.

The effect of fibre on the feline microbiome is less well studied. Research comparing the effects of three different types of fibre supplemented at 4% to an experimental diet—but not to an un-supplemented diet—found limited effects on microbiome structure (Barry et al. 2012). Supplementation with pectin resulted in increased Firmicutes, Chlorobi, Elusimicrobia and Proteobacteria compared to cellulose- or fructooligosaccharide (FOS)-supplemented diets. FOS-supplemented diets resulted in increased Actinobacteria compared to supplementation with cellulose or pectin. Although the inclusion of pectin or FOS did not greatly alter the functional potential of the faecal metagenome overall, some minor changes were present. For example, supplementation with FOS increased the proportion of genes for amino acid metabolism compared to the pectin-supplemented diet. The proportion of genes for nitrogen metabolism was increased in faeces of cats receiving a pectin- vs. FOS- or cellulose-supplemented diets.

Diet format may potentially represent the most extreme differences in the macro- and micronutrient profile of diets fed to dogs and cats and can dramatically alter the faecal microbiome. When a commercially available wet diet was compared to a dry diet, cats fed with dry diets had higher percentages of Actinobacteria (16.5% vs. 0.1%) and decreased percentages of Fusobacteria (0.3% vs. 23.1%) and Proteobacteria (0.4% vs. 1.1%) (Bermingham et al. 2013). Out of the 46 genera identified, 30 were significantly different in their percentage abundance. The strongest increases on the dry diet were *Lactobacillus* (31.8% vs. 0.1%), *Megasphaera* (23.0% vs. 0.0%) and *Olsenella* (16.4% vs. 0.0%), while *Bacteroides* (0.6% vs. 5.7%) and *Blautia* (0.3% vs. 2.3%) were decreased. An increase in Actinobacteria and a decrease in Fusobacteria were also observed in two other studies as an effect of increased carbohydrate and reduced protein content (Deusch et al. 2014; Hooda et al. 2013). This contrast may partly reflect differences in the protein content between the wet (42% of dry matter) and dry diet (33% of dry matter). While differences in phyla were similar between studies, differences were noted in bacterial genera indicating that factors other than protein content also affect the faecal microbiomes. In the manipulation of the faecal microbiota towards the enhancement of health, an understanding of the long-term effects of dietary influence on the microbiome and the impact on host physiology would be a significant advancement to the area of research.

17.2.7 The Gastrointestinal Microbiota in Disease

HTS studies have uncovered associations between the GI microbiota and numerous diseases in their human host (Kostic et al. 2015; Turnbaugh et al. 2006; Jackson et al. 2016; O'Mahony et al. 2015). Similarly such observations are starting to emerge from companion animal studies, and insights of veterinary importance are increasingly found. HTS studies of the microbiome describe bacterial groups previously considered infectious pathogens, to represent part of the normal microbiota in healthy cats and dogs (Handl et al. 2011; Tun et al. 2012; Chaban et al. 2012; Marks et al. 2011). This aligns with the estimated carrier rates for pathogenic species in cats and dogs varying widely, with the detection of such organisms inconsistently linked to clinical signs (Moser et al. 2001; Washabau and Day 2012; Lowden et al. 2015; Burnie et al. 1983). Since several species traditionally viewed as pathogenic are now considered putative commensals of the canine and feline GI tract, recent studies largely focus on the clinical endpoint to further understand GI diseases in veterinary medicine.

17.2.7.1 The Microbiota in Chronic and Acute Diarrhoea

Several studies have assessed the microbiota and metabolome during acute and chronic diarrhoeal disease in dogs and cats. Recently studies have revealed more intricate and complex interactions between the microbiota, bacterial products and metabolites than previously possible even following the advent of molecular analyses (Jia et al. 2010). Guard et al. (2015) used 454 pyrosequencing of the 16S rDNA, gas chromatography mass spectrometry (GCMS), ultra-performance liquid chromatography mass spectrometry (UPLC/MS) and high-performance liquid chromatography mass spectrometry (HPLC/MS) to assess the microbiota, short-chain fatty acid composition and metabolome of faeces from healthy dogs and those with acute diarrhoea. Lower levels of bacterial diversity were observed in dogs with acute diarrhoea as measured by species richness, chao1 and Shannon indices of microbial diversity. Specific microbial groups were also found to differ between diarrhoeic and control animals with Bacteroidetes, Faecalibacterium and Ruminococcaceae spp. under-represented and *Clostridium* spp. over-represented in the disease group compared to healthy controls. Propionic acid levels were also lower in dogs with diarrhoea, and these levels correlated with decreased numbers of *Faecalibacterium* spp. The predicted functional gene content based on 16S rDNA sequences detected using PICRUSt suggested an over-representation of genes for transposases and methyl accepting chemotaxis effectors in the diarrhoeal samples. In addition to these local effects, evidence of systemic changes in the canine metabolome were detected with reduced levels of 2-methyl-1H-indole and 5-methoxy-1H-indole-3-carbaldehyde in urine and of kynurenic acid in serum. Similarities exist in the findings of this study and research into human *C. difficile*-associated and *C. difficile*-negative nosocomial diarrhoea, where the microbiota detected in the distal gut demonstrated reduced phylogenetic diversity and species richness driven by changes in the Firmicutes phylum and particularly in the Ruminococcaceae, Lachnospiraceae and butyrate producers (Antharam et al. 2013). However, despite the increase in *Clostridia* spp. detected in diarrhoeic dogs (Guard et al. 2015) and in the microbiological culture era leading to the perception of *Clostridium* species as important enteropathogens in dogs and cats, researchers do not now believe *C. difficile and C. perfringens* are responsible for canine acute haemorrhagic diarrhoea syndrome (Allenspach 2015). This hypothesis is based on three main factors; firstly on the lack of association between the detection of clostridial enterotoxins or *Clostridium* spp. and clinical severity of disease (Busch et al. 2015), secondly on the range of clinical symptoms observed in dogs positive for clostridial enterotoxin (these being from mild self-limiting diarrhoea to severe and fatal acute haemorrhagic diarrhoea) (Marks et al. 2011) and finally on the detection rates of *Clostridium* sp. in the microbiota of healthy dogs (Suchodolski et al. 2012b). These confounding factors, however, fail to account for the concept of bacterial consortia being responsible for the degree of severity in such dysbioses. Antharam et al. 2016 used changes associated with lipid metabolism in the faecal metabolome of *C. difficile*-infected humans to correlate changes in the functional metabolome with the microbiota detected. They detected 63 taxa linked with high levels of the metabolite coprostanol, with 31 health-associated taxa and two *Veillonella* sp. linked with reduced levels of coprostanol and also associated with disease. The organisms clustered into 12 co-occurring subcommunities apparently acting as microbial consortia (Antharam et al. 2016). A recent study has also suggested a consortia-based approach to clostridial virulence and pathogenicity in dogs with an altered intestinal microbiota playing a key role in the activation of virulence (Minamoto et al. 2014). Minamoto et al. (2014) detected dysbiosis in diarrhoeic dogs with reduced abundance of Ruminococcaceae, Fusobacteria, *Blautia* and *Faecalibacterium* and increased levels of *Lactobacillus*, *Bifidobacterium* and *Escherichia coli* associated with disease. The dysbiosis was not linked to the detection of enterotoxigenic *C. perfringens* or enterotoxin, although the presence of *C. perfringens* enterotoxin as well as faecal dysbiosis was linked to GI disease. Thus an increased abundance of enterotoxigenic *C. perfringens* in disease was considered part of an intestinal dysbiosis associated with disease. Terminal restriction fragment length polymorphism (T-RFLP) analysis also previously demonstrated changes in the microbiota associated with acute episodes of diarrhoea in dogs with higher abundance of *Enterococcus faecalis*, *Enterococcus faecium* and *C. perfringens* detected (Bell et al. 2008).

The understanding of the microbiome in feline diarrhoeal conditions is less well studied. However a study of the 16S rDNA sequences using deep-sequencing and quantitative PCR approaches, acute or chronic (>21 days duration) feline diarrhoea compared to healthy controls, detected changes in the microbiota. Organisms from the *Burkholderiales* order, *Enterobacteriaceae* family and the *Collinsella* and *Streptococcus* genera were significantly higher in cats with diarrhoea compared to healthy controls. In healthy cats meanwhile bacteria from the Campylobacterales order, the *Bacteroidaceae* family and the *Megamonas*, *Roseburia* and *Helicobacter* genera were more abundant. *Bacteroidetes* species were significantly lower in cases of feline chronic diarrhoea, while organisms of the *Erysipelotrichia* class and *Lactobacillus* sp. were lower in acute feline diarrhoea. Analysis of the 16S rDNA sequences detected for prediction of the function gene content using PICRUSt suggested significant changes in genes encoding the metabolism of fatty acids, biosynthesis of glycosphingolipids, metabolism of biotin, tryptophan and ascorbate and aldarate by the microbiota in faeces of cats with diarrhoea (Suchodolski et al. 2015). Another study of the feline faecal microbiota during naturally occurring chronic diarrhoea using a 454-pyrosequencing approach to detect 16S rDNA content found significant correlations between faecal score and several bacterial taxa including Coriobacteriaceae, Enterobacteriaceae, *Slackia* spp., *Campylobacter upsaliensis*, *Raoultella* spp. and *Collinsella* spp. as well as unidentified organisms from the families Clostridiales, Lachnospiraceae, Aeromonadales and Succinivibrionaceae (Ramadan et al. 2014).

17.2.7.2 The Microbiota in Inflammatory Bowel Disease

Chronic diarrhoeal diseases may be diagnosed as inflammatory bowel disease (IBD); clinically IBD is defined as a group of chronic enteropathies characterised by recurrent and/or persistent GI symptoms such as diarrhoea and sickness, concurrent with inflammation of the GI tract (Simpson 2013). This group of diseases is considered to occur due to a loss in the regulation of the interaction between environmental triggers and immune and microbial factors brought about through host genetic susceptibility. As such the breed, sex and age of dog are considered in the development of diagnoses. In particular, Boxer and French bulldogs are susceptible to invasive *E. coli* isolates that are associated with clinical colitis, with the remission of disease coinciding with eradication of these organisms (Manchester et al. 2013; Packey and Sartor 2009; Simpson et al. 2006). German shepherd, Rottweiler, Border Collie, Boxer and Weimaraner are all considered breeds of dog susceptible to IBD with a higher incidence of disease observed in these populations (Kathrani et al. 2011c). The association of these breeds with such infections is suggestive of a genetic basis to susceptibility. Similarly a predisposition to antibiotic-responsive enteropathy, previously termed small intestinal bacterial overgrowth, is observed in the German shepherd dog (Batt Rm and Carter 1983). In the latter case, the disease is not associated with mucosally invasive organisms, but, like in humans, disease susceptibility is thought to be associated with genetic polymorphisms controlling the innate immune system. Canine breed studies have identified polymorphisms within the Toll-like receptor (TLR) genes TLR4 and TLR5 and differential expression of TLR4 and TLR2 to be associated with disease susceptibility (Allenspach et al. 2010; Kathrani et al. 2010; Mcmahon et al. 2010); these gene loci are also associated with human IBD (Fukata and Abreu 2007; Mcmahon et al. 2010). The TLRs are involved in signalling to induce innate immune responses including antimicrobial peptides, cellular proliferation and wound healing, in response to bacterial lipopolysaccharide (LPS), flagellin and lipopeptides (Akira et al. 2001). Polymorphisms in the nucleotide-binding oligomerization domain-containing protein 2 (NOD2) gene, the product of which activates nuclear factor kappa B (NFκB) in response to bacterial LPS, were also more frequently found in disease groups compared to healthy control dogs (Kathrani et al. 2011a). In addition to genetic polymorphisms and variations in gene expression, functional differences in the immune response have been observed in diseased dogs compared to healthy controls, with reduced levels of CD11+c cells detected in small intestinal and colonic endoscopic biopsies from dogs with IBD (Kathrani et al. 2011b). Taken together these findings are suggestive of a relatively nonspecific aetiology of disease for IBD, in which perhaps a generalised dysbiosis or a wide range of non-infectious organisms are able to elicit clinical symptoms.

Both the treatment and study of canine and feline IBD are, similarly to the human condition, complicated by the range of clinical presentations associated with disease. The treatment of disease is usually directed by the severity of the symptoms and most frequently involves dietary and immunomodulatory treatments (Malewska et al. 2011). Clinical investigation of presenting symptoms and histological analysis of intestinal biopsies may lead to diagnoses of minimal change enteropathy, granulomatous or neutrophilic IBD, lymphocyte and plasma cell predominant IBD, eosinophil predominant IBD or lymphangiectasia and crypt abscesses. The aetiology of disease may be different within each diagnosis category, despite the clinical and physiological signs of disease converging (Simpson 2013).

A cloning and sequencing approach to study the microbiota associated with small intestinal biopsies from German shepherd dogs with IBD identified differences in the 16S rDNA sequences detected in diseased dogs compared to healthy greyhounds, with an over-representation of bacteria from the class Bacilli and Erysipelotrichia and the orders Lactobacillales, Actinomycetales and Erysipelotrichales (Allenspach et al. 2010). Other cloning and sequencing studies of the microbiota associated with chronic intestinal inflammation in dogs have also identified changes in the mucosal populations associated with canine IBD. Dogs with active IBD had reduced total species richness and enrichment of *Enterobacteriaceae* (members of the *Gammaproteobacteria*) compared to healthy controls in one study (Xenoulis et al. 2008) and higher levels of Alphaproteobacteria, Betaproteobacteria and Gammaproteobacteria as well as lower abundance of clostridia in another (Suchodolski et al. 2010). In this latter study, organisms from the *Acinetobacter*, *Achromobacter*, *Brucella*, *Brevundimonas*, *Conchiformibious* and *Pseudomonas* genera were over-represented in cases of canine IBD. It is unclear whether the differences between studies are based on the breeds included or methods used in these analyses. More recently the advancement in HTS technologies has allowed a greater sequencing depth to be reached. Using a 454-pyrosequencing approach to analyse the 16S rDNA amplified from cDNA extracted from small intestinal mucosal biopsies, global differences were identified in the mucosal microbiota of canine IBD cases compared to healthy dogs (Suchodolski et al. 2012a). In this study higher levels of Bacteroidaceae, Fusobacteria, Prevotellaceae and Clostridiales were identified in healthy dogs compared to those with disease, while the genera *Diaphorobacter* and *Acinetobacter* from the Betaproteobacteria and Gammaproteobacteria classes, respectively, were more abundant or identified with increased frequency in dogs with clinical IBD. The Gammaproteobacteria class was also over-represented in cases of canine IBD compared to healthy control dogs in a 454-pyrosequencing study of the faecal microbiota, alongside lower bacterial diversity and lower levels of Erysipelotrichia, Clostridia and Bacteroidia (Minamoto et al. 2015). These findings show similarity to the current thinking in human IBD research where members of the phylum Proteobacteria are considered the aggressors in a dysbiosis caused by the breakdown of host-microbial mutualism (for a review, see Marchesi et al. 2016). The PICRUSt software was used to assess potential differences in functional gene content based on the 16S rDNA sequences detected, and this analysis identified increased bacterial transcription factors and secretion systems as well as a reduced capacity for amino acid metabolism in cases of canine IBD. In parallel with the analysis of the faecal microbiota, an untargeted metabolomics approach was used to detect systemic changes associated with disease. Several serum metabolites, including 3-hydroxybutyrate, hexuronic acid, ribose and gluconic acid lactone, were detected at higher levels in serum of dogs with clinical IBD; however, neither serum metabolites nor differences in faecal microbiota responded to treatment over a 3-week period, despite reduced clinical signs of disease. A further deep-sequencing study of the faecal microbiota associated with canine IBD compared to healthy controls and with dogs suffering from acute diarrhoeal diseases identified only significant reductions in the abundance of *Faecalibacterium* and Fusobacteria in dogs with clinical symptoms of IBD. The levels of these taxa were apparently descriptive of disease being subsequently increased when dogs experienced remission as measured by a clinical IBD activity index (CIBDAI) (Suchodolski et al. 2012b). It is possible that the mucosally associated microbiota may demonstrate profiles more descriptive of the disease phenotype in canine IBD as is found in humans (Gevers et al. 2014). However, a recent larger study of 85 healthy dogs and 65 dogs suffering from chronic GI disease and concurrent GI inflammation has demonstrated that microbiota profiles from both mucosal biopsy and faecal samples can discriminate between healthy and diseased groups (Vazquez-Baeza et al. 2016). In this study, *Gammaproteobacteria* and specifically the Enterobacteriaceae were significantly associated with IBD, with the Firmicutes such as *Clostridium* and *Ruminococcus*, which include SCFA producers, found to be associated with the healthy cohort. Similar to human IBD, the dysbiosis index was negatively correlated with measures of phylogenetic diversity. However, the authors described host species as being a greater determinant of the phylogenetic composition and predicted functional gene content than the presence of disease.

Deep-sequencing technologies have yet to be applied to the study of feline IBD. Targeted approaches are not expected to clearly identify associations between IBD and specific bacterial groups (with the exception of Enterobacteriaceae) due to their inability to account for the community effects of this apparently more generalised dysbiosis. Despite this, a small number of studies have used targeted approaches such as fluorescent in situ hybridisation to assess the microbiota in feline IBD (Inness et al. 2007; Janeczko et al. 2008). Inness et al. identified lower levels of total bacterial counts and reduced numbers of *Bifidobacterium* spp. and *Bacteroides* spp. in clinical feline IBD cases compared to healthy controls. Cats affected with IBD are also described to show higher levels of *Desulfovibrio* spp., known producers of hydrogen sulphide (Inness et al. 2007). In a second study, Enterobacteriaceae were present at higher levels in duodenal biopsy samples from cats with IBD compared to controls (Janeczko et al. 2008). The study also assessed the mucosal histology and showed that mucosally adherent bacterial numbers were associated with altered mucosal epithelial architecture and infiltration by macrophages and $CD3^+$ lymphocytes. Enterobacteriaceae including *E. coli* and *Clostridium* spp. were associated with mucosal abnormalities, cytokine mRNA upregulation and the clinical signs presenting. These differences may however be undetectable in the faecal microbiota. A dietary intervention study using identical methodologies demonstrated no detectable differences in the faecal populations of these groups in healthy cats and clinical cases of IBD (Abecia et al. 2010).

17.2.8 Summary

HTS technologies are ideally suited to provide deeper insight into the complexities of the GI and faecal microbiota and microbiome. The microbiota of cats and dogs appears relatively consistent with that of other monogastric mammals in terms of the phyla represented, although differences are described at lower taxonomic designations. Differences between individuals may be influenced by genetics and are accentuated between non-related individuals appearing more similar between littermates and genetically related dogs. Where specific pathogens were previously considered to be causal in disease, a more generalised dysbiosis is now understood to precede the development of clinical symptoms. While there is a complex crosstalk with host genetics, the adaptive and innate immune system appears to underlie the development of disease. Improved appreciation of the effect of nutrition on the host microbiome and the ability to modulate towards a healthy microbial composition remains a key research challenge in companion animal nutrition. Such refinement of nutritional strategies might be expected to have implications for promoting health and enhancing resistance to disease. Future advances in existing and other multi-omic-based approaches are expected to bring new insights to understand the relevance of microbial co-occurrence and metabolic crosstalk between microbial communities.

17.3 The Canine and Feline Oral Microbiome

The oral cavity is the first section of the GI tract and is where digestion is initiated. Distinct habitats exist in the oral cavity such as the mucosal surfaces (lips, cheek, tongue and palate) and the teeth. Unlike other parts of the body where mucosal surfaces are shed, teeth provide a surface for the formation of dental plaque. Dental plaque is a complex biofilm that is composed of a variety of microorganisms and their products. Microorganisms also exist in other parts of the oral cavity, the composition of which is influenced by the unique biological properties of each site and by the constant flow of saliva (Marsh 2000). The saliva of humans and companion animals differs in that human saliva has neutral pH, between 6.75 and 7.25 (Fejerskov and Kidd 2008), whereas canine saliva is more alkaline at around pH 8.5 (Lavy et al. 2012); such differences are expected to support the growth of different microorganisms. Saliva influences the microflora by coating the oral surfaces, especially the teeth, providing a site for attachment of microorganisms and delivering the primary source of nutrients (carbohydrates and proteins) for the inhabiting microflora (Marsh 2000). Another source of nutrients for the resident microflora is gingival crevicular fluid which is present in the gingival crevice, the area between the tooth and the gum.

The oral microbiome is believed to play an important role in health and disease, and, therefore, the main focus to date has been to characterise the healthy microbiome and to understand the microbial involvement in canine and feline oral diseases. Most attention has been directed at periodontal disease, which is the most commonly diagnosed oral disease of dogs (Butković et al. 2001; Hamp et al. 1984; Kortegaard et al. 2008; O'Neill et al. 2014b; Kyllar and Witter 2005) and cats (Girard et al. 2009; Lommer and Verstraete 2001; O'Neill et al. 2014a). There have also been a small number of studies of two other common oral diseases of cats: feline chronic gingivostomatitis (FCGS) and tooth resorption. These canine and feline oral diseases represent a major health issue, not only because of the number of animals affected but also because they can cause pain, loss of appetite, bad breath and, in severe cases, loss of teeth. They have also been associated with systemic conditions such as heart, liver and kidney disease (Glickman et al. 2009; Pavlica et al. 2008). Therefore given the number of animals affected and considering the serious consequences of these diseases, it is surprising that until recently, studies that sampled the microbial populations in sufficient depth and in the appropriate numbers of dogs or cats have not been performed. This was partly due to early studies being constrained by the number of animals and the limited number and type of bacteria studied due to reliance on culture- or molecular-based methods which targeted only organisms known to exist in the human oral cavity. However, over the last few years, there have been a number of studies that have further enhanced our understanding of the canine and feline oral microbiome through the advances in HTS technologies. This section provides a review of studies on the canine and feline oral microbiome by summarising results from early culture- and molecular-based studies but with a particular focus on new insights from HTS studies.

17.3.1 The Oral Microbiota in Health

Early estimates of the biodiversity in the oral cavity of healthy dogs and cats relied on the use of traditional culture techniques. The first large-scale surveys of the cultivable microflora from plaque and saliva of healthy dogs identified 84 phylotypes from 37 genera: approximately 50% of the phylotypes had not been previously described, and only 28% were deemed representatives of the human oral microbiota (Elliott et al. 2005). The genera commonly isolated from healthy dogs included *Actinomyces*, *Streptococcus* and *Granulicatella* in saliva and *Porphyromonas*, *Actinomyces* and *Neisseria* from plaque (Elliott et al. 2005). Other studies showed that the most prevalent bacteria cultured from the oral cavity and gingival sulcus of apparently healthy dogs were staphylococci, streptococci and *Bacillus* spp. (Ebrahimi et al. 2010). With respect to cats, the most frequently isolated species belonged to the phyla Bacteroidetes, Fusobacteria and Firmicutes (Ebrahimi et al. 2010; Harvey et al. 1995; Magaji et al. 2008). Although these studies provided preliminary insights into the oral microbiome, they vastly underestimated the richness (total number of species) and diversity (relative abundance of each species) of the bacterial populations. Further, they overestimated the contribution of species that are easily cultured and underestimated the role of difficult-to-culture organisms, which may be abundant and play a key role in the environment (Wade 2013). In fact it has been estimated that only about 3% of all bacteria can be cultured (Foster et al. 2012).

With the development of HTS technologies, extensive sequencing of microbial populations became possible, eliminating many of the inherent biases associated with conventional culture-based studies. Table 17.3 provides a summary of studies that have used HTS of the 16S rDNA gene to characterise the canine and feline oral microbiota. Analysis of gum and supragingival plaque samples from six clinically healthy client-owned dogs, using 454 pyrosequencing, to determine bacterial community membership, diversity and zoonotic potential resulted in the identification of 226 operational taxonomic units (OTUs, 97% identity), representing 181 genera from 13 bacterial phyla (Sturgeon et al. 2013). The phyla Bacteroidetes (60.2%), Proteobacteria (20.8%), Firmicutes (11.4%), Fusobacteria (4.7%) and Spirochaetes (1.7%) predominated. At the genus level, the majority of sequences (39.2%) were assigned to *Porphyromonas*. Other abundant genera included Fusobacterium (4.5% of sequences), Capnocytophaga (3.8%), Derxia (3.7%), Moraxella (3.3%) and Bergeyella (2.7%). There was evidence of a stable core microbiome, in that 56 OTUs were present in all samples. Although this core only comprised 5.6% of the total number of OTUs, it represented 80.9% of the total number of sequences.

Characterisation of the healthy feline oral microbiome of 11 clinically healthy cats using the Illumina MiSeq platform showed that gum, cheek and plaque samples comprised eight bacterial phyla representing 97.6% of the sequences: Proteobacteria (75.2%), Bacteroidetes (9.3%), Firmicutes (6.7%), SR1 (2.7%), Spirochaetes (1.8%), Fusobacteria (1.3%) and *Actinobacteria* (0.6%) (Sturgeon et al. 2014). At the genus level classification, 273 genera were identified, but 29.3% of the OTUs were unclassified at this level. The most abundant genera were an unclassified Pasteurellaceae (18.7%), Moraxella (10.9%), Thermomonas (6.9%), an unclassified Comamonadaceae (5.6%), Neisseria (4.9%) an unclassified Moraxellaceae (4.4%) and Pasteurella (4.3%). Only 0.6% of OTUs were present in all cats, but these represented 48.2% of sequences. The core comprised species from the genera *Capnocytophaga*, *Pasteurella*, *Bergeyella*

Table 17.3 Summary of publications that have used high-throughput sequencing technologies to characterise the canine and feline oral microbiota

	Canine microbiota					Feline microbiota		
Reference	Davis et al. (2013)	Sturgeon et al. (2013)	Holcombe et al. (2014)	Wallis et al. (2015)	Oh et al. (2015)	Sturgeon et al. (2014)	Harris et al. (2015)	Adler et al. (2016)
Number and breed of animals	223 client-owned dogs, various breeds	Six client-owned dogs, various breeds	12 Labrador retrievers	Miniature schnauzers; 30 periodontitis analysis, 39 gingivitis analysis	Four dogs (various breeds), their owners and two controls (non-dog owners)	11 client-owned cats	92 client-owned cats	Ten client-owned cats
Number and type of samples	Subgingival plaque, 72 healthy gingiva, 77 gingivitis, and 74 mild periodontitis	Composite oral sample (gums, tongue, teeth and cheeks); six healthy gingiva	24 supragingival plaque; 24- and 48-h post-descale and polish (dogs had mild gingivitis)	Subgingival plaque; 444 sample periodontitis analysis (47 progressing and 47 nonprogressing teeth) 955 samples for gingivitis analysis	Ten samples; composite oral sample from three oral sites (buccal, palatal, subgingival)	11 composite oral sample (gums, teeth and cheeks) from cats without oral or dental diseases	92 subgingival plaque samples (20 healthy teeth and gums, 50 gingivitis and 22 mild periodontitis)	Supragingival plaque
Sequencing platform	Roche 454 GS FLX Titanium	Roche 454 GS Junior	Roche 454 GS FLX Titanium	Roche 454 GS FLX Titanium	Roche 454 GS FLX Titanium	Illumina MiSeq	Roche 454 GS FLX Titanium	Illumina MiSeq
Number of sequence reads (after removal of sequencing error)	3,110,837	110,643	173,642	4,813,887	41,821	1,452,240	1,112,543	2,421,096

Number of OTUs	274 after removal of rare (<0.05%)	993 (624 singletons/doublets)	134 after removal of rare (<0.05%)	287 for periodontitis analysis; 290 for gingivitis analysis; after removal of rare (<0.05%)	246	10,177	267	411
Number of phyla	12	13	7	9	8	18	11	14
Number of genera	99	181	–	–	57	273	75	–
Predominant phyla (% of sequence reads):								
Bacteroidetes	26.5%	60.2%	23.9%	26.6%	19.7%	9.3%	21.8%	31.0%
Proteobacteria	17.4%	20.8%	32.1%	16.5%	25.7%	75.2%	16.7%	21.0%
Fusobacteria	3.7%	4.7%	6.0%	5.7%	12.3%	1.3%	3.6%	–
Firmicutes	28.5%	11.4%	20.9%	22.7%	19.3%	6.7%	30.0%	24.0%
Spirochaetes	1.9%	1.7%	–	10.6%	–	1.8%	7.4%	–
Actinobacteria	15.3%	–	14.9%	9.6%	21.0%	0.6%	8.2%	–
TM7	1.1%	–	2.2.%	–	–	–	–	–
SR1	<1%	–	–	–	–	2.7%	–	–
Chlorobi	<1%	–	–	–	–	–	3.2%	–

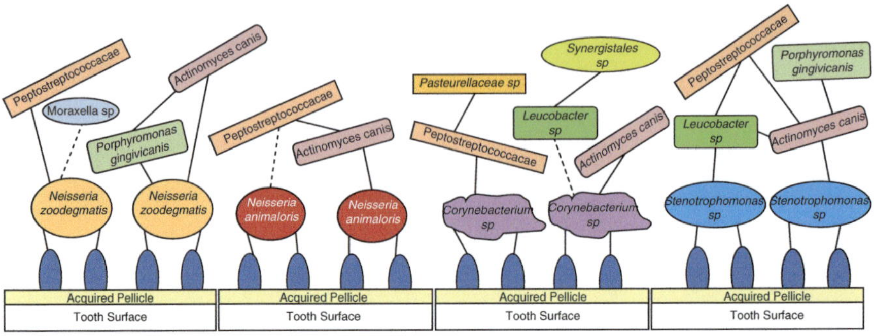

Fig. 17.2 Four hypothesised spatio-temporal model of *in vitro* early canine oral biofilm communities. Interaction networks for four primary coloniser species *Neisseria animaloris* COT-016, *Neisseria zoodegmatis* COT-349, *Corynebacterium* sp. 3105 and *Stenotrophomonas* sp. COT-224. Each community was identified using canine oral isolates and species-specific qPCR probes. Dotted lines represent tentative interactions. Reproduced from Holcombe et al. (2014)

and *Desulfomicrobium*. Cats could be separated by household and demonstrated intra-household similarity in bacterial community profiles, suggesting direct exchange of microorganisms through close contact and indicating that cohabiting animals could share disease risk. Alternatively these similarities could be due to environmental factors such as being fed the same diet. These HTS studies highlighted that the canine and feline oral cavity harbours a far more rich and diverse bacterial community than estimated by previous culture- and cloning-based studies.

Investigations into the initiation of plaque biofilm formation on the tooth surface demonstrated that the early colonisers of enamel in dogs were Gram-negative bacteria (Holcombe et al. 2014). The most abundant species in these early biofilms were *Bergeyella zoohelcum, Neisseria shayeganii* and an unclassified species of *Moraxella*. In dogs, streptococcal species, which are primary colonisers in human plaque biofilms, were rarely detected. *In vitro* simulations of canine biofilm formation resulted in the identification of five primary colonisers; three of these species belonged to the genus *Neisseria* (*N. zoodegmatis*, *N. animaloris*, *N. weaveri*) (Holcombe et al. 2014). This study led to the construction of a model detailing the development of the early canine plaque biofilm (Fig. 17.2). It also highlighted that, with the identification of 134 OTUs, the biofilm is already well formed 24 h after a professional cleaning. To the best of our knowledge, there have been no studies on biofilm formation in cats, but given the bacterial similarities between dogs and cats, it is likely that the process will be more similar to that in dogs than humans.

17.3.2 The Oral Microbiota in Disease

17.3.2.1 Periodontal Disease

Periodontal disease is initiated by the build-up of plaque on the tooth surface resulting in the gingiva becoming inflamed (gingivitis). Without an effective oral care regime, the host inflammatory response can cause the structures that support the tooth, such as the periodontal ligament and alveolar bone, to become inflamed and progressively destroyed (periodontitis) (Williams 1990). Discovering the aetiological influences responsible for the initiation and progression of the disease will lead to advancements in methods of diagnosis, treatment and prevention.

Early culture-based investigations into bacterial involvement in canine and feline periodontal disease focussed on identifying bacterial species with known associations to human periodontitis. Putative pathogens in human periodontitis include *Aggregatibacter actinomycetemcomitans*, *Porphyromonas gingivalis*, *Prevotella intermedia*, *Fusobacterium nucleatum*, *Tannerella forsythia*, *Eikenella corrodens* and *Treponema denticola* (Zarco et al. 2012). Overall, at the genus level, the majority of culturable bacteria in dogs show similarities with those from humans, but, at the species level, distinct differences exist (Dahlén et al. 2012; Hardham et al. 2005). The most frequently cultured bacterial species from subgingival plaque of dogs included *Porphyromonas crevioricanis*, *Fusobacterium canifelinum*, *Porphyromonas salivosa*, *Porphyromonas denticanis*, *Porphyromonas gulae*, *Porphyromonas macacae* and *F. nucleatum* (Dahlén et al. 2012; Hardham et al. 2005; Senhorinho et al. 2012). Many of these species were associated with periodontal disease, but some were also identified in subgingival plaque from dogs with healthy gingiva. Another study more frequently isolated previously uncultured bacteria from dogs with normal flora, *Bacteroides heparinolyticus*/*Pasteurella dagmatis* from dogs with gingivitis and *Actinomyces canis* from dogs with periodontitis (Riggio et al. 2011). Cultivation of spirochaetes from canine plaque resulted in the identification of *T. denticola*, *Treponema socranskii*, *Treponema vincentii*, *Treponema maltophilia*, *Treponema medium* and *Treponema pectinovorum* (Valdez et al. 2000).

The first survey of the microbial associations with feline periodontal disease, using standard culture techniques, showed that numbers of Gram-negative rods in the black-pigmented *Bacteroides* group and *Peptostreptococcus anaerobius* increased with increasing disease severity (Mallonee et al. 1988). In contrast, *Pasteurella multocida* appeared to decrease in numbers with increasing periodontal disease (Mallonee et al. 1988). Another study that evaluated the subgingival microbiota of cats with clinical signs of periodontal disease frequently identified three bacterial species: *P. gulae*, *Porphyromonas circumdentaria* and *F. nucleatum* (Perez-Salcedo et al. 2013). *P. gulae* was shown to be associated with increased tooth mobility, gingival recession and a tendency for deeper probing depths and loss of attachment (Perez-Salcedo et al. 2013). This finding was supported by a study of the oral microflora of cats and their owners where the prevalence of species associated with human periodontitis was investigated: *P. gulae* (86%), *P. gingivalis* (70%) and *T. forsythia* (90%) were found in cats with varying stages of periodontal disease (Booij-Vrieling et al. 2010). *T. forsythia* was isolated from both cats and owners, and the proportion was found to be higher in cats with periodontitis compared to those without the disease (Booij-Vrieling et al. 2010). Cultivation of spirochaetes from feline plaque resulted in the identification of *T. maltophilum* and *T. socranskii*, both of which have also been identified in canine plaque (Valdez et al. 2000).

Overall, it was found that as gingival inflammation increased, the anaerobic bacteria in subgingival plaque increased and the facultative aerobic species gradually decreased (Mallonee et al. 1988). Other studies, completed in dogs, have shown that Gram-negative anaerobes and Gram-positive aerobes predominate in supragingival and subgingival plaque in animals with periodontal disease (Ebrahimi et al. 2010; Syed et al. 1981; Forsblom et al. 1997; Harvey et al. 1995). This finding was supported by additional work which showed that Gram-positive bacterial species predominated in health, but as periodontal disease progressed, Gram-negative species prevailed (Hennet and Harvey 1991a, b, c). Based on the results of these early culture-based studies, which focussed on isolating bacterial species known to be associated with human periodontal disease, it was incorrectly concluded that there were no major differences in the bacterial taxa within subgingival plaque of dogs, cats and humans with periodontitis.

Later studies employed monoclonal antibodies, PCR, DNA-DNA hybridisation and cloning and sequencing of the 16S rDNA gene to identify bacterial species from dogs and cats with and without periodontal disease. However, these early molecular investigations were also biased in that they focussed on the identification of bacterial species associated with human periodontal disease. Using specific monoclonal antibodies for bacterial detection, associations of *T. denticola* and *T. socranskii* with canine periodontitis were identified (Riviere et al. 1996). PCR analysis detected *P. gingivalis*, *P. intermedia*, *T. forsythensis*, *F. nucleatum*, *Dialister pneumosintes*, *A. actinomycetemcomitans*, *Campylobacter rectus*, *E. corrodens* and *T. denticola* in subgingival samples from dogs with and without periodontitis (Nishiyama et al. 2007). In contrast, another study again using PCR identified *P. gulae* in the subgingival plaque of 92% of dogs with periodontitis and 56% of dogs without periodontitis (Senhorinho et al. 2011). Whole genomic DNA probes constructed for human periodontopathogens and a DNA-DNA hybridisation technique showed that the microbial counts and biofilm complexity differed, although not significantly, depending on the periodontal pockets depth (Papadimitriou 2016).

Following the development of molecular techniques based on sequencing of the 16S rDNA, *Pseudomonas* sp., *Porphyromonas gingivalis* and *Desulfomicrobium orale* were found to be the predominant species in normal, gingivitis and periodontitis samples from dogs, respectively (Riggio et al. 2011). A total of 353 taxa were identified by the generation of full-length 16S rDNA sequences from clone libraries derived from canine subgingival plaque samples. Of these, 80% were novel, and only 16.4% had previously been identified in the human oral cavity (Dewhirst et al. 2012). Similarly in cats, the sequencing of 16S rDNA clone libraries, and bacterial isolates cultured from the feline subgingival plaque, resulted in the identification of 171 taxa (Dewhirst et al. 2015). Not only did these studies indicate clear differences between the bacterial populations in the oral cavity of humans, dogs and cats, they also highlighted the lack of 16S rDNA reference sequences for dogs and cats in public DNA sequence databases. These canine and feline oral microbiome 16S rDNA gene reference sets were instrumental in enabling subsequent HTS studies by providing taxonomically curated full-length 16S rDNA gene sequences to which sequence reads could be mapped.

A number of HTS surveys have been undertaken to characterise the canine and feline oral microbiota in health and disease (Table 17.3). A large-scale study, using 454 pyrosequencing of the 16S rDNA gene, analysed subgingival plaque samples from dogs with healthy gingiva, gingivitis and mild periodontitis (223 dogs, approximately 70 samples per health state) (Davis et al. 2013). The most abundant phyla were Firmicutes (28.5%), Bacteroidetes (26.5%), Proteobacteria (17.4%), Actinobacteria (15.3%), Fusobacteria (3.7%), Spirochaetes (1.9%) and TM7 (1.1%). The bacterial community composition differed in health and disease: The phyla Proteobacteria and Bacteroidetes were frequently observed in dogs with healthy gingiva, whereas Firmicutes were more abundant in dogs with mild periodontitis. Across all health states, Porphyromonas was the most abundant genus, but it was more apparent in plaque from healthy dogs together with *Moraxella* and *Bergeyella*. In dogs with mild periodontitis, the most abundant genera were *Peptostreptococcus* and *Actinomyces* and the family Peptostreptococcaceae. *P. gingivalis* was the most abundant species across all health states and accounted for 7.4% of sequences. This is in contrast to humans in that Gram-negative aerobic bacterial species dominated in plaque from healthy dogs and Gram-positive anaerobic species predominated in disease. A subsequent longitudinal study underlined the temporal dynamics of the microorganisms within subgingival plaque by analysing samples from individual teeth, collected every 6 weeks for up to 60 weeks, of 52 Miniature Schnauzers (Wallis et al. 2015). This study showed that periodontal disease is characterised by a gradual decline of previously abundant, health-associated taxa, such as *B. zoohelcum* COT-186, *Moraxella* sp. COT-017, *N. shayeganii* COT-090 and *Pasteurellaceae* sp. COT-080, and, to a lesser extent, an increase in the relative abundance of a small number of species which predominantly belonged to the phylum Firmicutes. The bacterial diversity only slightly increased with periodontitis progression indicating that the community membership of dental plaque is relatively stable. *P. gingivalis* was the predominate species in all samples and was the most abundant overall representing over 7% of the total number of sequences supporting the findings of (Davis et al. 2013). It has been postulated that *P. gingivalis* is able to flourish in both health and disease because it has a complete protoporphyrin IX synthesis pathway potentially enabling it to synthesise its own haem (O'Flynn et al. 2015). In contrast, many of the genes from this pathway were absent in several of the porphyromonads associated with periodontal disease such as *P. gingivalis*. The authors hypothesised that *P. gingivalis* was able to predominate in the oral cavity of dogs due to its ability to produce haem, along with other compounds such as sirohaem and vitamin B12.

The most comprehensive HTS study of cats, whereby subgingival plaque samples from 92 client-owned cats with healthy gingiva ($n = 20$), gingivitis ($n = 50$) or mild periodontitis ($n = 22$) were analysed using 454 pyrosequencing, showed that seven phyla predominated: Firmicutes (30.0%), Bacteroidetes (21.8%), Proteobacteria (16.7%), Actinobacteria (8.2%), Spirochaetes (7.4%), Fusobacteria (3.6%) and Chlorobi (3.2%) (Harris et al. 2015). Overall 34 OTUs accounted for 50% of sequence reads. In contrast to dogs, where one species was particularly dominant, there were seven relatively abundant species in cats each representing between 2.0% and 2.9% of sequence reads (Peptostreptococcaceae bacterium FOT-028, *Moraxella* sp. FOT-087 and FOT-089, *Treponema* sp. FOT-201, Clostridiales bacterium FOT-072, *P. circumdentaria* FOT-102, *Filifactor villosus* FOT-044). Investigation of the core microbiome identified the most prevalent members as two species of *Porphyromonas* (*P. canoris* and *Porphyromonas* species FOT-110) together with Peptostreptococcaceae bacterium FOT-036 and *Filifactor* sp. FOT-129. The most abundant members of the healthy core microbiome were *Moraxella* sp. FOT-087, *B. zoohelcum* strain 357 FOT-329, *Fusobacterium* sp. FOT-120, Chlorobi bacterium COT-312, *P. circumdentaria* FOT-102, *Porphyromonas* sp. COT-290 and *Bacteroides* sp. FOT-113. With respect to the mild periodontitis samples, the bacterial species were more conserved, and the majority were within the class Clostridia (*Peptostreptococcaceae*, *F. villosus*, Lachnospiraceae, *Helcococcus* and Clostridiales species) and the genus *Treponema*. The species that were significantly more abundant in cats with healthy gingiva (>1%) compared to those with disease included Chlorobi bacterium COT-312, *P. circumdentaria* FOT-102, *Capnocytophaga* FOT-330, *Bacteroides* sp. FOT-113 and *B. zoohelcum* FOT-329. As was found for dogs, Peptostreptococcaceae were significantly more abundant in cats with periodontal disease compared to those that had healthy gingiva. Similar to dogs, approximately 75% of the subgingival bacterial species in healthy cats were Gram-negative bacteria, with the percentage of Gram-positive bacterial species increasing as periodontal health status declined. This study showed that the oral microbiota of cats is more similar to dogs than that of humans. One important finding was that the feline *P. gulae* strain, which had a similar occurrence to that in dogs, being only slightly more abundant in disease than health, was not considered to be involved in disease pathogenesis as has been suggested for the closely related *P. gingivalis* in humans (Hajishengallis et al. 2012). The number of OTUs and Shannon diversity did not significantly differ across health states suggesting, as was hypothesised for dogs, that the oral microbiota remains relatively stable, but the proportions of individual taxa change depending on health status (Harris et al. 2015).

To date, research into canine and feline periodontitis has focussed on the identification and characterisation of the bacterial communities present. However, other microorganisms are known to occupy the oral cavity and could also influence the disease process. Recently, a novel, broad spectrum 18S rDNA PCR was developed and used, in conjunction with HTS analyses, to characterise the protozoal composition of canine plaque and elucidate associations with periodontal disease. This led to the identification of two protists: *Trichomonas* sp. and *Entamoeba* sp. (Patel et al. 2016). The overall prevalence of trichomonads was 56.52% (52/92), and entamoebas was 4.34% (4/92). HTS of pooled healthy, gingivitis, early-stage periodontitis and severe periodontitis samples revealed the proportion of trichomonad sequences to be 3.51%, 2.84%, 6.07% and 35.04%, respectively, and entamoebae to be 0.01%, 0.01%, 0.80% and 7.91% respectively. This study showed that both genera of protists were significantly associated with periodontal disease. These findings provide the first conclusive evidence for the existence of oral protozoa in dog plaque and suggest a possible role for protozoa in the periodontal disease process.

17.3.2.2 Feline Chronic Gingivostomatitis

Feline chronic gingivostomatitis (FCGS) is a chronic inflammatory disease of the oropharyngeal mucosa and submucosa that causes pain and distress (Diehl and Rosychuk 1993; Niemiec 2008; Winer et al. 2016). The aetiology of FCGS is currently unknown, but the presence of bacteria is thought to be a major contributing factor. Viral and immunological causes have also been implicated (Perrone 2010). Cultivation of bacteria from oral swabs acquired from healthy ($n = 3$) and FCGS ($n = 5$) cats most frequently isolated *Pasteurella pneumotropica* and *P. multocida* subsp. *multocida* (Dolieslager et al. 2011). Further identification of bacteria using culture-independent methods (cloning and sequencing of the bacterial 16S rDNA gene) showed that, of the 158 clones obtained from oral swabs from healthy cats, *Capnocytophaga canimorsus* was the predominant species (10.8% of clones analysed). Of the clones analysed, approximately half were potentially novel species (43.7% of clones) or previously uncultured (8.2% of clones). *P. multocida* subsp. *multocida* was the most abundant species in cats with FCGS (51.8% of 253 clones analysed), and several were previously uncultured (8.7% of clones analysed) or potentially novel species (4.7% of clones analysed). From this small-scale study, the authors concluded that the oral microbiota in cats with FCGS appeared to be less diverse than that of healthy cats and that *P. multocida* subsp. *multocida* may be of aetiological significance in this disease (Dolieslager et al. 2011). A subsequent study, using the same samples, sequenced 54 clones, and 22 of these had <97% identity to known sequences and were therefore considered novel. The proportion of novel phylotypes in each group was 19.6% and 2.3% for cats with and without FCGS, respectively. *C. canimorsus* was the most prevalent species followed by uncultured bacterium, *Bergeyella* spp. and a Xanthomonadaceae bacterium. This study highlighted the shortcomings of culture-based studies in that many of the bacterial species identified by cloning could not be cultured. However, there were also a number of species identified using traditional culture-based techniques that were not identified molecular methods.

17.3.2.3 Tooth Resorption

Tooth resorption, formally known as feline odontoclastic resorptive lesion (FORL) and also referred to as cervical line lesions, neck lesions and feline caries, is characterised by the loss of dental tissue on the crown or the neck of the tooth. The aetiology and pathogenesis of tooth resorption remain to be fully determined but is almost certainly multifactorial. Thus, texture of the diet, abnormal calcium regulation, hypervitaminosis A, mechanical stress, anatomical abnormalities of the teeth, viral infections, plaque bacteria and periodontal disease have all been implicated in the disease (Gorrel 2015; Reiter et al. 2005).

A preliminary investigation of a small number of cats ($n = 21$) showed differences in the metabolic composition of saliva from healthy cats and cats with resorptive lesions (Ramadan et al. 2007). In cats with resorptive lesions, the levels of many organic and amino acids linked to microbial metabolism were increased, such as acetate, lactate, propionate, isovalerate, tryptamine and phenylalanine, suggesting the oral microflora is different in disease. The authors concluded that further studies are required to confirm this possible biomarker profile for tooth resorption using a larger number of cats.

17.3.3 Oral Microorganisms Shared Between Pets and Their Owners

To date there is only one example of 454 pyrosequencing being used to explore the differences between the oral microbiota of dogs and their owners (Oh et al. 2015). The study showed that in humans Firmicutes (57.6%), Proteobacteria (21.6%), Bacteroidetes (9.8%), Actinobacteria (7.1%) and Fusobacteria (3.9%) were the predominant phyla, whereas in dogs Proteobacteria (25.7%), Actinobacteria (21%), Bacteroidetes (19.7%), Firmicutes (19.3%), Fusobacteria (12.3%) and an unknown phylum (1.3%) were the most abundant. At the genus level, the oral samples from humans predominantly comprised *Streptococcus* (43.9%), *Neisseria* (10.3%), *Haemophilus* (9.6%), *Prevotella* (8.4%) and *Veillonella* (8.1%), whereas in dogs *Actinomyces* (17.2%), an unknown genera (16.8%), *Porphyromonas* (14.8%), *Fusobacterium* (11.8%) and *Neisseria* (7.2%) were the most abundant. The canine oral microbiota, although not significantly different, was richer and more diverse than that of humans. In total there were only 12 OTUs (4.9%) that were common to both human and canine oral plaque samples. As in previous studies, *P. gingivalis* was identified in all canine samples along with *T. forsythia* and *Streptococcus minor*. Overall this study showed that the oral microbiota of dogs is appreciably different to that of their owners.

17.3.4 Effect of Nutrition on the Oral Microbiota

Diet has been shown to have a significant effect on the overall diversity and abundance of specific bacterial species within the oral cavity (Adler et al. 2016). The oral microbiome of cats fed exclusively on dry diets has been reported to be more diverse than cats fed with a wet diet (canned and/or fresh meat combinations). The most enriched taxa in cats receiving dry diets included those that have previously been associated with oral health and disease: *Actinobacillus*, *Acholeplasma*, *Treponema* and *Porphyromonas* (Adler et al. 2016). Dental plaque from cats consuming a wet diet was significantly more abundant in Proteobacteria from the family Neisseriaceae (38%). Overall bacterial diversity was also shown to increase with age. Further studies are required to determine the influence of diet on the canine and feline oral microbiome.

17.3.5 Summary

HTS technologies have provided a deeper understanding of the complexity of the canine and feline oral microbiota. They have shown that the oral cavity of dogs and cats harbours a rich and diverse bacterial community that surpasses estimates obtained from early culture- and molecular-based studies. They have also revealed that the canine and feline oral microbiota are similar but distinctly different to the human oral microbiota. Key differences include the lack of canine and feline streptococcal species which suggests that the human oral colonisation process is not representative of what happens in dogs and cats. In addition, the high prevalence of *Porphyromonas*, and other genera associated with human disease, in clinically healthy dogs and cats suggests that different species of bacteria are potentially involved in the disease process. If we assume that the disease process is conserved between humans, dogs and cats, then the difference in bacterial species observed is most likely due to differences in the oral environment (e.g. saliva and diet). Conversely some commonality in the function of the oral bacteria must remain between humans, dogs and cats as the disease pathology is the same. Determining which species are involved in the various disease processes as opposed to adapting to the environment is more difficult to determine. Given these findings, it suggests that interventions designed to target human pathogenic species will not necessarily improve the oral health of cats and dogs. It is encouraging that preliminary studies suggest the potential for dietary intervention to alter the relative proportions of bacterial species in the oral cavity and potentially improve the oral health of cats and dogs.

Further improvements in HTS technologies will support future development and understanding of the canine and feline oral microbiome. This is vital to enable the role of oral bacteria in health and disease to be understood. To maximise progress, advances in understanding the oral microbiome need to be coupled to a deeper understanding of the interaction of oral bacteria with the immune response. Defining oral health and understanding what triggers disease will eventually help to provide veterinarians with the means to recognise and diagnose oral diseases and will enable the development of targeted treatment strategies.

17.4 The Canine and Feline Skin Microbiome

The skin could be regarded as a site which is relatively difficult for microbes to colonise, not least because the environment is dry and exposed and because of the rapid turnover of cells within the skin epithelium. Nonetheless, trillions of bacteria are known to colonise the human skin, and estimates have suggested that up to 1 billion bacteria per square centimetre are able to colonise this niche (Kong and Segre 2012). Further, there are a variety of different habitats on the skin where potential bacterial colonisation can occur, and colonisation dynamics can be influenced by solid structures (such as hair) and chemical composition (dictated by skin secretions produced by sweat and sebaceous glands). Thus significant diversity is expected and has indeed been reported for human skin (Findley et al. 2013).

As typified by the gastrointestinal tract, community stabilisation occurs at these various skin sites, disturbances of which are expected to have impacts for host health. The advent of HTS approaches has heralded a wealth of new information on the microbiome which inhabits the human skin (Grice 2015). Such insights, as with other regions of the body, were not possible using traditional culture-based approaches of the past. Moving forward, these technologies are expected to bring new insights to the canine and feline skin microbiome.

As with HTS studies in other specialties, the wealth of information on the various bacterial taxa, all the way from phylum, down to genus and species level, is immense. Such investigations demonstrate the tremendous diversity and variability of bacterial populations, detailed discussion of which, even from limited studies, is beyond the scope of this short overview. This section thus aims to briefly review the early culture-based work in the field of canine and feline skin microbiome and compare this to new insights brought by the small number of HTS studies that have been performed in companion animals. Current knowledge about skin microbial communities in companion animals is then briefly contextualised against the likely implications for health and disease in these species.

17.4.1 The Skin Microbiota in Health

Early studies describing the skin microbiota of cats and dogs involved culture-dependant techniques liable to gross underestimation of bacterial numbers and diversity due to inherent difficulties in ability to isolate, culture and characterise varied bacterial species using these methods. Example studies include culture-dependant studies in dogs which identified anywhere between 350 and 16,500 bacteria per square centimetre in healthy dogs versus those suffering from seborrhoeic skin lesions, a mild form of dermatitis (Ihrke et al. 1978). These studies identified *Staphylococcus*, *Micrococcus*, *Clostridium*, *Propionibacterium* and *Acinetobacter* as the dominant genera. Early culture-based studies described for cats are even less detailed, with some early studies limited to defining the presence of a limited subset of bacterial genera in small numbers of healthy cats (Krogh and Kristensen 1976). These studies also showed consistent presence of *Micrococcus* and *Acinetobacter* and presence of *Streptococcus*.

As with the intestinal microbiota, exploration of the canine and feline skin microbiomes using HTS approaches lags behind the human field and remains a relatively unexplored area, particularly in cats, where studies are largely absent. Pilot studies reported in conference proceedings in 2012 described between 38 and 110 different bacterial species per canine skin sample, highlighted significant differences both between human and canine microbial representation and showed significantly greater species diversity when compared with culture-dependant approaches (Sturgeon et al. 2012). Other recent studies describe the microbial sampling of dogs cohabiting with families. The primary focus of these investigations was to understand sharing of microbiota across species and to provide insights into the bacterial species representation on the foreheads and paw pads of dogs (Song et al. 2013). Illumina sequencing of samples from 36 dogs showed that compared to human skin, a greater diversity of taxa were found on the canine fur and paws and, also, as compared to humans, a more even mixture of bacterial taxa was found. Highest prevalence on the forehead was members of the Pasteurellaceae and Sphingomonadaceae families, while in the paws, Pseudomonadaceae and Sphingomonadaceae predominated (Song et al. 2013).

17.4.2 The Skin Microbiota in Disease

Rodriguez-Hoffman and colleagues studied swabs from healthy and allergic dogs from different regions of haired skin and mucosal surfaces using 454 pyrosequencing. Differences in species diversity were found in the 12 different skin sites that were analysed and showed high interregion and interindividual variation. Sampling regions included those from haired skin (including the dorsal lumbar and groyne areas) and those from mucosal surface (including such regions as the lips, paw pads and perianal regions). A total of 17 phyla were identified in all samples, with the most common phyla (in order of highest representation) being members of the Proteobacteria, Firmicutes, Actinobacteria, Bacteroidetes and Cyanobacteria families. At the genus level, the most abundant member was *Ralstonia* spp., an organism that is an emerging opportunistic pathogen that is well adapted to living in nutrient low conditions (Ryan and Adley 2014). However, the significance of its high occurrence remains unclear and may be the result of environmental contamination. When comparing healthy versus allergic dogs, the latter exhibited lower species diversity, a potential reflection of the use of antimicrobial agents in these animals to control the allergic response. However, lower bacterial diversity may simply be the result of inflammation reducing bacterial diversity as described in previous studies. In related studies, dogs with spontaneous atopic dermatitis were followed longitudinally by sampling at the axilla, concave pinna and groyne during periods of inflammation and remission. As with the Olivry study, these authors similarly showed decreased bacterial diversity versus healthy control dogs and demonstrated the propensity of the dermatitis group to show increased levels of *Staphylococcus* and *Corynebacterium* (Bradley et al. 2016).

Recent studies have also branched out to understand the role of cutaneous mycobiota in healthy and allergic dogs using HTS in cats (Meason-Smith et al. 2016). These studies have also demonstrated aberrant expression of certain fungal sequences such as from the Agaricomycetes and Sordariomycetes in allergic animals.

17.4.3 Summary

Dermatitis and allergy in companion animals remain an area in which further research is required. What is already clear is that current untargeted strategies using nonspecific antimicrobial treatments to resolve skin disorders may need to be reviewed in light of the importance of normal bacterial communities at the skin surface. Thus, untargeted antimicrobial approaches may not only suppress pathogenic but also commensal bacteria. The role of the microbiota in various other skin conditions, especially in companion animals, remains an area which should be explored.

17.5 Conclusion

Microbes inhabit varied environments from the oral cavity and intestinal compartments to the skin, where, in general, they live in symbiosis with the host. Extensive studies have demonstrated tremendous complexity and diversity in population and community structure in all these compartments with an increasing appreciation for their role in modulating health and disease.

Canine and feline microbial communities share some similarities but also some important differences versus their human counterparts where the microbiota is better characterised. Thus, while organisation of intestinal communities at the phylum level shows Firmicutes, Bacteroidetes and Fusobacteria to represent >95% of all bacteria in both humans and companion animals, variation is seen at the class, order, genus and species levels. Environment and genetic profiles also effect microbial populations. HTS approaches have already begun to yield insights into mechanisms of some disease processes, particularly into understanding intestinal microbial dysbiosis and implications for animal health. This review highlights the pressing need to better understand the species-specific community structure to enable improved insights into mechanisms of health and disease in which microbes are implicated.

17.6 Opinion

As more information emerges over the coming years about the community structure of the canine and feline microbiome, differing analysis methodologies will almost certainly hamper efforts towards gaining clarity on species representation in health and disease. Indeed, it is already clear that current molecular techniques and differences between laboratory approaches can introduce bias in community profiling, making comparison across different research studies and using different analysis platforms, a significant challenge. Future studies may focus on how such disparity can be rationalised; this may be using current platforms or using newer technologies on the horizon such as nanopore sequencing technologies.

The influence of geography, inter-animal variation, climate, diet and genetics (amongst other factors) are also likely to impact on species representation. Understanding this dysbiosis and how the microbiota produces harmful effects, as well as the influence of perturbations, is crucial to understanding both health and disease. Teasing apart the complex bacterial interactions from the host responses and then linking these to health and disease are and will remain a challenge.

While data from companion animals are still sparse, the advent of HTS approaches will almost certainly ensure rapid proliferation of data in this field. Insights from human studies, where cross-species observations have already identified microbial links with disease, should also not be ignored and will undoubtedly help in the assembly of the big picture.

Linking phylogeny to function remains a significant challenge. Thus, multi-omic approaches to understand not only the impact of bacterial representation but also the interface with genetic factors and interactions at the metabolomic and transcriptomic level would provide advancements to current research. This will require complex *in silico* analyses of multi-omic data sets, developments which are still in their infancy. Nevertheless, as research continues at pace, understanding host-microbial interactions, identifying core microbial populations and defining microbes associated with disease using such multi-omic strategies can be expected to bring new insights to animal health and may ultimately be the key to the targeted treatment of a number of diseases.

References

Abecia LHL, Khoo C, Frantz N, Mccartney AL (2010) Effects of a novel galactooligosaccharide on the faecal microbiota of healthy and inflammatory bowel disease cats during a randomized, double-blind, cross-over feeding study. Int J Probiotics Prebiotics 5:61–68

Adler CJ, Malik R, Browne GV, Norris JM (2016) Diet may influence the oral microbiome composition in cats. Microbiome 4:23

Akira S, Takeda K, Kaisho T (2001) Toll-like receptors: critical proteins linking innate and acquired immunity. Nat Immunol 2:675–680

Al-Asmakh M, Zadjali F (2015) Use of germ-free animal models in microbiota-related research. J Microbiol Biotechnol 25:1583–1588

Allenspach K (2015) Bacteria involved in acute haemorrhagic diarrhoea syndrome in dogs. Vet Rec 176:251–252

Allenspach K, House A, Smith K, Mcneill FM, Hendricks A, Elson-Riggins J, Riddle A, Steiner JM, Werling D, Garden OA, Catchpole B, Suchodolski JS (2010) Evaluation of mucosal bacteria and histopathology, clinical disease activity and expression of Toll-like receptors in German shepherd dogs with chronic enteropathies. Vet Microbiol 146:326–335

Antharam VC, Li EC, Ishmael A, Sharma A, Mai V, Rand KH, Wang GP (2013) Intestinal dysbiosis and depletion of butyrogenic bacteria in Clostridium difficile infection and nosocomial diarrhea. J Clin Microbiol 51:2884–2892

Antharam VC, Mcewen DC, Garrett TJ, Dossey AT, Li EC, Kozlov AN, Mesbah Z, Wang GP (2016) An integrated metabolomic and microbiome analysis identified specific gut microbiota associated with fecal cholesterol and coprostanol in Clostridium difficile infection. PLoS One 11:e0148824

Backhed F, Ley RE, Sonnenburg JL, Peterson DA, Gordon JI (2005) Host-bacterial mutualism in the human intestine. Science 307:1915–1920

Baele M, Decostere A, Vandamme P, Van Den Bulck K, Gruntar I, Mehle J, Mast J, Ducatelle R, Haesebrouck F (2008) *Helicobacter baculiformis* sp. nov., isolated from feline stomach mucosa. Int J Syst Evol Microbiol 58:357–364

Barry KA, Middelbos IS, Vester Boler BM, Dowd SE, Suchodolski JS, Henrissat B, Coutinho PM, White BA, Fahey GC Jr, Swanson KS (2012) Effects of dietary fiber on the feline gastrointestinal metagenome. J Proteome Res 11:5924–5933

Batt Rm NJ, Carter MW (1983) Bacterial overgrowth associated with a naturally occurring enteropathy in the German shepherd dog. Res Vet Sci 35:42–46

Becker AA, Hesta M, Hollants J, Janssens GP, Huys G (2014) Phylogenetic analysis of faecal microbiota from captive cheetahs reveals underrepresentation of Bacteroidetes and Bifidobacteriaceae. BMC Microbiol 14:43

Bell JA, Kopper JJ, Turnbull JA, Barbu NI, Murphy AJ, Mansfield LS (2008) Ecological characterization of the colonic microbiota of normal and diarrheic dogs. Interdiscip Perspect Infect Dis 2008:149694

Beloshapka AN, Dowd SE, Suchodolski JS, Steiner JM, Duclos L, Swanson KS (2013) Fecal microbial communities of healthy adult dogs fed raw meat-based diets with or without inulin or yeast cell wall extracts as assessed by 454 pyrosequencing. FEMS Microbiol Ecol 84:532–541

Benno Y, Nakao H, Uchida K, Mitsuoka T (1992) Impact of the advances in age on the gastrointestinal microflora of Beagle dogs. J Vet Med Sci 54:703–706

Bermingham EN, Young W, Kittelmann S, Kerr KR, Swanson KS, Roy NC, Thomas DG (2013) Dietary format alters fecal bacterial populations in the domestic cat (*Felis catus*). Microbiology 2:173–181

Biagi E, Nylund L, Candela M, Ostan R, Bucci L, Pini E, Nikkila J, Monti D, Satokari R, Franceschi C, Brigidi P, De Vos W (2010) Through ageing, and beyond: gut microbiota and inflammatory status in seniors and centenarians. PLoS One 5:e10667

Booij-Vrieling HE, Van Der Reijden WA, Houwers DJ, De Wit WE, Bosch-Tijhof CJ, Penning LC, Van Winkelhoff AJ, Hazewinkel HA (2010) Comparison of periodontal pathogens between cats and their owners. Vet Microbiol 144:147–152

Bradley CW, Morris DO, Rankin SC, Cain CL, Misic AM, Houser T, Mauldin EA, Grice EA (2016) Longitudinal evaluation of the skin microbiome and association with microenvironment and treatment in canine atopic dermatitis. J Invest Dermatol 136:1182–1190

Burnie AG, Simpson JW, Lindsay D, Miles RS (1983) The excretion of campylobacter, salmonellae and Giardia lamblia in the faeces of stray dogs. Vet Res Commun 6:133–138

Busch K, Suchodolski JS, Kuhner KA, Minamoto Y, Steiner JM, Mueller RS, Hartmann K, Unterer S (2015) Clostridium perfringens enterotoxin and Clostridium difficile toxin A/B do not play a role in acute haemorrhagic diarrhoea syndrome in dogs. Vet Rec 176:253

Butković V, Šimpraga M, Šehić M, Stanin D, Sušić V, Capak D, Kos J (2001) Dental diseases of dogs: a retrospective study of radiological data. Acta Vet Brno 70:203–208

Chaban B, Links MG, Hill JE (2012) A molecular enrichment strategy based on cpn60 for detection of epsilon-proteobacteria in the dog fecal microbiome. Microb Ecol 63:348–357

Claesson MJ, Cusack S, O'Sullivan O, Greene-Diniz R, De Weerd H, Flannery E, Marchesi JR, Falush D, Dinan T, Fitzgerald G, Stanton C, Van Sinderen D, O'Connor M, Harnedy N, O'Connor K, Henry C, O'Mahony D, Fitzgerald AP, Shanahan F, Twomey C, Hill C, Ross RP, O'Toole PW (2011) Composition, variability, and temporal stability of the intestinal microbiota of the elderly. Proc Natl Acad Sci U S A 108(Suppl 1):4586–4591

Clemente JC, Ursell LK, Parfrey LW, Knight R (2012) The impact of the gut microbiota on human health: an integrative view. Cell 148:1258–1270

Dahlén G, Charalampakis G, Abrahamsson I, Bengtsson L, Falsen E (2012) Predominant bacterial species in subgingival plaque in dogs. J Periodontal Res 47:354–364

Davis IJ, Wallis C, Deusch O, Colyer A, Milella L, Loman N, Harris S (2013) A cross-sectional survey of bacterial species in plaque from client owned dogs with healthy gingiva, gingivitis or mild periodontitis. PLoS One 8:e83158

De Godoy MR, Kerr KR, Fahey GC Jr (2013) Alternative dietary fiber sources in companion animal nutrition. Forum Nutr 5:3099–3117

Delgado S, Suarez A, Mayo B (2006) Identification of dominant bacteria in feces and colonic mucosa from healthy Spanish adults by culturing and by 16S rDNA sequence analysis. Dig Dis Sci 51:744–751

Desai AR, Musil KM, Carr AP, Hill JE (2009) Characterization and quantification of feline fecal microbiota using cpn60 sequence-based methods and investigation of animal-to-animal variation in microbial population structure. Vet Microbiol 137:120–128

Dethlefsen L, Huse S, Sogin ML, Relman DA (2008) The pervasive effects of an antibiotic on the human gut microbiota, as revealed by deep 16S rRNA sequencing. PLoS Biol 6:e280

Deusch O, O'Flynn C, Colyer A, Morris P, Allaway D, Jones PG, Swanson KS (2014) Deep Illumina-based shotgun sequencing reveals dietary effects on the structure and function of the fecal microbiome of growing kittens. PLoS One 9:e101021

Deusch O, O'Flynn C, Colyer A, Swanson KS, Allaway D, Morris P (2015) A longitudinal study of the feline microbiome identifies changes into early adulthood irrespective of sexual development, PLoS One 10:e0144881

Dewhirst FE, Klein EA, Thompson EC, Blanton JM, Chen T, Milella L, Buckley CMF, Davis IJ, Bennett ML, Marshall-Jones ZV (2012) The canine oral microbiome. PLoS One 7:e36067

Dewhirst FE, Klein EA, Bennett ML, Croft JM, Harris SJ, Marshall-Jones ZV (2015) The feline oral microbiome: a provisional 16S rRNA gene based taxonomy with full-length reference sequences. Vet Microbiol 175:294–303
Diehl K, Rosychuk RAW (1993) Feline gingivitis-stomatitis-pharyngitis. Vet Clin N Am Small Anim Pract 23:139–153
Dolieslager SM, Riggio MP, Lennon A, Lappin DF, Johnston N, Taylor D, Bennett D (2011) Identification of bacteria associated with feline chronic gingivostomatitis using culture-dependent and culture-independent methods. Vet Microbiol 148:93–98
Ebrahimi A, Oskuiezadeh K, Khoshnevisan R (2010) A study on the prevalent bacterial population in oral cavity of owned healthy dogs and cats. Intas Polivet 11:271–273
Eckburg PB, Bik EM, Bernstein CN, Purdom E, Dethlefsen L, Sargent M, Gill SR, Nelson KE, Relman DA (2005) Diversity of the human intestinal microbial flora. Science 308:1635–1638
Elliott DR, Wilson M, Buckley CMF, Spratt DA (2005) Cultivable oral microbiota of domestic dogs. J Clin Microbiol 43:5470–5476
Farris MH, Olson JB (2007) Detection of Actinobacteria cultivated from environmental samples reveals bias in universal primers. Lett Appl Microbiol 45:376–381
Fejerskov O, Kidd E (2008) Dental caries: the disease and its clinical management. Wiley-Blackwell, Chichester, West Sussex
Findley K, Oh J, Yang J, Conlan S, Deming C, Meyer JA, Schoenfeld D, Nomicos E, Park M, Program NIHISCCS, Kong HH, Segre JA (2013) Topographic diversity of fungal and bacterial communities in human skin. Nature 498:367–370
Forsblom B, Love DN, Sarkiala-Kessel E, Jousimies-Somer H (1997) Characterization of anaerobic, gram-negative, nonpigmented, saccharolytic rods from subgingival sites in dogs. Clin Infect Dis 25(Suppl 2):S100–S106
Foster JA, Bunge J, Gilbert JA, Moore JH (2012) Measuring the microbiome: perspectives on advances in DNA-based techniques for exploring microbial life. Brief Bioinform 13:420–429
Foster ML, Dowd SE, Stephenson C, Steiner JM, Suchodolski JS (2013) Characterization of the fungal microbiome (mycobiome) in fecal samples from dogs. Vet Med Int 2013:658373
Fredricks DN, Smith C, Meier A (2005) Comparison of six DNA extraction methods for recovery of fungal DNA as assessed by quantitative PCR. J Clin Microbiol 43:5122–5128
Fukata M, Abreu MT (2007) TLR4 signalling in the intestine in health and disease. Biochem Soc Trans 35:1473–1478
Garcia-Mazcorro JF, Suchodolski JS, Jones KR, Clark-Price SC, Dowd SE, Minamoto Y, Markel M, Steiner JM, Dossin O (2012) Effect of the proton pump inhibitor omeprazole on the gastrointestinal bacterial microbiota of healthy dogs. FEMS Microbiol Ecol 80:624–636
Garrett WS, Gordon JI, Glimcher LH (2010) Homeostasis and inflammation in the intestine. Cell 140:859–870
German AJ, Day MJ, Ruaux CG, Steiner JM, Williams DA, Hall EJ (2003) Comparison of direct and indirect tests for small intestinal bacterial overgrowth and antibiotic-responsive diarrhea in dogs. J Vet Intern Med 17:33–43
Gevers D, Kugathasan S, Denson LA, Vazquez-Baeza Y, Van Treuren W, Ren B, Schwager E, Knights D, Song SJ, Yassour M, Morgan XC, Kostic AD, Luo C, Gonzalez A, Mcdonald D, Haberman Y, Walters T, Baker S, Rosh J, Stephens M, Heyman M, Markowitz J, Baldassano R, Griffiths A, Sylvester F, Mack D, Kim S, Crandall W, Hyams J, Huttenhower C, Knight R, Xavier RJ (2014) The treatment-naive microbiome in new-onset Crohn's disease. Cell Host Microbe 15:382–392
Girard N, Servet E, Biourge V, Hennet P (2009) Periodontal health status in a colony of 109 cats. J Vet Dent 26:147–155
Glickman LT, Glickman NW, Moore GE, Goldstein GS, Lewis HB (2009) Evaluation of the risk of endocarditis and other cardiovascular events on the basis of the severity of periodontal disease in dogs. J Am Vet Med Assoc 234:486–494
Gorrel C (2015) Tooth resorption in cats: pathophysiology and treatment options. J Feline Med Surg 17:37–43
Greetham HL, Giffard C, Hutson RA, Collins MD, Gibson GR (2002) Bacteriology of the Labrador dog gut: a cultural and genotypic approach. J Appl Microbiol 93:640–646

Grice EA (2015) The intersection of microbiome and host at the skin interface: genomic- and metagenomic-based insights. Genome Res 25:1514–1520

Guard BC, Barr JW, Reddivari L, Klemashevich C, Jayaraman A, Steiner JM, Vanamala J, Suchodolski JS (2015) Characterization of microbial dysbiosis and metabolomic changes in dogs with acute diarrhea. PLoS One 10:e0127259

Hajishengallis G, Darveau RP, Curtis MA (2012) The keystone-pathogen hypothesis. Nat Rev Microbiol 10:717–725

Hamp SE, Olsson SE, Farsø-Madsen K, Viklands P, Fornell J (1984) A macroscopic and radiological investigation of dental diseases of the dog. Vet Radiol 25:86–92

Hand D, Wallis C, Colyer A, Penn CW (2013) Pyrosequencing the canine faecal microbiota: breadth and depth of biodiversity. PLoS One 8:e53115

Handl S, Dowd SE, Garcia-Mazcorro JF, Steiner JM, Suchodolski JS (2011) Massive parallel 16S rRNA gene pyrosequencing reveals highly diverse fecal bacterial and fungal communities in healthy dogs and cats. FEMS Microbiol Ecol 76:301–310

Hardham J, Dreier K, Wong J, Sfintescu C, Evans RT (2005) Pigmented-anaerobic bacteria associated with canine periodontitis. Vet Microbiol 106:119–128

Harris S, Croft J, O'Flynn C, Deusch O, Colyer A, Allsopp J, Milella L, Davis IJ (2015) A pyrosequencing investigation of differences in the feline subgingival microbiota in health, gingivitis and mild periodontitis. PLoS One 10:e0136986

Harvey CE, Thornsberry C, Miller BR (1995) Subgingival bacteria—comparison of culture results in dogs and cats with gingivitis. J Vet Dent 12:147–150

Hennet PR, Harvey CE (1991a) Aerobes in periodontal disease in the dog: a review. J Vet Dent 8:9–11

Hennet PR, Harvey CE (1991b) Anaerobes in periodontal disease in the dog: a review. J Vet Dent 8:18–21

Hennet PR, Harvey CE (1991c) Spirochetes in periodontal disease in the dog: a review. J Vet Dent 8:16–17

Holcombe LJ, Patel N, Colyer A, Deusch O, O'Flynn C, Harris S (2014) Early canine plaque biofilms: characterization of key bacterial interactions involved in initial colonization of enamel. PLoS One 9:e113744

Hooda S, Vester Boler BM, Kerr KR, Dowd SE, Swanson KS (2013) The gut microbiome of kittens is affected by dietary protein:carbohydrate ratio and associated with blood metabolite and hormone concentrations. Br J Nutr 109:1637–1646

Human Microbiome Project C (2012) A framework for human microbiome research. Nature 486:215–221

Ihrke PJ, Schwartzman RM, Mcginley K, Horwitz LN, Marples RR (1978) Microbiology of normal and seborrheic canine skin. Am J Vet Res 39(9):1487

Inness VL, Mccartney AL, Khoo C, Gross KL, Gibson GR (2007) Molecular characterisation of the gut microflora of healthy and inflammatory bowel disease cats using fluorescence in situ hybridisation with special reference to Desulfovibrio spp. J Anim Physiol Anim Nutr (Berl) 91:48–53

Jackson MA, Jeffery IB, Beaumont M, Bell JT, Clark AG, Ley RE, O'Toole PW, Spector TD, Steves CJ (2016) Signatures of early frailty in the gut microbiota. Genome Med 8:8

Janeczko S, Atwater D, Bogel E, Greiter-Wilke A, Gerold A, Baumgart M, Bender H, Mcdonough PL, Mcdonough SP, Goldstein RE, Simpson KW (2008) The relationship of mucosal bacteria to duodenal histopathology, cytokine mRNA, and clinical disease activity in cats with inflammatory bowel disease. Vet Microbiol 128:178–193

Jia J, Frantz N, Khoo C, Gibson GR, Rastall RA, Mccartney AL (2010) Investigation of the faecal microbiota associated with canine chronic diarrhoea. FEMS Microbiol Ecol 71:304–312

Johnston KLA, Batt RM (1993) An unexpected bacterial flora in the proximal small intestine of normal cats. Vet Rec 132:2

Johnston K, Lamport A, Batt R (1993) An unexpected bacterial flora in the proximal small intestine of normal cats. Vet Rec 132:362–363

Johnston KL, Swift NC, Forster-Van Hijfte M, Rutgers HC, Lamport A, Ballavre O, Batt RM (2001) Comparison of the bacterial flora of the duodenum in healthy cats and cats with signs of gastrointestinal tract disease. J Am Vet Med Assoc 218:48–51

Jumpstart Consortium Human Microbiome Project Data Generation Working G (2012) Evaluation of 16S rDNA-based community profiling for human microbiome research. PLoS One 7:e39315

Kathrani A, House A, Catchpole B, Murphy A, German A, Werling D, Allenspach K (2010) Polymorphisms in the TLR4 and TLR5 gene are significantly associated with inflammatory bowel disease in German shepherd dogs. PLoS One 5:e15740

Kathrani A, House A, Catchpole B, Murphy A, Werling D, Allenspach K (2011a) Breed-independent toll-like receptor 5 polymorphisms show association with canine inflammatory bowel disease. Tissue Antigens 78:94–101

Kathrani A, Schmitz S, Priestnall SL, Smith KC, Werling D, Garden OA, Allenspach K (2011b) CD11c+ cells are significantly decreased in the duodenum, ileum and colon of dogs with inflammatory bowel disease. J Comp Pathol 145:359–366

Kathrani A, Werling D, Allenspach K (2011c) Canine breeds at high risk of developing inflammatory bowel disease in the south-eastern UK. Vet Rec 169:635

Kong HH, Segre JA (2012) Skin microbiome: looking back to move forward. J Invest Dermatol 132:933–939

Kortegaard HE, Eriksen T, Baelum V (2008) Periodontal disease in research beagle dogs—an epidemiological study. J Small Anim Pract 49:610–616

Kostic AD, Gevers D, Siljander H, Vatanen T, Hyötyläinen T, Hämäläinen AM, Peet A, Tillmann V, Pöhö P, Mattila I, Lähdesmäki H, Franzosa EA, Vaarala O, de Goffau M, Harmsen H, Ilonen J, Virtanen SM, Clish CB, Orešič M, Huttenhower C, Knip M, Group DS, Xavier RJ (2015) The dynamics of the human infant gut microbiome in development and in progression toward type 1 diabetes. Cell Host Microbe 17:260–273

Krogh HV, Kristensen S (1976) A study of skin diseases in dogs and cats: II. Microflora of the normal skin of dogs and cats. Nord Vet Med 28:459–463

Kyllar M, Witter K (2005) Prevalence of dental disorders in pet dogs. Vet Med 50:496–505

Langille MG, Zaneveld J, Caporaso JG, Mcdonald D, Knights D, Reyes JA, Clemente JC, Burkepile DE, Vega Thurber RL, Knight R, Beiko RG, Huttenhower C (2013) Predictive functional profiling of microbial communities using 16S rRNA marker gene sequences. Nat Biotechnol 31:814–821

Lavy E, Goldberger D, Friedman M, Steinberg D (2012) Values and mineral content of saliva in different breeds of dogs. Isr J Vet Med 67:244–248

Ley RE, Bäckhed F, Turnbaugh P, Lozupone CA, Knight RD, Gordon JI (2005) Obesity alters gut microbial ecology. Proc Natl Acad Sci U S A 102:11070–11075

Ley RE, Hamady M, Lozupone C, Turnbaugh PJ, Ramey RR, Bircher JS, Schlegel ML, Tucker TA, Schrenzel MD, Knight R, Gordon JI (2008) Evolution of mammals and their gut microbes. Science 320:1647–1651

Lommer MJ, Verstraete FJ (2001) Radiographic patterns of periodontitis in cats: 147 cases (1998–1999). J Am Vet Med Assoc 218:230–234

Lowden P, Wallis C, Gee N, Hilton A (2015) Investigating the prevalence of Salmonella in dogs within the Midlands region of the United Kingdom. BMC Vet Res 11:239–245

Magaji AA, Saulawa MA, Salihu MD, Junaidu AU, Shittu A, Gulumbe ML, Chafe UM, Buhari S, Raji AA (2008) Oral microflora of stray domestic cats (Felis catus) found in the premises of two human hospitals in Sokoto, Nigeria. Sokoto J Vet Sci 7:9–12

Malewska K, Rychlik A, Nieradka R, Kander M (2011) Treatment of inflammatory bowel disease (IBD) in dogs and cats. Pol J Vet Sci 14:165–171

Mallonee DH, Harvey CE, Venner M, Hammond BF (1988) Bacteriology of periodontal disease in the cat. Arch Oral Biol 33:677–683

Manchester AC, Hill S, Sabatino B, Armentano R, Carroll M, Kessler B, Miller M, Dogan B, Mcdonough SP, Simpson KW (2013) Association between granulomatous colitis in French Bulldogs and invasive Escherichia coli and response to fluoroquinolone antimicrobials. J Vet Intern Med 27:56–61

Marchesi JR, Adams DH, Fava F, Hermes GD, Hirschfield GM, Hold G, Quraishi MN, Kinross J, Smidt H, Tuohy KM, Thomas LV, Zoetendal EG, Hart A (2016) The gut microbiota and host health: a new clinical frontier. Gut 65:330–339

Marks SL, Rankin SC, Byrne BA, Weese JS (2011) Enteropathogenic bacteria in dogs and cats: diagnosis, epidemiology, treatment, and control. J Vet Intern Med 25:1195–1208

Marsh PD (2000) Role of the oral microflora in health. Microb Ecol Health Dis 12:130–137

Mcmahon LA, House AK, Catchpole B, Elson-Riggins J, Riddle A, Smith K, Werling D, Burgener IA, Allenspach K (2010) Expression of Toll-like receptor 2 in duodenal biopsies from dogs with inflammatory bowel disease is associated with severity of disease. Vet Immunol Immunopathol 135:158–163

Meason-Smith C, Diesel A, Patterson AP, Older CE, Johnson TJ, Mansell JM, Suchodolski JS, Rodrigues Hoffmann A (2016) Characterization of the cutaneous mycobiota in healthy and allergic cats using next generation sequencing. Vet Dermatol 28:71

Mentula S, Harmoinen J, Heikkila M, Westermarck E, Rautio M, Huovinen P, Kononen E (2005) Comparison between cultured small-intestinal and fecal microbiotas in beagle dogs. Appl Environ Microbiol 71:4169–4175

Middelbos IS, Vester Boler BM, Qu A, White BA, Swanson KS, Fahey GC Jr (2010) Phylogenetic characterization of fecal microbial communities of dogs fed diets with or without supplemental dietary fiber using 454 pyrosequencing. PLoS One e9768:5

Minamoto Y, Hooda S, Swanson KS, Suchodolski JS (2012) Feline gastrointestinal microbiota. Anim Health Res Rev 13:64–77

Minamoto Y, Dhanani N, Markel ME, Steiner JM, Suchodolski JS (2014) Prevalence of Clostridium perfringens, Clostridium perfringens enterotoxin and dysbiosis in fecal samples of dogs with diarrhea. Vet Microbiol 174:463–473

Minamoto Y, Otoni CC, Steelman SM, Buyukleblebici O, Steiner JM, Jergens AE, Suchodolski JS (2015) Alteration of the fecal microbiota and serum metabolite profiles in dogs with idiopathic inflammatory bowel disease. Gut Microbes 6:33–47

Moser I, Rieksneuwohner B, Lentzsch P, Schwerk P, Wieler LH (2001) Genomic heterogeneity and O-antigenic diversity of Campylobacter upsaliensis and Campylobacter helveticus strains isolated from dogs and cats in Germany. J Clin Microbiol 39:2548–2557

Neish AS (2013) Redox signaling mediated by the gut microbiota. Free Radic Res 47:950–957

Neish AS, Jones RM (2014) Redox signaling mediates symbiosis between the gut microbiota and the intestine. Gut Microbes 5:250–253

Nielsen HB, Almeida M, Juncker AS, Rasmussen S, Li J, Sunagawa S, Plichta DR, Gautier L, Pedersen AG, Le Chatelier E, Pelletier E, Bonde I, Nielsen T, Manichanh C, Arumugam M, Batto JM, Quintanilha Dos Santos MB, Blom N, Borruel N, Burgdorf KS, Boumezbeur F, Casellas F, Dore J, Dworzynski P, Guarner F, Hansen T, Hildebrand F, Kaas RS, Kennedy S, Kristiansen K, Kultima JR, Leonard P, Levenez F, Lund O, Moumen B, Le Paslier D, Pons N, Pedersen O, Prifti E, Qin J, Raes J, Sorensen S, Tap J, Tims S, Ussery DW, Yamada T, Meta HITC, Renault P, Sicheritz-Ponten T, Bork P, Wang J, Brunak S, Ehrlich SD, Meta HITC (2014) Identification and assembly of genomes and genetic elements in complex metagenomic samples without using reference genomes. Nat Biotechnol 32(8):822

Niemiec BA (2008) Oral pathology. Top Companion Anim Med 23:59–71

Nishiyama SAB, Senhorinho GNA, Giosa MA, Avilo-Campos MJ (2007) Detection of putative periodontal pathogens in subgingival specimens of dogs. Braz J Microbiol 38:23–28

Norris CR, Marks SL, Eaton KA, Torabian SZ, Munn RJ, Solnick JV (1999) Healthy cats are commonly colonized with "Helicobacter heilmannii" that is associated with minimal gastritis. J Clin Microbiol 37:189–194

O'Flynn C, Deusch O, Darling AE, Eisen JA, Wallis C, Davis IJ, Harris SJ (2015) Comparative genomics of the genus Porphyromonas identifies adaptations for heme synthesis within the prevalent canine oral species Porphyromonas cangingivalis. Genome Biol Evol 7:3397–3413

O'Mahony SM, Clarke G, Borre YE, Dinan TG, Cryan JF (2015) Serotonin, tryptophan metabolism and the brain-gut-microbiome axis. Behav Brain Res 277:32–48

O'Neill DG, Church DB, PD MG, Thomson PC, Brodbelt DC (2014a) Prevalence of disorders recorded in cats attending primary-care veterinary practices in England. Vet J 202:286–291

O'Neill DG, Church DB, PD MG, Thomson PC, Brodbelt DC (2014b) Prevalence of disorders recorded in dogs attending primary-care veterinary practices in England. PLoS One 9:e90501

Oh C, Lee K, Cheong Y, Lee SW, Park SY, Song CS, Choi IS, Lee JB (2015) Comparison of the oral microbiomes of canines and their owners using next-generation sequencing. PLoS One 10:e0131468

Packey CD, Sartor RB (2009) Commensal bacteria, traditional and opportunistic pathogens, dysbiosis and bacterial killing in inflammatory bowel diseases. Curr Opin Infect Dis 22:292–301

Panasevich MR, Kerr KR, Dilger RN, Fahey GC Jr, Guerin-Deremaux L, Lynch GL, Wils D, Suchodolski JS, Steer JM, Dowd SE, Swanson KS (2015) Modulation of the faecal microbiome of healthy adult dogs by inclusion of potato fibre in the diet. Br J Nutr 113:125–133

Papadimitriou SA (2016) Detection and quantification of subgingival plaque bacteria in Beagle dogs. J Vet Sci Anim Welf 1:1–5

Papasouliotis K, Sparkes AH, Werrett G, Egan K, Gruffydd-Jones EA, Gruffydd-Jones TJ (1998) Assessment of the bacterial flora of the proximal part of the small intestine in healthy cats, and the effect of sample collection method. Am J Vet Res 59:48–51

Patel N, Colyer A, Harris S, Holcombe L, Andrew P (2016) The prevalence of canine oral protozoa and their association with periodontal disease. J Eukaryot Microbiol 64(3):286–292

Pavlica Z, Petelin M, Juntes P, Erzen D, Crossley DA, Skaleric U (2008) Periodontal disease burden and pathological changes in organs of dogs. J Vet Dent 25:97–105

Perez-Salcedo L, Herrera D, Esteban-Saltiveri D, Leon R, Jeusette I, Torre C, O'Connor A, Gonzalez I, Gonzalez I (2013) Isolation and identification of Porphyromonas spp. and other putative pathogens from cats with periodontal disease. J Vet Dent 30:208–213

Perrone JE (2010) Gingivostomatitis in cats. Vet Tech 31:2

Qin J, Li R, Raes J, Arumugam M, Burgdorf KS, Manichanh C, Nielsen T, Pons N, Levenez F, Yamada T, Mende DR, Li J, Xu J, Li S, Li D, Cao J, Wang B, Liang H, Zheng H, Xie Y, Tap J, Lepage P, Bertalan M, Batto JM, Hansen T, Le Paslier D, Linneberg A, Nielsen HB, Pelletier E, Renault P, Sicheritz-Ponten T, Turner K, Zhu H, Yu C, Li S, Jian M, Zhou Y, Li Y, Zhang X, Li S, Qin N, Yang H, Wang J, Brunak S, Dore J, Guarner F, Kristiansen K, Pedersen O, Parkhill J, Weissenbach J, Meta HITC, Bork P, Ehrlich SD, Wang J (2010) A human gut microbial gene catalogue established by metagenomic sequencing. Nature 464:59–65

Quince C, Lanzen A, Curtis TP, Davenport RJ, Hall N, Head IM, Read LF, Sloan WT (2009) Accurate determination of microbial diversity from 454 pyrosequencing data. Nat Methods 6:639–641

Rajilic-Stojanovic M, Smidt H, De Vos WM (2007) Diversity of the human gastrointestinal tract microbiota revisited. Environ Microbiol 9:2125–2136

Ramadan Z, Zhang P, Jacobs DM, Tavazzi I, Kochhar S (2007) An NMR- and MS-based metabonomic investigation of saliva metabolic changes in feline odontoclastic resorptive lesions (FORL)-diseased cats. Metabolomics 3:113–119

Ramadan Z, Xu H, Laflamme D, Czarnecki-Maulden G, Li QJ, Labuda J, Bourqui B (2014) Fecal microbiota of cats with naturally occurring chronic diarrhea assessed using 16S rRNA gene 454-pyrosequencing before and after dietary treatment. J Vet Intern Med 28:59–65

Reiter AM, Lewis JR, Okuda A (2005) Update on the etiology of tooth resorption in domestic cats. Vet Clin North Am Small Anim Pract 35:913–942. vii

Riggio MP, Lennon A, Taylor DJ, Bennett D (2011) Molecular identification of bacteria associated with canine periodontal disease. Vet Microbiol 150:394–400

Ritchie LE, Steiner JM, Suchodolski JS (2008) Assessment of microbial diversity along the feline intestinal tract using 16S rRNA gene analysis. FEMS Microbiol Ecol 66:590–598

Ritchie LE, Burke KF, Garcia-Mazcorro JF, Steiner JM, Suchodolski JS (2010) Characterization of fecal microbiota in cats using universal 16S rRNA gene and group-specific primers for Lactobacillus and Bifidobacterium spp. Vet Microbiol 144:140–146

Riviere GR, Thompson AJ, Brannan RD, Mccoy DE, Simonson LG (1996) Detection of pathogen-related oral spirochetes, Treponema denticola, and Treponema socranskii in dental plaque from dogs. J Vet Dent 13:135–138

Ryan MP, Adley CC (2014) Ralstonia spp.: emerging global opportunistic pathogens. Eur J Clin Microbiol Infect Dis 33:291–304

Senhorinho GN, Nakano V, Liu C, Song Y, Finegold SM, Avila-Campos MJ (2011) Detection of Porphyromonas gulae from subgingival biofilms of dogs with and without periodontitis. Anaerobe 17:257–258

Senhorinho GN, Nakano V, Liu C, Song Y, Finegold SM, Avila-Campos MJ (2012) Occurrence and antimicrobial susceptibility of Porphyromonas spp. and Fusobacterium spp. in dogs with and without periodontitis. Anaerobe 18:381–385

Simpson K (2013) Managing canine inflammatory bowel disease. Vet Focus 23:29–36

Simpson KW, Dogan B, Rishniw M, Goldstein RE, Klaessig S, Mcdonough PL, German AJ, Yates RM, Russell DG, Johnson SE, Berg DE (2006) Adherent and invasive Escherichia coli is associated with granulomatous colitis in boxer dogs. Infect Immun 74:4778–4792

Smith K, Mccoy KD, Macpherson AJ (2007) Use of axenic animals in studying the adaptation of mammals to their commensal intestinal microbiota. Semin Immunol 19:59–69

Sommer F, Backhed F (2013) The gut microbiota—masters of host development and physiology. Nat Rev Microbiol 11:227–238

Song SJ, Lauber C, Costello EK, Lozupone CA, Humphrey G, Berg-Lyons D, Caporaso JG, Knights D, Clemente JC, Nakielny S, Gordon JI, Fierer N, Knight R (2013) Cohabiting family members share microbiota with one another and with their dogs. Elife 2:e00458

Spor A, Koren O, Ley R (2011) Unravelling the effects of the environment and host genotype on the gut microbiome. Nat Rev Microbiol 9:279–290

Sturgeon AC, Costa M, Weese JS (2012) Preliminary evaluation of the bacterial microbiome of the skin and ear in dogs. Vet Dematol 23:68

Sturgeon A, Stull JW, Costa MC, Weese JS (2013) Metagenomic analysis of the canine oral cavity as revealed by high-throughput pyrosequencing of the 16S rRNA gene. Vet Microbiol 162:891–898

Sturgeon A, Pinder SL, Costa MC, Weese JS (2014) Characterization of the oral microbiota of healthy cats using next-generation sequencing. Vet J 201:223–229

Suchodolski JS (2011) Intestinal microbiota of dogs and cats: a bigger world than we thought. Vet Clin North Am Small Anim Pract 41:261–272

Suchodolski JS, Camacho J, Steiner JM (2008a) Analysis of bacterial diversity in the canine duodenum, jejunum, ileum, and colon by comparative 16S rRNA gene analysis. FEMS Microbiol Ecol 66:567–578

Suchodolski JS, Morris EK, Allenspach K, Jergens AE, Harmoinen JA, Westermarck E, Steiner JM (2008b) Prevalence and identification of fungal DNA in the small intestine of healthy dogs and dogs with chronic enteropathies. Vet Microbiol 132:379–388

Suchodolski JS, Xenoulis PG, Paddock CG, Steiner JM, Jergens AE (2010) Molecular analysis of the bacterial microbiota in duodenal biopsies from dogs with idiopathic inflammatory bowel disease. Vet Microbiol 142:394–400

Suchodolski JS, Dowd SE, Wilke V, Steiner JM, Jergens AE (2012a) 16S rRNA gene pyrosequencing reveals bacterial dysbiosis in the duodenum of dogs with idiopathic inflammatory bowel disease. PLoS One 7:e39333

Suchodolski JS, Markel ME, Garcia-Mazcorro JF, Unterer S, Heilmann RM, Dowd SE, Kachroo P, Ivanov I, Minamoto Y, Dillman EM, Steiner JM, Cook AK, Toresson L (2012b) The fecal microbiome in dogs with acute diarrhea and idiopathic inflammatory bowel disease. PLoS One 7:e51907

Suchodolski JS, Foster ML, Sohail MU, Leutenegger C, Queen EV, Steiner JM, Marks SL (2015) The fecal microbiome in cats with diarrhea. PLoS One 10:e0127378

Sunvold GD, Fahey GC Jr, Merchen NR, Bourquin LD, Titgemeyer EC, Bauer LL, Reinhart GA (1995) Dietary fiber for cats: in vitro fermentation of selected fiber sources by cat fecal inocu-

lum and in vivo utilization of diets containing selected fiber sources and their blends. J Anim Sci 73:2329–2339
Swanson KS, Dowd SE, Suchodolski JS, Middelbos IS, Vester BM, Barry KA, Nelson KE, Torralba M, Henrissat B, Coutinho PM, Cann IK, White BA, Fahey GC Jr (2011) Phylogenetic and gene-centric metagenomics of the canine intestinal microbiome reveals similarities with humans and mice. ISME J 5:639–649
Syed SA, Svanberg M, Svanberg G (1981) The predominant cultivable dental plaque flora of beagle dogs with periodontitis. J Clin Periodontol 8:45–56
Tap J, Mondot S, Levenez F, Pelletier E, Caron C, Furet JP, Ugarte E, Munoz-Tamayo R, Paslier DL, Nalin R, Dore J, Leclerc M (2009) Towards the human intestinal microbiota phylogenetic core. Environ Microbiol 11:2574–2584
Tun HM, Brar MS, Khin N, Jun L, Hui RK, Dowd SE, Leung FC (2012) Gene-centric metagenomics analysis of feline intestinal microbiome using 454 junior pyrosequencing. J Microbiol Methods 88:369–376
Turnbaugh PJ, Ley RE, Mahowald MA, Magrini V, Mardis ER, Gordon JI (2006) An obesity-associated gut microbiome with increased capacity for energy harvest. Nature 444:1027–1031
Turnbaugh PJ, Hamady M, Yatsunenko T, Cantarel BL, Duncan A, Ley RE, Sogin ML, Jones WJ, Roe BA, Affourtit JP, Egholm M, Henrissat B, Heath AC, Knight R, Gordon JI (2009) A core gut microbiome in obese and lean twins. Nature 457:480–484
Valdez M, Haines R, Riviere KH, Thomas DD (2000) Isolation of oral spirochetes from dogs and cats and provisional identification using polymerase chain reaction (PCR) analysis specific for human plaque Treponema spp. J Vet Dent 17:23–26
Vazquez-Baeza Y, Hyde ER, Suchodolski JS, Knight R (2016) Dog and human inflammatory bowel disease rely on overlapping yet distinct dysbiosis networks. Nat Microbiol 1:16177
Wade WG (2013) The oral microbiome in health and disease. Pharmacol Res 69:137–143
Wallis C, Marshall M, Colyer A, O'Flynn C, Deusch O, Harris S (2015) A longitudinal assessment of changes in bacterial community composition associated with the development of periodontal disease in dogs. Vet Microbiol 181(3–4):271–282
Wang X, Heazlewood SP, Krause DO, Florin THJ (2003) Molecular characterization of the microbial species that colonize human ileal and colonic mucosa by using 16S rDNA sequence analysis. J Appl Microbiol 95:508–520
Washabau RJ, Day MJ (2012) Canine & Feline Gastroenterology. VI Diseases of the Gastrointestinal tact, Bacterial infection, Hellicobacter. Elsevier Saunders 622–624
Williams RC (1990) Periodontal disease. N Engl J Med 322:373–382
Winer JN, Arzi B, Verstraete FJ (2016) Therapeutic management of feline chronic gingivostomatitis: a systematic review of the literature. Front Vet Sci 3:54
Xenoulis PG, Palculict B, Allenspach K, Steiner JM, Van House AM, Suchodolski JS (2008) Molecular-phylogenetic characterization of microbial communities imbalances in the small intestine of dogs with inflammatory bowel disease. FEMS Microbiol Ecol 66:579–589
Zarco MF, Vess TJ, Ginsburg GS (2012) The oral microbiome in health and disease and the potential impact on personalized dental medicine. Oral Dis 18:109–120
Zoetendal EG, Akkermans ADL, Akkermans-van Vliet WM, de Visser JGM, de Vos WM (2001a) The host genotype affects the bacterial community in the human gastronintestinal tract. Microb Ecol Health Dis 13:129–134
Zoetendal EG, Ben-Amor K, Akkermans AD, Abee T, De Vos WM (2001b) DNA isolation protocols affect the detection limit of PCR approaches of bacteria in samples from the human gastrointestinal tract. Syst Appl Microbiol 24:405–410
Zoetendal EG, Cheng B, Koike S, Mackie RI (2004) Molecular microbial ecology of the gastrointestinal tract: from phylogeny to function. Curr Issues Intest Microbiol 5:31–48

Recent Advances in the Metagenomics of Marine Mammals Microbiome

18

Chandra Shekar Mootapally, Paresh Poriya, Neelam Mustakali Nathani, Balu Alagar Venmathi Maran, and Indra Ramjibhai Gadhvi

Abstract

In recent years, marine life has gotten highly popularized through research, education, and audio-visual programs especially in exploring marine mammals. Marine mammals are warm-blooded animals, evolved from three different terrestrial groups and adapted to marine environment; few of them spend their entire lifespan in the sea and few come out to sea shore at a particular stage. Further, they act as an indicator of environmental change and ocean health. Seawater and different biological niches provide the base to marine mammals for harboring a rich microbiome that plays a major role in host nutrition, tissue differentiation, health, disease, and immune responses. Microbiota has been observed to be species-specific, related to the evolutionary divergence and host phylogeny. Several researchers are pursuing studies to better understand the marine microbial association, as these megafauna are under threat due to hunting, habitat degradation, and infectious diseases. This expanding knowledge will help in developing strategies for marine mammal's health improvement and their conservation. In this book chapter, we emphasize the microbe-host association in marine mammals and recent metagenomic studies untapping the marine host-specific microbial diversity. It also portrays the unique evolutionary lineages of marine mammals and provides baseline information on normal as well as pathogenic microbiota.

C.S. Mootapally (✉) • P. Poriya • I.R. Gadhvi
Department of Marine Science, Maharaja Krishnakumarsinhji Bhavnagar University, Bhavnagar, Gujarat 364002, India
e-mail: chandu.avi@gmail.com

N.M. Nathani
Department of Life Sciences, Maharaja Krishnakumarsinhji Bhavnagar University, Bhavnagar, Gujarat 364002, India

B.A. Venmathi Maran
Fisheries Science Institute, Chonnam National University, Yeosu 59626, Republic of Korea

R.P. Singh et al. (eds.), *Understanding Host-Microbiome Interactions - An Omics Approach*, DOI 10.1007/978-981-10-5050-3_18

Keywords
Marine • Host-microbiome • Molecular advancements • Omics

18.1 Introduction

Bacteria, viruses, fungi, and protozoa constitute the normal microbiota that is part of the environment. These microbes are also observed residing in the marine mammals that are one of the important members of the marine food chain as well as act as an indicator of marine environment stability (Estes et al. 2011; Wells et al. 2004). Despite numerous conservation efforts, the exploitation of marine niche is leading to extinction of several mammals. Mutual relationship or coexistence of this microbiota with marine mammals is sometimes beneficial for many physiological processes of host including host nutrition, tissue differentiation, and colonization as well as immune development. However, normal microbiota could become opportunistic pathogens when the animals are stressed, immunosuppressed, or under antimicrobial medical treatment (Avalos-Téllez et al. 2010). Also, up to now the studies characterizing the marine mammals' microbiome are very meager in comparison to those on the terrestrial mammals (Bik et al. 2016).

On account of being underwater creatures, many physiological, anatomical and ecological adaptations evolved in marine mammals like the gastrointestinal tract of marine mammals is comparatively longer than terrestrial mammals (Mårtensson et al. 1998). Some marine mammals have uncommon metabolic adaptations, as they can reduce elevated metabolic rates considerably during foraging dives (Hurley and Costa 2001; Williams et al. 2001); still the role of microbial symbionts in digestion remains unsolved in marine mammals. Anaerobic bacteria have been observed in the forestomach of baleen whales and small-toothed whales (Olsen et al. 1994), indicating that microbial fermentation occurs here as it does in the rumen, while the presence of gut microflora in dugong and manatee enables them to digest cellulose and other fibrous carbohydrates (Nishiwaki and Marsh 1985; Lanyon and Marsh 1995). These all may have special effects on composition of gut microbes, but on the whole very few studies have been carried on gut microbes of marine mammals.

The metagenomics era is now changing the outlook of studying the marine mammalian microbiome too, with the latest 2016 news on the huge proportion of novel bacteria observed in the US Navy dolphins by David Relman group, which will be described in detail in the following sections. PubMed search (as of November 2016) using the term "Marine mammal AND metagenomics" shows about 65 publications with 48 in the past decade; manual check revealed that almost 50% of the publications just referred the words in context of their upcoming importance in the field or were used in new software releases using the marine samples as validation. Thus, there are very few studies in limited areas across the globe involved in metagenomics of marine mammals inhabiting microbes. Deagle et al. (2009) explained the utility and advancement of metagenomic technology in dietary studies of marine mammals and suggested that the use of this technology can increase sample sizes and the speed of dietary analysis as well as provide insights into the ecology of prey species in marine and terrestrial ecosystems. Using the advanced approaches, it is also becoming feasible to explore the viral communities in the marine mammals. The viral flora of the sea lion feces has been assessed using the metagenomic approach from the wild and captive California sea lions (*Zalophus californianus*) (Li et al. 2011). Earlier reports have revealed several viruses in the sea lions such as San Miguel sea lion virus (SMSV) (Smith et al. 1973), canine distemper virus (CDV) from the brain tissue (Barrett et al. 2004; Pomeroy et al. 2008), and many others as summarized by the authors including astrovirus (AstV), polyomavirus, anellovirus, gammaherpesvirus, parapoxvirus, retrovirus, and adenovirus (Li et al. 2011). The feces data revealed that the majority of the eukaryotic viruses detected belonged to DNA and RNA viral species not previously reported for marine mammals with astrovirus, bocavirus, and rotavirus dominating the niche. Thus, this study by elucidating the fecal virome of marine mammal added several viruses like sapeloviruses, sapoviruses, noroviruses, bocavirus, and dependovirus in the list of possible mammal-infecting viruses. The sea lions who share the coastal areas could be infected by the terrestrial animals viruses, and the data would further allow a study of these viruses' role as commensals or pathogenic in the host life cycle.

Another fecal virome was very recently analyzed using Ion torrent and Illumina for the fur seal species *Arctocephalus australis* and *Arctocephalus tropicalis* from the Southern Hemisphere (Kluge et al. 2016). The deceased fur seals along the coast were assessed to study the etiology of diseases in marine mammals by means of their viral metagenomics. The fecal was dominated by bacteriophages, and majority of the sequences were found to be lacking similarity with the databases, which are expected to be quite updated at these past years due to increased viral sequencing-based studies. The outcomes revealed a novel species in the *Sakobuvirus* genus, which is the first genome of the genus from marine mammal host, with one member earlier reported from cat (Ng et al. 2014). These two studies seem to be the only virome-characterizing studies by the advanced approaches in the marine mammals till date, and it gives important directions as the two studies involved geographic, host, etc., variations. The results also well depicted that few viruses for instance, sapovirus, were infecting along a huge geographic range infecting the pinniped populations (Li et al. 2011; Kluge et al. 2016; Ng et al. 2009a).

As the relationship of the animal hosts and microbes is complex and includes several factors affecting the microbial composition, metagenomics will allow us to understand the factors such as species specificity of the microbes, geographical influence, organ colonization by selected microbes, disease manifestation, pathogenicity/symbiotic relationship, and their taxonomic-functional variability compared to the sea/sediment microbiomes, if any.

18.2 Marine Mammal-Microbiome Mutualism

Some past studies (Preen 1995) on the fecal analysis of dugongs from Moreton Bay in tropical Australia revealed that it fed primarily on seagrasses, ascidians, colonial ascidians, polychaetes, etc., which explains the omnivory of the Moreton Bay animals. Using the basic knowledge about the marine mammals from culture-based studies and performing the recent metagenomic approach would provide more insight into such kind of evolutionary information and lead to explanations on the relationship of dietary food and associated microbial community in the marine mammals. Mammals serve as multi-genomic as they are composed not only of their own genes but also those of all of their associated microbes (Ley et al. 2008). Metagenomic study can unveil coevolution of the mammals and their indigenous microbial communities.

Recently, Eigeland et al. (2012) explored and characterized a novel microbial ecosystem, viz., the hindgut of dugong, which is the only strict herbivorous medium-sized hindgut fermenting marine mammal. This study indicated marked differences between the hindgut bacterial communities of wild and captive dugongs and influence of ontogeny and diet on hindgut microbial composition and diversity in wild dugongs. Thus, the microbial niche may be unique and have an important role in seagrass fiber degradation in the dugong hindgut. Eigeland et al. (2012) characterized for the first time few of the microbial genera like *Roseburia*, *Clostridium*, *Bacteroides*, *Sedimentibacter* and *Ruminococcus* in the dugong hindgut, which are commonly found in terrestrial herbivores (Brulc et al. 2009; Hess et al. 2011; Nathani et al. 2015). Metagenomics and phylogenetic study can thus unveil the microbiome and its relationship with the host and can also help in understanding of the dugong evolution and adaptation to the marine environment.

Another such recent study on the gut microbiota of southern elephant seals (*Mirounga leonina*) and leopard seals (*Hydrurga leptonyx*), inhabiting Antarctica, suggested that the gut microbiota has coevolved with the wild mammals (Nelson et al. 2013). Also, the results highlighted that the factors like age, captivity, diet, and host species were modulating the gut microbiota composition, along with the impact of other differences like the gut physiology to a lower extent. Prey consumed by the seals also contributed to the alterations in the gut microbiota. The slightly anaerobic gut of the southern elephant seal seemed to be more suitable for the phyla Bacteroidetes compared to other marine hosts. The same group also performed a comparative analysis of the marine and terrestrial mammals and observed that both the niche had different bacterial community structures, with the marine gut more rich in diversity, with *Fusobacteria* as the dominant phyla than the latter mammals.

Metagenomic studies by sequencing of microbial community DNA and targeted sequencing of bacterial 16S ribosomal RNA genes of gut microbiota in different mammalian species including human found that the adaptation of the microbiota to diet is similar across different mammalian lineages (Muegge et al. 2011). Symbiotic relationship between mammals and these bacteria may have contributed in evolution of Mammalia, allowing them to radiate in large numbers to occupy a variety of habitat (Collinson et al. 1991; Ley et al. 2008). Another group of researchers conducted metagenomic study of the hindgut microbiota in the Florida manatees which indicated the presence of novel microbial community in the hindgut of Florida manatees (Merson et al. 2014) and also pointed out the evolution of the sirenians with respect to their gut microbiota.

DNA-based techniques are useful for defining trophic links in a variety of ecosystems, and recently developed metagenomic technologies provide new opportunities for dietary studies in marine mammals (Deagle et al. 2009). A novel *Lactobacillus* species with a new strain of *Lactobacillus salivarius* was isolated with potential for veterinary probiotic applications (Diaz et al. 2013). This type of study will enable the understanding of symbiotic members of the marine mammals' microbiota. Apart from the gut, the microbial niche also resides in the other organs of the marine mammals, and the community composition of different organs such as the skin, mouth, different regions of the gut, etc. is observed to be different among the organs and also from the microbial communities in the surrounding waters (Apprill 2011).

Among the mammalian organs, the skin has the largest surface area and fur to its hardiness serves as the protective barrier against the microbial entry. However, the skin is susceptible to lesions, which allows the inhabitation of microbes on the skin, but the role of microorganisms in the skin is still less studied. Recent study indicated that two genera of bacteria (Bacteroidetes genus *Tenacibaculum* and Gammaproteobacteria genus *Psychrobacter*) were found to be cosmopolitan and abundant associates on humpback whale (*Megaptera novaeangliae*) skin (Apprill et al. 2014a, 2014b). It is possible that these bacterial communities may have some means to maintain their residence on the whale skin and could provide benefits to their host (Nelson et al. 2015). This was assumed as the population of this core microbiome varied in the healthy and the starved whales, thus assuming that the functional microbiome in the gut was also impacting the exogenous microflora composition. Such interpretations lead to prediction that the two genera could be used as a microbial health indicator in the baleen whales.

18.3 Marine Mammal-Microbiome and Pathogenesis

Microbiome, apart from being beneficial to the host, at times turns out to be harmful to the host leading to infections and diseases. Being an integral part of the biosphere, marine parasites are of immense ecological, medical, and economic importance. Bacterial infections associated to pneumonia are frequently developed by free-living cetaceans by the bacterial species belonging to the genera such as *Aeromonas*, *Edwarsiella*, *Klebsiella*, *Pseudomonas*, *Pasteurella*, *Salmonella*, and *Vibrio* recognized to cause strong parasitism under captivity and free-living dolphins (Dunn et al. 2001). A large number of parasitic groups/species have been reported from marine mammals including protozoans, ciliates, flagellates, Nematoda, Trematoda, Cestoda, Insecta, Copepoda, etc. Some protozoans like *Sarcocystis* sp., *Toxoplasma* sp., and *Eimeria phocae* have been reported from numerous marine mammals including pinnipeds, cetaceans, sea otters, and manatee (Dailey 2005). Many zoonotic pathogens are reported from marine mammals including influenza (Geraci et al. 1982; Ohishi et al. 2003; Anthony et al. 2012), *Brucella* (Garner et al. 1997; McDonald et al. 2006; Nymo et al. 2011), *Toxoplasma gondii* (Lapointe et al. 1998; Miller et al. 2001), and *Leptospira interrogans* (Stamper et al. 1998; Stevens et al. 1999). The disease "brucellosis," caused by bacteria *Brucella* spp., is best known for causing abortion in domestic livestock and is also isolated from seals, sea lions, whales, porpoises, and dolphins and appears to be endemic in many marine mammal populations globally. Maratea et al. (2003) isolated an evident infection of *Brucella* spp. from harbor seals (*Phoca vitulina*) and harp seals (*Phocagroenlandica*) stranded along the coast of Southern New England. Molecular characterizations of *Brucella* spp. isolated from marine mammals indicated that they are distinct from those commonly isolated from terrestrial mammals, specifically *B. abortus*, *B. melitensis*, and *B. ovis* (Bricker et al. 2000). Recent metagenomic techniques can highlight more in this regard and can help to guard reproductive physiology of marine mammals. Similarly, Morbillivirus and canine distemper virus (CDV) are causative agents of serious infection in dogs and many other carnivores (Appel 1987), also reported from Lake Baikal seal (*Phoca sibirica*) population (Butina et al. 2010).

Contamination of coastal waters can carry pathogens and toxins that cause human origin disease in marine mammals. Stewart et al. (2014) isolated *Vibrio*, *Escherichia coli*, *Shewanella putrefaciens*, *Pseudomonas fluorescens/putida*, and *Staphylococcus aureus* from bottlenose dolphins in the Southeastern USA. However, all these strains were antibiotic resistant and were observed among wild animals that have not been given antibiotics. Marine environment is well known as the source of novel antibiotics as well as the antibiotic resistant genes (ARGs) (Nathani et al. 2016). These strains associated with marine mammals may have a functional role as antibiotic producers or possess inherent resistance that could be answered by metagenomic techniques. Many parasitological studies concerning marine mammals are being addressed by scientists from different regions of the world. A study by the Relman laboratory is initiated by the group to assess the bacterial population in the dolphins used by the US Navy for purpose of mining and detecting water intruder locations (Bik et al. 2016). The study involved oral bacterial community detection and their comparison with that of the trained sea lions, both kept under same environmental conditions and fed same diet. But both showed varied communities with the dolphins possessing more of the novel unknown bacteria compared to the sea lions, whose microbiota in contrast was more similar to big cats such as tigers, lions, leopards etc., despite similar survival conditions with dolphins. Such studies are believed to provide a path in developing a diagnostic approach (by means of indicative bacterial species) to assess the impact of environmental changes on the mammal health condition, similar to that proposed for indicative microbial community structure change in marine sediments on account of uranium pollution (Suriya et al. 2017). Ecological, physiological, and evolutionary studies of marine mammal parasites in interdisciplinary research can provide huge scientific information in the field of biological oceanography. However, these aspects have been less studied in marine mammals, perhaps due to insufficient sampling or problems of sampling in gigantic ocean. Mortalities caused by parasites also have not been well recognized in the wild for marine mammals, may be due to huge size of ocean.

The higher emerging rate of novel diseases in marine mammals especially viral infections and the persistent difficulties in their characterization due to the methodological boundaries in identifying novel viruses in diseased animals poses an incessant challenge to the scientific groups. With the recent intervention of -omic technologies in this specific area, i.e. viral metagenomics, effective methodologies are now being standardized for identification of novel viruses in diseased animals. In-depth understanding obtained from viral metagenomics-based studies and with the technological advancements, Fahsbender et al. (2015) developed the enzyme-linked immunosorbent assay (ELISA) for detection of *Zalophus californianus anellovirus* (ZcAV) pathogenesis in live sea lions which enabled the option to understand the epidemiology of ZcAV as well as it allowed measurement of the ZcAV pathogenicity during the period of seroconversion in wild and captive sea lions. One of the other important case studies involving viral metagenomics involves the concern regarding the mortality episode of harbor seals in the year 2000 at California. The outcomes revealed that the cause of mortality was through a novel seal anellovirus (SealAV), which was later also predicted as the cause for seals death in the year 2008. However, in 2009, the SealAV genome was fully studied by metagenomics of the stored lung sample of the mortal seals at California (Ng et al. 2011). The SealAV was observed in a comparatively lesser seal samples of the event, and also no confirmation on the virus SealAV comprising seals being positive for necropsies was observed; hence, the hypothesis of this anellovirus being the causative agent for the deaths was not proved. But as both studies showed its presence, it was concluded that anellovirus commonly infects marine mammals impacting their health and immune function. Both the works unveiled the importance of determining the variants and taxonomy of SealAV to understand the hazardous infections in marine mammals (Ng et al. 2011). With the use of -omic technology, novel single-stranded DNA virus (STTV1) from a sea turtle fibropapilloma was identified, and its genome described was characterized. It was the second ssDNA viruses known in reptiles (Ng et al. 2009b).

In a surveillance investigation considering the effect of urbanization levels on the marine mammals during the past decade, Greig et al. (2014) revealed low prevalence of the *Salmonella*, *Campylobacter*, *Giardia*, and *Cryptosporidium* genera in the feces of stranded and wild-caught seals. However, it has been reported that in the same study, *Toxoplasma gonidii, Sarcocystis neurona*, and type A influenza were only detected in the wild-caught harbor seals, whereas antibody titers to *Leptospira* spp. were detected in stranded and wild-caught seals. The study highlights the role of marine mammals like harbor seals as sentinel species for zoonotic and terrestrial pathogens.

The emergence of new strains of microbes and natural mutations are of great concern and challenges nowadays. Anthony et al. (2012) reported the emergence of an avian influenza virus (H3N8) in New England harbor seals (*Phoca vitulina*) that had naturally acquired mutations and were known to increase transmissibility and virulence in mammals. The genomic study of this case also indicated that this avian influenza virus was similar to a virus circulating in North American waterfowl but with mutations that indicate recent adaption to mammalian hosts. Thus, the emergence of this virus may be a sign of the appearance of an H3N8 influenza clade with potential for persistence and cross-species transmission (Anthony et al. 2012).

18.4 Conclusion and Future Perspectives

The research works compiled in this chapter clearly indicate the dearth of information on the marine mammalian microbiota using the advanced metagenomic approaches. The chapter also indicates the importance and vital role of newer molecular techniques and bioinformatics in unraveling the microbial ecology in the different organs of marine mammals. Studies on the gut and fecal species reveal the host-specific bacterial populations to be involved in feed utilization depending on herbivorous or carnivorous diet. Viral metagenomics is also an area of interest for elucidating the etiology of disease in marine mammals and the viral taxa involved in the same. Further studies on the functional gene profile of the microbes will give detailed information on the mechanisms involved in either commensalism or pathogenicity of the microbe with the marine mammal. Conservation strategies can also be planned by using the species as health or threat indicators, as many of these mammals are under extinction threat. The marine mammal host-microbiome interaction is highly attracting conservationists and ecologists to get insight into the marine mammals' unique microbial niche.

Acknowledgment The authors are thankful to the Science and Engineering Research Board, Government of India, for support in the form of Early Career Research Award-National Post Doctoral Fellowship to CSM (Grant No. PDF/2016/001239) and NMN (Grant No. PDF/2016/000190).

References

Anthony SJ, Leger JAS, Pugliares K et al (2012) Emergence of fatal avian influenza in New England harbor seals. mBio 3(4):e00166-12. doi:10.1128/mBio.00166-12

Appel MJG (1987) Canine distemper virus. In: Appel MJ (ed) Virus infections of carnivores. Elsevier, New York, NY, pp 133–159

Apprill A (2011) Humpback whales harbor a combination of specific and variable skin bacteria. Environ Microbiol Rep 3:223–232

Apprill A, Robbins J, Eren AM et al (2014a) Humpback whale populations share a core skin bacterial community: towards a health index for marine mammals? PLoS One 9(3):e90785. doi:10.1371/journal.pone.0090785

Apprill A, Robbins J, Eren AM et al (2014b) Humpback whale populations share a core skin bacterial community: towards a health index for marine mammals? PLoS One 9(3):e90785. doi:10.1371/journal.pone.0090785

Avalos-Téllez R, Suárez-Güemes F, Carrillo-Casas EM et al (2010) Bacteria and yeast normal microbiota from respiratory tract and genital area of bottlenose dolphins (*Tursiops truncatus*). In: Mendez-Vilas A (ed) Current research, technology and education topics in applied microbiology and microbial biotechnology. Formatex Pub, Malaga, pp 666–673

Barrett T, Wohlsein P, Bidewell CA et al (2004) Canine distemper virus in a Californian sea lion (*Zalophus californianus*). Vet Rec 154:334–336

Bik EM, Costello EK, Switzer AD et al (2016) Marine mammals harbor unique microbiotas shaped by and yet distinct from the sea. Nat Commun 7:10516. doi:10.1038/ncomms10516

Bricker BJ, Ewalt DR, Mac Millan AP et al (2000) Molecular characterization of Brucella strains isolated from marine mammals. J Clin Microbiol 38:1258–1262

Brulc JM, Antonopoulos DA, Berg Miller ME et al (2009) Gene-centric metagenomics of the fiber-adherent bovine rumen microbiome reveals forage specific glycoside hydrolases. Proc Natl Acad Sci 106:1948–1953. doi:10.1073/pnas.0806191105

Butina TV, Denikina NN, Belikov SI (2010) Canine distemper virus diversity in Lake Baikal seal (*Phoca sibirica*) population. Vet Microbiol 144:192–197

Collinson ME, Hooker JJ, Skelton PW et al (1991) Fossil evidence of interactions between plants and plant-eating mammals. Philos Trans R Soc Lond B Biol Sci 333:197–208. http://rstb.royalsocietypublishing.org/content/333/1267/197.abstract

Dailey MD (2005) Parasites of marine mammals. In: Rohde K (ed) Marine Parasitology. Cisro Publishing, CAB International, Wallingford, pp 408–414

Deagle BE, Kirkwood R, Jarman SN (2009) Analysis of Australian fur seal diet by pyrosequencing prey DNA in faeces. Mol Ecol 18:2022–2038. doi:10.1111/j.1365-294X.2009.04158.x

Diaz MA, Bik EM, Carlin KP et al (2013) Identification of Lactobacillus strains with probiotic features from the bottlenose dolphin (*Tursiops truncatus*). J App Microbiol 115:1037–1051. doi:10.1111/jam.12305

Dunn LJ, Buck JD, Robeck TR (2001) Bacterial diseases of cetaceans and pinnipeds. In: Dierauf LA, Gulland FMD (eds) Handbook of marine mammal medicine: health, disease, and rehabilitation. CRC Press, Boca Raton, FL, pp 309–335

Eigeland KA, Lanyon JM, Trott DJ et al (2012) Bacterial community structure in the hindgut of wild and captive dugongs (*Dugong dugon*). Aquat Mamm 38(4):402–411. doi:10.1578/AM.38.4.2012.402

Estes JA, Terborgh J, Brashares JS et al (2011) Trophic downgrading of planet Earth. Science 333:301–306

Fahsbender E, Rosario K, Cannon JP et al (2015) Development of a serological assay for the sea lion (*Zalophus californianus*) Anellovirus, ZcAV. Sci Rep 5:9637. doi:10.1038/srep09637

Garner MM, Lambourn DM, Jeffries SJ et al (1997) Evidence of Brucella infection in Parafilaroides lungworms in a Pacific harbor seal (*Phoca vitulina richardsi*). J Vet Diagn Invest 9:298–303

Geraci JR, DJ ST Aubin IK, Barker RG et al (1982) Mass mortality of harbor seals: Pneumonia associated with influenza A virus. Science 215:1129–1131

Greig DJ, Gulland FMD, Smith WA et al (2014) Surveillance for zoonotic and selected pathogens in harbor seals *Phoca vitulina* from central California. Dis Aquat Org 111:93–106

Hess M, Sczyrba A, Egan R et al (2011) Metagenomic discovery of biomass-degrading genes and genomes from cow rumen. Science 331:463–467. doi.org/10.1126/science.1200387

Hurley JA, Costa DP (2001) Standard metabolic rate at the surface and during trained submersions in adult California sea lions (*Zalophus californianus*). J Exp Biol 204:3273–3281

Kluge M, Campos FS, Tavares M et al (2016) Metagenomic survey of viral diversity obtained from feces of subantarctic and south american fur seals. PLoS One 11(3):e0151921. doi:10.1371/journal.pone.0151921

Lapointe JM, Duignan FJ, Marsh AE et al (1998) Meningoencephalitis due to a *Sarcocystis neurona*-like protozoan in Pacific Harbor Seals (*Phoca vitulina richardsi*). J Parasitol 84:1184–1189

Lanyon JM, Marsh H (1995) Digesta passage time in the dugong. Aust J Zool 43:119–127

Ley RE, Hamady M, Lozupone C et al (2008) Evolution of mammals and their gut microbes. Science 320(5883):1647–1651

Li L, Shan T, Wang C et al (2011) The fecal viral flora of california sea lions. J Virol 85(19):9909–9917

Maratea J, Ewalt DR, Frasca S Jr et al (2003) Evidence of *Brucella* sp. infection in marine mammals stranded along the coast of southern New England. J Zoo Wild Med 34:256–261

Mårtensson PE, Nordøy ES, Messelt EB et al (1998) Gut length, food transit time and diving habit in phocid seals. Polar Biol 20:213–217

Miller MA, Crosbie FR, Sverlow KW et al (2001) Isolation and characterization of *Sarcocystis* from brain tissue of a free living southern sea otter (*Enhydra lurris nereis*) with fatal meningoencephalitis. Parasitol Res 87:252–257

McDonald WL, Jamaludin R, Mackereth G et al (2006) Characterization of a *Brucella* sp. strain as a marine-mammal type despite isolation from a patient with spinal osteomyelitis in New Zealand. J Clin Microbiol 44:4363–4370

Merson SD, Ouwerkerk D, Gulino LM et al (2014) Variation in the hindgut microbial communities of the Florida manatee, *Trichechus manatus latirostris* over winter in Crystal River, Florida. FEMS Microbiol Ecol 87:601–615. doi:10.1111/1574-6941.12248

Muegge BD, Kuczynski J, Knights D et al (2011) Diet drives convergence in gut microbiome functions across mammalian phylogeny and within humans. Science 332(6032):970–974. doi:10.1126/science.1198719

Nathani NM, Patel AK, Chandra Shekar M et al (2015) Effect of roughage on rumen microbiota composition in the efficient feed converter and sturdy Indian Jaffrabadi buffalo (*Bubalus bubalis*). BMC Genomics 16:1116. doi:10.1186/s12864-015-2340-4
Nathani NM, Chandra Shekar M, Dave BP (2016) Sedimentomics – exploring the microbial treasures in deep marine environments. J Investig Genomics 3(2):00047. doi:10.15406/jig.2016.03.00047
Nelson TM, Rogers TL, Carlini AR et al (2013) Diet and phylogeny shape the gut microbiota of Antarctic seals: a comparison of wild and captive animals. Environ Microbiol 15:1132–1145. doi:10.1111/1462-2920.12022.
Nelson TM, Apprill FA, Mann J (2015) The marine mammal microbiome: current knowledge and future directions. Microbiol Aust 36:8–13. doi:10.1071/MA15004
Ng TFF, Suedmeyer WK, Wheeler E et al (2009a) Novel anellovirus discovered from a mortality event of captive California sea lions. J Gen Virol 90:1256–1261. doi:10.1099/vir.0.008987–0
Ng TFF, Charles M, Kelly B (2009b) Discovery of a novel single-standard DNA virus from a sea turtle fibropapilloma by using viral metagenomics. J Virol 83:2500–2509. doi:10.1128/JVI.01946-08
Ng TFF, Elizabeth W, Denise G et al (2011) Metagenomic identification of a novel anellovirus in pacific harbor seal (*Phoca vitulina richardsii*) lung samples and its detection in samples from multiple years. J Gen Virol 92:1318–1323
Ng TFF, Mesquita JR, Nascimento MSJ et al (2014) Feline fecal virome reveals novel and prevalent enteric viruses. Vet Microbiol 171:102–111. doi:10.1016/j.vetmic.2014.04.005
Nishiwaki M, Marsh H (1985) Dugong. In: Ridgway SH, Harrison R (eds) Handbook of marine mammals. Academy, New York, NY, pp 1–32
Nymo IH, Tryland M, Godfroid J (2011) A review of *Brucella* infection in marine mammals, with special emphasis on *Brucella pinnipedialis* in the hooded seal (*Cystophora cristata*). Vet Res 42:93. doi:10.1186/1297-9716-42-93
Ohishi K, Ninomiya A, Kida H et al (2003) Influenza virus infection in seal (Phocidae): seroepidemiological survey of influenza virus in Caspian seals (*Phoca caspica*). Comp Immunol Microb 28:39–44
Olsen MA, Nordoy ES, Blix AS et al (1994) Functional anatomy of the gastrointestinal system of the northeastern atlantic minke whale (*Balaenoptera acutorostrata*). J Zool London 234:55–74
Pomeroy LW, Bjornstad ON, Holmes EC (2008) The evolutionary and epidemiological dynamics of the *paramyx oviridae*. J Mol Evol 66:98–106
Preen A (1995) Diet of dugongs: are they omnivores? J Mammal 76(1):163–171
Smith AW, Akers TG, Madin SH et al (1973) San Miguel sea lion virus isolation, preliminary characterization and relationship to vesicular exanthema of swine virus. Nature 244:108–110
Stevens E, Lipscomb TP, Gulland FMD (1999) An additional case of leptospirosis in a harbor seal. J Wildl Dis 35:150
Stewart JR, Townsend FI, Lane SM et al (2014) Survey of antibiotic-resistant bacteria isolated from bottlenose dolphins *Tursiop struncates* in the southeastern USA. Dis Aquat Org 108:91–102. doi:10.3354/dao02705
Suriya J, Chandra Shekar M, Nathani NM et al (2017) Assessment of bacterial community composition in response to uranium levels in sediment samples of sacred Cauvery River. Appl Microbiol Biotechnol 101(2):831–841. doi:10.1007/s00253-016-7945-2
Wells RS, Rhinehart HL, Hansen LJ et al (2004) Bottlenose dolphins as marine ecosystem sentinels: developing a health monitoring system. Eco health 1:246–254
Williams TM, Haun J, Davis RW et al (2001) A killer appetite: metabolic consequences of carnivory in marine mammals. Comp Biochem Physiol A Mol Integr Physiol 129:785–796

Advancements in Molecular Techniques and Bioinformatics for Understanding the Rumen Microbiome

19

Neelam M. Nathani, Amrutlal K. Patel, Chandra Shekar Mootapally, Ramesh K. Kothari, and Chaitanya G. Joshi

Abstract

Microbial diversity in the forestomach is of major importance to the ruminants. Rumen microbiome diversity and gene tag predictions are important areas of scientific interests due to their global impacts including methane emission control and ruminant health. Earlier the knowledge of rumen biodiversity was entirely dependent on the anaerobic culturing approach. However, advancements in molecular techniques have revealed hitherto uncovered features of the rumen microbiome. Next-generation sequencing in combination with conventional techniques has provided significant opportunities to understand animal performance in context to rumen ecology and subsequent nutrient manipulation. Metagenomic sequencing generates enormous data making its informatics

N.M. Nathani (✉)
UGC-CAS Department of Biosciences, Saurashtra University, Rajkot 360 005, Gujarat, India

Department of Life Sciences, Maharaja Krishnakumarsinhji Bhavnagar University, Bhavnagar, Gujarat 364002, India
e-mail: neelam.nathani@yahoo.com

A.K. Patel
Hester Biosciences Limited, Pushpak, Panchvati Circle, Ahmedabad 380 006, Gujarat, India

C.S. Mootapally
Department of Animal Biotechnology, College of Veterinary Science & Animal Husbandry, Anand Agricultural University, Anand 388 001, Gujarat, India

Department of Marine Science, Maharaja Krishnakumarsinhji Bhavnagar University, Bhavnagar, Gujarat 364002, India

R.K. Kothari
UGC-CAS Department of Biosciences, Saurashtra university, Rajkot 360 005, Gujarat, India

C.G. Joshi
Department of Animal Biotechnology, College of Veterinary Science & Animal Husbandry, Anand Agricultural University, Anand 388 001, Gujarat, India

R.P. Singh et al. (eds.), *Understanding Host-Microbiome Interactions - An Omics Approach*, DOI 10.1007/978-981-10-5050-3_19

handling and analysis intricate due to the immense complexity of the data. Bioinformatic resources are rapidly emerging to outwit the intricacies. In this book chapter, we highlight the advances in the metagenomic analysis that help to improve the genomic aspect of the tree of life by getting an in-depth understanding of the *Candidatus* uncultured population of the rumen microbiome focusing on correlation between the microbial functioning and animal efficiency. It also describes the strategies and specific bioinformatic approaches that can be applied to metagenomic sequences for accurate and in-depth metagenomic analysis, data binning, and genome reconstruction of individual microbes.

Keywords
Microbial diversity • Ruminants • Meta-omics • Next generation sequencing

19.1 Introduction

Ruminants, considered as one of the highest successful clusters among the herbivorous beings of the globe, comprise of ~200 species. These are widespread across the world as millions of wild and domestic animals (Hackmann and Spain 2010). Ruminant animals are grouped based on their way of feed digestion, which includes the forestomach (aka the rumen) that facilitates the partial digestion of feed by the persistent microbial diversity in rumen, before it enters the true stomach. Ruminants are unable to produce the degradation enzymes to utilize the complex polysaccharides of the feed material; it is the rumen that imparts the niche to the huge microbiota which in turn fulfill the feed digestion by their interactive metabolic potentials. The metabolic processes lead to formation of varied volatile fatty acids (VFAs) which are used by the ruminants as one of the major energy/nutrient source. These majorly contribute to the animal health, homeostasis, and productivity (Hendersen et al. 2015). Thus, the rumen-possessing group of animals since earlier times has undergone evolutionary adaptations at anatomical and behavioral levels such that the rumen-residing microbes can utilize a variety of different feedstuffs (Hofmann 1989). This host-microbe relationship has also been the core reason behind the innumerable benefits of their domestication including milk, meat and other products of animal origin; also their religious importance cannot be neglected.

19.2 Rumen Microbes

Ruminal microbes, viz., prokaryotic: bacterial and archaeal populations and eukaryotic: protozoa and fungal populations, viruses, and bacteriophage populations are vital to the host's life span (Nathani et al. 2013). And on account of their major function being digestion of feeds, this microbial population of rumen has been categorized based on their associations with the feed particles such as the cellulolytic microbes, amylolytic, proteolytic, lipolytic, or hemicellulolytic, overall accounting as the bio-hydrogenating rumen population. Also, this microbial population is observed to be diverse based on several variable factors including the type of diet, breed of the ruminant host, age group, concentration of dietary components, etc. Specifically, the complexity of the substrates in the feed uptake of ruminants leads to difference in the microbial community depending on the animal and their respective diet conditions. This subsequently leads to microbiota variation; however, in contrast to other niches, the rumen microbial diversity is observed to be higher at the lower taxa with high strain variability and lesser at the higher phylum-like levels. Thus, along with its high micro-diversity levels, a vast range of substrate availability makes it a continuous fermenter, with high number of microbes producing a huge amount of fermentation intermediates and end products. The degradation is carried out in a systematic manner with a set of core microbes actively participating as primary degraders by carrying out breakdown of the complex feed biomolecules, and the intermediate molecules hence produced are then utilized by different set of microbes; both these groups of microbial species are most often different in their abundance (Wallace 2008). Thus, in-depth knowledge of the rumen microbiome as a whole, including the detailed study of the not so dominant yet important members of the microbial taxa, is reforming the rumen research studies. The advancements in the sequencing and molecular techniques have not left apart this field of microbial ecological studies. These techniques have led to huge number of data inflow on the ruminant anaerobes that are difficult to cultivate due to their stringent requirements (Nathani et al. 2015a, b, c; Patel et al. 2014). More and more understanding of the rumen microbiome will help in addressing several environmental questions, including the methane gas emission, and will also help in understanding the ruminant host-microbiota relationship to improve the feed intake, digestibility, and the overall feed efficiency (Firkins and Yu 2015). Hence, rumen microbial community studies are key to understanding conversions of feed molecules to ruminant products and fecal waste for continuous developments in the nutrient utilization efficiency by ruminant hosts.

19.2.1 Rumen Prokaryotic Population

Rumen can be characterized as being dominantly comprising of bacteria with almost >10^{11} bacteria per gram of the rumen content, constituting almost 50% of the cell biomass. Traditional and basic knowledge on the rumen bacteria is totally relied upon the highly specific microbiological techniques developed for cultivation of the strict anaerobes. The extremely diverse population of bacteria in rumen is well adapted to the highly specific conditions naming few such as the low to neutral pH (5.0–7.0), temperature around 40 °C, anoxic environment amidst the feed particles, and continuously produced fermented products. Due to all these strict conditions, the culturing-based isolation of rumen microbes has not been as much successful as it has been for other niches. The efforts of the large number of research groups in the 1990s and the following decade had formed the base for rumen bacteria knowledge and understanding. Starting with the very first mass culture attempt of rumen cellulolytic bacteria by von Tappeiner in the 1880s till successful isolation of numerous bacteria from bovine rumen by Hungate were the outcomes that paved the path for understanding the existent rumen bacterial diversity and their metabolic capacities based on morphological, microscopic, and culturing approaches. This with time led to several media component standardization for rumen microbiologists to pursue their studies with accuracy and ease.

Among the rumen prokaryotic microbiota, the ruminal archaea account for 0.3–3.3% (Janssen and Kirs 2008) and are well targeted by the researchers due to their methane production process. Methane emitted from ruminants is responsible for increasing the global warming problem to a huge extent (Kumar et al. 2012). And hence to find a strategy to reduce the methane emission from rumen sources, in-depth understanding of the archaeal community is extremely vital (Kumar et al. 2015). Rumen methanogens genera are often observed to be falling within the orders Methanobacteriales, Methanomicrobiales, and Methanosarcinales of the phylum Euryarchaeota. Genus Methanobrevibacter is the most dominantly observed falling under Methanobacteriales. Similar to the bacterial community, many members of methanogens that are observed by culture-independent advanced approaches have outnumbered those studied by the traditional approaches (Zhou et al. 2010).

19.2.2 16S rDNA-Based Bacterial Diversity

A number of genera and species of rumen origin have been increasing on account of the several studies being performed on rumen bacteria isolation (Creevey et al. 2014; Denman et al. 2015). Creevy and colleagues in a survey based on five culture collections had reported that the number of genera described by culturing of rumen microbes has reached >85 and these are falling under nine different phylas (Creevey et al. 2014). Though Bacteroidetes have been reported to be the dominant phylum in rumen, the cultivation-based reports revealed that the majority of the members belonged to Firmicutes, thus, indicating that Bacteroidetes are more difficult to isolate and need more understanding of their survival in rumen for developing strategies to grow them in vitro. Nevertheless, in the past few years, emphasis on rumen bacterial studies has relied on the culture-independent methods like the 16S rRNA gene thus expanding the rumen microbes' database by great amount of data (Kamra 2005). Small subunit 16S rDNA sequences are commonly used for bacterial community evaluation in various niches because of its dual characteristic having conserved and variable regions. Numerous studies by the use of 16S rDNA sequencing studied the bacterial community composition of ruminants (Brulc et al. 2009; Pitta et al. 2010; Hess et al. 2011; Jami and Mizrahi 2012). The results of all these studies could spread light on the fact that, similar to other niches, in rumen also the majority of the bacteria remain uncultured, and these novel, hitherto not studied, bacterial species could be identified and added to the taxonomic tree of the rumen microbes using advanced approached. Few of the earlier studies on cattle rumen fluid showed that the major 16S clones from the source were best clustering with the low G + C

Gram-positive bacteria, most of them highly related to the *Clostridium* genus (Whitford et al. 1998). The other major group involved the species falling under the *Prevotella* and *Bacteroides* genus. These implications during the tenure could be well attributed to the then available database, which was more reliant on the cultured data.

Earlier, due to the lack of sequencing platforms, the studies on 16S amplicon involved the tedious and lengthy cloning procedure. But with recent advancements and cost reduction in the sequencing techniques, 16S-based amplicon analysis is a common approach chosen to study the bacterial community composition of rumen animals and also to check the community shifts under different variables like the animal host, breeds, age group, environmental rearing conditions, dietary treatments, and forage content. The bacterial colonization in the rumen of goats from 80 to 110 day of age was performed using the V4–V5 hypervariable regions of 16S rRNA by high-throughput next-generation sequencing (NGS) (Han et al. 2015). The study reported a striking observation that the Bacteroidetes abundance was increased with the increase in age, while Firmicutes and Synergistetes constituted to be the most abundant phyla during the lower age groups. Also, the group observed that the abundance of unclassified/uncultured bacteria was significantly lowering with increase in age, suggesting that the adult goat rumen bacteria are more studied compared to the younger ones. Thus, the sequencing ease has led to determination of rumen bacteria to a broader perspective including the knowledge of the abundance of the difficult to cultivate bacteria. However, the 16S amplicon-based approach could reveal the structure and dynamics of the bacterial species, but the question on who does what remained unclear for the uncultured population which was answered by the shotgun genomics and metagenomics approach (sequence-based or function-based) to study the rumen bacteria.

19.2.3 Rumen Eukaryotes

Apart from the prokaryotic organisms, fungi and protozoa are also well known for their role in the host homeostasis maintenance. Also, the highly intricate genetic architecture of the rumen microbiome involves the interactions/associations of the four major microbial groups viz., the bacteria, fungi, archaea, and protozoa, for proper rumen functioning and ruminant health. Rumen fungi are lower in numbers with about 10^6 cells per milliliter of rumen content. The rumen-residing anaerobic fungi are also reported to play substantial role in the digestion of feed fibers by means of the vast array of degradation enzymes encoded by their genomes (Dagar et al. 2011; Gruninger et al. 2014). The feed fibers are primarily colonized by the fungi, which further facilitates the colonization of other microbes, viz., the bacteria, protozoa, etc. (Sehgal et al. 2008). This well predicts that the fungal population in the rumen has a great impact on the abundance of bacteria and other fiber-degrading microbes (Kittelmann et al. 2012). Most of the rumen fungi identified from rumen sources are observed to be falling under the phylum Neocallimastigomycota (Liggenstoffer et al. 2010).

Rumen protozoa, apart from the fungus, are also observed in the eukaryotic population. These participate in the carbohydrate and protein digestion. Several ciliate species are observed in the rumen, constituting ~50% of the total microbial biomass, and reported to digest one fourth to half of the total fiber content in the feed intake (Shah et al. 2015). There is very meager data on rumen protozoa due to their complex polymorphic nature making their cultivation difficult. Bacteria, fungi, and protozoa together are involved in the degradation of host indigestible feed material, and the metabolic process releases hydrogen molecules (Akin et al. 1988) that are in turn utilized by the ruminal archaea aka hydrogen scavengers (Janssen and Kirs 2008). These microbial interactions and reliance on each other had quite earlier been proved by coculture studies (Joblin et al. 1989) and were again verified in the recent study on fiber degradation wherein the hydrogen utilization by methanogens was observed as a metabolism key step (Piao et al. 2014).

19.3 Advancements in Microbial Community Assessment Using -Omics Approaches

The composition and role of rumen bacteria are now being deeply assessed using the whole rumen microbial DNA study using the NGS techniques. Metagenomics and metatranscriptomics revealed the overall and active functional profiles of bacteria in the rumen content. During the development of metagenomics as an emerging field with vast applications, the softwares used to analyze the sequences are needed to be validated for the actual representation of the bacterial abundance in the rumen as revealed by the software output. One such study was conducted to check the statistical correlation between the metagenomic MG-RAST software and the convention real-time PCR outcomes (Nathani et al. 2013). Forty-eight samples of rumen content (24 fluid and solid content each) from eight river buffalos (Bubalus bubalis) fed on varied diets were assessed for the abundance of four bacterial species using both the MG-RAST-based taxonomic profiling of shotgun metagenomic sequences and absolute quantification using the qPCR. The outcomes of the study indicate that the shotgun sequencing approach to describe metagenome analysis/annotation by MG-RAST was reliable with statistically similar outcomes of both the approaches. Few recent studies using advanced techniques based on molecular markers, such as rRNA gene/ITS regions, have also added information to the rumen fungal taxonomic data. Several novel taxonomic clades unable to be classified in the available taxonomic rank also indicate that the fungal population of the rumen is still in dearth of information due to less cultured representatives available (Liggenstoffer et al. 2010; Kittelmann et al. 2012).

Thus, the sequence-based direct analyses of nucleic acid are providing deeper insights into the microbiome functioning and the subsequent impact on the host. In rumen, the microbiome is known to be affected by the change in type of diet and also based on the host animal. One such study well depicts the correlation between microbes and their functional dynamics with increased roughage incorporation in the diet. Four animals of the strong Jaffrabadi buffalo breed of India were fed with green forage and four with dry forage content under three dietary treatments as roughage to concentrate ratios 50:50, 75:25, and 100:0 for 6 weeks each (Nathani et al. 2015a). The results revealed significant variations in the bacterial phyla and genus levels between the liquid and solid portions of the rumen fluid, and the bacterial genera in the three-dietary treatment were also specific to each treatment. High roughage content influenced the population by increasing the fibrolytic phyla members and reducing the Bacteroidetes population. One of the important findings was in the study on the correlations between the phylum in both green and dry roughage, wherein the abundance of Bacteroidetes was observed to positively influence the higher abundance of bacterial phylum Verrucomicrobiae, Proteobacteria, and Tennericutes. Thus, this data would add knowledge to the microbial phyla interactions and their efficient degrading action in symbiosis. The diet also influenced the functional genes involved in volatile fatty acid production and methane formation; thus this confirmed the influence of diet on the host microbiota (Shanks et al. 2011).

Similar effect of the diet variation was studied on the carbohydrate enzyme profile in the Mehsana breed of buffalo (Patel et al. 2014). The outcomes clearly revealed the impact of roughage in diet on the enzyme profile with oligosaccharide-degrading enzymes in increased proportions as compared to low roughage-fed animals, and pectin lyase was observed to be the most abundant gene tag found in the green forage-fed diet samples. Both the studies well describe the alterations in the rumen microbiota and their corresponding functional genetic repertoire in response to dietary composition changes. Apart from these, the knowledge of individual species genomic data will provide better understanding of each microbes' role in the feed digestion, and their mutual relationship can be hypothesized based on analysis of their substrate utilization profile and metabolic potentials.

19.4 Rumen Microbial Genomics and In Silico Binning of Genomes

Genomic analysis of microorganism provides the details of its genetic basis behind its morphology and metabolic abilities. Hence, genomic studies are highly carried out since the past few years. Hosts like humans and environments such as soil are well studied with a huge number of representative microbial information in the respective databases. However, rumen niche still needs exploration, and with this aim, several projects like the Hungate 1000 genome project had been initiated to improve the database of rumen microbial genomes (Creevey et al. 2014). The major objectives of this project include performing methane mitigation and rumen feed improvement technologies based on the genomic data developed through the study and also to use the genome-enabled research for understanding the rumen function. The project has resulted into >200 genomic sequences of rumen bacteria and archaea being added to the database, which also includes several novel-type strain data.

Apart from these, several rumen bacteria and archaea genomes are sequenced and have revealed the efficacy of these microbes in utilizing the feed particles in host rumen. The genome sequence of *Fibrobacter succinogenes* S85 revealed its mechanism in degrading cellulose (Suen et al. 2011). To access the cellulose, it is important to first remove the hemicellulose and pectins, and the genome clearly indicated the presence of a huge array of enzymes involved in hemicellulose degradation. Thus, the unique pathway of the bacteria gave an idea of its role in animal health and performance. Another study has reported a novel species of *Clostridium* and subspecies of *Bacillus neaslsoni* isolated from rumen having a gene repertoire for utilizing polysaccharides (Nathani et al. 2014, 2015b). Also, the first genome of rumen methanogen *Methanobrevibacter ruminantium* revealed insights into the cellular processes of rumen methanogen which further was proposed for use as vaccine-mediated inhibition of rumen methanogens to reduce the ruminant-based greenhouse gas emissions (Leahy et al. 2010).

In continuation to the above studies that included the genomic analysis from cultured rumen isolates, now the advancements in bioinformatics are allowing genomic study of individual uncultured genomes from metagenomics data. Bioinformaticians around the globe, to a great extent, are currently involved in developing algorithms and user-friendly softwares for reconstructing individual species genomes from metagenomic data. These advancements in the assembly and other sequence-dependent parameters are bringing about a cutting-edge improvement in the field of metagenomics study of ecological niches. Due to the large micro-diversity, persistent in the rumen niche, practically at this point, it is as of yet difficult to apply the strategies of sequence parameters like GC, K-mer frequency, etc., to the rumen data for separation of individual genomes, as these parameters would not be able to bin/distinguish the huge diversity at lower taxonomic ranks like subspecies/strain. Yet, there are finger-countable number of studies which have initiated and succeeded in reconstructing genomes from rumen data. Hess et al. (2011) could reconstruct 15 genomes of hitherto uncultured rumen microbes from 268 Gb of cattle metagenomic data. Nathani et al. (2015c) also reconstructed *Prevotella ruminicola* genomes from cattle and buffalo rumen metagenomic datasets by homology-based approach and observed significant niche-adapted genetic features in the species in comparison with the same species from another source. Thus, genomics, metagenomics, and metatranscriptomics are leading a way ahead in rumen host-microbiome research understanding. Major implications of the data would include the use of efficient enzymes in agricultural and industrial areas and reduction of methane emission as a way of environmental hazard prevention strategy.

19.5 Gene Mining of Agro-industrially Important Enzymes from Rumen-Omics Data

Ruminants have the uniqueness in their biodegradation ability and converting the feed particles to the high-quality products for human benefits including the meat and nutritious milk. All this is possible only by the highly efficient microbiota that degrade the feedstuffs, which the host cannot otherwise utilize by itself (Flint et al. 2008). The microbiota does this by their systematic enzyme machinery which is very well enhanced compared to the same enzymes from other niches. These facts about the enzymes from the natural continuous fermenter are seeking the attention of numerous scientists across the globe to mine the genes encoding these enzymes for use in the feed industry, biofuel, textile, and enumerable other industrial applications (Selinger et al. 1996; Wang et al. 2010).

In this advanced era of -omic technologies, mining gene is extremely an important aspect as it allows screening and characterization of enzymes from the 85% of the uncultured rumen bacteria (Morgavi et al. 2013). Many metagenomics and metatranscriptomic studies have mined and characterized lignocellulose-degrading enzymes from the rumen microbiome (Dai et al. 2015). The enzymes studied and the former genes mined from the studies are expanding the fibrolytic gene database such as the CAZy (Carbohydrate-Active Enzymes) database by the novel (having lesser percent homology/nucleotide similarity with database available gene sequences) enzyme-coding gene sequences with unique structural domains and biochemical properties from the bacteria and fungus inhabiting a specialized niche like the rumen (Ribeiro et al. 2016). As far as this application of -omics technology to understand the host microbiome relationship is concerned, it is obvious that the metatranscriptome-based gene mining would give the actual representation of the active enzymes in the rumen ecosystem, providing accurate insight into the metabolic potential of the rumen, which was also observed by several groups (Qi et al. 2011; Dai et al. 2015). The information also provides an insight into the rumen microbial community shifts.

These studies are also helping to again improve the rumen digestibility by incorporating the efficiently expressing enzyme from one host to that potentially limiting in that enzyme activities by means of feed uptake. Also, knowledge about the lack of specific-enzyme families from the rumen source provides knowledge that can be used to perform feed trials of the absent enzyme from other relevant sources to ruminants to check their impact. For example, studies reveal that combination of rumen-originating and aerobic cellulases such as the glycoside hydrolase families 7 and 12 resulted into release of glucose + xylose from the barley and hay feedstuff materials (Badhan et al. 2014; Riley et al. 2014).

In addition to the lingo-cellulosic enzymes, rumen also harbors the phytase-like enzymes, which make phosphorous available by acting on the phytate molecule and which are usually absent in the monogastric animals (Konietzny and Greiner 2004). Mootapally et al. (2016) mined a phytase gene of 1251 bp length and 417 amino acids from the rumen metagenomics assembly data of Mehsana breed of buffalo. The gene showed 92% similarity with the histidine acid phosphatase (HAP) domain of *Prevotella* sp. The enzyme was characterized for its optimum requirements for highest expression activity, and based on the results that showed a 52,000-fold higher enzyme activity to the commercially available enzyme from plant origin, the enzyme proved to be a potential candidate for use as additional feed supplement in monogastric (birds, horses, rabbits etc.) animal farming.

19.6 Rumen Microbiome Deep Sequencing and Methanogenesis

Rumen digestive function is known to release significant concentration methane gas leading to global concerns (Denman et al. 2015) and is considered as the second highest anthropogenic promoter for greenhouse threats. This process also has a negative impact on the host as it leads to energy of almost up to 10% (Johnson and Johnson 1995). In context to this, several attempts are being done by researchers to develop nutritional strategies that lead to metabolic pathways yielding lower methane. The shifts in microbiota composition in relation to high- or low-methane yield provide more and more information to improve the feed strategies for reduction of methane emission.

The main target for the studies is the methanogens and in small ruminant-like sheep, there is an advantage to distinguish them based on their amount of methane yield. A metagenomic and metatranscriptomic study on the low- and high-yielding sheep revealed different methanogenic enrichments in the two types with lactate formation as the chosen alternative in the less methane-yielding hosts (Kamke et al. 2016). The two samples showed clear difference in the bacterial abundance, with high-yield sheep rumen microbiota dominated by the Lachnospiraceae and Ruminococcaceae families, while in contrast the low-yield sheep rumen microbiota revealed almost tenfold higher abundance of the genus *Sharpea* falling under the Erysipelotrichaceae family. Also, the low-yield animals seemed to be highly expressing the genes for the pyruvate » lactate » volatile fatty acids (butyrate/propionate) as compared to the high methane-yielding animals. Thus, confirming the theory that a heterofermentative environment (here small rumen of sheep) leads to lower methane emissions. And as the results pointed toward the specific bacterial families that lead to this, the knowledge about their metabolic pathways can be further brought to practical applications for manipulating the microbiota in other ruminants by nutrient modification. Similar metagenomic shotgun study on the gene-level functional diversity in the methane formation pathways in the rumen of Mehsana breed buffalo fed on different roughage proportions led to significant observations on the archaea population that can be used to provide roughage-specific diet to mitigate the methane emission caused by ruminants (Singh et al. 2015). Euryarchaeota was the most dominant phyla observed among the five phyla including Crenarchaeota, Thaumarchaeota, Koraarchaeota, and Nanoarchaeota found to be representing the 0.8–1.8% of the archaea population of the Indian breed of buffalo. Overall, the group observed that hydrogenotrophic methanogens were more abundant in the buffalo rumen with few acetogenic bacteria also observed to be existent in the samples.

Another study including the application of basic concepts undertook the rumen microbial metagenomic analysis posttreatment of goats by a halogenated methane analog for inhibition of methanogenesis. The administration of the anti-methanogenic compound to the goats resulted into the increase in the gene expression level of propionate production via succinate pathway as compared to the acrylate pathway as observed in the precious case of sheep rumen (Denman et al. 2015). Also, the analysis depicted the metagenomic sequences for the pathway dominantly belonged to the *Prevotella* and *Selenomonas* genus. Thus, the reduction in methane directly or indirectly associates with the microbial composition shifts and subsequent changes in the metabolic gene content.

Conclusion

In this book chapter, we have compiled the advancements in the approach of studying the rumen microbiome ecology and functions. Since long, the rumen microbiome has since been known as an excellent source of agro-industrially important microbial enzymes. Advancements in the molecular techniques are allowing study of the uncultured rumen microbial population, and this is leading to enormous data outputs, that need to be properly stored and are distributed for a wider livestock-related personnel to be able to use and interpret it for improving the host health and productivity. Bioinformatics is developing at a very rapid rate, and this is allowing easier storage, retrieval, and analysis of rumen microbiome -omics data. The data is giving insight into the efficient fermentation pathways, the microbial dynamics, and symbiosis with effect to dietary, host, and environmental variables as well as the knowledge about how animal traits and genotype of the microbial communities shape the methane emission in the host will allow developing strategies to reduce ruminant-caused green gas release. More efforts toward applying the advanced algorithms and statistics to the sequence data will further enhance the rumen microbial database filling the gaps in the taxonomic tree of rumen microbes.

Acknowledgements The authors are thankful to Department of Science and Technology (DST), Government of India for providing support to NMN in the form of INSPIRE fellowship.

References

Akin D, Borneman W, Windham W (1988) Rumen fungi: morphological types from Georgia cattle and the attack on forage cell walls. Biosystems 21:385–391. doi:10.1016/0303-2647(88)90037-8
Badhan AY, Wang R, Gruninger D et al (2014) Formulation of enzyme blends to maximize the hydrolysis of alkaline peroxide pretreated alfalfa hay and barley straw by rumen enzymes and commercial cellulases. BMC Biotechnol 14:31. doi:10.1186/1472-6750-14-31
Brulc JM, Antonopoulos DA, Miller ME et al (2009) Gene-centric metagenomics of the fiber-adherent bovine rumen microbiome reveals forage specific glycoside hydrolases. Proc Natl Acad Sci U S A 106:1948–1953
Creevey CJ, Kelly WJ, Henderson G et al (2014) Determining the culturability of the rumen bacterial microbiome. Microb Biotechnol 7(5):467–479
Dagar SS, Kumar S, Mudgil P et al (2011) D1/D2 domain of large-subunit ribosomal DNA for differentiation of *Orpinomyces spp.* Appl Environ Microbiol 77:6722–6725. doi:10.1128/AEM.05441-11
Dai X, Tian Y, Li J et al (2015) Metatranscriptomic analyses of plant cell wall polysaccharide degradation by microorganisms in the cow rumen. Appl Environ Microbiol 81:1375–1386. doi:10.1128/AEM.03682-14
Denman SE, Fernandez GM, Shinkai T et al (2015) Metagenomic analysis of the rumen microbial community following inhibition of methane formation by a halogenated methane analog. Front Microbiol 6:1087
Firkins JL, Yu Z (2015) Ruminant nutrition symposium: how to use data on the rumen microbiome to improve our understanding of ruminant nutrition. J Anim Sci 93:1450–1470. doi:10.2527/jas2014-8754
Flint HJ, Bayer EA, Rincon MT et al (2008) Polysaccharide utilization by gut bacteria: Potential for new insights from genomic analysis. Nat Rev Microbiol 6:121–131. doi:10.1038/nrmicro1817
Gruninger RJ, Puniya AK, Callaghan TM (2014) Anaerobic fungi (phylum *Neocallimastigomycota*): advances in understanding their taxonomy, life cycle, ecology, role and biotechnological potential. FEMS Microbiol Ecol 90:1–17. doi:10.1111/1574-6941.12383
Hackmann TJ, Spain JN (2010) Invited review: ruminant ecology and evolution: perspectives useful to ruminant livestock research and production. J Dairy Sci 93:1320–1334

Han X, Yang Y, Yan H et al (2015) Rumen bacterial diversity of 80 to 110-day-old goats using 16S rRNA sequencing. PLoS One 10(2):e0117811. doi:10.1371/journal.pone.0117811

Hendersen G, Cox F, Ganesh S et al (2015) Rumen microbial community composition varies with diet and host, but a core microbiome is found across a wide geographical range. Sci Rep 5:Article ID: 14567. doi:10.1038/srep14567

Hess M, Sczyrba A, Egan R et al (2011) Metagenomic discovery of biomass-degrading genes and genomes from cow rumen. Science 331:463–467

Hofmann RR (1989) Evolutionary steps of ecophysiological adaptation and diversification of ruminants: a comparative view of their digestive system. Oecologia 78:443–457

Jami E, Mizrahi I (2012) Composition and similarity of bovine rumen microbiota across individual animals. PLoS One 7:e33306

Janssen PH, Kirs M (2008) Structure of the archaeal community of the rumen. Appl Environ Microbiol 74:3619–3625. doi:10.1128/AEM.02812-07

Joblin K, Campbell GP, Richardson A et al (1989) Fermentation of barley straw by anaerobic rumen bacteria and fungi in axenic culture and in co-culture with methanogens. Lett Appl Microbiol 9:195–197. doi:10.1111/j.1472-765X.1989.tb00323.x

Johnson KA, Johnson DE (1995) Methane emissions from cattle. J Anim Sci 73:2483–2492

Kamke J, Kittelmann S, Soni P et al (2016) Rumen metagenome and metatranscriptome analyses of low methane yield sheep reveals a Sharpea-enriched microbiome characterised by lactic acid formation and utilization. Microbiome 4:56. doi:10.1186/s40168-016-0201-2

Kamra DN (2005) Rumen Microbial Ecosystem. Curr Sci 89(1):124–135

Kittelmann S, Naylor GE, Koolaard JP et al (2012) A proposed taxonomy of anaerobic fungi (Class *Neocallimastigomycetes*) suitable for large-scale sequence-based community structure analysis. PLoS One 7:e36866. doi:10.1371/journal.pone.0036866

Konietzny U, Greiner R (2004) Bacterial phytase: potential application, in vivo function and regulation of its synthesis. Braz J Microbiol 35:11–18

Kumar S, Dagar SS, Puniya AK (2012) Isolation and characterization of methanogens from rumen of Murrah buffalo. Ann Microbiol 62:345–350. doi:10.1007/s13213-011-0268-8

Kumar S, Indugu N, Vecchiarelli B et al (2015) Associative patterns among anaerobic fungi, methanogenic archaea, and bacterial communities in response to changes in diet and age in the rumen of dairy cows. Front Microbiol 6:781. doi:10.3389/fmicb.2015.00781

Leahy SC, Kelly WJ, Altermann E et al (2010) The genome sequence of the rumen methanogen *Methanobrevibacter ruminantium* reveals new possibilities for controlling ruminant methane emissions. PLoS One 5(1):e8926. doi:10.1371/journal.pone.0008926

Liggenstoffer AS, Youssef NH, Couger M et al (2010) Phylogenetic diversity and community structure of anaerobic gut fungi (phylum *Neocallimastigomycota*) in ruminant and non-ruminant herbivores. ISME J 4:1225–1235. doi:10.1038/ismej.2010.49

Mootapally CS, Nathani NM, Patel AK et al (2016) Mining of ruminant microbial phytase (RPHY1) from metagenomic data of Mehsani buffalo breed: identification, gene cloning, and characterization. J Mol Microbiol Biotechnol 26(4):252–260. doi:10.1159/000445321

Morgavi DP, Kelly WJ, Janssen PH et al (2013) Rumen microbial (meta)genomics and its application to ruminant production. Animal 7:184–201. doi:10.1017/S1751731112000419

Nathani NM, Patel AK, Dhamannapatil PS et al (2013) Comparative evaluation of rumen metagenome community using qPCR and MG-RAST. AMB Express 3(1):55. doi:10.1186/2191-0855-3-55

Nathani NM, Duggirala SM, Bhatt VD et al (2014) Genomic analysis of a novel strain of *Bacillus nealsonii*, isolated from Surti buffalo rumen. Adv Biosci Biotechnol 5(3):235–245. doi:10.4236/abb.2014.53030

Nathani NM, Patel AK, Chandra Shekar M et al (2015a) Effect of roughage on rumen microbiota composition in the efficient feed converter and sturdy Indian Jaffrabadi buffalo (*Bubalus bubalis*). BMC Genomics 16:1116. doi:10.1186/s12864-015-2340-4

Nathani NM, Duggirala SM, Chandra Shekar M et al (2015b) Isolation of chitinolytic *Clostridium* sp. NCR from Mehsani buffalo rumen, its genomic analysis and potential role in rumen. Genomics Data 5:109–111. doi:10.1016/j.gdata.2015.05.017

Nathani NM, Kothari RK, Patel AK et al (2015c) Functional characterization reveals novel putative coding sequences in *Prevotella ruminicola* genome extracted from rumen metagenomic studies. J Mol Microbiol Biotechnol 25:292–299

Patel DD, Patel AK, Parmar NR et al (2014) Microbial and carbohydrate active enzyme profile of buffalo rumen metagenome and their alteration in response to variation in the diet. Gene 545:88–94. doi:10.1016/j.gene.2014.05.003

Piao H, Lachman M, Malfatti S et al (2014) Temporal dynamics of fibrolytic and methanogenic rumen microorganisms during in situ incubation of switchgrass determined by 16S rRNA gene profiling. Front Microbiol 5:307. doi:10.3389/fmicb.2014.00307

Pitta DW, Pinchak E, Dowd SE et al (2010) Rumen bacterial diversity dynamics associated with changing from bermuda grass hay to grazed winter wheat diets. Microb Ecol 59:511–522

Qi M, Wang P, O'Toole N et al (2011) Snapshot of the eukaryotic gene expression in musk-oxen rumen - A metatranscriptomic approach. PLoS One 6:e20521. doi:10.1371/journal.pone.0020521

Ribeiro GO, Gruninger RJ, Badhan A et al (2016) Mining the rumen for fibrolytic feed enzymes. Anim Front 6(2):20–26. doi:10.2527/af.2016-0019

Riley R, Salamov AA, Brown DW et al (2014) Extensive sampling of basidiomycete genomes demonstrates inadequacy of the white-rot/brown-rot paradigm for wood decay fungi. Proc Natl Acad Sci U S A 111:9923–9928. doi:10.1073/pnas.1400592111

Sehgal J, Jit D, Puniya A et al (2008) Influence of anaerobic fungal administration on growth, rumen fermentation and nutrient digestion in female buffalo calves. J Anim Feed Sci 17:510–518

Selinger LB, Forsberg CW, Cheng K-J (1996) The Rumen: A unique source of enzymes for enhancing livestock production. Anaerobe 2(5):263–284

Shah RK, Patel AK, Shah TM et al (2015) Analysis of community structure and species richness of protozoa enriched rumen metagenome from Indian Surti by shotgun sequencing. Curr Sci 111(1):184–191

Shanks OC, Kelty CA, Archibeque S et al (2011) Community structures of fecal bacteria in cattle from different animal feeding operations. Appl Environ Microbiol 77(9):2992–3001. doi:10.1128/AEM.02988-10

Singh KM, Patel AK, Shah RK et al (2015) Potential functional gene diversity involved in methanogenesis and methanogenic community structure in Indian buffalo (*Bubalus bubalis*) rumen. J Appl Genet 56:411–426. doi:10.1007/s13353-015-0270-0

Suen G, Weimer PJ, Stevenson DM et al (2011) The complete genome sequence of *Fibrobacter succinogenes* S85 reveals a cellulolytic and metabolic specialist. PLoS One 6(4):e18814. doi:10.1371/journal.pone.0018814

Wallace RJ (2008) Gut microbiology–broad genetic diversity, yet specific metabolic niches. Animal 2:661–668

Wang L, Mavisakalyan V, Tillier ERM, Clark GW, Savchenko AV, Yakunin AF, Master ER (2010) Mining bacterial genomes for novel arylesterase activity. Microb Biotechnol 3(6):677–690

Whitford MF, Forster RJ, Beard CE et al (1998) Phylogenetic analysis of rumen bacteria by comparative sequence analysis of cloned 16S rRNA genes. Anaerobe 4(3):153–163

Zhou M, Hernandez-Sanabria E, Guan LL (2010) Characterization of variation in rumen methanogenic communities under different dietary and host feed efficiency conditions, as determined by PCR-denaturing gradient gel electrophoresis analysis. Appl Environ Microbiol 76(12):3776–3786

Part V

Marine Microbiome and Microbial Interaction

Multi-omic Approaches for Mapping Interactions Among Marine Microbiomes

20

Shubhrima Ghosh, Rameshwar Tiwari, R. Hemamalini, and S.K. Khare

Abstract

The marine ecosystems, teeming with microscopic life, provide an interesting avenue for studying host-microbe interactions. Microbes are mostly found to interact with other organisms as a community, and such relationships range from predation and symbiosis to pathogenesis. Their interactions with their immediate environmental niche also deserve considerable attention, as the microbiome is the source of major energy and carbon flux in the ocean biosystem. New-generation multi-omic approaches help us gather and analyze large amounts of data for comprehensive understanding. While metagenomics helps us know the taxonomic profile, metatranscriptomics and metaproteomics help us derive the functional angle. Metabolomics of microbiomes on the other hand has been employed to study the resulting metabolic products of such interactions. This diverse research have involved various consortia of institutions in recent times, resulting in the discovery of important linkages between nutrient exchanges, cell signaling, and biogeochemical processes. The article focuses on the current developments and challenges in the vast area of omic-based marine research.

Keywords

Marine microbiome • Host-microbe interaction • Metagenomics • Metatranscriptomics • Metaproteomics • Metabolomics

S. Ghosh • R. Tiwari • R. Hemamalini • S.K. Khare (✉)
Enzyme and Microbial Biochemistry Laboratory, Department of Chemistry, Indian Institute of Technology Delhi, Hauz Khas, New Delhi 110016, India
e-mail: skkhare@chemistry.iitd.ac.in; skhare@rocketmail.com

R.P. Singh et al. (eds.), *Understanding Host-Microbiome Interactions - An Omics Approach*, DOI 10.1007/978-981-10-5050-3_20

20.1 Introduction

Microorganisms are the most ubiquitous of all organisms, present in all environments ranging from polar caps to deep-sea vents. They can be free-living or associated with host organisms, behaving as a single functional unit. Many of these microbes have actually taken the same evolutionary route as their host. Ambitious projects such as Tara Oceans and Global Ocean Sampling have provided us with unprecedented amounts of information about the distribution and variety of microbes present in the oceans (Moran 2015). Thus, it is pertinent to look at these interactions through the omic approaches of metagenomics, metaproteomics, metatranscriptomics, and metabolomics, which provide useful information about their correlations and their effect on the immediate environment. The article focuses on some major marine microbiomes.

20.1.1 Coral Microbiome

Coral reefs are one of the most diverse terrains in the marine ecosystem. They provide habitats for a number of organisms such as bacteria, sea anemones, fishes, and turtles. The interactions between coral and microbes mostly form a symbiotic, mutualistic relationship. Some of these microbes have been shown to promote coral health and protect against pathogens. Under stress condition such as temperature rise and acidification, these bacteria are also affected (Fisher et al. 2012). Studies show that beneficial bacteria such as *Pseudoalteromonas sp.* ZJ6102 and *Pseudoalteromonas euthinica* which live on the surface of healthy *Acropora cervicornis*, when challenged with pathogenic *V. shiloi*, produced an antibacterial response.

The dinoflagellate *Symbiodinium* is the most widely known eukaryotic microbe to have a symbiotic association with corals, sea anemones, sponge, flatworms, mollusks, and ciliates. They enter the host cell through phagocytosis, persist as intracellular symbionts, reproduce, and distribute to the environment. Corals grow in warm, nutrient-deficit waters, and *Symbiodinium* provides them with 90% of photosynthetic carbon (Fisher et al. 2012). Though perceived as a single species, study of the genetic markers has shown diversity of species (Rowan and Powers 1992).

20.1.2 Sponge Microbiome

Sponges belong to the phylum Porifera and are repositories of diverse microbial communities, which constitute about 35% of the sponge biomass. About 25 distinct phyla of bacteria have been reported from sponges only (Webster and Taylor 2012). Most of this influx comes from the constant filter-feeding activities. Major genera found in the sponge microbiome revealed though 16S rRNA profiling are *Actinobacteria*, *Gammaproteobacteria*, *Alphaproteobacteria*, *Betaproteobacteria*, *Deltaproteobacteria*, *Cyanobacteria*, *Chloroflexi*, and *Acidobacteria*, among others

(Webster and Taylor 2012). They are remarkable models to study the host-metazoan interactions in the sea. Genomic approaches using next-generation sequencing into the host and the symbiont genome have enabled the study of these intimate partnerships (Hentschel et al. 2012). There is little similarity to be found in the species structure; however, sponge-specific symbionts are more common rather than opportunists (Thomas et al. 2016). Photosynthetically fixed cyanobacteria provide 50% of food requirements for the sponges. In addition, they are also responsible for a number of diseases in sponges, the incidence of which has increased with elevated temperatures due to climate change. Mukherjee et al. (2009) found a collagenolytic enzyme from a pathogen in sponges of the Great Barrier Reef, which has the capability to degrade the sponge. The capacity of these organisms has also been explored in recent years for their potential for pharmacological compounds.

20.1.3 Seaweed Microbiome

Seaweeds or macroalgae are diverse photosynthetic organisms which play the role of primary producers in the marine ecosystem. They harbor a large number of epiphytic microbial communities which are essential for their morphological development, algal health, and protection against secondary colonization (Singh and Reddy 2014). Thus, they interact as a unified functional entity, what is known as the holobiont. Various factors such as algal metabolites, levels of oxygen and carbon dioxide, and pH determine the nature of surface environment available to the microbes. Bacteria of the genus *Cyanobacteria*, *Bacillus*, *Marinomonas*, *Vibrio,* and *Pseudoalteromonas* have been found to occur in macroalgal microbiomes (Egan et al. 2013). Transcriptomic studies into the algae *Laurencia dendroidea* revealed the occurrence of major bacterial categories of nitrogen-fixing *Cyanobacteria* and aerobic heterotrophic *Proteobacteria* (de Oliveira et al. 2012). Lachnit et al. (2015) also studied the interactions of viruses with the red macroalga, *Delisea pulchra.* Virome sequencing revealed the presence of dsDNA viruses of the genus *Totivirus*, which infects plant pathogenic fungi, and ssRNA virus of *Picornavirales* genus, a pathogen of marine diatoms. They could possibly play the role of algal pathogens too.

20.1.4 Marine Mammal Microbiome

The number of known marine mammal totals to about 130 living and near-extinct species, distributed across the vast oceans and seas of the world. They play the roles of predators and primary and secondary consumers. Microbes have a close association with marine mammals and the major locations where they have been studied include the gut, respiratory tracts, oral cavity, and skin (Nelson et al. 2015). Bik et al. (2016) studied bacterial communities in dolphins and sea lions, through 16S rRNA sequencing. The widest variety have been recorded in oral, gut and chuff specimens. *Proteobacteria*, *Bacteroidetes*, *Actinobacteria*, *Tenercutes*, *Spirochaetes*, and *Fusobacteria* make up the major taxa. Interestingly, this group found that the microbiota of the surrounding seawater is quite different from what is found on the body of the mammals. The microbes also play a vital role in the hosts' disease manifestation and immune response often causing a decrease in the population. For example, the respiratory tract of cetaceans is found to be a hotbed of pathogenic bacterial genera such as *Plesiomonas*, *Aeromonas*, *Escherichia*, *Clostridium and Pseudomonas*, *Burkholderia*, *Mycobacterium*, *Haemophylus*, *Streptococcus*, and *Staphylococcus* (Nelson et al. 2015).

20.2 Interactions in Marine Microbiomes

The marine systems work as a mutualistic cooperative consortium, whole of which sustains and evolves together under natural selection processes. Due to this interconnectedness, it is often impossible to obtain culturable bacteria from marine samples. Thus, cultivation-independent techniques such as high-throughput genomics, proteomics, and metabolomics could be employed to study the molecular basis for these interactions. The different types of marine microbiome interactions can be broadly classified as follows:

20.2.1 Symbiosis

Symbiosis is defined as the close long-term relationship between two different species and is one of the major interactions found in the marine environment. Such interactions are either beneficial, detrimental, or not significantly affecting in terms of both the species. Factors, including host-derived nutrients, chemicophysical characteristics, and host properties (like immune response), determine the composition and structure of symbiont communities. Symbiosis can be further distinguished into mutualism (both species are benefitted), commensalism (one species is benefitted, while the other is not significantly affected), and parasitism (one species is benefitted, other is negatively affected).

One classic example of symbiosis in marine ecosystems is cnidarian-algal-protist interactions. The dinoflagellates such as *Symbiodinium* exist in the endoderm of cnidarians such as corals, jellyfish, and sea anemones. They provide photosynthetic products to the cnidarian species, while the microbe gets inorganic molecules and a place to harbor in exchange (Baker 2003; Iglesias-Prieto et al. 1992).

Hermit crabs living inside the shell of gastropods is an instance of commensalism. The larvae of roundworm often reside in fishes in a parasitic relationship. Another example is *Phronima*, a tiny amphipods which use salps, a species of zooplankton, as their hosts. *Phronima* eat the animal and burrow into the center of the gelatinous shell, creating a living barrel, in which the *Phronima* live, raising their young and eating food that floats by (Laval 1978).

20.2.2 Pathogenesis

There is prevalence of toxic alga and harmful bacteria in the marine systems. They affect corals, fishes, bivalves, and other fauna. Marine pathogenesis is also important for humans as they often cause diseases in organisms consumed as seafood and can be a major health concern. *Vibrio* is the biggest genus of naturally occurring marine pathogens. It is expected that due to climate change, the levels of marine pathogens will increase causing ramifications in coral diseases and subsequent bleaching. Turner et al. (2016) exposed the mussels, *Perna viridis*, to simulated climate change and subjected them to harmful bacteria and/or toxin-producing dinoflagellates. They found increase in the toxin-pathogen load with significant relation to climate change. Influx of pollutants in oceans results in the release of toxins from organisms such as algae, a condition also known as harmful algal blooms. Paralytic shellfish poisoning (PSP), caused by *Alexandrium fundyense*, is a significant harmful algal threat. Often sewage-borne bacteria such as *Escherichia coli* and *Enterococcus* species also end up in oceans from effluent channels.

20.2.3 Predation

Predation as an interaction is most explored in case of mammals and eukaryotic microbes. However, bacteria of the genus *Halobacteriovorax* have been found to be a predator found on the surface of coral reefs. In a library of 198 16S rRNA samples spanning three coral genera, 79% were positive carriers of *Halobacteriovorax* (Welsh et al. 2016).

Studies have also explored the role of marine viruses as predators and parasites. Their major involvement results from their capability to induce genetic changes in the host organisms, affecting their life cycles as well as evolutionary outcomes. They affect a number of phytoplanktons, while virally encoded proteins modify the marine host genomes which influence their photosynthetic, phosphate recycling, carbon metabolism, and apoptotic processes (Rohwer and Thurber 2009).

20.3 Omics Approaches to Study the Marine Microbiome

The vast knowledge that the ocean microbiome upholds has been mined in recent years through high-throughput approaches of metagenomics, metatranscriptomics, metaproteomics, and metabolomics. With the advent of 16S rRNA sequencing and culture-free methods in the last four decades, scientists now have an access to gigantic sets of data and equally equipped analysis techniques to annotate and derive information from them. Some of these state-of-the-art techniques are discussed below.

20.3.1 Metagenomics Approaches for Studying Interactions in Marine Microbiome

The interaction between marine microbiome is presented by metagenomic approach by various researchers. A wide array of microbial population including algae, fungi, archaea, bacteria, and protozoa are detected in high abundance. The DNA-based approach demonstrated the richness and evenness of the microbial population in marine holobiont system. The researchers demonstrated that the microbial community comprised of marine holobiont is completely different from the neighboring water body. This study revealed the coevolved symbiotic relationship between microorganisms and marine habitat (Dimijian 2000).

The dominance of the microbial groups is found to be very specific with their host coral symbiont, and this research has been extended with the development of molecular tools for DNA sequencing particularly the next-generation parallel sequencing. The operational taxonomic units (OTUs)-based determination of coral microbiome revealed low richness and evenness of coral-associated microbial population as compared with surrounding seawater bodies. This trend describes coral holobionts as a very selective microbial population for a particular coral. However, it is difficult to distinguish the specific coral microbial composition on the basis of coral environmental conditions like temperature, pH, and nutrient availability. Moreover, some symbiotic microbes are very common throughout the coral species from different marine environments. For example, metagenomic-based approach showed that genus *Endozoicomonas* belonging to Gammaproteobacteria has been found in reef-building stony corals like gorgonian coral *Eunicella cavolini* (Bayer et al. 2013a), Red Sea coral *Stylophora pistillata* (Bayer et al. 2013b; Neave et al. 2016), and *Pocillopora verrucosa* (Neave et al. 2016). In *E. cavolini*, *S. pistillata* and *P. verrucosa* corals the highly dominating genus *Endozoicomonas* accounts for 60–90% of sequence read compared with other native microbiomes. Bayer et al. (2013b) designed a genus-specific DNA probe to recognize the localization of genus *Endozoicomonas* in coral tissue through fluorescence in situ hybridization (FISH) technique. The dense cell aggregation of *Endozoicomonas* bacteria was detected in endoderm of *S. pistillata* coral tissue which reflects its established relationship with the corals worldwide. This localization was further explained by Neave et al. (2016), which uses more advanced catalyzed reporter deposition-fluorescence in situ hybridization (CARD-FISH) technique. This technique explained highly close and coral-specific localization of *Endozoicomonas*. This genus lies in the gastrodermis region of *P. verrucosa* and tentacles of *S. pistillata*. The role of *Endozoicomonas* as a major coral holobiont is still unknown. However, some hypotheses like coral metabolic product utilization, prey acquisition, and nutrient-managed quorum sensing are assumed by various researchers. Ultimately, the metagenomic-based studies have proven a unique interaction between *Endozoicomonas* and different coral animals.

Many reports on metagenomic information of coral holobionts have already been published, for example, taxonomic and functional abundance of different coral microbiome including *Fungia echinata* (Badhai et al. 2016), *Millepora alcicornis*

(Santos et al. 2016), *Muricea californica*, and *Muricea fruticosa* (Holm and Heidelberg 2016). From the last decade, the main application of metagenomic approach is to explain the decline of coral reefs worldwide. The metagenomic-based taxonomic analysis decodes the coral holobiont shifting from beneficial to disease-causing microbial community (Ainsworth and Gates 2016; Pandolfi et al. 2005). Thurber et al. (2009) showed that different marine stresses like high temperature, eutrophication, and low pH cause coral disease. The pyrosequencing data analysis revealed that healthy coral microbial community has been abundant in *Cyanobacteria*, *Proteobacteria*, and *Symbiodinium*; however, in stressed coral microbial community, it has shifted toward *Bacteroidetes*, *Fusobacteria*, and fungi of microbes often found on diseased corals (Ainsworth and Gates 2016). Moreover, functional elucidation of metagenomic data through SEED and KEGG analysis revealed that genes involved in antibiotic and toxic compound resistance, stress resistance, oxidative stress, DNA repair, N and S metabolism, secondary metabolism, membrane transport, and signal transduction were found highly abundant in stressed coral holobiont as compared to the healthy coral (Badhai et al. 2016; Thurber et al. 2009).

Since the 1970s, Clive Wilkinson and Jean Vacelet have initiated work on sponge microbiology, and further studies revealed that 40% of the total sponge volume is comprised of microbial community (Wilkinson and FAY 1979; Vacelet and Donadey 1977). With the advancement of molecular techniques, marine microbial community is been unwrapped through metagenomic approach (Taylor et al. 2007). All three domains of microbial life are present in sponge microbial community encompassing diverse interactions ranging from commensalistic to mutualistic and exploitative (Webster and Taylor 2012). Decoding of sponge-microbe interactions displayed that most of the sponges were sharing a few common microbial consortia which again coevolved in nature as sponges and are one of the oldest metazoans (Thomas et al. 2010). For example, the cyanobacterial clade of *Candidatus synechococcus spongiarum* group is highly abundant and distributed worldwide as marine sponge symbiont (Burgsdorf et al. 2015). The global ocean metagenomic survey (Nealson and Venter 2007) revealed many specific functional signatures found in sponge symbiotic cyanobacterial clad (Fan et al. 2012; Thomas et al. 2010). Clusters of orthologous groups (COGs) and KEGGs annotation of various draft genomes retrieved from the metagenomic data demonstrated high abundance of transposable insertion and horizontal gene transfer elements which is a common signature of host-microbe interaction and took part in evolutionary adaptation of microbial symbionts to their specific host (Burgsdorf et al. 2015; Fan et al. 2012). Another specific and important functional signature in sponge-associated microbial community is the prevalence of CRISPR-Cas systems which indicates a high selective pressure against the high load of viruses and phages inside the sponges (Horn et al. 2016). Some eukaryotic functional elements involved in ankyrin repeats, tetratricopeptide repeats, leucine-rich domains, and protein-protein interaction domains are hypothesized as host-specific fingerprints of sponge microbiome (Tian et al. 2014; Gao et al. 2014; Fan et al. 2012).

The most recent single-cell genomic approach also revealed various functional characteristics of sponge-associated microbiota (Raghunathan et al. 2005; Hentschel et al. 2012). For example Siegl et al. (2011) sequenced an exclusive marine sponge candidate, the mixotrophic bacterium *Poribacteria*. The sequencing result displayed several symbiotic factors including tetratricopeptide repeat domain, polyketide synthases (PKSs), ankyrin, and fibronectin type III which proves symbiotic interaction of *Poribacteria* with sponge.

The global climate changes also alter the marine sponge microbial community. Fan et al. (2013) have shown the effect of elevated temperature on the microbial interaction with ecologically important sponge *Rhopaloeides odorabile*. The elevated temperature causes necrosis in sponge, and eventually microbial shift has been found. The necrotic sponge microbial community encompasses all the bacterial groups but with lower abundance as compared with healthy sponge. The abundance of *Vibrionaceae*, *Pseudoalteromonas*, *Colwelliaceae*, *Ferrimonas*, *Oceanospirillaceae*, *Endozoicomonas*, *Arcobacter*, *Marinifillum*, and Fusibacter taxa in necrotic sponge microbial community showed clear difference from healthy sponge (Simister et al. 2012). In conclusion, these alterations in coevolving sponge-microbe symbiotic interactions are a sensitive display of global warming, and further efforts to explore these interaction shifts may be useful.

Seaweed-microbial interaction is also one of the important marine associations between microalgae and microbial community. The seaweed provides substratum for the settlement of microbes, and secreted algal polysaccharides act as carbon and energy source for the growth of bacteria mostly in the form of biofilm (Singh and Reddy 2014). Various attempts have been made to explore seaweed-microbial interactions. Seaweed-associated microbial community may be affected by various factors like host species, host life cycle, geographical locations, and different seasons. Furthermore, limited information is available regarding the metagenomic information on the seaweed-associated microbial community. Burke et al. (2011) revealed that unlike coral or sponge microbial community, seaweed green alga *Ulva australis* is lacking a taxonomically and functionally consistent and common core microbial community. The high richness showed that a large number of bacterial taxa are able to make biofilm on green alga. The author predicted the redundancy and lottery hypothesis behind this uncommon core microbial community. This was earlier proposed by Naeem (1998) in their redundancy hypothesis where functions of ecosystems presented specific roles governed by various species. In seaweed-microbial interaction, the OTUs with high abundance act as a macroorganism which may perform a very selective mechanism in a particular ecosystem. Secondly, the metagenomic data of seaweed-microbial interaction can by interrelated by the lottery hypothesis (Sale 1976). According to this hypothesis, the high abundance of similar bacterial species on the seaweed biofilm from an ecosystem corresponded to the first-come-first-win lottery for space. Furthermore, the invasion of microbial community into green macroalga *Caulerpa racemosa* was investigated by metagenomics (Aires et al. 2013). The identified OTUs of endophytic bacterial community were involved in various functional illustrations such as nitrate-reducing betaproteobacterium, legume nodule forming N_2 fixing bacteria like Burkholderia and sulfur-cycling *Desulfobacteraceae*. This endophytic bacterial community may be considered as "meta-organism" or "holobiont" acting synergistically with the host green macroalga *Caulerpa racemosa*.

20.3.2 Metatranscriptomics Approaches for Studying Interactions in Marine Microbiome

The metatranscriptomic analysis of any environmental sample allows us to understand the functional RNA framework which may not be efficiently covered by the metagenomic approach. In marine ecosystems, the microbial interaction with their respective hosts like coral, sponge, seaweed, etc. has been studied by various researchers. The interaction between coral animal algal symbiont *Zooxanthellae* and its associated microbial community was examined by RNA-based metatranscriptomic analysis (Gust et al. 2014). The metatranscriptomic analysis study revealed that accumulation of emerging marine pollutant 1,3,5-trinitro-1,3,5-triazine (RDX) into coral holobiont was displayed by increasing the expression of xenobiotic detoxification pathways, cytochrome P450s, and UDP glucuronosyltransferase 2 family enzyme. Moreover, the genes involved in photosynthetic energy metabolism and carbon fixation were found to reduce their expression in coral holobiont in the presence of the marine pollutant. The stress-related studies have also been conducted by Pinzón et al. (2015), where coral holobiont displayed reduced immune response-related genes with the effect of continuous bleaching in coral. The impact of diseased coral on the microbial community of coral holobiont was studied by Daniels et al. (2015). The bacterial community displayed 645 differentially abundant gene functions including DNA restriction-modification, phage-associated genes, multidrug efflux pumps. and type II secretion system. Moreover, the alteration in the expression of genes involved in bacterial expression and retrons showed the microbial adaptation toward the changing environmental conditions.

The metatranscriptomic analysis of sponge-microbe interaction revealed various microbial functional responses under different environmental conditions. The identification of bacterial mRNA tags of marine cold-water sponge *Geodia barretti* described high turnover of nitrification involving genes from ammonia-oxidizing archaea (Radax et al. 2012). Moreover, denitrification and anaerobic ammonia oxidation were also detected in the metatranscriptomic data which suggested that complete nitrogen cycle has been governed by the sponge-microbe interaction. The metatranscriptomic analysis of two sponges, namely, *Stylissa carteri* and *Xestospongia testudinaria*, also showed the core sponge-microbe interaction (Ryu et al. 2016). The abundance of fibronectin type III domain, xylose isomerase-like TIM barrel domain, HicB family domain, PIN domain, virus-related domain, mycoplasma protein of unknown function domain, and transposases is responsible for various functional aspects such as maintenance of host-microbe interaction, symbiotic association, niche invasion, toxin-antitoxin production, pathogenicity, viral defense, and genetic mobility across different species. This specialized functional abundance revealed the mechanisms of host-microbe interactions in marine ecosystems (Hentschel et al. 2012).

20.3.3 Metaproteomics

Mainly metagenomic and metatranscriptomic approach has been adopted to mine the unculturable functional microbial potential of any environment. One step ahead, metaproteomics has been established as an encouraging tool to establish microbial activity-based intrinsic metabolic function display. Despite DNA- or RNA-based analysis, metaproteomics represented the "actual active building blocks" in the microbial environment. Therefore, it might be a very essential tool to decipher the interaction effect of marine host-microbe relationships. Only few studies have been conducted regarding the marine-based host microbial interaction, although metaproteomic approach can be used as a fundamental tool to design marine microbial functional network (Wang et al. 2014; Williams and Cavicchioli 2014). The symbiotic associations between chemosynthetic bacteria and marine animals like *Olavius algarvensis* that bloom under the nutrient-deficient environment in deep sea have been studied by Kleiner et al. (2012). This interaction was studied and metabolic pathways of host and symbionts were reconstructed by the metaproteomic approach. The highly expressed proteins, were particularly related with transporters, host waste cycling, and energy and carbon conservation including various pathways like uptake of different substrates, utilization of inorganic energy sources and host waste material, and sulfate reduction, and CO_2 fixation. The metaproteomic approach shows metabolic interaction between the host (*Olavius algarvensis*) and microbial symbionts under oligotrophic environment. The sponge holobiont relationship was also evaluated by metaproteomic analysis where various transporters and phagocytosis resistance domains were found to be abundant (Liu et al. 2012). The effect of nutrient concentrations on the oceanic microbiome has also been studied through comparative membrane metaproteomic approach. This study revealed shifts in nutrient utilization and energy transduction along the environmental nutrient gradient from low-nutrient areas to high-productivity regions. TonB-dependent transporters (TBDTs) were the dominant membrane proteins identified, with around 19% abundance. Archaeal ammonia monooxygenase proteins were also found in the nutrient-rich regions, suggesting the involvement of archaea as nitrifiers (Morris et al. 2010).

20.3.4 Metabolomic Approaches for Studying Interactions in Marine Microbiome

Metabolomics, by definition, captures the global metabolite profile, in a given condition, mostly of low molecular weight compounds (<1500 Da) (Gordon and Leggat 2010). Thus, it presents an important and useful approach for studying marine microbiome interactions. This is due to the fact that co-occurring organisms release various chemicals which influence the other species in terms of growth, metabolic pathways, as well as defense mechanisms. Metabolomics can actually substantiate and complement other omic approaches correlating the genotype with the phenotype. With emerging issues of climate change and coral bleaching, it becomes even more pertinent to know the chemical moieties the species are interacting with in their niche.

Goodacre et al. (2004) classified metabolomics approaches to studying living systems as:

1. *Metabolite target analysis*—Study of metabolites restricted to a particular enzyme system
2. *Metabolomics*—Whole metabolome analysis under a given condition
3. *Metabolite profiling*—Analysis of a group of metabolites from a particular pathway
4. *Metabolic profiling*—Mostly used in the clinical sense to understand the fate of a drug or metabolite
5. *Metabolic fingerprinting*—Analyzing samples based on their biological origin
6. *Metabonomics*—To understand the biochemical profile in a disease condition or under the influence of a drug or toxin

Most metabolomics approaches center around the identification of the compounds using a mass spectroscopy (MS) paired with a liquid chromatography (LC) or gas chromatography modules for separation, resolution, and selectivity. In MS modalities, electrospray ionization (ESI), coupled to TOF or Fourier transform ion cyclotron resonance (FT-ICR) MS, is employed. Complex biological fluids can also be analyzed using ^{1}H, ^{13}C, and ^{31}P in nuclear resonance imaging (NMR), providing essential structural information (Goodacre et al. 2004). Sample preparation includes extraction, enrichment, and derivatization. The analysis pipeline is composed of preprocessing, statistical analysis, and machine-learning techniques for pattern recognition with available databases (Aguiar-Pulido et al. 2016). Links of some popular databases, helpful in metabolomics are compiled in Tables 20.1 and 20.2.

Marine bacteria are most widely studied using metabolomic profiling approach. This has resulted in information which has revealed newer species of bacteria. Metabolite profiling of genera, such as *Pseudoalteromonas* and *Vibrionaceae*, has led to the discovery of secondary metabolites and a number of bioactive compounds (Goulitquer et al. 2012).

Table 20.1 List of some relevant MS and NMR databases and their links

Mass spectral/NMR databases	Links
Metlin	https://metlin.scripps.edu/
MMCD	http://mmcd.nmrfam.wisc.edu/
LipidMaps	http://www.lipidmaps.org/
GMD (Golm Metabolome Database)	http://gmd.mpimp-golm.mpg.de/
FiehnLib (Fiehn Metabolome library)	http://fiehnlab.ucdavis.edu/projects/FiehnLib/
MassBank	http://www.massbank.jp/
NIST	https://www.nist.gov/

Table 20.2 List of some relevant metabolite databases and their links

Database of metabolites	Links
PubChem	https://pubchem.ncbi.nlm.nih.gov/
LipidMaps	http://www.lipidmaps.org/
ChemSpider	http://www.chemspider.com/
ChEBi (Chemical Entities of Biological Interest)	http://www.ebi.ac.uk/chebi/
SWMD (Seaweed Metabolite Database	http://www.swmd.co.in/
MarinLit	http://pubs.rsc.org/marinlit/
MMCD	http://mmcd.nmrfam.wisc.edu/
BMRB (Biological Magnetic Resonance Data Bank)	http://www.bmrb.wisc.edu/metabolomics/

Unicellular algae have been found to contribute to about 50% of total carbon fixation of the world. Thus, their role in the marine ecosystem cannot be undermined and have been found to be intertwined with their interactions with other species, especially that of bacteria. Earlier research has found that compounds such as dimethyl sulfide (DMS), cyanogen bromide and high molecular weight proteins trigger responses in phytoplanktons (Wolfe et al. 1997; Yamasaki et al. 2007; Vanelslander et al. 2012). Most of these studies are complex due to the difficulty in creating the marine ecological niches in laboratory environs. Co-culture setups with membrane diffusion have been used to study the chemically mediated interactions between bacteria and diatoms. Paul et al. (2013) studied the diatom *Thalassiosira pseudonana* and bacterium *Dinoroseobacter shibae* by growing them in a specially designed two-flask setup with a PVDF membrane for diffusion of the metabolites. LC-MS measurements were carried out with a UPLC paired to a Q-TOF micromass spectrometer. They also conducted analysis of the heptadienal (GC-MS), nitrate (UV spectrophotometer), and DMSP (UPLC-MS). This study found the upregulation of intracellular amino acids in *T. pseudonana* in the presence of *D. shibae*. However, this effect on metabolism was not affected in growth, probably due to adaptive response of the diatom.

The role of metabolites in nutrition and diel cycles in symbiotic relationships between the photosynthetic dinoflagellate, *Symbiodinium* sp., and invertebrates such as corals has been discussed by Gordon and Leggat (2010) using labeled carbon. Soluble products of photosynthesis, along with other lipids and amino acids released by *Symbiodinium* were utilized by their coral hosts.

Symbiotic relationship between tunicates such as *Lissoclinum patella* and the cyanobacterium *Prochloron didemni* was explored through metabolomics by Muller et al. (2010). They found that *L. patella* was essential for the survival of *P. didemni* as photosynthetic exchange has revealed. The cyanobacterium influences the lipid composition of the animals by synthesizing sterols and an unusual lipid with biofuel potential. Also, *L. patella* harbors a great number of other bacterial groups that contribute secondary metabolic products to the symbiosis (Donia et al. 2011).

Computational approaches such as Predicted Relative Metabolomic Turnover (PRMT) have been used to correlate metagenomic data with metabolite information. Such information has been successfully used to predict relationships between CO_2, iron, orthophosphate, nitrate, and chlorophyll with marine primary production. Utilization of organophosphorus and chitin could also be predicted (Larsen et al. 2011).

Major challenges associated with metabolomics in marine systems are the presence of high amount of salts which polymerize the ion source creating adducts which interfere with the MS procedure. This can be prevented by diluting the sample prior to MS procedure. Another major challenge is the presence of interfering compounds such as polysaccharides in seaweeds, which reduces the reproducibility of the extraction procedure. Cryo-grinding has been suggested to overcome this problem. Similarly, majority of pigments in algal samples can be removed by hexane/acetone extraction (Goulitquer et al. 2012). More studies are required in metabolomics of marine microbiome interactions for elucidating the signaling pathways involved, using inputs from a systems biology approach which can help decipher the complete picture.

An integrated solution comprising of all these meta-omic approaches supplemented with computational predictions could be beneficial in understanding the complex interactions of marine microbes with their hosts. They may help identify key structures and functions and their correlations in marine microbial community-wide networks.

20.4 Conclusion and Future Perspectives

Most of the studies in involving an omic approach to study the interactions in the marine microbiome depend largely on the samples collected from a particular marine niche. Owing to the vastness of the ocean ecosystems, this might not be enough to represent the complete data and their interrelations. Thus, it is necessary to develop innovative and noninvasive techniques for collection of relevant samples from the organisms.

Omic approaches present the means to mine vast amounts of data from the marine ecosystems and integrate them for useful information. These provide us the interconnection between the genotype and the phenotype of such communities. Understanding these correlations is crucial for the impact they have on environmental integrity, climate change, industrial practices, and human health.

References

Aguiar-Pulido V, Huang W, Suarez-Ulloa V, Cickovski T, Mathee K, Narasimhan G (2016) Metagenomics, metatranscriptomics, and metabolomics approaches for microbiome analysis. Evol Bioinformatics Online 12(Suppl 1):5

Ainsworth TD, Gates RD (2016) Corals' microbial sentinels. Science 352(6293):1518–1519

Aires T, Serrão EA, Kendrick G, Duarte CM, Arnaud-Haond S (2013) Invasion is a community affair: Clandestine followers in the bacterial community associated to green algae, *Caulerpa racemosa*, track the invasion source. PLoS One 8(7):e68429

Badhai J, Ghosh TS, Das SK (2016) Composition and functional characterization of microbiome associated with mucus of the coral *Fungia echinata* collected from Andaman Sea. Front Microbiol 7:936

Baker AC (2003) Flexibility and specificity in coral-algal symbiosis: diversity, ecology, and biogeography of *Symbiodinium*. Annu Rev Ecol Evol Syst 24:661–689

Bayer T, Arif C, Ferrier-Pagès C, Zoccola D, Aranda M, Voolstra CR (2013a) Bacteria of the genus *Endozoicomonas* dominate the microbiome of the Mediterranean gorgonian coral *Eunicella cavolini*. Mar Ecol Prog Ser 479:75–84

Bayer T, Neave MJ, Alsheikh-Hussain A, Aranda M, Yum LK, Mincer T, Hughen K, Apprill A, Voolstra CR (2013b) The microbiome of the Red Sea coral *Stylophora pistillata* is dominated by tissue-associated *Endozoicomonas* bacteria. Appl Environ Microbiol 79(15):4759–4762

Bik EM, Costello EK, Switzer AD, Callahan BJ, Holmes SP, Wells RS, Carlin KP, Jensen ED, Venn-Watson S, Relman DA (2016) Marine mammals harbor unique microbiotas shaped by and yet distinct from the sea. Nat Commun 7:10516

Burgsdorf I, Slaby BM, Handley KM, Haber M, Blom J, Marshall CW, Gilbert JA, Hentschel U, Steindler L (2015) Lifestyle evolution in cyanobacterial symbionts of sponges. MBio 6(3):e00391-00315. doi:10.1128/mBio.00391-15

Burke C, Thomas T, Lewis M, Steinberg P, Kjelleberg S (2011) Composition, uniqueness and variability of the epiphytic bacterial community of the green alga *Ulva australis*. ISME J 5(4):590–600

Daniels C, Baumgarten S, Yum LK, Michell CT, Bayer T, Arif C, Roder C, Weil E, Voolstra CR (2015) Metatranscriptome analysis of the reef-building coral *Orbicella faveolata* indicates holobiont response to coral disease. Front Mar Sci 2:62

Dimijian GG (2000) Evolving together: the biology of symbiosis, part 1. Proc (Baylor Univ Med Cent) 13(3):217

Donia MS, Fricke WF, Partensky F, Cox J, Elshahawi SI, White JR, Phillippy AM, Schatz MC, Piel J, Haygood MG (2011) Complex microbiome underlying secondary and primary metabolism in the tunicate-Prochloron symbiosis. Proc Natl Acad Sci 108(51):E1423–E1432

Egan S, Harder T, Burke C, Steinberg P, Kjelleberg S, Thomas T (2013) The seaweed holobiont: understanding seaweed–bacteria interactions. FEMS Microbiol Rev 37(3):462–476

Fan L, Reynolds D, Liu M, Stark M, Kjelleberg S, Webster NS, Thomas T (2012) Functional equivalence and evolutionary convergence in complex communities of microbial sponge symbionts. Proc Natl Acad Sci 109(27):E1878–E1887

Fan L, Liu M, Simister R, Webster NS, Thomas T (2013) Marine microbial symbiosis heats up: the phylogenetic and functional response of a sponge holobiont to thermal stress. ISME J 7(5):991–1002

Fisher P, Malme M, Dove S (2012) The effect of temperature stress on coral–*Symbiodinium* associations containing distinct symbiont types. Coral Reefs 31(2):473–485

Gao Z-M, Wang Y, Tian R-M, Wong YH, Batang ZB, Al-Suwailem AM, Bajic VB, Qian P-Y (2014) Symbiotic adaptation drives genome streamlining of the cyanobacterial sponge symbiont "*Candidatus Synechococcus spongiarum*". MBio 5(2):e00079-00014

Goodacre R, Vaidyanathan S, Dunn WB, Harrigan GG, Kell DB (2004) Metabolomics by numbers: acquiring and understanding global metabolite data. Trends Biotechnol 22(5):245–252

Gordon BR, Leggat W (2010) *Symbiodinium*—invertebrate symbioses and the role of metabolomics. Mar Drugs 8(10):2546–2568

Goulitquer S, Potin P, Tonon T (2012) Mass spectrometry-based metabolomics to elucidate functions in marine organisms and ecosystems. Mar Drugs 10(4):849–880

Gust KA, Najar FZ, Habib T, Lotufo GR, Piggot AM, Fouke BW, Laird JG, Wilbanks MS, Rawat A, Indest KJ (2014) Coral-zooxanthellae meta-transcriptomics reveals integrated response to pollutant stress. BMC Genomics 15(1):1

Hentschel U, Piel J, Degnan SM, Taylor MW (2012) Genomic insights into the marine sponge microbiome. Nat Rev Microbiol 10(9):641–654

Holm JBH, Heidelberg KB (2016) Microbiomes of *Muricea californica* and *Muricea fruticosa*: comparative analyses of two co-occurring Eastern Pacific octocorals. Front Microbiol 7:917

Horn H, Slaby BM, Jahn MT, Bayer K, Moitinho-Silva L, Förster F, Abdelmohsen UR, Hentschel U (2016) An enrichment of CRISPR and other defense-related features in marine sponge-associated microbial metagenomes. Front Microbiol 7:1751

Iglesias-Prieto R, Matta JL, Robins WA, Trench RK (1992) Photosynthetic response to elevated temperature in the symbiotic dinoflagellate *Symbiodinium microadriaticum* in culture. Proc Natl Acad Sci 89(21):10302–10305

Kleiner M, Wentrup C, Lott C, Teeling H, Wetzel S, Young J, Chang Y-J, Shah M, VerBerkmoes NC, Zarzycki J (2012) Metaproteomics of a gutless marine worm and its symbiotic microbial community reveal unusual pathways for carbon and energy use. Proc Natl Acad Sci 109(19):E1173–E1182

Lachnit T, Thomas T, Steinberg P (2015) Expanding our understanding of the seaweed holobiont: RNA viruses of the Red Alga *Delisea pulchra*. Front Microbiol 6:PMC4705237

Larsen PE, Collart FR, Field D, Meyer F, Keegan KP, Henry CS, McGrath J, Quinn J, Gilbert JA (2011) Predicted relative metabolomic turnover (PRMT): determining metabolic turnover from a coastal marine metagenomic dataset. Microb Inform Exp 1(1):4. doi:10.1186/2042-5783-1-4

Laval P (1978) The barrel of the pelagic amphipod *Phronima sedentaria* (Forsk.) (Crustacea: Hyperiidea). J Exp Mar Biol Ecol 33(3):187–211

Liu M, Fan L, Zhong L, Kjelleberg S, Thomas T (2012) Metaproteogenomic analysis of a community of sponge symbionts. ISME J 6(8):1515–1525

Moran MA (2015) The global ocean microbiome. Science 350(6266):aac8455

Morris RM, Nunn BL, Frazar C, Goodlett DR, Ting YS, Rocap G (2010) Comparative metaproteomics reveals ocean-scale shifts in microbial nutrient utilization and energy transduction. ISME J 4(5):673–685. http://www.nature.com/ismej/journal/v4/n5/suppinfo/ismej20104s1.html

Mukherjee J, Webster N, Llewellyn LE (2009) Purification and characterization of a collagenolytic enzyme from a pathogen of the Great Barrier Reef sponge, *Rhopaloeides odorabile*. PLoS One 4(9):e7177

Muller J, Szklarczyk D, Julien P, Letunic I, Roth A, Kuhn M, Powell S, von Mering C, Doerks T, Jensen LJ, Bork P (2010) KeggNOG v2.0: extending the evolutionary genealogy of genes with enhanced non-supervised orthologous groups, species and functional annotations. Nucleic Acids Res 38(Database issue):D190–D195

Naeem S (1998) Species redundancy and ecosystem reliability. Conserv Biol 12(1):39–45

Nealson KH, Venter JC (2007) Metagenomics and the global ocean survey: what's in it for us, and why should we care? ISME J 1(3):185

Neave MJ, Rachmawati R, Xun L, Michell CT, Bourne DG, Apprill A, Voolstra CR (2016) Differential specificity between closely related corals and abundant *Endozoicomonas* endosymbionts across global scales. ISME J 11(1):186–200

Nelson TM, Apprill A, Mann J, Rogers TL, Brown MV (2015) The marine mammal microbiome: current knowledge and future directions. Microbiol Aust 36(1):8–13

de Oliveira LS, Gregoracci GB, Silva GGZ, Salgado LT, Amado Filho G, Alves-Ferreira M, Pereira RC, Thompson FL (2012) Transcriptomic analysis of the red seaweed *Laurencia dendroidea* (Florideophyceae, Rhodophyta) and its microbiome. BMC Genomics 13(1):1

Pandolfi J, Jackson JBC, Baron N, Bradbury R, Guzman H, Hughes T, Kappel C, Micheli F, Ogden J, Possingham HP (2005) Are US coral reefs on the slippery slope to slime? Science 307(5716):1725–1726

Paul C, Mausz MA, Pohnert G (2013) A co-culturing/metabolomics approach to investigate chemically mediated interactions of planktonic organisms reveals influence of bacteria on diatom metabolism. Metabolomics 9(2):349–359

Pinzón JH, Kamel B, Burge CA, Harvell CD, Medina M, Weil E, Mydlarz LD (2015) Whole transcriptome analysis reveals changes in expression of immune-related genes during and after bleaching in a reef-building coral. R Soc Open Sci 2(4):140214

Radax R, Rattei T, Lanzen A, Bayer C, Rapp HT, Urich T, Schleper C (2012) Metatranscriptomics of the marine sponge *Geodia barretti*: tackling phylogeny and function of its microbial community. Environ Microbiol 14(5):1308–1324

Raghunathan A, Ferguson HR, Bornarth CJ, Song W, Driscoll M, Lasken RS (2005) Genomic DNA amplification from a single bacterium. Appl Environ Microbiol 71(6):3342–3347

Rohwer F, Thurber RV (2009) Viruses manipulate the marine environment. Nature 459(7244):207–212

Rowan R, Powers DA (1992) Ribosomal RNA sequences and the diversity of symbiotic dinoflagellates (zooxanthellae). Proc Natl Acad Sci 89(8):3639–3643

Ryu T, Seridi L, Moitinho-Silva L, Oates M, Liew YJ, Mavromatis C, Wang X, Haywood A, Lafi FF, Kupresanin M (2016) Hologenome analysis of two marine sponges with different microbiomes. BMC Genomics 17(1):1

Sale PF (1976) Reef fish lottery. Nat Hist 85:60–65

Santos HF, Carmo FL, Martirez N, Duarte GA, Calderon EN, Castro CB, Pires DO, Rosado AS, Peixoto RS (2016) Cyanobacterial and microeukaryotic profiles of healthy, diseased, and dead *Millepora alcicornis* from the South Atlantic. Dis Aquat Org 119(2):163–172

Siegl A, Kamke J, Hochmuth T, Piel J, Richter M, Liang C, Dandekar T, Hentschel U (2011) Single-cell genomics reveals the lifestyle of Poribacteria, a candidate phylum symbiotically associated with marine sponges. ISME J 5(1):61–70

Simister R, Taylor MW, Tsai P, Fan L, Bruxner TJ, Crowe ML, Webster N (2012) Thermal stress responses in the bacterial biosphere of the Great Barrier Reef sponge, *Rhopaloeides odorabile*. Environ Microbiol 14(12):3232–3246

Singh RP, Reddy CRK (2014) Seaweed–microbial interactions: key functions of seaweed-associated bacteria. FEMS Microbiol Ecol 88(2):213–230

Taylor MW, Hill RT, Piel J, Thacker RW, Hentschel U (2007) Soaking it up: the complex lives of marine sponges and their microbial associates. ISME J 1(3):187–190

Thomas T, Rusch D, DeMaere MZ, Yung PY, Lewis M, Halpern A, Heidelberg KB, Egan S, Steinberg PD, Kjelleberg S (2010) Functional genomic signatures of sponge bacteria reveal unique and shared features of symbiosis. ISME J 4(12):1557–1567

Thomas T, Moitinho-Silva L, Lurgi M, Björk JR, Easson C, Astudillo-García C, Olson JB, Erwin PM, López-Legentil S, Luter H (2016) Diversity, structure and convergent evolution of the global sponge microbiome. Nat Commun 7:11870

Thurber RV, Willner-Hall D, Rodriguez-Mueller B, Desnues C, Edwards RA, Angly F, Dinsdale E, Kelly L, Rohwer F (2009) Metagenomic analysis of stressed coral holobionts. Environ Microbiol 11(8):2148–2163

Tian RM, Wang Y, Bougouffa S, Gao ZM, Cai L, Bajic V, Qian PY (2014) Genomic analysis reveals versatile heterotrophic capacity of a potentially symbiotic sulfur-oxidizing bacterium in sponge. Environ Microbiol 16(11):3548–3561

Turner LM, Alsterberg C, Turner AD, Girisha S, Rai A, Havenhand JN, Venugopal M, Karunasagar I, Godhe A (2016) Pathogenic marine microbes influence the effects of climate change on a commercially important tropical bivalve. Sci Rep 6:32413

Vacelet J, Donadey C (1977) Electron microscope study of the association between some sponges and bacteria. J Exp Mar Biol Ecol 30(3):301–314

Vanelslander B, Paul C, Grueneberg J, Prince EK, Gillard J, Sabbe K, Pohnert G, Vyverman W (2012) Daily bursts of biogenic cyanogen bromide (BrCN) control biofilm formation around a marine benthic diatom. Proc Natl Acad Sci 109(7):2412–2417

Wang D-Z, Xie Z-X, Zhang S-F (2014) Marine metaproteomics: current status and future directions. J Proteome 97:27–35

Webster NS, Taylor MW (2012) Marine sponges and their microbial symbionts: love and other relationships. Environ Microbiol 14(2):335–346

Welsh RM, Zaneveld JR, Rosales SM, Payet JP, Burkepile DE, Thurber RV (2016) Bacterial predation in a marine host-associated microbiome. ISME J 10(6):1540–1544. doi:10.1038/ismej.2015.219

Wilkinson C, FAY P (1979) Nitrogen fixation in coral reef sponges with symbiotic cyanobacteria. Nature 279:527–529. doi:10.1038/279527a0

Williams TJ, Cavicchioli R (2014) Marine metaproteomics: deciphering the microbial metabolic food web. Trends Microbiol 22(5):248–260

Wolfe GV, Steinke M, Kirst GO (1997) Grazing-activated chemical defence in a unicellular marine alga. Nature 387(6636):894–897

Yamasaki Y, Nagasoe S, Matsubara T, Shikata T, Shimasaki Y, Oshima Y, Honjo T (2007) Allelopathic interactions between the bacillariophyte *Skeletonema costatum* and the raphidophyte *Heterosigma akashiwo*. Mar Ecol Prog Ser 339:83–92

MIX
Papier aus verantwortungsvollen Quellen
Paper from responsible sources
FSC® C105338

If you have any concerns about our products,
you can contact us on
ProductSafety@springernature.com

In case Publisher is established outside the EU,
the EU authorized representative is:
Springer Nature Customer Service Center GmbH
Europaplatz 3, 69115 Heidelberg, Germany

Printed by Libri Plureos GmbH
in Hamburg, Germany